Dynamic Aspects of Detonations

Edited by
A. L. Kuhl
Lawrence Livermore National Laboratory
El Segundo, California

J.-C. Leyer
Université de Poitiers
Poitiers, France

A. A. Borisov
Russian Academy of Sciences
Moscow, Russia

W. A. Sirignano
University of California
Irvine, California

Volume 153
PROGRESS IN ASTRONAUTICS AND AERONAUTICS

A. Richard Seebass, Editor-in-Chief
University of Colorado at Boulder
Boulder, Colorado

Technical papers from the Thirteenth International Colloquium on Dynamics of Explosions and Reactive Systems, Nagoya, Japan, July 1991, and subsequently revised for this volume.

Published by the American Institute of Aeronautics and Astronautics, Inc.,
370 L'Enfant Promenade SW, Washington, DC 20024-2518

Copyright © 1993 by the American Institute of Aeronautics and Astronautics, Inc. Printed in the United States of America. All rights reserved. Reproduction or translation of any part of this work beyond that permitted by Sections 107 and 108 of the U.S. Copyright Law without the permission of the copyright owner is unlawful. The code following this statement indicates the copyright owner's consent that copies of articles in this volume may be made for personal or internal use, on condition that the copier pay the per-copy fee ($2.00) plus the per-page fee ($0.50) through the Copyright Clearance Center, Inc., 21 Congress Street, Salem, Massachusetts 01970. This consent does not extend to other kinds of copying, for which permission requests should be addressed to the publisher. Users should employ the following code when reporting copying from this volume to the Copyright Clearance Center:

1-56347-057-8/93 $2.00 + .50

Data and information appearing in this book are for informational purposes only. AIAA is not responsible for any injury or damage resulting from use or reliance, nor does AIAA warrant that use or reliance will be free from privately owned rights.

ISSN 0079-6050

Progress in Astronautics and Aeronautics

Editor-in-Chief
A. Richard Seebass
University of Colorado at Boulder

Editorial Board

Richard G. Bradley
General Dynamics

Allen E. Fuhs
Carmel, California

George J. Gleghorn
*TRW Space
and Technology Group*

Dale B. Henderson
Los Alamos National Laboratory

Carolyn L. Huntoon
NASA Johnson Space Center

John L. Junkins
Texas A&M University

Daniel P. Raymer
*Conceptual Research
Corporation*

Martin Summerfield
*Princeton Combustion Research
Laboratories, Inc.*

Charles E. Treanor
*Arvin/Calspan
Advanced Technology Center*

Jeanne Godette
Director
Book Publications
AIAA

Preface

The four companion volumes on *Dynamic Aspects of Detonation and Explosion Phenomena* and *Dynamics of Gaseous and Heterogeneous Combustion and Reacting Systems* present 111 of the 230 papers given at the Thirteenth International Colloquium on the Dynamics of Explosions and Reactive Systems held in Nagoya, Japan from July 28 to August 2, 1991.

These four volumes are included in the Progress in Astronautics and Aeronautics series published by the American Institute of Aeronautics and Astronautics, Inc. *Dynamics of Gaseous Combustion* (Volume 151) and *Dynamics of Heterogeneous Combustion and Reacting Systems* (Volume 152) span a broad area, encompassing the processes of coupling the exothermic energy release with the fluid mechanics occurring in various combustion processes. *Dynamic Aspects of Detonations* (Volume 153) and *Dynamic Aspects of Explosion Phenomena* (Volume 154) principally address the rate processes of energy deposition in a compressible medium and the concurrent nonsteady flow as it typically occurs in explosion phenomena. The Colloquium, in addition to embracing the usual topics of explosions, detonations, shock phenomena, and reactive flow, includes papers that deal primarily with the gasdynamic aspects of nonsteady flow in combustion systems, the fluid mechanic aspects of combustion (with particular emphasis on turbulence), and diagnostic techniques.

In this volume, *Dynamic Aspects of Detonations*, papers have been arranged into chapters on gaseous detonations, initiation of detonation waves, and nonideal detonations and boundary effects. Although the brevity of this Preface does not permit the editors to do justice to all of the papers, we offer the following highlights of some of the especially noteworthy contributions.

In Chapter I, *Manson and Dabora* present an extended review of research and publications on detonation waves covering the period from 1920 to 1950. This is a sequel to their chronological review of such research prior to 1920, which was published in the previous Proceedings (Volume 133 of the Progress in Astronautics and Aeronautics series).

In Chapter II, Gaseous Detonations, *Bourlioux and Majda* present high-resolution numerical simulations of unstable detonations, using a second-order Godunov code. *Dremin* explores the origin of detonation front instabilities during the transition from an overdriven detonation to the Chapman-Jouguet (CJ) state. This chapter also contains a number of articles on the cellular structure of gaseous detonations (e.g., *Lefebvre et al., Lee et al.*, and *Huang and Van Tiggelen*), galloping detonations by *Aksamentov et al.*, and cylindrical detonations in methane mixtures by *Aminallah and Brossard*.

Chapter III covers recent progress on the initiation of detonation waves. This includes articles on the initiation of gaseous detonations by turbulent jets and obstacle-generated transverse waves from the McGill group, reignition behind Mach stems (by *Oran and co-workers*) and reflected shocks (by *Takano*), and the failure of classical dynamic parameter relationships in cellular structures by

Desbordes and co-workers.

Chapter IV, Nonideal Detonations and Boundary Effects, investigates the fundamental propagation mechanisms of nonideal or quasidetonations that propagate at a fraction of the CJ velocity. This includes articles on detonations in a porous medium (by *Makris et al.*) and in tube bundles (by *Laberge et al.*). Also described are nonideal detonations in hybrid or two-phase mixtures by *Khasainnov and Veyssière* and *Gois et al.*

The companion volumes, *Dynamics of Gaseous Combustion* (Volume 151), *Dynamics of Heterogeneous Combustion and Reacting Systems* (Volume 152), and *Dynamic Aspects of Explosion Phenomena* (Volume 154), include papers on gas explosions, dust explosions, vapor explosions, and nonsteady flows; papers on the behavior of propagating premixed flames, ignition dynamics, diffusion flames and their structure, nonsteady flames and combustion in shear layers; as well as papers on the dynamics of turbulent combustion, combustion in dust-air mixtures, droplet combustion, pulsed jet combustion, and internal combustion engines.

These four volumes will, we trust, help satisfy the need first articulated in 1966 and will continue the tradition of augmenting our understanding of the dynamics of explosions and reactive systems begun the following year in Brussels with the first colloquium. Subsequent colloquia have been held on a biennial basis: 1969 in Novosibirsk, 1971 in Marseilles, 1973 in La Jolla, 1975 in Bourges, 1977 in Stockholm, 1979 in Göttingen, 1981 in Minsk, 1983 in Poitiers, 1985 in Berkeley, 1987 in Warsaw, 1989 in Ann Arbor, and 1991 in Nagoya. The Colloquium has now achieved the status of a principal international meeting on these topics, and attracts contributions from scientists and engineers throughout the world.

To provide an enduring focal point for the administrative aspects of the ICDERS, the organization was formally incorporated in the state of Washington under the name Institute for Dynamics of Explosions and Reactive Systems (IDERS). Professor J. R. Bowen is serving as the current president. Communications may be sent to:

Dean J. R. Bowen
President, IDERS
College of Engineering FH-10
University of Washington
Seattle, Washington 98195
USA

Papers from the first six colloquia have appeared as a part of the journal *Acta Astronautica*, or its predecessor, *Astronautica Acta*. With the publication of the Seventh Colloquium, selected papers have appeared as part of the Progress in Astronautics and Aeronautics series published by the American Institute of Aeronautics and Astronautics (AIAA). These are the last Dynamics of Explosions and Reactive Systems Colloquium papers to appear in the Progress in Astronautics and Aeronautics series.

Acknowledgments

The Thirteenth Colloquium was held under the auspices of Nagoya University from July 28 to August 2, 1991. Local arrangements were organized by Professors T. Fujiwara and A. K. Hayashi. Publication of selected papers from the Colloquium was made possible by grants from the National Science Foundation and the Defense Nuclear Agency of the United States. Generous financial support for the meeting was received from the following organizations: Aichi Machine Industry Company, Aichi Prefecture, Aishin AW, Canon Sales Company, Central Japan Nagoya Airport, Central Japan Nagoya Station, Chubu Aeronautics and Space Technology Development Association, Chubu Electric Power Company, Daikin Industry, DAIKO Foundation, ENGAKU, Haruki (Mr.), ET Planning, FUJIMA Sohke School of Kabuki Dances, Gifu Auto Body Industry Company, Hitachi, Honda Motor Company, IBM Japan, Ishikawajima-Harima Heavy Industries, Isuzu Motor Company, Japan Gas Association, KATO Ryutaro Foundation, Kawasaki Heavy Industries, Kobe Steel, Matsushita Graphic Communication Systems, Mazda Motor Company, Meitec Corporation, Mitsubushi Heavy Industries, Nagoya City, Nippon Denso, Nippon Oil and Fats Company, Nippon Sanso, Nippon Steel Corporation, Nissan Motor Company, Rinnai Corporation, Science Research Fundings from the Ministry of Education, Science, and Culture (Profs. K. Abe, T. Fujiwara, and K. Takayama), Shachihata Industrial Company, Sogo Solvent Company, Takashimaya-Nippatsu Kogyo Company, Toho Gas, Tokai Bank, Toshiba Corporation (Chubu Branch), Toyoda Automatic Loom Works, Toyoda Gosei Company, Toyoda Machine Tools, Toyota Central Research and Development Laboratory, Toyota Motor Company, and Toyota Techno Service Company.

A. L. Kuhl
J.-C. Leyer
A. A. Borisov
W. A. Sirignano
May 1993

Table of Contents

Preface

Chapter I

Chronology of Research on Detonation Waves: 1920–1950 3
N. Manson, *Ecole Nationale Supérieure de Mécanique et d'Aérotechnique, Poitiers, France*, and E. K. Dabora, *University of Connecticut, Storrs, Connecticut*

Chapter II. Gaseous Detonations

High Resolution Numerical Simulations for Two-Dimensional Unstable Detonations 43
Anne Bourlioux and Andrew J. Majda, *Princeton University, Princeton, New Jersey*

Simulation of Cellular Structure in a Detonation Wave 64
M. H. Lefebvre, *Royal Military Academy, Brussels, Belgium*, E. S. Oran and K. Kailasanath, *Naval Research Laboratory, Washington, DC*, and P. J. Van Tiggelen, *Université Catholique de Louvain, Louvain-la-Neuve, Belgium*

Mach Reflection of Detonation Waves 78
J. Meltzer, J. E. Shepherd, R. Akbar, and A. Sabet, *Rensselaer Polytechnic Institute, Troy, New York*

Formation and Propagation of Photochemical Detonations in Hydrogen-Chlorine Mixtures 95
Norihiko Yoshikawa, *Toyohashi University of Technology, Toyohashi, Japan*, and John H. Lee, *McGill University, Montreal, Quebec, Canada*

Mechanism of Unstable Detonation Front Origin 105
A. N. Dremin, *Russian Academy of Sciences, Moscow, Russia*

Numerical Modeling of Galloping Detonation 112
S. M. Aksamentov, V. I. Manzhaley, and V. V. Mitrofanov, *Lavrentyev Institute of Hydrodynamics, Novosibirsk, Russia*

Experimental Study of the Fine Structure in Spin Detonations 132
Z. W. Huang and P. J. Van Tiggelen, *Université Catholique de Louvain, Louvain-la-Neuve, Belgium*

Influence of Fluorocarbons on H_2O_2 Ar Detonation: Experiments and Modeling 144
  M. H. Lefebvre, E. Nzeyimana, and P. J. Van Tiggelen, *Université Catholique de Louvain, Louvain-la-Neuve, Belgium*

Oxidation of Gaseous Unsymmetrical Dimethylhydrazine at High Temperatures and Detonation of UDMH/O_2 Mixtures 162
  Said Abid, Gabrielle Dupré, and Claude Paillard, *National Center of Scientific Research and University, Orléans, France*

Digital Signal Processing Analysis of Soot Foils 182
  J. J. Lee, D. L. Frost, J. H. S. Lee, and R. Knystautas, *McGill University, Montreal, Quebec, Canada*

Cylindrical Detonations in Methane-Oxygen-Nitrogen Mixtures 203
  Miloud Aminallah and Jacques Brossard, *Université d'Orléans, Bourges, France*, and A. Vasiliev, *Siberian Academy of Sciences, Novosibirsk, Russia*

Chapter III. Initiation of Detonation Waves

Structure of Reaction Waves Behind Oblique Shocks 231
  C. Li, K. Kailasanath, and E. S. Oran, *Naval Research Laboratory, Washington, DC*

Ignition in a Complex Mach Structure 241
  E. S. Oran and J. P. Boris, *Naval Research Laboratory, Washington, DC*, D. A. Jones, *Materials Research Laboratory, Victoria, Australia*, and M. Sichel, *University of Michigan, Ann Arbor, Michigan*

Photographic Study of the Direct Initiation of Detonation by a Turbulent Jet 253
  M. Inada, J. H. Lee, and R. Knystautas, *McGill University, Montreal, Quebec, Canada*

Transition from Fast Deflagration to Detonation Under the Influence of Wall Obstacles 270
  R. S. Chue, J. H. Lee, T. Scarinci, A. Papyrin, and R. Knystautas, *McGill University, Montreal, Quebec, Canada*

Simulations for Detonation Initiation Behind Reflected Shock Waves ... 283
  Yasunari Takano, *Tottori University, Tottori, Japan*

Limiting Tube Diameter of Gaseous Detonation 298
  S. M. Frolov and B. E. Gelfand, *Russian Academy of Sciences, Moscow, Russia*

Effect of Flame Inhibitors on Detonation Characteristics of Fuel-Air Mixtures 312
  A. A. Borisov, V. V. Kosenkov, A. E. Mailkov, V. N. Mikhalkin, and S. V. Khomik, *Russian Academy of Sciences, Moscow, Russia*

Propagation of Gaseous Detonations Through Regions of
Low Reactivity 324
    T. Engebretsen, *Norwegian Defense Construction Service, Oslo, Norway*,
    D. Bjerketvedt, *Christian Michelsen Institute, Bergen, Norway*, and
    O. K. Sønju, *Norwegian Institute of Technology, Trondheim, Norway*

Failure of the Classical Dynamic Parameters Relationships in Highly
Regular Cellular Detonation Systems 347
    D. Desbordes, C. Guerraud, L. Hamada, and H. N. Presles, *Ecole Nationale Supérieure de Mécanique et d'Aérotechnique, Poitiers, France*

Chapter IV. Nonideal Detonations and Boundary Effects

Mechanisms of Detonation Propagation in a Porous Medium 363
    A. Makris, A. Papyrin, M. Kamel, G. Kilambi, J. H. S. Lee, and R. Knystautas, *McGill University, Montreal, Quebec, Canada*

Propagation and Extinction of Detonation Waves in Tube Bundles 381
    S. Laberge, R. Knystautas, and J. H. S. Lee, *McGill University, Montreal, Quebec, Canada*

Simultaneous Strong and Quasi-Chapman-Jouguet Detonation
Wave Propagation 397
    Roger Chéret, *Commissariat à l'Energie Atomique, Paris, France*

Structure and Velocity Deficit of Gaseous Detonation in Rough Tubes ... 405
    A. Teodorczyk, *Warsaw University of Technology, Warsaw, Poland*

Possible Method for Quenching of Gaseous Detonations 425
    J. Bakken and O. K. Sønju, *Norwegian Institute of Technology, Trondheim, Norway*, D. Bjerketvedt, *Christian Michelsen Institute, Bergen, Norway*, and T. Engebretsen, *Norwegian Defense Construction Service, Oslo, Norway*

Effect of Losses on the Existence of Nonideal Detonations in
Hybrid Two-Phase Mixtures 447
    B. A. Khasainov, *Russian Academy of Sciences, Moscow, Russia*, and
    B. Veyssière, *Ecole Nationale Supérieure de Mécanique et d'Aérotechnique, Poitiers, France*

Effect of Hollow Heterogeneities on Nitromethane Detonation 462
    C. Gois, H. N. Presles, and P. Vidal, *Ecole Nationale Supérieure de Mécanique et d'Aérotechnique, Poitiers, France*

Author Index for Volume 153 471

List of Series Volumes 473

Table of Contents for Companion Volume 151

Preface

Chapter I

Behavior of Propagating Flames in Premixed Media 3
Toshisuke Hirano, *University of Tokyo, Tokyo, Japan*

Chapter II. Ignition Dynamics

Numerical Simulation of Ignition Processes and Combustion Wave Propagation in H_2O_2 Reaction Systems 27
H.-J. Weber, A. Mack, and P. Roth, *Universität Duisburg, Duisburg, Germany*

Detailed Numerical Simulation of H_2O_2 Ignition in Two-Dimensional Geometries 39
Ulrich Maas and Jürgen Warnatz, *Universität Stuttgart, Stuttgart, Germany*

Simulation of "Hot-Spot" Ignition in H_2O_2 and CH_4-Air Mixtures: A Parametric Study 59
G. Goyal, Ulrich Maas, and Jürgen Warnatz, *Universität Stuttgart, Stuttgart, Germany*

Analysis of Self-Ignition for Nonunit Lewis Number 77
K. L. Henderson and J. W. Dold, *University of Bristol, Bristol, United Kingdom*

Chapter III. Diffusion Flames and Their Structure

Temperature Measurement of an Axisymmetric Flame Using Phase Shift Holographic Interferometry with Fast Fourier Transform 97
S. M. Tieng and W. Z. Lai, *National Cheng Kung University, Taiwan, Republic of China*

Effect of Gas-Phase Radiation on Flame Speed in Counterflow Premixed Flames 114
Suk H. Chung, Joon S. Lee, and Jong S. Lee, *Seoul National University, Seoul, Korea*

Two-Dimensional Simulation of a Methane-Air Premixed Flame near Stoichiometry 128
Nílson Kunioshi, Seishiro Fukutani, and Hiroshi Jinno, *Yoshida-honmachi, Sakyo-ku, Kyoto, Japan*

NOx Emission Characteristics of Rich Methane-Air Flames 141
Makihito Nishioka and Tadao Takeno, *Nagoya University, Nagoya, Japan*, Shigeto Nakagawa, *Toho Gas Company, Ltd., Tokai City, Japan*, and Yoshihiro Ishikawa, *Rinnai Company, Aichi-ken, Japan*

OH Radical Distribution in the Cold Zone of C_3H_8-Air Flame 163
A. A. Konnov and I. V. Dyakov, *Kazakh Interdisciplinary Scientific and Technical Center of Self-Propagating High-Temperature Synthesis, Alma-Ata, Kazakhstan*

Dynamics of Laminar Counterflow Hydrogen-Air Diffusion Flames near Extinction and Ignition Limits 173
N. Darabiha and S. Candel, *Centre National de la Recherche Scientifique, Ecole Centrale Paris, Chatenay-Malabry, France*

Modeling and Computation of Strained Laminar Diffusion Flames with Thermal Radiation 188
Y. Liu and B. Rogg, *University of Cambridge, Cambridge, United Kingdom*

Systematically Reduced Kinetic Mechanisms: Sensitivity Analysis 202
B. Rogg, *University of Cambridge, Cambridge, United Kingdom*

Quenching Corrected Laser Saturated Fluorescence Measurements of OH Concentration at High Pressure 228
Pascale Desgroux, Eric Domingues, Douglas A. Feikema, Annie Garo, and Marie-Joseph Cottereau, *Centre National de la Recherche Scientifique, Mont Saint Aignan, France*

Chapter IV. Nonsteady Flames

Dynamics of Flames near the Rich-Flammability Limit of Hydrogen-Air Mixtures 247
K. Kailasanath, K. Ganguly, and G. Patnaik, *Naval Research Laboratory, Washington, DC*

Stability of Nonadiabatic Cellular Flames near Extinction 263
L. Sinay and F. A. Williams, *University of California, San Diego, La Jolla, California*

Numerical Simulations of Interactions of Flamelets with Shock Waves in the Premixed Gas 274
Shiro Taki, *Hiroshima University, Higashi-Hiroshima, Japan*

Behavior of Propagating Flame in a Rotating Flowfield 284
Satoru Ishizuka and Toshisuke Hirano, *University of Tokyo, Tokyo, Japan*

Flame Propagation and Extinction in a Closed Channel with Cold Sidewalls 307
Georgii M. Makhviladze and V. I. Melikhov, *Russian Academy of Sciences, Moscow, Russia*

Experimental Determination of the Laminar Burning Velocity of Iso-Octane-Air Mixtures by Means of a Spherical Combustion Vessel 323
T. Kageyama, F. Fisson, and T. Ludwig, *Ecole Nationale Supérieure de Mécanique et d'Aérotechnique, Poitiers, France*

New Flamelet Approach to Model the Transient Phenomena Following Ignition in a Turbulent Diffusion Flame 331
F. Fichot, D. Schreiber, F. Lacas, D. Veynante, and B. Yip, *Centre National de la Recherche Scientifique, Ecole Centrale Paris, Chatenay-Malabry, France*

Detailed Analysis of Tulip Flame Phenomenon Using Numerical Simulation 344
M. Gonzalez, R. Borghi, and A. Saouab, *Université de Rouen, Mont Saint Aignan, France*

Chapter V. Combustion in Shear Layers

Study of Combustion Dynamics for Passive and Active Control 365
K. C. Schadow, E. Gutmark, and T. P. Parr, *Naval Air Warfare Center, China Lake, California*

Proposed Discrete Vortex Model for Vortex Pairing 389
A. Umemura and S. Kachi, *Yamagata University, Yonezawa, Japan*

Three-Dimensional Calculation of a Hydrogen Jet Injection into a Supersonic Air Flow 402
A. Koichi Hayashi and Masahiro Takahashi, *Nagoya University, Nagoya, Japan*

Compressibility, Exothermicity, and Three Dimensionality in Spatially Evolving Reactive Shear Flows 413
F. F. Grinstein and K. Kailasanath, *Naval Research Laboratory, Washington, DC*

Author Index for Volume 151 437

List of Series Volumes 439

Table of Contents for Companion Volume 152

Preface

Chapter I. Dynamics of Turbulent Combustion

Amplification of a Pressure Wave by Its Passage Through a Flame Front..................... 3
    T. Scarinci and J. H. Lee, *McGill University, Montreal, Quebec, Canada,* and G. O. Thomas, R. Bambrey, and
    D. H. Edwards, *University of Wales, Aberystwyth, Dyfed, United Kingdom*

Flame Curvature and Flame Speed of a Turbulent Premixed Flame in a Stagnation Point Flow.......... 25
    Yuji Yahagi, Toshihisa Ueda, and Masahiko Mizomoto, *Keio University, Yokohama, Kanagawa, Japan*

Near-Field CARS Measurements and the Local Extinction of Turbulent Jet Diffusion Flames........... 37
    Fumiaki Takahashi and Marlin D. Vangsness, *University of Dayton, Dayton, Ohio*

Correlation of Temporal and Spatial Data in Turbulent Premixed Bunsen Flames..................... 56
    Y. Zhang and K. N. C. Bray, *University of Cambridge, Cambridge, United Kingdom*

Numerical Simulation and Statistical Aspects of a Simple Model for "Hole Dynamics" in Turbulent
Diffusion Flames... 70
    L. J. Hartley, J. W. Dold, and D. Green, *University of Bristol, Bristol, United Kingdom*

Modeling of Autoignition in Nonpremixed Turbulent Systems: Closure of the Chemical-Source Terms.... 87
    Y. Zhang, B. Rogg, and K. N. C. Bray, *University of Cambridge, Cambridge, United Kingdom*

Chapter II. Combustion in Dust-Air Mixtures

Shock-Wave Induced Combustion of Dust Layers.. 105
    Marek Wolinski and Piotr Wolanski, *Warsaw University of Technology, Warsaw, Poland*

Some Fundamental Characteristics of Cornstarch Dust-Air Flames................................. 119
    Józef Jarosinski, *Institute of Aeronautics, Warsaw, Poland,* Yi Kang Pu, *Chinese Academy of Sciences, Beijing,*
    China, Elżbieta M. Bulewicz, *Technical University of Cracow, Cracow, Poland,* and C. W. Kauffman and
    Vincent G. Johnson, *University of Michigan, Ann Arbor, Michigan*

Combustion of Single Nonspherical Cellulosic Particles... 136
    P. J. Austin, C. W. Kauffman, and M. Sichel, *University of Michigan, Ann Arbor, Michigan*

Mechanism of Flame Propagation in Dust-Air and Hybrid Mixtures................................ 155
    R. Klemens, *Warsaw University of Technology, Warsaw, Poland*

Boron Particle Ignition and Liquid Film Rupture Because of Surface Tension Effects................ 178
    M. Konczalla, *University of Bielefeld, Bielefeld, Germany*

Quenching of Rich Dust Flames.. 192
    Guy Joulin, *Laboratoire d'Energétique et de Détonique, Poitiers, France*

Experiments on Turbulent Flame Propagation in Dust-Air Mixtures............................... 211
    F. Rzal and B. Veyssière, *Laboratoire d'Energétique et de Détonique, Poitiers, France,* and Y. Mouilleau and
    C. Proust, *CERCHAR-INERIS, Verneuil-en-Halatte, France*

Chapter III. Droplet Combustion

Liquid Vaporization from Fine Metal Slurry Droplets 235
    Rakesh Bhatia and William A. Sirignano, *University of California, Irvine, California*

**Numerical Simulation of Fuel Droplet Evaporation and Ignition Under High
Temperature and High Pressure** 263
    T. Tsukamoto, *Tokyo University of Mercantile Marines, Tokyo, Japan*, and T. Niioka, *Tohoku University, Sendai, Japan*

Euler System Modeling Vaporizing Sprays 280
    Lionel Sainsaulieu, *E.N.P.C., La Courtine, Noisy-le-Grand, France*

Ignition Process of Compound Spray Combustible Mixtures 306
    Masataka Arai, *Gunma University, Kiryu, Japan*, Hajime Yoshida, *Maritime Safety Academy, Wakaba-cho, kure, Japan*, and Hiroyuki Hiroyasu, *University of Hiroshima, Higashi-Hiroshima, Japan*

Chapter IV. Pulsed Jet Combustion

Augmentation of Combustion in a Chamber by a Small Hydrogen-Air Jet Flame 319
    Kazunori Wakai, *Gifu University, Gifu, Japan*, and Makoto Nagai, *NGK Spark Plug Company, Ltd., Aichi, Japan*

Performance of a Pulsed Jet Combustion System in a Swirl and a Turbulent Field 332
    S. I. Abdel-Mageed, T. Lezanski, and P. Wolanski, *Warsaw University of Technology, Warsaw, Poland*

Numerical Simulation of Pulsed Jet Plume Combustion 343
    Manabu Hishida and A. Koichi Hayashi, *Nagoya University, Nagoya, Japan*

Chapter V. Internal Combustion Engines

Thermodynamics of Combustion in an Enclosure 365
    A. K. Oppenheim and J. A. Maxson, *University of California, Berkeley, California*

Large Eddy Simulation of the Premixed Flame in an Engine 383
    Ken Naitoh, *NISSAN Research Center, Kanagawa, Japan*, and Kunio Kuwahara, *Institute of Space and Astronautical Science, Kanagawa, Japan*

Mechanism for Inhomogeneity in Ignition of Compressed Mixtures 392
    Satoshi Kadowaki, Yasuhiko Ohta, and Ko Terada, *Nagoya Institute of Technology, Nagoya, Japan*

Ionization of Compression Ignition Low-Temperature Flames 403
    Masahiro Furutani, Yasuhiko Ohta, and Kenji Komatsu, *Nagoya Institute of Technology, Nagoya, Japan*

**Examination of the Degree of Sudden Compression Required to Produce Detonation for
Thermally Sensitive Chemistry** 414
    J. W. Dold, *University of Bristol, Bristol, United Kingdom*, and A. K. Kapila, *Rensselaer Polytechnic Institute, Troy, New York*

Author Index for Volume 152 431

List of Series Volumes 433

Table of Contents for Companion Volume 154

Preface

Chapter I. Gas Explosions

Modeling of Turbulent Unvented Gas-Air Explosions 3
  Francesco Tamanini, *Factory Mutual Research Corporation, Norwood, Massachusetts*

Dynamics of Flame Propagation in Multichamber Systems 31
  R. H. Abdullin, A. V. Borisenko, and V. S. Babkin, *Institute of Chemical Kinetics and Combustion, Novosibirsk, Russia*

Fuel and Obstacle Dependence in Premixed Transient Deflagrations 51
  A. T. Cates and S. J. Bimson, *Shell Research Limited, Chester, United Kingdom*

Corrections to Zel'dovich's "Spontaneous Flame" and the Onset of Explosion via Nonuniform Preheating 59
  M. Short and J. W. Dold, *University of Bristol, Bristol, United Kingdom*

Numerical and Experimental Studies of Flame Propagation Through a Grid 75
  G. O. Thomas and R. J. Bambrey, *University of Wales, Aberystwyth, Dyfed, United Kingdom*, and B. H. Hjertager, T. Solberg, and J.-E. Forrisdahl, *Telemark Institute of Technology and Telemark Innovation Centre, TMIH Kjolnes, Porsgrunn, Norway*

Experimental Study of Large-Scale Unconfined Fuel Spray Detonations 95
  V. I. Alekseev, S. B. Dorofeev, V. P. Sidorov, and B. B. Chaivanov, *I. V. Kurchatov Institute of Atomic Energy, Moscow, Russia*

Investigation on Blast Waves Transformation to Detonation in Two-Phase Unconfined Clouds 105
  V. I. Alekseev, S. B. Dorofeev, V. P. Sidorov, and B. B. Chaivanov, *I. V. Kurchatov Institute of Atomic Energy, Moscow, Russia*

Dynamics of Gas Explosions in Vented Vessels: Review and Progress 117
  Vladimir Molkov, Anatoly Baratov, and Alexander Korolchenko, *All-Russia Scientific Research Institute for Fire Protection, Balashikha-6, Moscow, Russia*

Chapter II. Dust Explosions

Detonation Processes in Dusty Mixtures of Different Oxygen Contents 135
  Marek Wolinski, Marek Kapuscinski, and Piotr Wolanski, *Warsaw University of Technology, Warsaw, Poland*

Measurements of Cellular Structure in Spray Detonation 148
  J. Papavassiliou, A. Makris, R. Knystautas, and J. H. S. Lee, *McGill University, Montreal, Quebec, Canada*, and C. K. Westbrook and W. J. Pitz, *Lawrence Livermore National Laboratory, Livermore, California*

Experimental Investigations of Accelerating Flames and Transition to Detonation in Layered Grain Dust 170
  Y.-C. Li, C. G. Alexander, P. Wolanski, C. W. Kauffman, and M. Sichel, *University of Michigan, Ann Arbor, Michigan*

Enhancement and Generation of Detonations Using Dust Layers 185
  J. Sheng, C. W. Kauffman, M. Sichel, P. Wolanski, and N. A. Tonello, *University of Michigan, Ann Arbor, Michigan*

Detonability of Organic Dust-Air Mixtures. 195
    F. Zhang and H. Grönig, *Stosswellenlabor, RWTH Aachen, Germany*

Two-Head Detonation Structure in Cornstarch-Oxygen Mixtures 216
    F. Zhang, P. Greilich, A. v. d. Ven, and H. Grönig, *Stosswellenlabor, RWTH Aachen, Germany*

Detonation Wave Propagation in Combustible Mixtures with Variable Particle Density Distributions 228
    Shmuel Eidelman and Xiaolong Yang, *Science Applications International Corporation, McLean, Virginia*

Structure of Detonation Waves in a Vacuum with Propellant Particles 252
    Sergei A. Zhdan, *Lavrentyev Institute of Hydrodynamics, Novosibirsk, Russia*

Effect of Inert Particle Evaporation on the Chemical Reaction in a Combustible Medium. 263
    S. M. Frolov, *Russian Academy of Sciences, Moscow, Russia*, and J. M. Timmler and P. Roth,
        Universität Duisburg, Duisburg, Germany

Ignition Mechanism of Coal Suspension in Shock Waves 278
    V. M. Boiko, A. N. Papyrin, and S. V. Poplavski, *Russian Academy of Sciences, Novosibirsk, Russia*

Chapter III. Vapor Explosions

Developments of the CULDESAC Physical Explosion Model 293
    D. F. Fletcher, *Atomic Energy Authority Technology, Oxfordshire, United Kingdom*

Behavior of Free-Falling Boiling Spheres with Relation to Vapor Explosion Phenomena 322
    F. S. Gunnerson and P. R. Chappidi, *University of Central Florida, Orlando, Florida*

Effect of Fluid Flow Velocity on the Fragmentation Mechanism of a Hot Melt Drop. 334
    G. Ciccarelli and D. L. Frost, *McGill University, Montreal, Quebec, Canada*

Implications for the Existence of Thermal Detonations from Equilibrium Hugoniot Analysis 362
    D. L. Frost and G. Ciccarelli, *McGill University, Montreal, Quebec, Canada*

Flash X-Ray Visualization of the Steam Explosion of a Molten Metal Drop 388
    D. L. Frost, G. Ciccarelli, and P. Watts, *McGill University, Montreal, Quebec, Canada*

Onset of Boiling Liquid Expanding Vapor Explosion 421
    C. K. Chan and K. N. Tennankore, *Whiteshell Laboratories, Pinawa, Manitoba, Canada*, and C. A. McDevitt
    and F. R. Steward, *University of New Brunswick, Fredericton, New Brunswick, Canada*

Models of Rapid Evaporation in Nonequilibrium Mixtures of Tin and Water 432
    S. McCahan and J. E. Shepherd, *Rensselaer Polytechnic Institute, Troy, New York*

Shock Waves by Sudden Expansion of Hot Liquid. 449
    S. P. Medvedev, A. N. Polenov, B. E. Gelfand, and S. A. Tsyganov, *Russian Academy of Sciences,
        Moscow, Russia*

Thermal Detonation in Molten Sn-Water Suspension 459
    B. E. Gelfand, A. M. Bartenev, S. M. Frolov, and S. A. Tsyganov, *Russian Academy of Sciences,
        Moscow, Russia*

Chapter IV. Nonsteady Flows

Analysis of Combustion Processes in a Mobile Granular Propellant Bed 477
    Tony W. H. Sheu and Shi-Min Lee, *National Taiwan University, Taiwan, Republic of China*, Ming-Yih Chen,
        Tamkang University, Republic of China, and Vigor Yang, *Pennsylvania State University, University Park,
        Pennsylvania*

Unstable Wall Layers Created by Shock Reflections 491
    A. L. Kuhl, *Lawrence Livermore National Laboratory, El Segundo, California*, and R. E. Ferguson, K.-Y. Chien,
        and P. Collins, *Naval Surface Warfare Center, Silver Spring, Maryland*

Numerical Prediction of Mechanism on Oscillatory Instabilities in Shock-Induced Combustion.........516
Akiko Matsuo and Toshi Fujiwara, *Nagoya University, Nagoya, Japan*

Influence of Nonequilibrium Processes on Gasdynamic Parameters of Nonstationary Supersonic Jets...532
T. V. Bazhenova, V. V. Golub, A. V. Emelyanov, A. V. Eremin, A. M. Shulmeister, O. D. Miloradov, and V. T. Ziborov, *Russian Academy of Sciences, Moscow, Russia*

Shock Waves in Self-Propagating High-Temperature Synthesis Research............................539
Yury Gordopolov and Alexander Merzhanov, *Russian Academy of Sciences, Moscow, Russia*

Author Index for Volume 154 561

List of Series Volumes 563

Chapter I

Chronology of Research on Detonation Waves: 1920–1950

N. Manson*
Ecole Nationale Supérieure de Mécanique et d'Aérotechnique, Poitiers, France
and
E. K. Dabora†
University of Connecticut, Storrs, Connecticut 06269

Abstract

For the 12th ICDERS Proceedings we presented a paper entitled "A Chronology of the Early Research on Detonation Waves" in which we listed the early research, prior to 1922, that led to the discovery of the detonation wave and to the development of the theory related to this phenomenon, namely, the Chapman-Jouguet theory, which has since been considered as the classical theory of detonation. The purpose of this paper is to provide a list of the research that followed on detonation waves in homogeneous explosives and continued during the period of 1922-1950. We choose these dates because they cover an era in which new studies were conducted due to new motivations based on either technological needs or desires for fundamental understanding. Thus between 1922 and 1945, for instance, numerous studies on detonation waves were motivated by the desires: 1) to understand and eliminate "knock" in piston engines and 2) to better predict the characteristics of detonation waves and the products through calculations, especially in the case of high explosives. After 1948-50 many research investigations were conducted for the purpose of designing supersonic combustors as well as initiating new approaches to the study of the properties of matter at elevated temperatures and pressures. In highlighting some special studies published between 1920 and 1950, we intend to show to what extent these studies ushered this era.

Nomenclature

BG = burned gases
CE = condensed explosive
CJ = Chapman-Jouguet state

Copyright © 1993 by the American Institute of Aeronautics and Astronautics, Inc. All rights reserved.
*Dean Emeritus, deceased February 11, 1993
† Professor of Mechanical Engineering

d	=	tube diameter
D	=	detonation velocity
DDT	=	deflagration to detonation transition
D_S	=	shock wave velocity
D_{th}	=	"theoretical" DW velocity
DW	=	detonation wave
EOS	=	equation of state
ES	=	electric spark
GEM	=	gaseous explosive mixture
$PbEt_4$	=	tetraethyl lead
p	=	pressure
SI	=	spark ignition piston engine
SW	=	shock wave
T	=	temperature
T_S	=	shock temperature
ρ	=	density

Introduction

In our previous paper which we presented at the 12th ICDERS we covered the following: 1) Studies that led to the discovery of detonation waves in condensed explosives by F. A. Abel (1869-1874) and in gaseous mixtures by M. Berthelot and P. Vieille (1881) were cited. 2) Conclusions based on the outcome of the research performed on these waves up to 1920-1921 which was aimed at clarifying their properties regarding their application (e.g., use of explosives in coal mines, for prevention or, at least, reduction of the effects of explosions, etc.) were listed. 3) The different steps leading to the development of the gasdynamic theory of these waves as independently pursued by V. A. Michelson (1890), D. L. Chapman(1899), P. Vieille (1900), and E. Jouguet (1905-1917), were enumerated.

In the present communication, we propose to point out, in the same chronological order, studies aimed at the understanding of the detonation phenomenon over a period from 1920 and 1950. As in our previous paper, we 1) summarize the main conclusions of these studies without separating the experimental from theoretical ones; and 2) attempt to highlight the motivation of these studies which became increasingly more important with respect to the development of internal combustion engines, the exploration of possible supersonic combustion chambers designs and the necessity to learn more about the properties of matter at high temperature and pressure. To achieve the latter objective, we have chosen to cite works performed after 1950, as we believe that in doing so we can bring out the avant-garde character of certain studies of the 1920-1950 period.

On the other hand, we may well be criticized, and for good reason, for not having sufficiently described other studies (in particular, theoretical ones) which at the time of their publication and immediately afterward were considered important. It is not our intention to deny this, but if we mentioned them only briefly, it is because their results have been noted and discussed in survey papers that have been published not only during the 1920-1950s, but thereafter as well. Among such papers we cite in the text and list in the references, we mention those due to: Bone and Townsend[10], Bowden and Joffe[16b], Courant and Friedrich[31], Evans and Ablow[45], Fickett and Davis[50], Johansson and Persson[63], Jost[67], Lewis and von Elbe[81], Sokolik[128], Soloukhine[130], Taylor[132], Zeldovitch and Kompaneetz.[151]

Finally, at the end we present photographs of those researchers who contributed in a more systematic way to the improvement of our knowledge of detonation waves.

Chronology

1921

Woodbury et al.[146] conclude from their experiments on C_2H_2 and ether - air mixtures (contained in a 305-mm-long cylinder with d = 101.6 mm at initial pressures and temperatures up to 5.1 atm and 160°C, and ignited by means an ES) that autoignition of the compressed "end gas" ahead of the flame may account for engine knock.

They did not observe DDT except in rich C_2H_2 or in ether with oxygen enriched air mixtures, and noted that the pressure oscillation frequencies are identical to those of alternating brightness of the combustion of end-gases which were already observed by Dixon et al.[36,37] in 1911-1914 in a rapid compression machine equipped with a spark plug (see Comments 1 and 2).

1922

Campbell[24] observes that in mixtures of various combustible gases (H_2, C_2N_2, CS_2) with O_2 (eventually diluted in N_2) at ambiant p and T contained in tubes of various diameters (5-9 mm) the velocity D: 1) decreases with d, especially when the latter is small and when the mixture is either rich or diluted with N_2; and 2) undergoes a rapid change when the DW transits to a tube of a smaller or a larger d, and moreover, in the larger one, the DW can either be quenched and re-established if the tube is long enough, or definitely destroyed.

1923-1925

Payman and Walls[103] demonstrate that the "law of speeds" established for the uniform movement of flames applies, within the limits of experimental error, to the rate of DW, in mixtures of CH_4 and H_2 in which O_2 is present in sufficient quantity to burn the flammable gas completely to CO_2 and H_2O or to H_2O.

By means of chronophotographic records Laffitte[75a-75d] demonstrates the following: 1) For GEM contained in tubes (5 < d < 55 mm) the DDT occurs usually after a "predetonation distance" which decreases with increasing tube diameter and tube roughness and also when the igniter (an ES placed at one end of the tube) energy is increased, and 2) In spherical glass containers (20-26 cm i.d.) filled with stoichiometric O_2-C_2S or O_2-H_2 mixtures a spherical divergent DW, initiated by 1g Hg-fulminate placed at the center of the balloon, propagates with the same velocity as in tubes (see Comment 3).

He also confirmed earlier findings, on the dependence of the DW velocity of loading density and diameter of CE cartridges and demonstrated that the luminous SW launched by a DW into a gas-filled glass tube which extends along the CE cartridge may have a significantly higher velocity D_S than that of the DW.

However, this D_s drops quickly, but more slowly than the velocity of the BG that follow, at an increasing distance from the front of the SW.

Laffitte also studied the luminosity emitted by the SW as related to the nature of the gas (air, O_2, H_2, C_2H_2...) in glass tubes. His conclusions were examined and discussed later on by Perrot and Gawthorp.[107] Patry[96] (see also Refs. 78b, and 97) and his experiments were repeated very systematically in 1934-1939 by MICHEL-LEVY and MURAOUR[87a, 87b, 87c], who are currently considered the originators of the detonation driven argon flash lamp.[88]

Wendland[142] measured by means of a ballistic galvanometer chronograph, the DW velocities in 5-9 m long tubes of d = 21 mm, filled with H_2-air and CO-O_2 mixtures at ambient pressure and temperature. Away from the stoichiometric composition of these mixtures the DW velocity decreases at first slightly, then more and more and finally ceases to be uniform near the "detonation limit."

New a priori calculation of DW velocities in H_2-O_2 and CO-O_2 mixtures performed by Jouguet[68c] with more recent values of specific heat of gases at high temperature, convinced him of the necessity to consider the dissociation in the BG in a more rigorous manner.

1926, 1927

On the basis of their experiments in C_2H_2 and C_5H_{12} - air mixtures (more and less diluted by N_2, Ar, CO_2) at initial temperatures and pressures up to 230°C and 10 atm respectively in tubes of d = 9 or 16 mm and. 1.5 m long, A. EGERTON and S. GATES[44] concluded that 1) the DDT is promoted by pressure increase but delayed by temperature increase; 2) an addition of up to 1% Pb Et$_4$ has no effect on DDT; and 3) the knocking in the SI is not due to an onset of DW, but is connected with a vibratory combustion.

Laffitte and Dumanois[77] demonstrated that in the GEM contained in tubes, the DDT occurs at distances from the igniter spark that diminish as the initial pressure of the GEM is increased.

Campbell and Woodhead[25a] ascertained that for CO-O_2 mixtures contained at ordinary initial p and T in long (up to 3 m) tubes of d =15 mm the following occurred. 1) The velocity of the DW created by means of a DW in $2H_2$ + O_2 mixture falls rapidly to less than half the normal value but is re-established after the wave has travelled about 1 - 2 m. 2) During the period that immediately precedes the onset of the DW, the unburnt gaseous medium had a marked translational velocity in front of advancing flame and several separate flames appear in front of this main flame. 3) The addition to the CO-O_2 mixture of more than 1% of H_2 allows the DW to proceed immediately at its full rate. 4) Some of the chronophotographic records shows marked horizontal bands "striae" (striations) in the BG behind the front of the DW.

Later on, in another paper, Campbell and Woodhead[25b] note that close examination of some of Dixon,[37] records published in 1903, discloses also "striae" in the BG and describe new experiments they performed in order to determine the conditions required for the production of an apparently undulatory form of the DW.

According to the experiments that they performed, they concluded the following. 1) This undulatory form of the DW seems to be characteristic of DW in CO-O_2 and in some gases with O_2 and with O_2-N_2 mixtures. 2) The "undulation" of DW front in CO-O_2 mixtures are eliminated if these mixtures are saturated with H_2O vapor at ordinary temperature or contains more than 6% of H_2. 3) The distance between the "undulations" depends directly on the i.d. of the tube. 4) An increase of initial pressure to 3 atm does not seem to change this distance. 5) An interpretation based on the assumption that the DW propagates along a helical path, may explain the dependence of the length of each undulation on the tube diameter.

The book by Bone and Townsend[10] entitled: "Flame and Combustion in Gases" was published in 1927.

1928

Campbell and Finch[26] examined several possible explanations of the DW front "undulations" and of the related "striae" visible on the chronophotographic record. After numerous ingenious experiments they concluded that the BG behind the DW rotate; the wave front seems to "spin" with a pitch about three times the tube i.d., and the period of the undulation and the frequency of the striae are characteristics of the phenomenon especially if the tube is narrow. However, it remained unclear how the rotation of the BG is initiated.

Jones[64] made improved chronophotographic records of DW in cartridges (d = 32 and 76 mm) of several high explosives (i.e., nitroglycerine, dynamite) and demonstrated that the front of the DW is curved and can propagate not only with a constant normal (high) velocity but also with a low (< 3000 m/s) one.

Payman[99,100] reported several observations concerning the DDT, based on experiments that made use of a newly developed "wave speed camera" utilizing the Schlieren method and thus making it possible to photograph shock and compression waves in gases. More importantly, he showed that the pressure waves in front of the flame are not due to the igniter spark. These waves propagate at speeds greater than that of sound and appear to have their origin in the gases behind the flame front.

In 1935, after an improvement of the "wave speed camera" accomplished by Woodhead,[104a] Payman and Titman[102] studied the initiation of detonation waves in C_2H_4/O_2 and CO/O_2 mixtures and pointed out that 1) "Detonations may be set up either ahead or within the flame front due to the effect of shock waves travelling in front or behind the flame, the collision or overtaking of wave and flame or wave and wave..."; 2) "The energy of waves overtaking the flame will be added to that of detonation..."; and 3) "... The cause of production of these waves is still, however, a matter of conjuncture" (see Comment 3).

The First Symposium on Combustion[156] was held in connection with meetings of the American Chemical Society at Swampscott, Massachusetts, September 1928.

1929

In 1929, Rumpf[122] described an improved version of Jones' chronograph (with condensor and ballistic pendulum) and an optical, drum chronograph especially suitable for DW velocity measurements (with in an uncertainty of about ± 2%) in CE charges, 5-10 cm long for the first and about 1 m for the second.

Further improvements of the electrical chronograph were made in 1933 by Roth[119a] which allowed measurement of the DW velocity in CE over distances of 15-30 mm but with an uncertainty of about ± 3%.

More accurate measurements become possible in 1936 thanks to the rotating mirror camera built by Fraser (see Ref. 11 and Comment 4) and later on (1940-1945) due to the development of electronic chronographs.

1930

Lewis and Friauff[82] computed the DW velocity in numerous H_2-O_2 mixtures diluted with N_2, He, and Ar at ordinary p and T. In their calculations, they took into account the most recent (Eastman's 1929) values of the heat capacities and of the dissociation energies at high temperature of the BG components (considered ideal gases) and assumed that sound velocities are those of the "frozen" composition (i.e., corresponding to quenched equilibrium composition of the BG).

The agreement between the calculated and the measured D in tubes of d =19 mm, by means of direct and Schlieren chronophotographic records, was excellent (1 to 3% discrepancy) for undiluted mixtures and somewhat lesser (up to 6-7 %) for some diluted ones. In all cases the agreement was significantly better than that previously obtained by Jouguet[68b,68c] for these mixtures (see Comment 5).

Lewis[80] tried to calculate the DW velocity in stoichiometric H_2-O_2 mixtures by assuming that the reaction inside the DW occurs as a chain reaction in which an active product (OH) of one reaction is continuously used and regenerated. However, several discussions that followed this attempt to explain the initiation and the propagation of DW in explosives by means of chain reaction theory indicated (see e.g., Ref. 150) that the mechanism of the propagation of reactive chains and that of the DW were of quite different nature to be directly correlated. As noted by Andreev and Khariton[2] for the priming of a self-sustained reaction, what was needed was not a unique "reactive center," but a group of such centers localized in space and time.

1931, 1932

In 1931, the Institute of Chemical Physics was founded in Leningrad and headed by N. N. Semenov. There, a laboratory for studying explosive systems was organized by J. B. Khariton. Later (1939), the Institute became an Institute of the SSSR Academy of Sciences. In 1943, it was transferred to Moscow.

For a better understanding of the initiation and propagation mechanisms of DW in CE, Taylor and Weale[135a] performed several "drop tests" (fall hammer tests) under carefully controlled conditions. The results of their experiments led them to conclude that the old (1871) statement by M. Berthelot (see Ref. 5, for instance) on the fundamental role in this mechanism of "layer by layer" heating by means of mechanical shock, was to be expanded in that the necessary heating to prime the exothermic reactions might be provided by other mechanical actions, especially friction.

This study became one of the starting points of a great deal of scientific research on the initiation and propagation mechanism of DW in CE (particularly, as it will be pointed out later, by Bowden[13]) and on the practical test on sensitivity of an explosive to detonate (see e.g., Ref. 86).

In the mean time Taylor and Weale[135b] performed a new study on this problem, and their conclusions (as well as conclusions of other researchers) were the subject of a discussion during a meeting of the Faraday Society in 1938.

To clarify to what extent the velocity of a DW along a tube of constant diameter is independent of the nature of the material of the tube as previously noted in pioneering studies[5], Campbell et al.[28] used a drum camera to record the behavior of DW passing through a section of rubber tube (3.5 to 100 cm in length and varied wall thickness 0.25 to 5 mm) connecting two glass sections of the same 15-mm diameter that were long enough to ascertain whether D was constant.

The experiments performed with various GEM (made with H_2, CH_4, C_2H_2, C_2H_4, and O_2 or N_2O pure or diluted by CO_2, SO_2), initially at ordinary pressure and temperature, showed two distinct behaviors. In one class of mixtures (e.g., $2H_2 + O_2 + N_2$, $CH_4 + 2O_2$), streak records showed that the horizontal bright and dark striae were undistinguishable, and that DW passes through a considerable length of (even thin) rubber tubing without any apparent change in velocity. In another class of mixtures (e.g., $2CO + O_2$, $2H_2 + O_2 + CO_2$, $CH_4 + 7 O_2$), the striations were obvious, and the velocity as the DW passed through the rubber section showed marked reduction that depended upon such factors as the thickness and length of the rubber section. According to Campbell[28], this phenomenon may be due to pressure release of BG behind the front.

Campbell et al.[27] performed experiments on the breakup of copper foils (0.1 to 0.3 mm thick) by DW in different GEM (H_2, C_2H_4, C_2H_2, CH_4, CO with O_2), at ordinary p and T. With the copper foils calibrated by means of static air pressures, these experiments seemed to provide a method of measuring the pressure existing behind the front of the DW. The measured values agreed fairly well with the pressures calculated by the Jouguet's method.[68b] However, as noted by Gordon,[54] who in 1948 made measurements of the pressure behind the DW in H_2/O_2 and H_2/air mixtures by means of piezoelectric gauges, the Campbell et al.[27] results were somewhat fortuitous due to the uncertainties of the mechanism of the diaphragm bursting that is not exactly the same under dynamic and static conditions (occurrence and lack of wave reflection); and the accuracy of his own results had to be further improved by eliminating uncertainties in the calculation of the crystals and providing a better interpretation of the oscillographic records.

New, significantly improved measurements of the BG pressure were performed after 1948 in gases by means of the Hopkinson pressure bar (see Refs. 34 and 61) and new tourmaline gauges (see Refs. 7, 72, and 90), and in condensed explosives by means of measurement of free surface velocity.[53,43] The results obtained were often used in the discussion about the validity of the ZND theory and more especially concerning the question as to where the chemical reaction starts inside the DW, i.e., whether inside the leading SW (in a relaxation zone), or clearly behind (after a time delay) as assumed by Vieille[137] and later by Zeldovitch.[150]

1933

Pursuing his sustained efforts to clarify and specify the assertion made in the earliest studies of DW in GEM (see Ref. 5), Campbell performed with C. Whithworth and D. W. Woodhead (Ref. 29) experiments to verify that, after some propagation distance of the DW, its velocity became independent from the device used to initiate the DW in GEM contained in tubes. Using as ignitor an electrical spark surrounded by small amounts of various mixtures more or less detonable (such as $2H_2 + O_2$ and $CH_4 + 6O_2$) at one end of a long (3 to 5 m) metal (copper, lead) tube that was extended by a 2.25 m long glass tube of the same diameter (15 mm), he observed from streak records that D in CO/O_2 mixtures (saturated with H_2O vap.) initially at ordinary pressure and temperature, was constant (within ± 1%) and independent of the mixture around the igniter spark, (except for the lean and rich CO/O_2 mixtures), and that this velocity appears indeed to be mainly determined by the caloric value and density of the CO/O_2 mixture.

Careful systematic measurements of DW velocity in several condensed explosives were performed by Roth[119] and Friedrich.[52] The results obtained by the latter, on the variation of D versus charge density, were extensively used as reference until about 1960. However, the value of the temperature of the detonation products that he deduced from his measurement of D raised a controversy with Schmidt concerning the molecular mechanisms inside the DW (see Refs. 52 and 123b).

1934

Draper[39] demonstrated that the noise of a knocking SI engine is a consequence of mainly transverse vibrations of the gas in the combustion chamber, set on by the autoignition of the end gas. The frequency of these vibrations can be predicted by Rayleigh's accoustical theory[112] and is in agreement with pressure oscillations and hence with the dark and bright bands chronophotographically recorded already (see Refs. 37 and 146) in closed vessels and later on by Withrow and Boyd[144] and Withrow and Rassweller.[145] (See Comments 1 and 2).

With 1934 also came the publication of a first draft of a book by A. S. Sokolik, *Combustion and Detonation in Gases*.[128a] A more extensive version was published in 1960 and translated into English in 1963[126b.] In this version, studies performed at the Institute of Chemical Physics of the SSSR Academy of Science since its foundation, on the DW in gases are discussed.

Shtelkine and Sokolik[127] studied the DW in mixtures of H_2 and of CH_4 with oxygen and in the H_2/Cl_2 mixture (contained in long tubes to insure constancy at D) initially at ordinary temperature, and at pressures different from atmospheric down to the minimum at which DW occurs. They concluded that 1) in $CH_4 + 2 O_2$ and $2 H_2 + O_2$, D increases with p in a similar way as calculated by Jouguet's method when the effect of dissociation is taken into account; 2) in the case if the H_2/Cl_2 mixture, D is independent of the initial pressure; and 3)the assumption according to which D varies with p because of dissociation of BG behind the front of the DW is confirmed.

1935

In 1935, Scorah[125] established that the available energy of the BG in the CJ state, calculated in reference to the initial condition of the explosive, is maximum. In that same year, Schmidt[123a] resumed the a priori calculation of the DW characteristics (i.e., of the state of BG and D) using the EOS established in 1880 by Noble and Abel, namely,

$$p(v-b) = RT$$

as did Taffanel and Dautriche in 1912 for dynamite at low load density ρ (see Refs. 5 and 68b). He computed the temperature T_b of BG using newer values of the mean specific heats at constant volume.

A comparison of computed D with the measured one for different load densities ρ led Schmidt to conclude that 1) the covolume b is not constant, but is pressure (or specific volume) dependent and likely to be temperature dependent as well; and 2) the experimentally measured D versus ρ values for some typical explosives can advantageously be employed for the determination of the EOS of BG at high pressures and elevated temperatures, which can be used to compute the parameters of DW in other CE by means of the CJ theory.

This last assertion summarized the principle of the <u>inverse method</u> and became a starting point of several studies including that of Roth,[119b,119c] who had already attempted in 1939-1941 to take into account the dissociations of some of BG's components. As we will note again it was employed later by Landau and Staniukovitch,[79] Jones, Cook,[30] and several others (see Refs. 49, 85d, and 85e).

1936

In 1936, Jouguet[68d] set forth a theory that described the characteristics of the SW generated by a DW when the latter reached the end of cylindrical high explosive cartridge. The calculation performed by means of the relations developed in this theory allowed the inference of some details such as the velocity of SW being higher than that of the DW and the reason for the luminosity of the SW in different gases observed by Laffitte,[75d] and Perrot, Gawthorp,[107] Patry,[96,97] and Michel-Levy and Muraour[87] (see also Ref. 85b).

After examining the detailed photographic records of Fraser, Wheeler, and Bone[11b,12] on "spinning detonations," Jouguet concluded the following. 1) The DW could not spin, since the alternate bright and dark bands and the ondulations of the front of the wave were registered on the chronophotograph, no matter what the internal section of the tube happened to be (circular, rectangular, or even if provided with a rod along its axis or along its wall). 2) In the same GEM contained in circular tubes, one could observe one, two, or even three systems of bands having different frequencies (e.g., in a CH_4/O_2 mixture, $f_1 = 68$, $f_2 = 110$ and, $f_3 = 221$ kH, depending on the internal diameter of the tube). 3) The "spinning" DW would actually be a coupling, more or less stable, of an SW with a "reactive localized head of intense combustion" that "travels spiral wise as a wave front through otherwise stationary medium."

"Spinning detonations" were acknowledged as characteristic of DW in GEM close to the detonation limits, in composition by Breton[18,76] and in initial pressure, and in inert dilution by Rivine and Sokolik[116a,b] (see also Ref. 128b).

These latter underscore that it is strictly necessary to distinguish the "explosion limits" (i.e., the initial composition and pressure of the GEM, where the DDT is just possible), from the classical "detonation limits" as determined by Wendlant[142] and Breton[18] (i.e., the initial conditions of the GEM where a constant velocity DW can propagate). Following their experiments (performed in the 30-m-long lead tube of 18 mm in diameter equipped with windows for chronographic records), they ascertained that the "explosion domain" 1) in H_2-air mixtures is much narrower than the "detonation domain" and may at initial pressure lower than 1 atm split in two separate domains; and 2) in HC-air (Ref. 116b) mixtures even initially at 1 atm this domain may not exist (except for C_2H_2 and C_2H_4).

They also noted that Crussard and Jouguet (see Refs. 5 and 68b) had already tried to formulate criteria for the "explosion limit" by taking into account the ability of the SW to heat the GEM to its self-ignition temperature, which seemed to predict qualitatively the composition limit. But inasmuch as the movement of the flame was never fully independent from the power of the igniter, it was not easy to cleary distinguish the explosion limit from the detonation one.

It is worth noting that subsequent investigators who were interested in the detonation limit (e.g. Mooradian and Gordon[90]) seemed to completely ignore the problem set forth by Rivine and Sokolik. The latter recognized that close to the detonation limit the reaction zone became wider behind the SW (Ref. 116b) (see also Ref. 141), and confirmed that the spin characteristics were well-defined near the limit, and that as a consequence, a detonation limits theory must in fact be the same as that of a spinning detonation.[110b]

Also, later striae, characteristic of SW spinning detonations, have been observed on Schlieren streak records, in the compressed unburnt gases between the reaction zone and the SW (Ref. 21).

1937

In 1937, new measurements of the minimum pressure at which a DDT occurs and of the velocity of laminar flame at these pressures in several combustible mixtures (H_2, C_2H_2, and CH_4 with O_2 or air) were performed by Bresker, Rivine, and Sokolik (see Ref. 128b) in a 30-m-long tube of 18 mm i.d.

Shtelkine and Sokolik[127] conducted experiments in mixtures of pentane-air with or without a small amount of Pb Et$_4$, contained in a 220-cm-long, 28-mm-i.d. glass tube at pressures of 50 to 500 mm Hg. They noted that after a preheat to a temperature of about 50°C, the DDT distance was clearly shorter than in the case of no preheat and without Pb Et$_4$. If the preheat temperature was higher, reaching the temperature at which a cold flame occurs, this distance diminished further. It exhibited a minimum that depended on the preheat temperature and the time elapsed between the end of the preheat and the igniting spark. The possible role of cool flames had been re-examined already by Shtelkine.[126a]

The Second Symposium on Combustion[159] was held in 1937 in connection with the meeting of the American Chemical Society in Rochester, New York. Most of the papers dealt with knock in SI engines.

1938 - 1940

Two classical books were published during this time period, including the first edition of *Combustion Flames and Explosions* by Lewis and Elbe[81] (a much more extended edition of this text appeared in 1951) and *Explosion und Verbrennungsforgange in Gasen* by Jost.[67] (An English translation of this was published in 1946 by Mc Graw Hill, New York).

At this time, Apine[3a] explained the "layer by layer" propagation of the DW in CE, theorizing that a small volume undergoes an explosive combustion which generates "microjets" which start the combustion of adjacent volumes. Later on,[3b] taking into account the conclusions of studies on the onset of DW (performed in the mean time by Belaev and Andreev), he specified the stages of the mechanism he proposed and reviewed with Bobolev[9] the properties of CE that have to be taken into account for the understanding of the ability of the CE to detonate.

Rosing and Khariton[118] developed a criteria for determining the critical diameter d_{cr} of CE charge by comparing the reaction time τ_r to the time $\tau_{exp} = d/2a_s$ (where a_s = the speed of sound) needed for an expansion wave (in the reaction zone behind the front of the DW) to reach the axis of the charge. They asserted that, to a first approximation, the self-sustained DW could be produced if $\tau_r > \tau_{exp}$, i.e., if $d < d_{cr} = 2\ \tau_r a_s$, and that in order to determine exact values of d_{cr} it was necessary to take into account not only the properties of the CE (load density in particular) but the inertia and the resistance of the casing (confinement) as well. Later on this problem was re-examined, by Bobolev[9] among others.

Extending the theory of Vieille[137b] regarding the coupling between the chemical reaction and the SW inside the DW and taking into account the frictional and heat losses to the tube wall inside the reaction zone behind the SW and the exponential variation of the induction time τ_i of the reaction with the shock temperature T_s, Zeldovitch[150a,150b] developed a theory of the structure of the DW and derived an approximate relation for the detonation velocity in a tube of diameter, d:

$$D = D_\infty(1 - e^*/d)$$

where D_∞ is the value of D in the unconfined explosive (i.e., when d is infinite) and the theoretical width of the reaction zone and $e^* = \tau_i D_\infty$ (i.e. the distance between the front of the SW and the CJ plane) that depends on T_s.

This relation allowed ZELDOVITCH to discuss the detonation limits and was used later to compare the measured values of the DW velocity with those calculated a priori, first by Kistiakovsky et al.[72a] and then by many others.

Some developments on the phenomena within the DW were also performed (in 1942) by Neumann[92] and (in 1943) by Doring.[38] These two were not aware of Vieille's theory nor of Zeldovitch's work.

All three men, Zeldovitch, von Neumann, and Doring noted the existence of a maximum pressure (that is usually called the "ZND-SPIKE") whose value is greater than the pressure in the BG in the CJ plane (i.e., the CJ pressure) as theorized already by Vieille in 1900 who computed its order of magnitude quite accurately. However, despite a great number of attempts to verify the existence of the ZND spike in either GEM or in CE, it still remains questionable.

Shtelkine[126b,126c] completed his first study concerning the effect of metallic spirals placed in contact with the internal surface of glass tubes (1.2 to 1.5 m long, 18 mm i.d.) on the DDT and D values in various gaseous hydrocarbon-oxygen mixtures and especially in hexane-air mixture at initial ordinary temperatures and pressures of 0.2 - 1 atm. Based on this study and of those published in 1945 (Ref. 125e), he concluded that the presence of the spiral reduces: 1) appreciably the predetonation distance, and makes it possible to obtain DDT in C_6H_{14} -air mixtures at ordinary p and T; and 2) the value of D, depending on the ratio of the diameter of the spiral wire to the tube diameter, the spiral pitch and composition of the mixture, with the effect being more pronounced for lean or rich mixtures than for the stoichiometric. These conclusions would be confirmed and extended later by Shtelkine[126d,126e,126f] himself and also by Guenoche.[59,60]

Zeldovitch[150c] examined the extent to which a DW stabilized in a supersonic combustion chamber could be used for ramjet propulsion and concluded that from a thermodynamic point of view, this process is of lower interest than that of a subsonic combustion chamber.

1941, 1942

At this time, Sokolik[128c,128d] studied with M. A. Rivine the variation of the lean detonation limit with initial temperatures of up to 500°C of H_2-air mixtures at ordinary pressure in 1.5-m-long, 20-mm-i.d. glass tubes initiated by means of an SW produced by a DW in $2H_2 + O_2$ mixture. Sokolik concluded that as long as the initial temperature is lower than 150°C, the onset of the DW is somewhat difficult, but when this temperature exceeds 200°C, it is clearly easier (and the H_2 limit in the mixture drops from 21% at 20°C to 17% at higher T. These conclusions, however, were vigorously contested by Rivine[115a,115b] who pointed out the inadequacy of the techniques used.

Zeldovitch and Ratner[153] described methods for approximate and for rigorous computations of DW characteristics in GEM which took into account the best available thermodynamic data of gases at high temperatures (and, especially, the dissociation energies). By comparing the measured D with the calculated one for different possible values of dissociation energies of some of the BG components of the CO/O_2 and C_2N_2/O_2 mixtures, they demonstrated the ability to choose the most appropriate values of these energies. This procedure was significantly extended later by Kistiakovsky et al.[72a,72b](see also Ref. 41).

By comparing the increase of D with the density of CE (PETN and TNT at $0.2 < \rho < 1.0$ g/cm^3) and of GEM ($2H_2 + O_2$ at ordinary temperature and pressures such that $0.2 < \rho < 0.5$ g/cm^3), Schmidt[123c] noted that the variation is practically similar (somewhat more than linear) and concluded that the thermogasdynamic theory in respect to the main characteristics of the DW and more especially D, is equally valid for both systems.

Kistiakovsky and Wilson[73] summarized the main features of the CJ theory and the properties of the rarefaction and of the SW. They also drew attention to the a priori calculations of the characteristics of DW in CE, recently studied in cooperation with R. S. Halford, and those in the Bureau of Mines (by

Brown[23] and as noted in (Ref. 73) by Mac Dougal and Epstein). The E.O.S. of the BG used was the one proposed in 1921-1922 by Becker (see Ref. 5) and the required values of the constants needed in this E.O.S, were determined following E. SCHMIDT's[183] "inverse method," i.e., in using the measured D vs ρ in some "representative" CE. Similar, but improved, calculations were performed in 1942 by Brinkley and Wilson[19] and later with further improvement at Los Alamos by several others (Fickett and Wood,[49] Cowan and Fickett,[32] and Mader[83]).

Zeldovitch[150d] developed a theory of nonplanar DW: cylindrical and spherical. He demonstrated especially that, unlike Jouguet's[68,a,68b] conclusion made in 1907 such waves when they are divergent, exhibit the same uniform propagation velocity as the planar one, but the condition of their existence is more severe. The same conclusions were reached, independently at the same time, by Taylor,[133] but published several years later after declassification.

1943-1945

Unaware of Jouguet's theoretical study (Ref. 68d) of the phenomena that occurs when a DW reaches the end of an explosive cartridge, Grib[56] advanced a similar theory on the movement (assumed to be one-dimensional and isentropic) of the BG and of the SW in the surrounding atmosphere. His results are in accordance with the main conclusions of Jouguet.

Pokrovsky and Staniukovitch[108] as well as Landau and Staniukovitch[79] consider that in CE the variation of D with the load density ρ may be written as:

$$D = D_1 \rho^m$$

where D_1 is the measured D value for $\rho = 1$ and the exponent m can vary from 0.3-0.8 (about 0.79 ± 0.01 for the usual high explosives such as TNT, PETN, picric acid). By assuming that the E.O.S. of the BG is:

$$p = AT\rho + p'$$

where A is a coefficient depending on the degrees of freedom of the components of the BG, and p' is the component of pressure which is independent of the temperature, they show that

$$p\rho^{-n} = \mathrm{const}$$

where the values of n (about 3) can be determined by taking into account Eq. (1), and indicate that from this, the main characteristics of the expansion of the BG behind the DW can be calculated.

Goranson[53] reported on the methods developed at the Los Alamos National Laboratory, for measurements of the speed of target plates located at the boundary of CE charges. These methods were used by Goranson and Malin to determine the dynamic properties of various materials (solids and liquids) at high pressures (up to 600 kB) as well as the characteristics of the detonation products.

It may be interesting to note that many of the results of these determinations (e.g., Ref. 43) were published before Goranson's report was declassified in 1974 and that ever since, the methods remain applicable to research on the properties of materials at high pressures and elevated temperatures. These methods were also used in many experiments, to determine the "thickness" of DW in CE (Jacobs[62] and Duff[41]).

Zeldovitch[150f] (see also Ref. 151) offered an explanation of the spinning detonation proposed some months earlier by Shtelkine,[126g] based on the assumption that the front of such DW is not plane but divided in two parts (one of which being oblique) by a "break" that moves along a circumference inducing rotation of the BG and triggering the chemical reaction. However, after pointing out that it is impossible to explain the rotatory motion of the BG by means of fluid dynamics laws, Zeldovitch[150f] proposed first to assume that only the oblique part of the frontal SW ensures the ignition of the GEM, because this part is in fact a strong DW and its velocity is about 30% higher than that of the front. By means of some calculations he then demonstrated this oblique part moves spirally along the circumference with rotational frequency of the order of the experimentally observed frequency.

Later (in 1950) Brodsky and Zeldovitch[22] performed in the framework of the preceding Shtelkine - Zeldovitch theory of spin, new calculations of the characteristics of BG and specified the development of the chain reactions behind the oblique and the straight part of the spinning detonation front in H_2-air mixtures. They indicated that their calculations demonstrated that spinning detonations are possible only in diluted mixtures and that it is possible to evaluate approximately the composition detonation limits and their widening with the tube diameter. However, as they noted, the results concerning the influence of the initial pressure on the limits were contradictory to the experimental observations, and the phenomena in noncircular tubes remained unexplained.

An explanation of the recorded phenomena, the clear and dark bands characteristic of "spinning detonation" in circular as well as in rectangular tubes, was advanced at nearly the same time by Manson[85a,85b] who was then unaware of the Shtelkine-Zeldovitch theory. By assuming that the BG movement is not strictly one-dimensional but proceeds also in the transverse direction as well, he analyzed the "spin" phenomena taking into account a superposition of a vibratory motion, described by Rayleigh's acoustical theory,[112] on the longitudinal unidimensional motion as assumed in the classical CJ theory. Because of the supersonic value of D and the CJ condition, only the transverse components of the vibratory motion of the BG were significant, and it appeared that the characteristic frequencies of this motion were in very good agreement with those measured by Campbell et al.[26] as well as by Bone et al.[11,12] whatever the form of the tube section.

The same results based on the same assumptions were obtained independently, first in 1948 by G. I. Taylor who, in addition, determined the frequencies of the vibratory motion in tubes whose sections were changed by means of rods of different diameter [this was not published until 1954 (Ref. 134)], and then in 1952 by Fay.[47] However, at that time none of these investigators had provided an explanation on how this transverse vibration was set and how it was sustained (see Comment 6).

1947

To verify that the spin appears definitely just before the limit of propagation of a DW is reached, Rakipova et al.,[110a] made several chronophotographic records of the DW in H_2-O_2 and H_2-air mixtures contained in glass tubes (1.5-1.8 m long, and 15 - 17.5 mm i.d.) and ignited by means of a small PETN charge. They concluded that their experiments confirm Breton[18], and Rivine and Sokolik's[116] statements concerning the spin appearance as the mixture composition tends to the detonation limit, and that the theory of these limits must be that of spinning detonation.

Aivazov and Zeldovitch[1] made numerous measurements of D in GEM contained in tubes of decreasing diameter. In fact, their experiments partly repeated those performed by Campbell[24] in 1922. However, concerning the velocity of strong DW that are produced when the DW transits to a tube of smaller diameter and the characteristics of the DW reflection at the closed end of the tube, the information they provided is much more instructive.

To explain the dependency of D on the cartridge diameter, Jones[65a] developed a theory assuming that the main cause of this dependency is the lateral expansion of the medium due to the high pressure behind the leading SW. According to him, this expansion could be described as a Prandtl-Meyer one, and using other simplifying assumptions, he established that

$$D^{-2} \approx D_{th}^{-2} (1 + \alpha \, d^{-n})$$

where D_{th} is the velocity of the DW when d is infinite, n = 2, and α is a constant related to the reaction zone thickness and the confinement characteristics.

Shtelkine[126e] reported the results of new systematic experiments concerning the effect of the tube roughness on the DDT. He used easily detonable GEM in smooth glass and metallic tubes of $10 \leq d \leq 37.5$ mm, equipped with spirals, of varied wire diameter and pitch, placed about 8-10 diameters from the closed end of the tube at which the ignition is initiated by a hot wire. He concluded that the predetonation length 1) was higher in smooth tubes than in "rough" tubes by a factor of 2.5-10 times, depending on GEM composition, its initial pressure, and the spiral wire diameter; 2) decreases as the initial pressure increases whether the tube is smooth or "rough"; 3) increases if the spiral is placed farther away from the igniter; and 4) that a theory of the DDT must take into account all gasdynamic phenomena that occur in unburnt gases in front of the flame.

Zeldovitch and Rozlovsky[154] conducted experiments on flame propagation of H_2-O_2 (with CS_2 added for increased luminosity) in 15-cm-diameter spherical vessels. Their streak records showed that at p = 10 atm, DDT occured after the flame travelled about 1/6 of the vessel radius. Although they were unable to measure D, they reasoned that at DDT the Reynolds number based on flame velocity was unexpectedly high. They thus reasoned that on the basis of the Landau's self-turbulization theory, such a value indicates the existence of some other mechanism for flame (detonation) stabilization (see Comment 7).

Ratner[111] discussed the computations of the DW characteristic performed before 1946. However, he was unaware of those performed at this time in the

United States by Kistiakovsky et al[19,73], and in his review he underscored chiefly the gaps of the EOS of BG considered by Schmidt,[123a] Roth[119c] and Landau and Staniukovitch.[79] However, he noted the pioneering attempts made by Becker[5].

At about the same time several new studies dealing with the a priori computations of CE were reported by Cook[30], Jones and Miller,[66] and Paterson.[98] Cook used the

$$pV = nRT + b(V)p$$

EOS, and, after a comparison of his own results with those of Brinkley and Wilson,[19] concluded that the only detonation BG property which could provide experimental information on the accuracy of various EOS was the "detonation temperature". Jones and Miller (see also Ref. 132) used the

$$pV = nRT + b(T)p + b'(T)p^2 + b''(T)p^3$$

EOS and obtained (despite simplified BG composition) D variation with load density in a quite good agreement with the experiment. Paterson used the Boltzmann's EOS and obtained numerous results useful for comparing different CE from the point of view of practice.

Later on, as already noted, many new EOS were developed on different bases (see, e.g., Ref. 69), and the comparisons between the calculated and measured values of the DW characteristics became more and more instructive (see Refs. 62 and 122b).

1948

In 1948, the now classical monograph "Supersonic Flow and Shock Waves" by Courant and Friedrichs[31] was published. The same year, the Third Symposium on Combustion and Flame and Explosion phenomena was held in Madison, Wisconsin.

To verify some conclusions concerning the effect of losses at the tube walls, Kogarko and Zeldovitch[74] tried to make pressure measurements by means of copper crushers placed at the end of a 12.2-m-long and 305-mm-diam detonation tube. Their results were very questionable, but they observed that for H_2-air mixtures, initially at room temperature and pressure, if the diameter of the tube is increased, the detonation domain between the rich and lean limits increases; and that near these limits the DW remains "spinning."

Kistiakovsky[71] reviewed and discussed the studies on the initiation of detonation of explosives (mostly CE) performed more frequently during 1939-1945 (World War II). In particular, he underscored the finding of Bowden et al.[13-15] who performed many studies, including some with Yoffe[16a,16b] on CE, and summarized the state of our knowledge concerning 1) the primordial role of the hot spots; 2) the increase of sensitivity of CE to mechanical impact when the melting temperature of grit added to the explosive is above a specific temperature (e.g. for PETN, 625 K); 3) the three factors, i.e., the adiabatic compression of gas bubbles entrapped in CE, the heating by friction between the particles and by the viscous flow at the impacting surfaces; 4) the existence of two phases for all detonations by impact, the first during which a deflagration propagates with moderate velocity, and the second abrupt where a change from deflagration to a high velocity detonation occurs.

Note that at this same time in the SSSR some similar features were studied and discussed (see, e.g., Khariton[70a,70b]).

By referring to Schmidt's suggestion[123a] on the use of D correlation when the EOS of BG is unknown, Jones[65b] deduced from the main relations of the CJ theory quite simple expressions for the pressure, p_{CJ}, and the density, ρ_{CJ}, of the BG in the CJ state. However, as indicated later by Ref. 109 because of the approximations needed for the numerical calculations, the derived expressions were highly questionable.

1949, 1950

During these years, Cybulsky et al.[33] reported a study performed from 1943 to 1946 by means of a high-speed rotating mirror camera, on the influence of different initial parameters of various charges (up to 3.17 cm for d and up to 40 cm in length CE for picric acid, tetryl, etc., and especially cast TNT) on D. Among the realized findings (such as the limit diameter and the diameter above which D remains constant), the authors noted that differences may exist in the reaction rate along different crystallographic axes, and therefore the propagation of DW can be stable along one axis and unstable along another.

During a meeting at the Royal Society of London held under the chairmanship of W. G. Penney,[106] reports were presented on theoretical and experimental investigations performed in the last few years on physical and chemical processes occurring in a DW (including the just declassified work). In particular, the already noted works [of Bowden[13] on detonation pressure in GEM,[34] of Jones,[65c] and also of others (e.g. on high and low velocity[131])] were discussed.

Eyring et al.[46] presented a theory of the variation of D in CE with d and the casing inertia. By taking into account the curvature of the DW front which they assumed to be due to the lateral expansion of the material (behind the front), they demonstrated that:

$$D \approx D_{th}(1 + a^* d^{-n})$$

where D_{th} is the "ideal" DW velocity (i.e., such as its value is defined in the CJ theory), and a^* and n are some coefficients that depend on the reaction zone length e_r and the characteristic of the casing. They concluded that in agreement with experimental results, 1) the coefficient a^* may be considered as a measure of the width of the reaction zone; 2) the exponent n can be taken in practice to be equal to unity if the charge is unconfined or is very heavily confined and 3) for intermediate casing, a^* is proportional to the ratio of the mass of explosive to that of the casing per unit length of the charge.

Later on the problem of the DW curvature was reconsidered in detail by Wood and Kirkwood.[148]

Zeldovitch and Shliapintokh[155] resumed an early unfinished study undertaken by Zeldovitch and Leipunsky[152] on the triggering of chemical reaction by SW. They performed experiments on the ignition and the subsequent phenomena that occur when a supersonic (< 1.8 km/s initial velocity) rifle bullet is fired into a tube filled with different GEM, and also when the bullet is set into a supersonic jet (flow). From shadow photography they noted a "striated structure" of the movement of the gases around the bullet and concluded that 1) the frequency of the striae is approximately the same as that observed on streak camera records of spinning detonation in the same GEM and 2) the increase of the luminosity of BG near the axis of the tube in the wake seems to be due to the crossing of Mach waves.

Experiments on DW produced by means of rifle bullet fired in GEM were resumed by several investigators later on (e.g., Ruegg[121], and Behrens et al.[6]).

Reviewing the Shtelkine-Zeldovitch[150f] theory of the spin (Ref. 129), Voinov[139] concluded that the triggering of the exothermic reactions can occur behind the leading SW of the DW only if there were also some transverse SW. These waves move back and forth and the structure of a spinning DW appears to be that of an unstable coupling of the SW with the reaction zone.

A very detailed survey of the studies performed thus far by several investigators on the DDT phenomena in GEM contained in tubes led Kistiakovsky[71b] to underscore that 1) the nearly discontinuous initiation of detonation was due to an autoignition that occurs between the SW and the front of the original deflagration; 2) the shortening of the predetonation run up distances in tubes with macroscopically rough tube walls (see Shtelkine[126f]) suggested that the acceleration of the deflagration was not only due to the development of turbulent movement of GEM between the SW and the reaction zone (flame front) but also to the reflection of SW at oblique angles by the uneven walls; 3) the cause of pulsating detonation observed already in 1926-1927 by Campbell et al.[29] in CO-O_2 mixtures was not the detonation spin. On the contrary, this latter was an effect of the former and was due to reflections of oblique shocks formed by the autoignition.

Several of these statements were amplified later on, particularly by Greifer et al.[55] and by Oppenheim et al.,[94,95] and especially by Urtiew.[136]

Reingold and Viaud[113] resumed Roy's proposal[120] to use detonation for propulsion, and they described a design on the stabilization of combustion in supersonic flow.[114] The completed facility allowed him to stabilize by means of an SW system (created thanks to an appropriate injector) the combustion of H_2 (and later on that of kerosene) in a Mach = 2.5 flow of "vitiated air" (i.e., burnt gases of about 900 K stagnation temperature produced in a turbojet combustion chamber and slightly enriched with O_2).

Later on, several teams of researchers (Gross[57,58] and Nicholls et al.[93]) pursued the studies of standing detonation waves with the idea of using such waves for scramjets.

Comments

1) In contrast to Nernst's[91] (see also Ref. 5) opinion in 1905 that the "anomaleous combustion" in SI engines was due to buildup of DW, Ricardo[117] (1912-1920) believed that knock was due to autoignition of "end gas."

At that time the temperature history of these gases was already known, thanks to Hopkinson[61] and Mache,[84] as well as the antiknock effects of additives such as Pb Et$_4$, thanks to Midgley Jr. (see Ref. 17).

2) Following the 1928-1930 noteworthy improvements of cracking and reforming processes of fuels and the setting up of the CFR test engine that allowed a better characterization of these fuels by means of a global antiknock property, namely the "octane number," numerous new experimental studies were undertaken (see, e.g., Refs. 67, 81, 85f, and 128b).

Besides, according to several investigators such as Draper (see Ref. 85f), the knock of SI engines appeared to be a consequence of the explosive selfignition of "end gas" as stated by Ricardo[117]; according to others such as Voinov and Sokolik,[138] the knock was a consequence of a DDT. Although the occurrence of such DDT in the conditions of p and T that arise in the SI combustion chamber is unlikely as shown by Egerton and Gates,[44] these investigators interpreted the bands perpendicular to the time axis on the chronophotographic records, not as due to transverse vibratory phenomena, but as shock waves that move back and forth in the chamber after DDT.

Finally, to combine both points of view, in 1946 Miller[89] suggested that the characteristic vibratory phenomena of knocking are of a particular kind which he called "detonation wave type vibration."

3) At this time Laffitte's assertion[75b] of the existence of a spherical divergent DW that propagated with the same velocity as the DW in tubes was in disagreement with the conclusions of Jouguet's theoretical studies.[68] Later on, however, it not only appeared to be in agreement with a theory developed independently by Zeldovitch[150d] and Taylor,[133] but it was also fully comfirmed by experiments performed with numerous combustible gases with oxygen mixtures (H$_2$/O$_2$, C$_2$H$_2$/O$_2$, C$_2$H$_4$/O$_2$, etc.) contained in large plastic balloons (0.5-1 m diameter), first by Ferie and Manson[48,85c] and Freiwald and Uhde,[51] and afterwards by several others. Also, 1) after that, Cheret and Verdes (see Ref. 20) demonstrated in 1970 that a spherical divergent DW propagated also in solid explosives, and a comparison of the DW velocities in GEM and in CE (solid and liquid) in tubular, cylindrical, and spherical charges were made by Brochet et al.[20] (2) Some experiments were performed that showed evidence of the spherical DDT, more successfully by Rakipova[110] and also by Zeldovitch and Rozlovsky.[154]

4) During the development of the wave speed camera, Payman[99], with Robinson (in 1926) and with W. C. F. Shepherd (in 1927) performed several studies related to the safety of explosives used in coal mines. Once improved (with Woodhead[104]), this camera enabled Payman et al.[104b,105] to systematically study explosions and SW sent out by detonators, as well the DDT in GEM,[101a] and also (in 1946) with Shepherd[101b] the phenomena inside a shock tube, the first built after that of Paul Vielle[137]; see also Ref. 40.

5) In the 1950's, Berets et al.[7] performed the same calculation of D as Lewis and Friauff[82] but with much more improved thermodynamics data of the BG components at high temperature. They stated that the discrepancy noted by Lewis and Friauff[82] could be due to effects that were not taken into account in the classical CJ theory of DW, or were due to the lack of energy equipartition inside of BG components.

The Lewis and Friauff[82] choice of the "frozen" velocity of sound was at the origin of numerous discussions, despite the fact that the calculated D by different investigators as well as the accuracy of measurement of D were within 1-2%. However, this was not always the case for other characteristics (pressure, density, etc.) of BG behind the front of the DW (see Duff et al.[42] and White,[143] for instance).

Wood and Salzburg[149] concluded from their extensive study of the CJ-ZND theories that the calculations of the values of the BG parameters in the CJ state could only be done after assuming that this state was an equilibrium state and not a frozen one, since, it had clearly been demonstrated by Voitsekhovsky[140] in 1957, and by Denissov and Trochine[35] in 1959, that the structure of the DW was multidimensional (i.e., appeared to be a coupling between the SW and reaction zones). The utility of the discussion on which value of the sound velocity, "frozen" or "equilibrium," that should be considered in the CJ state became moot.

6) A straightforward explanation of the vibration phenomena in BG of a spinning detonation was given in 1959 only after analyzing the back and forth movement of the refracted branch of the leading SW. This occured, however, after the following developments.

a) In 1950 Voinov[139] on the one hand and Kistiakovsky[71b] on the other hand independently pointed out the probable importance of transverse oblique SW in the birth of autoignition.

b) In 1956, by assuming that the front of the spinning DW behaved very much like a rotating heat source that generates pressure waves, Boa Teh Chu[8] demonstrated that the waves propagated out of a nondecaying wave train if their frequency was not lower than the natural frequency of transverse vibration of gases in a tube behind the DW front.

c) In 1957-1958, Voitsekhovsky[140a] ascertained (thanks to a technique[110] that allowed him to obtain stationary photographs of the DW) that the "break" considered in the Shtelkine-Zeldovitch[126g,150f] theory was related to Voinov's[139] transverse SW wave; he also analyzed the state of the gases at the triple point where the crossing of the two SW and the DW occurred.

d) In 1959 Denissov and Trochine[35] demonstrated (by means of the soot track method) the changes of the structure and of the characteristics (number of "pulsation" and the size of "cells") of the spinning DW, as affected by the GEM composition, its initial pressure, and the tube size.

e) Also in 1959 Soloukhine and Topchian (see Ref. 128) demonstrated the mechanism of the maintenance (support) of the transverse vibration in the tail of the DW. Concerning the structure and its continuous change, however, the ideas put forward by White[143] on the possible role of the turbulence left the problem open. Only the structure of the pure spinning DW in round tubes became, thanks to Schott,[124] a very instructive description.

7) Concerning the Zeldovitch and Rozlovsky[154] conclusions and the Rakipova et al.[110a] experiments on spherical DDT in plastic balloons as noted by Sokolik,[128b] the explanation of DDT by "selfturbulization" is questionable, and the absence of an accelerating period of a spherical flame similar to that

observed in tubes indicated the impossibility of an accumulation of compression waves that build up into a spherical SW.

Acknowledgment

Among the publications that we have cited, there are several that we would not have been able to examine or note, if it were not for the assistance of many colleagues. B. A. Khasainov of the Institute of Chemical-Physics, SSSR Academy of Sciences, was of particular help. Professor P. Bauer, who coauthored the first paper on the Chronology of Detonation Waves, has again assisted us during the initial stage of this paper. Finally, A. Barreau, and L. Hockla patiently assisted to bring this manuscript to its final form. To all of the above and many others, we would like to extend our gratitude and thanks.

W. BONE
1870-1939

C. CAMPBELL
1886-1953

S.Ch. DRAPER
1901-1987

A. EGERTON
1886-1959

Marcelin BERTHELOT
1827-1907

Paul VIEILLE
1854-1934

Ernest MALLARD
1833-1894

Henry LE CHATELIER
1850-1936

Harold DIXON
1852-1930

Vladimir MICHELSON
1860-1928

CHRONOLOGY OF DETONATION WAVES

W. JOST
1903-1988

G.B. KISTIAKOVSKY
1900-1982

P. LAFFITTE
1898-1981

J. von NEUMAN
1903-1957

W. PAYMAN
1896-1946

A. SCHMIDT
1881-1970

A.S. SOKOLIK
1899-1969

K.I. SHCHELKINE
1911-1968

G.I. TAYLOR
1886-1975

J.B. ZELDOVITCH
1914-1987

Alfred NOBEL
1833-1896

Frederic ABEL
1827-1902

References

[1] Aivazov, B. V., and Zeldovitch, Ja. B., "Obrasovanie perejatoi detonazoinoi volni v sujaushtoi trubke," (Formation of Strong Detonation Wave in a Constricted Tube), *Zhurnal Experimentalnoi i Teoritishekeskoi Fiziki. Moscow*, Vol. 17, 1947, pp. 889-900.

[2] Andreev, K. K., and Khariton, Ju B., "Nekotorie Soobrajenia o zmekhanisme Samorasprostronianushtikh Reakzii," (Some Reflexion on the Mechanism of Self-Sustained Reactions), *Doklady of the Acad. of Sciences SSSR*, Vol. 1, 1934, pp. 402-404.

[3] Apine, A. J., a) "O mekanisme vzrivshatova razlojenia tetrila," (On the Mechanism of Explosive Decomposition of Tetryl), *Doklady of the Acad. of Sciences SSSR*, Vol. 24, 1939, pp. 223-225; b) "O detonazi i vzrivnom goreni vzrivzhatikh vestshestv," (On the Detonation and Explosive Combustion of Explosives), *Doklady of the Acad. of Sciences SSSR*, Vol. 50, 1945, pp. 285-288.

[4]Apine, A. J. and Bobolev V. K., "Vlianie fisisheskoï strukturi vzrivshatikh vesshestv na ikh detonazionosposobnost," (Influence of the Physical Structure of Explosives on their Ability to Detonate), *Zhurnal Fizisheskoi Khimii*, Vol. 20, 1946, pp. 1367-1370.

[5]Bauer, P., Dabora, E. K., and Manson, N., "Chronology of the Early Research on Detonation Wave," *Dynamics of Detonations and Explosions: Detonations*, Progress in Astronautics and Aeronautics, edited by A. L. Kuhl et al., Vol. 133, AIAA, Washington, DC, 1991, pp. 3-18.

[6]Behrens, H., Struth, W., and Wecken, F., "Studies of Hypervelocity Firings into Mixtures of Hydrogen With Air or Oxygen," *Xth Symposium (International) on Combustion*, The Combustion Institute, 1965, pp. 245-252.

[7]Berets, D. J., Greene, E. F., and Kistiakovsky, G. B., "Gaseous Detonations. I: Stationary Waves in Hydrogen-Oxygen Mixtures. II: Initiation by Shock Waves," *Journal of the Chemical Society*, Vol. 72, 1950, pp. 1080-1086, 1086-1091.

[8]Boa Teh Chu, "Vibration of the Gaseous Column Behind a Strong Detonation." *Proceeding of the Gas Dynamics Aerothermochemistry*, Northwestern Univ., Evanston, IL., 1956, pp. 95-111.

[9]Bobolev, V. K., "O Predelnikh diametrov gomoguenikh zariadov khimitsheki odnorodnikh vzrivshatikh veshestv" (On the Critical Diameter of Chemically Homogeneous Explosive Charges), *Doklady of the Acad. of Sciences SSSR*, Vol. 57, 1947, pp. 784-792.

[10]Bone, W. A., and Townsend, D. T. A., *Flame and Combustion*, Longmans, Green and Co., London, 1927.

[11]Bone, W.A., and Fraser, R. P., a) "A Photographic Investigation of Flame Movements in Carbonic Oxygen Explosion, Part III," *Philosophical Transactions of the Royal Soc.*, Vol. 223, 1929, pp. 197-234; b) "Photographic Investigation of Flame Movements in Gaseous Explosions. Parts IV, V and VI," *Philosophical Transactions of the Royal Soc.*, Vol. 230, 1932, pp. 363-384.

[12]Bone, W. A., Fraser, R. P., and Wheeler, W. H., "A Photographic Investigation of Flame Movements in Gaseous Explosives, Part VII," *Philosophical Transactions of the Royal Soc.*, Vol. 235, 1936, pp. 29-67.

[13]Bowden, F. P., "The Initiation of an Explosion and its Growth to Detonate," *Proceedings of the Royal Soc. of London*, Vol. A204, 1950, pp. 20-25.

[14]Bowden, F. P., Stone, M. A., and Tudor, E. K., "Hot Spots on Rubbing Surfaces and the Detonation of Explosives by Friction," *Proceedings of the Royal Soc. of London*, Vol. A188, 1947, pp. 329-349.

[15]Bowden, F. P., Mulcahy, M. F. R., Vines, R. G. and Yoffe, A. D., "The Detonation of Liquid Explosives by Impact. The Effect of Gas Spaces," *Proceedings of the Royal Soc. of London*, Vol. A188, 1947, pp. 291-328.

[16]Bowden, F. P. and Yoffe, A., a) "Hot Spots and Initiation of Explosions," *Third Symposium on Combustion*, Williams and Wilkins Co, Baltimore, MD, 1949, pp. 552-560; b) *Initiation and Growth of Explosion in Liquid and Solids*, University Press, Cambridge, UK, 1953.

[17]Boyd, T. A., "Pathfinding in Fuel and Engines," *SAE Quarterly Transactions*, Vol. 4, 1950, pp. 12-19.

[18]Breton, J., "Recherches sur la Detonation des Melanges Gazeux," (Research on the Detonation of Gaseous Mixtures), Dr Ing. Thesis, Nancy 1936. Also in *Annales de l'Office des Combustion Liquides*, 1936, pp. 487-559.

[19]Brinkley, S. R. and Wilson, E. B., "Revised Method of Predicting the Detonation Velocity of Solid Explosives," OSRD Rept. 905, 1942.

[20]Brochet, C., Brossard, J., Manson N., Cheret, R., and Verdes, G., "A Comparison of Spherical, Cylindrical and Plane Detonation Velocities in Some Condensed and Gaseous Mixtures," *Vth International Symposium on Detonation*, Pasadena, CA, 1970, pp. 41-46.

[21]Brochet, C., Leyer, J. C., and Manson N., "Phénomènes Vibratoires dans les Détonations Dissociées," (Vibratory Phenomena in Uncoupled Detonations), *Comptes-Rendus de l'Acad. des Sciences, Paris, France*, Vol. 253, 1961, pp. 621-623.

[22]Brodsky, A. M., and Zeldovitch, J. B., "O detonazii vodorodno vosdushnikh s messei," (On the Detonation of Hydrocarbon-Air Mixtures), *Zhurnal Fizischeskoi Khimii*, Vol. 24, 1950, pp. 778-785.

[23]Brown, F. W., "Theoretical Calculations for Explosives," *U.S. Bureau of Mines*, Rept. 652, 1941, and Rept. 653, 1942.

[24]Campbell, C., "The Propagation of Explosion Waves in Gases Contained in Tubes of Varying Cross-Section," *Journal of the Chemical Soc.*, 1922, pp. 2483-2498.

[25]Campbell, C., and Woodhead, D. W., a) "The Ignition of Gases by an Explosive Wave," *Journal of the Chemical Soc.*, 1926, pp. 3010-3021; b) "Striated Photographic Records of Explosion Waves," *Journal of the Chemical Soc.*, 1927, pp. 1572-1578.

[26]Campbell, C., and Finch, A. C., "Striated Photographic Records of Explosion Waves. An Explanation of the Striae," *Journal of the Chemical Soc.*, 1928, pp. 2094-2106.

[27]Campbell, C., Littler, W. B., and Whitworth, C., "The Measurements of Pressure Developed in Explosion Waves," *Proceedings of the Royal Society of London*, Vol. A137, 1932, pp. 380-396.

[28]Campbell, C., King, A., Whitworth, C., "The Propagation of Explosion Waves Through a System of Glass and Rubber Tubes," *Transactions of the Faraday Society*, Vol. 28, 1932, pp. 681-688.

[29]Campbell, C., Whitworth, C., and Woodhead, D. W., "The Rates of Detonation in Carbon Monoxyde-Oxygen Mixtures," *Journal of the Chemical Soc.*, 1932, pp. 380-396.

[30]Cook, M. A., "An Equation of State for Gases at Extremely High Pressures and Temperature from the Hydrodynamic Theory of Detonation," *Journal of Chemical Physics*, Vol. 15, 1947, pp. 518-524.

[31]Courant, R. and Friedrichs, K. O., *Supersonic Flow and Shock Wave*, Interscience, New York, 1948.

[32]Cowan, R. D., and Fickett, W., "Calculation of the Detonation Properties of Solid Explosives with the Kistiakovsky-Wilson Equation of State," *Journal of Chemical Physics*, Vol. 24, 1956, pp. 432-439.

[33]Cybulski W. B., Payman, W., and Woodhead, D. W., "Explosion Waves and Shock Waves, VIII, The Velocity of Detonation in TNT" (work performed in 1943-46), *Proceedings of the Royal Society of London*, Vol. A197, 1949, pp. 51-72.

[34]Davies, R. M., Owen, J. D., Edwards, D. H., and Thomas, D. E., "Pressure Measurements in Detonating Gases," *Proceedings of the Royal Society of London*, Vol. A204, 1950, pp. 17-19.

[35]Denisov, N. Ju, and Trochine, Ju. K., "Pulsirovskaia i spinovaia detonazia gasovikh smessei n turbakh," (Pulsating and Spinning Detonation of Gaseous Mixtures in Tubes), *Doklady of the Acad. of Sciences SSSR*, Vol. 125, 1959, pp. 110-113.

[36]Dixon, H. B., a) "Presidential Address: The Initiation and Propagation of Explosions," *Journal of the Chemical Society,* Vol. 99, 1911, pp. 588-599. b) "On the Movement of the Flame in the Explosion of Gases," *Philosophical Transactions of the Royal Soc.,* Vol. A200, 1903, pp. 315-352.

[37]Dixon, H. B., Bradshaw, L., and Campbell, C., "The Firing of Gases by Adiabatic Compression," *Journal of the Chemical Society,* Vol. 105, 1914, pp. 2027-2035, 2036-2053.

[38]Doring, W., "Uber den Detonationsvorgang in Gasen," (On the Detonation Process in Gases), *Annalen der Physick,* 5e Folge, Vol. 43, 1943, pp. 421-436.

[39]Draper, S., "The Physical Effect of Detonation in a Closed Cylindrical Chamber," Ph.D. Thesis, Massachusetts Institute of Technology, 1934; NACA Rept. 493, 1936.

[40]Dryden, H. L., Murnaghan, F. D. and Bateman, H., Report of the Committee on Hydrodynamics, Division of Physical Sciences, *Bulletin of the National Research Council,* Rept. 24, 1932, pp. 551-559.

[41]Duff, R. E., "Equation of State and Thermodynamic Functions," AGARDOgraph 41: *Fundamental Data Obtained from Shock tube Experiments,* Edited by A. Ferri, Pergamon Press, 1961, Chap. VIII, pp. 291-319.

[42]Duff, R. E., Knight, H. T. and Rink, J. B., Precision Flash X-Ray Determination of Density Ratio in Gaseous Detonation," *Physics of Fluids,* Vol. 1, 1958, pp. 393-398.

[43]Duff, R. E. and Houston, E., "Measurement of the Chapman-Jouguet Pressure and Reaction Zone Length in a Detonating High Explosive," *Journal of Chemical Physics,* Vol. 23, 1955, pp. 1268-1273.

[44]Egerton A., and Gates, S. F., a) "Detonation of Gaseous Mixtures of Acetylene and of Pentane," *Proceedings of the Royal Society of London,* Vol. A124, 1926, pp. 137-151. b) "Detonation of Gaseous Mixtures at High Initial Pressures and Temperatures," *Proceedings of the Royal Society of London,* Vol. A124, 1926, pp. 152-160. c) "Further Experiments on Explosions in Gaseous Mixtures of Acetylene, of Hydrogen and of Pentane," *Proceedings of the Royal Society of London,* Vol. A125, 1927, pp. 516-529.

[45]Evans, M. W. and Ablow, C. M., "Theories of Detonation," *Chemical Review,* Vol. 61, 1961, pp. 129-178.

[46]Eyring, H., Powell, R. E., Duffey, C. H., and Parlin, R. B., "The Stability of Detonation," *Chemical Review,* Vol. 45, 1949, pp. 69-181.

[47]Fay, J., "Mechanical Theory of Spinning Detonations," *Journal of Chemical Physics,* Vol. 20, 1952, pp. 942-950.

[48]Ferie, F., and Manson, M., a) "Sur les ondes explosives sphériques dans les mélanges gazeux," (On Spherical Explosion Waves in Gaseous Mixtures), *Comptes-Rendus de l'Acad. des Sciences,* Paris, France, Vol. 235, 1952, pp. 139-141; b) "Contribution to the Study of Spherical Detonation Waves," *4th Symposium (International) on Combustion,* Ed. Williams & Wilkins, 1953, pp. 486-494.

[49]Fickett, W. and Wood, W. W., "A Detonation Product Equation of State Obtained from Hydrodynamic Data," *Physics of Fluids,* Vol. 1, 1958, pp. 528-534.

[50]Fickett, W. and Davis, W. C., *Detonation,* Univ. of California Press, Berkeley, CA, 1979.

[51]Freiwald, H., and Uhde, H., a) "Sur les ondes explosives sphériques dans les mélanges d'acétylène et d'air," (On Spherical Explosion Waves in Mixtures of Acetylene and Air), *Comptes-Rendus de l'Acad. des Sciences, Paris,* Vol. 126, 1953, pp. 1741-1743; b) "Uber die Initiierung Kugelformiger Detonation swellen in

Gasgemischen," (On the Initiation of Spherical Detonation Waves in Gas Mixtures), *Zeitschrift fur Elektrochemie,* Vol. 50, 1955, pp. 910-913.

[52]Friedrich, W., "Uber die Detonation der Sprengstoffe," (On the Detonation of Explosives), *Zeitschrift fur Gesamte Schiess und Sprengstoffwesen,* 1933, pp. 2-6, 51-53, 80-83, 113-116, 213-215, and 244-247.

[53]Goranson, R. W., "A Method for Determining E.O.S. and Reaction Zones of High Explosives and its Application to Pentolite, Composition B, Buretol and TNT," *Los Alamos Rept.* 487, 1946 (unclassified in 1974).

[54]Gordon, W. E., "Pressure Measurements in Gaseous Detonation by Means of Piezo-Electric Gauges," *Third Symposium on Combustion, Flame and Explosion Phenomena,* The Williams & Wilkins Co, Baltimore, 1949, pp. 579-586.

[55]Greifer B., Cooper, J. C. and Mason, C. M., a) "Combustion and Detonation in Gases," *Journal of Applied Physics,* 1956, pp. 289-294; b) "Studies on Gaseous Detonation," *Second Symposium on Detonation,* U.S. Naval Ordinance Lab., White Oak, MD, 1955, pp. 165-175.

[56]Grib, A. A., a) "O rasprostranenie ploskoi udarnoi volni pri obiknovenim vzrive", (On the Propagation of Ordinary Shock Wave Produced by Means on Ordinary Explosion), *Prikladnaia Mekhanika i Teoritisteskaia Fizika,* Vol. 8, 1944, pp. 169-180. b) "Vlianie mesta inizirovania na parametri vosdushnoi udarnoi volni pri detonazi vzrivshatikh gazovikh smesei," (Influence of the Position of the Initiator of the Detonation in Gaseous Mixtures, on the Parameters of the Aerial Shock Wave), *Prikladnaia Mekhanika i Teoritisteskaia Fizika,* Vol. 8, 1944, pp. 273-290.

[57]Gross, R. A., "Research on Supersonic Combustion," *A.R.S. Journal,* Vol. 29, 1959, pp. 63-64.

[58]Gross, R. P., and Chinitz, W., "A Study of Supersonic Combustion," *Journal of Aerospace Sciences,* Vol. 27, 1964, pp. 517-534.

[59]Guenoche, H., "Recherches sur la détonation et les déflagrations dans les mélanges gazeux," (Investigations of Detonation and Deflagrations in Gaseous Mixtures), *Revue de l'Institut Français du Pétrole,* Vol. 4, 1949, pp. 15-36 and 48-69.

[60]Guenoche, H. and Laffitte, P., "Sur les variations de la vitesse de détonation des mélanges gazeux combustibles," (On the Variation of Detonation Velocities in Combustible Gaseous Mixtures), *Comptes-Rendus de l'Acad. des Sciences, Paris,* Vol. 224, 1947, pp. 1224-1227.

[61]Hopkinson, B., "Explosion of Coal Gas Mixtures with Air," *Proceedings of the Royal Society of London,* Vol. A77, 1906, pp. 387-413.

[62]Jacobs, S. J., "Recent Advances in Condensed Media Detonation," *A.R.S. Journal,* Vol. 30, 1960, pp. 151-154.

[63]Johansson, H., and Persson, P.A., *Detonics of High Explosives,* Academic Press, New York, 1970.

[64]Jones, E., "Photographic Study of Detonation in Solid Explosives. Part I: The Development of a Photographic Method for Measuring Rates of Detonation," *Proceedings of the Royal Society of London,* Vol. A120, 1928, pp. 603-619.

[65]Jones, H., a) "A theory of Dependence of the Rate of Detonation of Solid Explosives on Diameter of the Charge," *Proceedings of the Royal Society of London,* Vol. A189, 1947, pp.415-426; b) "The Properties of Gases at High Pressure which can be Deduced from Explosive Experiments," *Third Symposium on Combustion Flame and Explosion Phenomena,* The Williams Wilkins Co, Baltimore, 1949, pp. 540-544; c) "Theoretical Considerations Relating to the Detonation of Explosives," *Proceedings of the Royal Society of London,* Vol. A204, 1950, pp. 9-12.

[66]Jones, H., and Miller, A. R., "The Detonation of Solid Explosives," *Proceedings of the Royal Society of London,* Vol. A194, 1949, pp. 480-507.

[67]Jost, W., *Explosionsvorgänge in Gasen*, Springer Berlin, 1939 (English translation by McGraw-Hill, New York, 1946).

[68]Jouguet, E., a) "Sur les ondes de choc et de combustion sphériques," (On Spherical Shock and Combustion Waves), *Comptes-Rendus de l'Acad. des Sciences, Paris*, Vol. A120, 1907, pp. 632-634. b) *La Mécanique des Explosifs*, Doin, Paris, 1917. c) "Comparaison de la théorie de l'onde explosive avec quelques expériences récentes," (Comparison of the Theory of Explosion Wave with Some Recent Experiments), *Comptes-Rendus de l'Acad. des Sciences, Paris*, Vol. A162, 1925, pp. 546-548; d) "Sur les ondes de choc produites dans un gaz par un explosif solide," (On Shock Waves Produced in a Gas by a Solid Explosive), *Comptes-Rendus de l'Acad. des Sciences, Paris*, Vol. A202, 1936, pp. 1225-1228, 336-1338.

[69]Kahara, T. and Hikita, T., "Equation of State for Hot Dense Gases and Molecular Theory of Detonation," *Fourth Symposium (Int.) on Combustion*, Williams Wilkins Co, Baltimore, MD, 1953, pp. 458-464.

[70]Khariton, Ju. B., a) "K voprossou o detonazï ot oudara," (On the Detonation Produced by Means a Shock), *Problems of Theory of Explosives*, Ac. Sc. SSSR, and Oborongiz, Moscow, 1940; b) "O detonazionnoï spossonosti vzrivtshatikh vertchestv," (On the Detonability of Explosives), *Problems of the Theory of Explosives*, Acad. Sc. SSSR, Vol. 1, Moscow, 1947.

[71]Kistiakovsky, G. B., a) "Initiation of Detonation of Explosives", *Third Symposium on Combustion, Flames and Explosion in Gases*, Williams Wilkins Co, Baltimore, 1949, pp. 560-565; b) "Initiation of Detonation in Gases," *O.N.R. Technical Report, NR053-094*, 1950 and *Industrial Engineering Chemistry*, Vol. 43, 1951, 2794-2792.

[72]Kistiakovsky, G. B., Knight, H. T., and Malin, M. E., a) "Gaseous Detonation III: Dissociation Energies of Nitrogen and Carbon Monoxide," *Journal of Chemical Physics*, Vol. 20, 1952, pp. 876-883. b) "Gaseous Detonation V: Non Steady Waves in $CO-O_2$ Mixtures," *Journal of Chemical Physics*, Vol. 20, 1952, pp. 994-1000.

[73]Kistiakovsky, G. B. and Wilson, E. B., "The Hydrodynamic Theory of Detonation and Shock Waves," *O.S.R.D. Rept.* 114, 1941.

[74]Kogarko, S. M. and Zeldovitch, J. B., "O detonazii gasovikh smessei," (On the Detonation of Gaseous Mixtures), *Doklady Acad. of Sciences*, SSSR, Vol. 63, 1948, pp. 553-555.

[75]Laffitte, P., a) "Sur la formation de l'onde explosive," (On the Formation of an Explosion Wave), *Comptes-Rendus de l'Acad. des Sciences, Paris* Vol. 176, 1923, pp. 1393-1396; b) "Sur la propagation de l'onde explosive sphérique," (On the Propagation of a Spherical Explosion Wave), *Comptes-Rendus de l'Acad. des Sciences, Paris*, Vol. 177, 1923, pp. 178-180. c) "Sur l'onde explosive," *Comptes-Rendus de l'Acad. des Sciences, Paris*, Vol. 179, 1924, pp. 1393-1394; d) "Recherches expérimentales sur l'onde explosive et l'onde de choc," (Experimental Investigations on Explosion Waves and Shock Waves), *Annales de Physique*, Vol. 4, 1925, pp. 587-695; e) "Influence de la température sur la formation de l'onde explosive," (Influence of Temperature on the Formation of Explosion Waves), *Comptes-Rendus de l'Acad. des Sciences, Paris*, Vol. 186, 1928, pp. 951-953.

[76]Laffitte, P., and Breton, J., "Sur les limites de détonation de quelques mélanges gazeux," (On the Limit of Detonation in some Gaseous Mixtures), *Comptes-Rendus de l'Acad. des Sciences, Paris,*, Vol. 199, 1934, pp. 146-148.

[77]Laffitte, P., and Dumanois, P., a) "Influence de la pression sur la formation de l'onde explosive," (Influence of Pressure on the Formation of Explosion Waves), *Comptes-Rendus de l'Acad. des Sciences, Paris*, Vol. 184, 1926, pp. 284-286; b) "La

vitesse de l'onde explosive," (The Velocity of Explosion Waves), *Comptes-Rendus de l'Acad. des Sciences, Paris*, Vol. 186, 1928, pp. 146-148.

[78]Laffitte P., and Patry, M., a) "Sur la détonation des explosifs solides," (On the Detonation of Solid Explosives), *Comptes-Rendus de l'Acad. des Sciences, Paris*, Vol. 191, 1930, pp. 1335-1337. b) "Sur la détonation des explosifs solides," (On the Detonation of Solid Explosives), *Comptes-Rendus de l'Acad. des Sciences, Paris*, Vol. 192, 1931, pp. 744-746.

[79]Landau, L. D., and Sstaniukovitch, K. P., "Ob i zoutsheni detonazi kondensirovanikh vzrivshatikv veskstshestu," (On the Study of the Detonation of Condensed Explosives), *Doklady of the Acad.of Sciences*, SSSR, Vol. 46, 1945, pp. 399-402.

[80]Lewis, B., "A Chain Reaction Theory of the Rate of Explosion in Detonating Gas Mixture," *Journal of the American Chemical Society*, Vol. 52, 1930, pp. 3120-3127.

[81]Lewis, B., and von Elbe, G., *Combustion, Flames and Explosion of Gases*, 1st ed., Cambridge Univ. Press, 1938 (2nd ed., Academic Press, 1951).

[82]Lewis, B., and Friauf, J. B., "Explosion in Detonating Gas Mixtures," *Journal of the American Chemical Society*, Vol. 52, 1930, pp. 3905-3920.

[83]Mader, Ch. L., *Detonation Performance Calculation Using the Kistiakovsky-Wilson Equation of State*, Los Alamos Press, CA, pp. 26-29, 1961.

[84]Mache, H., *Die Physik der Verbrennungserscheinigungen*, Ed. Veit u. Co, Leipzig, 1918.

[85]Manson, N., a) "Sur la structure des ondes explosives hélicoidales," (On the Structure of Helical Explosion Waves), *Comptes-Rendus de l'Acad. des Sciences, Paris*, Vol. 222, 1945, pp. 46-48; b) "Propagation des détonations et des déflagrations dans les mélanges gazeux," (Propagation of Detonations and Deflagrations in Gaseous Mixtures), Dr. Sc. Thesis, University of Paris, 1946, ONERA and IFP, Paris 1947, (English translation: Technical Information and Library of Science, Ministery of Supply, USA, 1954); c) "Formation et célérité des ondes explosives sphériques dans les mélanges gazeux," (Formation and Velocity of Spherical Explosion Waves in Gaseous Mixtures), *Revue de l'Institut Français du Pétrole*, Vol. IX, 1954, pp. 133-143; d) "Une nouvelle relation de la théorie hydrodynamique des ondes explosives," (A New Relation of the Hydrodynamic Theory of Explosion Waves), *Comptes-Rendus de l'Acad. des Sciences, Paris*, Vol. 246, 1958, pp. 2860-2862; e) *Détermination par la Méthode Inverse des Caractéristiques des Ondes Explosives*, (Determination of Characteristics of Explosion Waves by the Inverse Method), Publications Scientifiques et Techniques du Ministère de l'Air, Rept. 365, Paris 1960; f) "Contribution de Ch.S. Draper à la connaissance des phénomènes vibratoires dans les foyers des moteurs et des réacteurs," (Contribution of Ch. Draper to the Understanding of Vibratory Phenomena in Motor and Jet Engine Combustors), *Acta Astronautica*, Vol. 21, 1990, pp. 349-353.

[86]Medard, L., *Les Explosifs Occasionnels*, Techniques et Documentations, Paris, 1978 (2 volumes).

[87]Michel-Levy, A., and Muraour, H., a) "Sur la luminosité des ondes de choc," (On the Luminosity of Shock Waves), *Comptes-Rendus de l'Acad. des Sciences, Paris*, Vol. 198, 1934, pp. 1499-1501; b) "Influence de la nature du gaz environnant sur la luminosité qui accompagne la détonation des explosifs," (The Influence of the Nature of the Boundary Gas on the Luminosity which Accompanies the Detonation of Explosives), *Mémorial des Poudres et Salpêtres*, Vol. 26, 1937, pp. 171-182; c) "Sur l'utilisation des rencontres d'ondes de choc dans l'argon comme source de lumière brève et puissante," (On the Utilization of Colliding Shock Waves in Argon as an Intense, short Duration Light Source), *Mémorial de l'Artillerie Française*, Vol. 29, 1947, pp. 105-123.

[88] Michel-Levy, Vassy, E., and A., Muraour, H., "Source de lumière très brève pour l'usage photographique," (A Short Duration Light Source for Photographic Applications), *Revue d'Optique*, Vol. 20, 1941, pp. 161-165.

[89] Miller, C. D., "Relation Between Spark Ignition Engine Knock, Detonation Waves and Auto-Ignition Shown by High Speed Photography," *NACA Rept.*, 855, 1946.

[90] Mooradian, A. J. and Gordon, W. E., "Gaseous detonation. I: Initiation of Detonation," *Journal of Chemical Physics*, Vol. 19, 1951, pp. 1166-1172.

[91] Nernst, W., "Physikalisch-Chemische Betrachtrungen uber den Verbrennung- prozess in den Gasmotoren," (Physico-Chemical Considerations of the Combustion Process in Gas-Fueled Motors), *Zeitschrift V.D.I.*, Vol. 49, 1905, pp. 1426-1431.

[92] Neumann, J. von, "Progress Report on the Theory of Detonation Wave," *O.S.R.D. Rept.*, Vol. 549, 1942, and Coll. Works, Macmillan, New York, 1963.

[93] Nicholls, J. A., Dabora, E. K. and Gealer, R. L., "Studies in Connection with Stabilized Gaseous Detonation Waves", *Seventh Symposium (International) on Combustion*, Combustion Institute, Butterworths, London, 1959, pp. 766-772.

[94] Oppenheim, A. K., and Stern, R. A., "On the Development of Gaseous Detonation. Analysis of Wave Phenomena," *Seventh Symposium (International) on Combustion*, Butterworths, London, 1959, pp. 837-859.

[95] Oppenheim A. K., Laderman A. J., and Urtiew, P. A., "On Generation of a Shock Wave by Flame in an Explosive Gas," *Ninth Symposium (International) on Combustion*, Pergamon Press, New York, 1963, pp. 265-274.

[96] Patry, M., "*Recherches Expérimentales sur la Combustion et la Détonation des Substances Explosives*," Dr. Thesis, University of Nancy, France, 1933.

[97] Patry, M., and Laffitte, P., "Sur la détonation des explosifs solides," (On the Detonation of Solid Explosives), *Comptes-Rendus de l'Acad. des Sciences, Paris*, Vol. 191, 1930, pp. 1335-1352.

[98] Patterson, J., "The Hydrodynamic Theory of Detonation. Part II: On Calculations for Condensed Explosives," *Research*, Vol.1, 1948, pp. 221-233.

[99] Payman W., "The Pressure Wave Sent out by an Explosive," (Part I with H. Robinson, Part II with W. C. F. Shepherd), *Safety in Mines Res. Board Papers*, No. 18, 1926; and No. 29, 1927.

[100] Payman, W., "The Detonation Wave in Gaseous Mixtures and the Predetonation Period," *Proceeding of the Royal Society of London*, Vol. A120, 1928, pp. 90-109.

[101] Payman, W. and Shepherd, W. C. F., a) "Explosion Waves and Shock Waves. IV: Quasi-Detonation in Mixtures of Methane and Air," *Proceeding of the Royal Society of London*, Vol. A158, 1937, pp. 348-363; b) "Explosion Waves and Shock Waves. VI: The Disturbance Produced by Bursting Diaphragms with Compressed Air" (written in 1940), *Proceeding of the Royal Soc. of London*, Vol. A186, 1946, pp. 293-321.

[102] Payman, W., and Titman, H., "Explosion Waves and Shock Waves. III: The Initiation of Detonation in Mixtures of Ethylene and Oxygen and of Carbon Monoxyde and Oxygen," *Proceeding of the Royal Society of London*, Vol. A152, 1935, pp. 418-445.

[103] Payman, W., and Walls, N. S., "The Rate of Detonation in Complex Gaseous Mixtures," *Journal of the Chemical Society*, Vol. 123, 1923, pp. 420-426.

[104] Payman, W., and Woodhead, D. W., a) "Explosion Waves and Shock Waves. Part I : The Wave Speed Camera and its Application to the Photography of

Bullet in Flight," *Proceeding of the Royal Society of London,* Vol. A132, 1931, pp. 200-213; b) "Explosion Waves and Shock Waves. V: The Shock Wave and Explosion Products from Detonating Solid Explosives," *Proceeding of the Royal Society of London,* Vol. A163, 1937, pp. 575-592.

[105]Payman, W., Woodhead, D.W. and Titman, H., "Explosion Waves and Shock Waves. Part II: The Shock Wave and Explosion Products Sent out by Blasting Detonation," *Proceedings of the Royal Society of London,* Vol. A148, 1935, pp. 604-622.

[106]Penney, W. G., "A Discussion on Detonation," *Proceeding of the Royal Society of London,* Vol. A204, 1950, pp. 1-33.

[107]Perrot, G. St. J. and Gawthrop, D. B., "Propagation of Detonation Across an Air-Gap Between Two Cartridges of Explosive,"*Journal Franklin Institute,* 1927, pp. 387-406.

[108]Pokrovsky, G. I. and Staniukovitch, K. P., "K voprosou o naprevlenia vzriva" (On the Direction of the Explosion), Izv. Acad. Naouk, SSSR serie Fiz., Vol. 8, 1944, p. 214.

[109]Pujol, Y., and Manson, N., "Estimation de la Pression de Détonation des Explosifs Condensés," (Estimate of the Pressure of Condensed Explosives), *Comptes-Rendus de l'Acad. des Sciences, Paris,* Vol. 254, 1962, pp. 3173-3175.

[110]Rakipova, Kh. A., Trochine, Ja, and Shtelkine, K. I., a) "Izmerenie normalkh skorostei plameni azetilenokislorodnikh smessei", (Measurements of Normal Flame Velocities in Acetylene-Oxygen Mixtures), *Zh. Tekhnisheskoi Frziki,* Vol. 17, 1947, pp. 1397-1408. b) "Spin u priedelov detonazii" (Spin Near the Detonation Limits), *Zh. Tekhnisheskoi Fiziki,* Vol. 17, 1947, pp. 1409, 1410.

[111]Ratner, S. B., "Problema oustanovlenia sviazi mejdou kharakteristikami detonazionikh voln i ikh razchteta dlia kondensirovanikh vestechstv", (Establishment of Relations Between Characteristics of Detonation Wave and the Calculations of the Values of these Characteristics in the Case of Condensed Explosives), *Problems of Theory of Explosives,* Ed. Acad. Sc. SSSR, Vol. 1, Moscow, 1947.

[112]Rayleigh, W., *Theory of Sound,* Second Edition, reprinted by MacMillan, London, 1926-1929.

[113]Reingold, L., and Viaud, L., "Perfectionnements apportés aux foyers à circulation interne supersonique," (Improvements Attained in Internal Supersonic Flow Combustors), *ONERA French Patent,* 1008 660, Feb. 1952.

[114]Reingold, L., "Recherches sur les combustions permanentes apportées aux foyers à circulation interne supersonique," (Investigations of Steady Combustion Attained in Internal Supersonic Flow Combustors), *ONERA TN2,* Department of Energy and Propulsion, Study 728-E, March 1950.

[115]Rivine, M. A., a) "K voprossu o vlianie temperaturi na predel detonazii smessei vodoroda s vosdukhom," (About the Influence of Temperature on the Detonation Limit of H_2-Air Mixtures), *Zhurnal Fizisheskoi Khimii,* Vol. 15, 1941, pp. 534-550. b) "Mekhanism gasovoi detonazii," (Mechanics of Gaseous detonation), *Ouspekhi Khimii,* Vol. 20, 1951, pp. 473, 474.

[116]Rivine, M. A. and Sokolik, A. S., a) Vzrivnie predeli vodorodno vosdushnikh smessei," (Explosion Limits of Hydrogen-Air Mixtures), *Zhurnal Fizisheskoi Khimii,*Vol. 7, 1936, pp. 571-582; b) "Detonazionnie predeli uglevodoroda vosdusnikh smessei," (Detonation Limits of Hydrocarbon-Air Mixtures), *Zhurnal Fizisheskoi Khimii,* Vol. 10, 1937, pp. 692-699.

[117]Ricardo, H. R., "Paraffin as Fuel," *Automotive Engineering,* Vol. 9, 1919, pp. 2-5; "Recent Research Work on Internal Combustion Engines," *SAE Journal,* Vol. 10, 1922, pp. 300-312.

[118] Rosing, V., and Khariton, J., "Prekrashtenie detonazi vzrivskatikh vesteshtv pri malikh diametrov zariada," (Quenching of the Detonation of Explosives in Small Diameter Charges), *Doklady of the Acad. of Sciences SSSR*, Vol. 26, 1940, pp. 360-363.

[119] Roth J.F., a) "Eine Laboratoriumsmethod zur Bestimmung der Detonationgeschwindigket mit Kurzen Metrecken," (A Laboratory Method for the Determination of Detonation Velocity Using Short Measurement Intervals), *Zeitschrift fur Gesamte Schiess und Sperengstoffwesen*, Vol. 28, 1935, pp. 42-46; b) "Uber dass Covolumen und den Zustand der Schwaden in der Detonationszone brisanter Sprengstoffe," (On the Co-Volume and the State of the Gas Cloud in the Detonation Zone of High Explosives), *Zeitschrift fur Gesamte Schiess und Sprengstoffwesen*, Vol. 34, 1939, pp. 193-197; c) "Die praktische Ausfuhrung der Berechnung von Sprengtechnischen Grossen," (The Practical Calculation of Explosions of Technical Scale), *Zeitschrift fur Gesamte Schiess und Sprengstoffwesen*, Vol. 35, 1940, pp. 193, 196, 220, 240, 243, 245, and Vol. 36, 1941, pp. 88, 52, 160.

[120] Roy, M., "Propulsion par Statoréacteur à Détonation," (Detonation Ramjet Propulsion), *Comptes-Rendus de l'Acad. des Sciences, Paris*, Vol. 222, 1946, pp. 31-32.

[121] Ruegg, F. W., "A Technique for Studying Supersonic Combustion in the Vicinity of a Hypersonic Missile," *NBS Technical News Bulletin*, Vol. 44, 1960, pp. 181, 182.

[122] Rumpf, H., a) "Der Kondensator-Chronograph," The Condenser Chronograph), *Zeitschrift fur Gesamte Schies und Sprengstoffwesen*, Vol. 24, 1929, pp. 13-17; b) "Der Optische Chronograph," (The Optical Chronograph), *Zeitschrift fur Gesamte Schiess und Sprengstoffwesen*, Vol. 24, 1929, pp. 55-57.

[122b] Schall, R., "Methoden und Ergebnisse der Detonationdruck bestimmung bei festen Sprengstoffen Zeitschrift fur Elektrochemie," (Methods and Results on the Determination of Detonation Pressure in solid Explosives), Vol. 61, 1957, pp. 629-635.

[123] Schmidt, A., a) "Uber die Detonation von Sprengstoffen und die Beziehungen zwischen Dichte und Detonationsgeschwindigkeit," (On the Detonation of Explosives and the Relationships Between Density and Detonation Velocity), *Zeitschrift fur Gesamte Schiess und Sprengstoffwesen*, Vol. 30, 1936, pp. 8, 38, 80, 114, 148, 183, 218, 240, 284, 322; b) "Explosion und detonation in molecular und reaktions kinetischer betrachtung," (Molecular and Reaction-Kinetic Consideration of Explosion and Detonation), *Zeitschrift fur Gesamte Schiess und Sprengstoffwesen*, Vol. 33, 1938, pp. 121-125; c) "Uber den Nachweiss der Gultigkeit der Hydrodynamischer Theorie der Detonation fur feste und flussige Sprengstoffe," (On the Proof of Validity of the Hydrodynamic Theory of Detonation for Solid and Liquid Explosives), *Zeitschrift fur Physikalische Chemie*, Vol. A183, 1941, pp. 88-94.

[124] Schott, G. L. a) "Observation of the Structure of Spinning Detonation," *Physics of Fluids*, Vol. 8, 1965, pp. 850-865; b) "Structure, Chemistry and Instability of Detonation in Gases,"*4th Symposium on Detonation*, O.N.R. Washington, 1965, pp. 67-77.

[125] Scorah, R. I., "On the Thermodynamic Theory of Detonation," *Journal of Chemical Physics*, Vol. 3, 1935, pp. 425-430.

[126] Shtelkine, K. I., a) "Detonation in Air Mixtures of Pentane and Hexane in Tubes," *Doklady of the Academy of Sciences, SSSR*, Vol. 22, 1939, pp. 111-112; b) "On the Theory of the Development of Detonation in Gases," *Doklady of the Acad. of Sciences, SSSR*, Vol. 23, 1939, pp. 636-640; c) "Vlianie sherokhovatosti trubi na voznikovenie i rasprostronemie detonazi v gasakh," (Influence of the Tube Walls

Roughness on the Onset and Propagation of Detonation in Gases), *Zhurnal Experimentalnoi i Teoritiskeskoi Fiziki*, Vol. 10, 1940, pp. 823-827; d) "Bistroe gorenie i spinovaia detonazia gasov," (Rapid Combustion and Spinning Detonation in Gases), Min. of the Army SSSR, Moscow, 1944; e) "Decrease of Detonation Velocity in Rough Tubes," *Acta Physicachimika*, SSSR, Vol. 20, 1945, pp. 305-306; f) "Voznikovenie detonazii v sherokhovatikh trubakh," (Onset of the Detonation in Rough Tubes), *Zhurnal Tekhnisheskoi Fiziki*, Vol. 17, 1947, pp. 613-618; (English translation by Fed. Sc. and Techn. Information); g) "On a Theory of the Phenomenon of Spin in Detonation," *Doklady of the Acad. of Sciences, SSSR*, Vol. 47, 1945, pp. 482-484 (in english).

[127]Shtelkine, K. I. and Sokolik, A. S., "Detonazia v gazovikh smessei, III: vlianie tetraethilsvinza na obrazovanie detonazionoi volni," (Detonation in Gaseous Mixtures. III Influence of Tetraethyl Lead on the Formation of the Detonation Wave), *Zhurnal Fizisheskoi Khimii*, Vol. 10, 1937, pp. 479-483.

[128]Sokolik, A. S., a) "Gorenie i detonazia v gasakh," (Combustion and Detonation in Gases), Gos. Izdat. Moscow, 1934; b) "Samovosplamenie, plamia in detonazia v gasakh", (Auto-ignition, Flame and Detonation in Gases), Academy of Sciences, SSSR, Moscow, 1960 (English translation, I.P.S.T., Jerusalem, 1963); c) "Vlianie temperaturi na detonaziornii predel vodorodno vosdushnikh smessei," (Influence of Temperature on the H_2-O_2 Mixture Detonation Limits), *Zhurnal Fizisheskoi Khimii*, Vol. 13, 1939, pp. 1031-1039 d) "O statie M.A. Rivina: Vlianie temperaturi n a predeli detonazii smessei vodoroda c vosdukhim," (On the M.A. Rivin's Paper About the Question of the Influence of Temperature on the Detonation Limit of H_2-O_2 Mixtures), *Zhurnal Fizisheskoi Khimii*, Vol. 15, 1941, pp. 551-555.

[129]Sokolik, A. S. and Shtelkine, K. I., "Detonation in Gas Mixtures: the Influence of Pressure on the Velocity of a Detonation Wave," *Acta Physicokhimika*, SSSR, Vol. 1, 1934, pp. 311-317.

[130]Soloukhine, R. I., "Udarnie volni i detonazia v gazakh," (Shock Waves and Detonation in Gases), State Edition of Physics and Mathematics Studies, Moscow, 1963 (English translation edited by Mono Book Corp. Baltimore, MD, 1966).

[131]Taylor, J., "High and Low Detonation Velocity Regimes in Condensed Explosives." *Proceedings of the Royal Society of London*, Vol. A204, 1950, pp. 30, 31.

[132]Taylor, J., *Detonation in Condensed Explosives*, The Clarendon Press, Oxford, UK, 1952.

[133]Taylor, G. I ., "The dynamics of Combustion Products Behind Plane and Spherical Detonation Front in Explosives," *Proceedings of the Royal Society of London*, Vol. A200, 1950, pp. 235-247.

[134]Taylor, G. I. and Tankin, R. S., "Gasdynamics Aspects of Detonations, Fundamentals of Gasdynamics," *High Speed Aeronautics and Jet Propulsion*, Vol. 3, Princeton Univ. Press, 1958.

[135]Taylor, W., and Weale, A., a) "The Mechanism of the Initiation and Propagation of Detonation of Solid Explosives," *Proceedings of the Royal Society of London*, Vol. A138, 1932, pp. 92-116; b) "Conditions for the Initiation and Propagation of Detonation in Solid Explosives," *Transactions of the Faraday Society*, Vol. 34, 1936, pp. 995-1003.

[136]Urtiew, P. A. and Oppenheim A. K., "Experimental Observation on the Transition to Detonation in an Explosive Gas," *Proceedings of the Royal Society of London*, Vol. A295, 1966, pp. 13-21.

[137]Vieille, P., a) "Sur les discontinuités produites par la détente brusque de gaz confinés," (On the Discontinuities Produced by the Sudden Expansion of Confined

Gases), *Comptes-Rendus de l'Acad. des Sciences, Paris*, Vol. 129, 1899, pp. 1228, 1230; b) "Sur le rôle des discontinuités dans les phénomènes de propagation," (On the Role of Discontinuities in Propagation Phenomena), *Mémorial des Poudres et Salpêtres*, Vol. 10, 1899-1900, pp. 177-200.

[138] Voinov and Sokolik, A. S., "Detonazia v motore s iskrovim vosplameneniem," (Knocking in Spark Ignition Engine), *Tekhnika Vozdushnova Flota*, No. 3, 1936, pp. 24-40; and Izvest. A. N. SSSR, No. 1, 1937, p. 137.

[139] Voinov, A. N., "O mekhanisme vozniknovenia detonazionogo spina," (On the Mechanism of Formation of the Spinning Detonation), *Doklady of the Academic of Sciences*, SSSR, Vol. 73, 1950, p. 125-128.

[140] Voitsekhovsky, B. V., a) "O Spinovoi Detonazii," (On Spinning Detonation), *Doklady of the Academic of Sciences, SSSR*, Vol. 114, 1957, pp. 717-720; b) "Isledovanie Strukturi Fronta Spinovoi Detonazii," (Study of the Structure of the Front of the Spinning Detonation), *Research on Physics and Radiotechnic*, Moscow Physics Technical Institute, State Edition of Defense Industry, Moscow, 1958, pp. 81-91.

[141] Wagner, H. G., "Gaseous Detonations and Structure of a Detonation Zone," AGARDOgraph 4, Fundamental Data obtained from Shock Tube Experiments, Edited by A. Ferri, Pergamon Press, Oxford, 1961, pp. 320-385.

[142] Wendland, R., a) "Experimentalle Untersuchungen zur Detonationsgreuze gasformiger Gemische," (Experimental Investigations of the Detonation Limits of Gaseous Mixtures), *Zeitschrift fur Physikalische-Chemie*, Vol. 110, 1924, pp. 637-655; b) "Die Detonationsgrenze in explosiven gas Gemischen," (Detonation Limits in Explosive Gaseous Mixtures), *Zeitschrift fur Physikalische-Chemie*, Vol. 116, 1925, pp. 227-260.

[143] White, D., "Turbulent Structure of Gaseous Detonation," *Physics of Fluids*, Vol. 4, 1961, pp. 465-480.

[144] Withrow, L., and Boyd, T. A., "Photographic Flame Studies in the Gasoline Engine," *Industrial and Engineering Chemistry*, Vol. 23, 1931, pp. 539-543.

[145] Withrow, L., and Rassweller, G., "Engine knock," *Automotive Engineering*, Vol. 24, 1934, pp. 281-284.

[146] Woodbury, C. A., Lewis, H. A. and Canby, A. T., "The Nature of Flame Movement in a Closed Cylinder," *SAE Journal*, Vol. 8, 1921, pp. 209-218.

[147] Wood, W. W., and Ficket, W., "Investigation of the Chapman-Jouguet Hypothesis by the Inverse Method," *Physics of Fluids*, Vol. 6, 1963, pp. 648-652.

[148] Wood, W. W., and Kirkwood, J. W., "Diameter Effect in Condensed Explosives: the Relation Between Velocity and Radius of Curvature," *Journal of Chemical Physics*, Vol. 22, 1954, pp. 1920-1924.

[149] Wood, W. W., and Salsburg, Z. W., "Analysis of Steady State Supported One Dimensional Detonations and Shocks," *Physics of Fluids*, Vol. 3, 1960, pp. 549-556.

[150] Zeldovitch, J. B., a) "K Teori rasprostranenia detonazi v gasoobrasnikh systemakh," (On the Theory of the Propagation of Detonation in Gaseous Systems), *Zhurnal Experimentalnoi i Teoritiskeskoi Fiziki*, Vol. 10, 1940, pp. 543-568. (English translation NACA TM 1261, 1960); b) "Teoria Gorenia i Detonazii Gasov," (Theory of Combustion and Detonation of Gases), Academy of Sciences, SSSR, Moscow, 1944 (English translation: TR GDAM, Ag-T-45, Air Material Command); c) "K voprossu ob energetitsheskom ispolzovani detonazionovo gorenia," (On the Use of Detonative Combustion for Energy Production), *Zhurnal Tekhnishesnor Fiziki*, Vol. 10, 1940, pp. 1453-1461; d) "O raspredeleni davlenia i skorosti v produktakh

detonazionovo vzriva, v tshastnosti pri sferitsheskom rasprostraneni detonazionoi volni," (On the Pressure and Velocity in Products of Detonative Explosion, and More Especially in Those of a Detonation Wave), *Zhurnal Experimentalnoi i Teoritiskeskoi Fiziki*, SSSR, Vol. 12, 1942, pp. 389-406; f) "On the Theory of Detonation Spin," *Doklady Academic of Sciences*, SSSR, Vol. 52, 1946, pp. 147-150.

[151]Zeldovitch, J. B., and Kompaneez, A. S., "Theoria Detonazii," (Theory of detonation), State Edition of Technical and Theoretical Literature, Moscow 1963 (English translation by Academic Press, New York, 1960).

[152]Zeldovitch, J.B., and Leipunsky, O. I., a) "Study of Chemical Reactions in Shock Waves: Theory of the Method and Preliminary Results," *Acta Physico-Chimica*, Vol. 18, 1943, pp. 167-172; b) "O rasprostraneni udarnikh voln v vode," (On Shock Wave Propagation in Water), *Zhurnal Experimentalnoi i Teoritiskeskoi Fiziki*, SSSR, Vol. 13, 1943, pp. 181, 182.

[153]Zeldovitch, J. B., and Ratner, S. B., "Rashtet skorosti detonazii v gasakh," (Calculation of the Detonation Velocity in Gases), *Zhurnal Experimentalnoi i Teoritiskeskoi Fiziki*, SSSR, Vol. 11, 1941, pp. 170-183.

[154]Zeldovitch, J. B., and Roslovsky, A. I., "Ob usloviakh vosniknovenia neustoishivasti normalnovo gorenia; perekhod sferisheskovo plameni v detonaziu," (On the Onset of Unstable Normal Burning: Transition of a Spherical Flame to Detonation), *Doklady Academic of Sciences*, SSSR, Vol. 57, 1947, pp. 365-368.

[155]Zeldovitch, J. B., and Shliapintokh, V. J., "Vosplamenemie vzrivshatikh smessei v udarnikh volnakh," (Ignition of Explosive Mixtures in Shock Waves), *Doklady Academic of Sciences*, SSSR, Vol. 65, 1949, pp. 871-874.

[156]Symposia on Combustion, "Ind. and Eng. Chem.," Vol. 20, 1928, pp. 998-1057; "Chem. Rev.," Vol. 21, 1937, p. 209 and 22, 1938, pp. 1-310.

Chapter II. Gaseous Detonations

High Resolution Numerical Simulations for Two-Dimensional Unstable Detonations

Anne Bourlioux* and Andrew J. Majda†
Princeton University, Princeton, New Jersey 08544

Abstract

A new high resolution numerical method is developed for the computation of unstable detonations in two-space dimensions. The new scheme combines an unsplit higher-order Godunov method for the robust computation of flows with discontinuities with an explicit conservative tracking of the leading detonation front and an automatic adaptive mesh refinement procedure . This numerical scheme is used to document the fact that a wide range of phenomena observed in experiments in gaseous and condensed phases can be found in the Euler equations for reacting gas flow with simple one-step chemistry as parameters such as the heat release or the activation energy are varied.

In particular, a series of numerical experiments are reported, describing transition from very weak pulsating instabilities to strong Mach stem structures. These numerical experiments are the first work to systematically compare the theory of instabilities for detonations with large-scale numerical simulations. A new regime of weak pulsating instabilities with perturbations on the length scale of the half-reaction length is reported. Good agreement between theory and numerical results is observed, both for the spatial and temporal structure of the flow. As the overdrive ratio is decreased away from the neutral stability curve in the unstable parameter domain, larger amplitude perturbations are sharply captured by the numerical scheme, including single-head and triple-point transverse wave structures and even trailing multiple shock configurations. These last re-

---
Copyright © 1991 by the American Institute of Aeronautics and Astronautics, Inc. All rights reserved.
* Graduate student, Program in Applied and Computational Mathematics.
† Professor, Program in Applied and Computational Mathematics and Mathematics Departement.

sults are very similar to the experimental results for gases with complex chemistry. As the detonation becomes more unstable, the full complexity of two-dimensional turbulence is observed in the wake of the front instability.

Introduction

The classical theory of von Neumann, Zeldovich, and Doring (the ZND theory) from the early 1940s postulates that detonation waves are steady traveling waves with a quasi-one-dimensional structure consisting of an ordinary fluid dynamic shock followed by a reaction zone. In contrast, experiments beginning in the 1960s demonstrate that the detonations observed in many circumstances are, instead, rather complicated unstable wave patterns in reacting gases as well as reacting liquids and solids. The nature of these instabilities ranges from complex transverse Mach stems in gases to regular and chaotically irregular pulsating fronts in both gaseous and condensed phases under appropriate circumstances (see Oppenheim and Soloukhin,[1] Lee and Moen,[2] and Fickett and Davis[3] for surveys of these phenomena).

In the last fifteen years, there have been many important contributions to our understanding of unstable detonations through direct numerical simulation of two-dimensional problems. Taki and Fujiwara[4,5] pioneered these efforts and conclusively demonstrated that contemporary numerical methods can reproduce the regular Mach stem cell structure typically observed in experiments with gaseous phases. In their numerical work, Oran et al.,[6] Kailasanath et al.,[7] Guirguis et al.,[8] and Boris[9] have also demonstrated several other important physical effects in the cellular structure for unstable detonations such as the formation of unreacted gas pockets and also some aspects of irregular cell structures. The goal of the work presented here by the authors and elsewhere in collaboration with Roytburd and Colella[10-16] is to develop numerical methods that are capable of detecting a variety of detonation instabilities besides the cellular structure and also of computing with higher resolution such important quantities as vorticity in the wake of the cellular structure to gain further physical insight into the complex phenomena of detonation instability in both gaseous and condensed phases. The approach also involves an interdisciplinary interaction with modern applied mathematics where simple qualitative models and asymptotic theories for various types of detonation instabilities are combined with systematic numerical test problems to uncover subtle numerical issues as well as validate both the numerical methods and the theories.

The present work in two-space dimensions is a continuation of preliminary systematic studies by the authors on a simplified model for high Mach number combustion[10,11,13,17,18] and on one-dimensional "galloping" detonation instabilities.[11-13] A very important conclusion of these papers was the need for a new high resolution numerical method to study detonation instability. For the first problem[11,13] a series of shock capturing methods including both classical (Godunov and Lax-Wendroff) and contemporary (random choice, piecewise parabolic method (PPM), flux corrected transport method (FCT), essentially non oscillatory method (ENO), and Superbee) schemes were tested on a model problem consisting of Burgers equation coupled with a chemical energy release equation. The methods were used to compute steady traveling ZND wave solutions to the equations and also to compute the initiation to a fully sustained wave from an initial pulse. Even for such a simple problem, dramatic numerical artifacts were observed for most of the methods. Certainly the most spectacular effect was the existence of a numerical "weak" detonation wave for methods with excessive numerical viscosity. Of all the methods tested, the best results were obtained with the random choice method, PPM and FCT.

In the instability study for one-space dimension,[12] the three methods above were tested on the classical problem of the galloping detonation, as predicted by Erpenbeck[19] and computed by Fickett and Wood.[20] Acceptable accuracy was obtained with PPM for a very fine mesh, whereas the two other methods failed to converge convincingly for any reasonable level of mesh refinement. Even for PPM, the number of mesh points needed to achieve the desired accuracy would make the scheme prohibitively expensive for two-space dimension calculations. For that reason, a new scheme was developed and successively tested quantitatively for the one space-dimension instability problem. The new scheme consists of the basic PPM scheme along with explicit tracking of the front and a specialized adaptive mesh refinement procedure. Detonation instabilities are essentially generated at the interface created by the leading detonation front between the unburnt gas at rest and the burnt gas in the wake of the front. Most of the important non-linear interactions responsible for those instabilities take place in a very short region behind the front where the hydrodynamic waves interact with the chemical heat release. Those features are well taken into account by the two procedures above: the front tracking treats the detonation front as an explicit interface across which Rankine-Hugoniot jump conditions are used to both predict accurately the motion of the interface and to compute the fluxes across it; the adaptive mesh procedure concentrates the computational effort right around the leading shock, where small length scales are generated by the interaction with the chemistry. By using the new scheme, the authors were able to confirm quantitatively the exact transition overdrive ratio, below which the detonation wave becomes unstable, as predicted by an asymptotic non-linear stability theory. Remarkable agreement was also observed while comparing the spatial profiles from the asymptotic theory and the direct simulation. Farther away from the transition, the numerical simulations also displayed complex multimodes and even chaotic instabilities. These last results support the earlier computational results of Moen et al.,[21] where irregular pulsation instabilities were observed in one-dimensional numerical simulations with much finer meshes.

The present paper is organized as follows. We first state the basic equations and summarize the linear stability theory for planar detonations. Erpenbeck's formulation[19] is implemented into an efficient numerical algorithm as described by Lee and Stewart[22] in one-space dimension. The linear stability predictions are used as guides in selecting interesting physical parameters for the model; the unstable eigenmodes are added to the basic ZND profile as a systematic way of exciting the instabilities with very small amplitude perturbations. For cases very close to the transition, where a small perturbation approach is relevant, comparison between the theoretical predictions and the numerical calculations provides a very severe test for the numerical method (see 12 for this type of comparison in one-space dimension and 15 for a description of the non-linear two-dimensional theory). Next, the numerical scheme in two-space dimensions is described in more detail. Finally, we report on a series of numerical experiments. The tests are meant to illustrate both the accuracy of the new numerical method and the capability of the simple mathematical model to represent complex unstable structures as observed in experiments. The accuracy of the method is verified both for very small amplitude perturbations by comparison with theoretical results as mentioned above, and also for very large perturbations as observed in cellular instabilities with complex multiple Mach stem structures. In this last case, the numerical results are verified qualitatively by comparison with experimental results.

By varying systematically the physical parameters of the model, we are able to demonstrate the existence of several distinct interesting regimes of detonation instability. Close to the stability transition, a single-headed wave, traveling transverse to the leading shock, is observed with perturbations on the half-reaction length scale. For larger amplitude perturbations, a wide range of cellular structures with triple-point configurations are observed. The weakest of those structures generates small and very regular vorticity in the wake. As the degree of instability is increased, roll-up of vortex sheets is observed, producing very regular large-scale paired eddies. Finally, in the most unstable situation, the wake region exhibits the full complexity of two-dimensional turbulence for incompressible flows. These last results confirm the ideas of Lee[23] that the instability of detonations can be an additional source of significant turbulence. A more extensive detailed comparison between theories for detonation instability and additional computational results as described here will be presented elsewhere in the near future by the authors.[16]

Preliminaries

Basic Equations

We study solutions of the Euler equations for inviscid reacting gas flow so we neglect all dissipation mechanisms. For the chemical interaction we consider the simplest model. There are only two species present, the reactant and the product, and the reactant is converted to the product by a one-step irreversible chemical reaction governed by Arrhenius kinetics. With these simplifying assumptions, the equations for reacting flow are given by

$$\rho_t + \nabla \cdot (\rho \underline{v}) = 0$$
$$(\rho \underline{v})_t + \nabla \cdot (\rho \underline{v}\underline{v}) + \nabla p = 0 \quad (1)$$
$$(\rho E)_t + \nabla \cdot (\rho \underline{v} E + \underline{v} p) = 0$$
$$(\rho Z)_t + \nabla \cdot (\rho \underline{v} Z) = -w$$

with

$$E = e + q_o Z + \frac{v^2}{2}$$
$$p = (\gamma - 1)\rho e \quad (2)$$
$$T = \frac{p}{\rho} e$$
$$w = K\rho Z \exp(-E^+/T)$$

In these equations, $p, \rho, T, \underline{v}, E$, and Z are, respectively, the pressure, density, temperature, velocity, specific energy, and reactant mass fraction. The variables have been made dimensionless by reference to a fixed constant state. The dimensionless parameters appearing above are the specific heat ratio γ, the heat release parameter q_o, and the activation energy E^+.

The above equations have explicit traveling wave profiles consisting of a precursor ordinary fluid dynamics shock followed by a chemical reaction. These profiles are the ZND waves and are readily computed by quadrature of a single non-linear ordinary differential equation.[2] Given a fixed constant prestate, there is a minimum speed for the ZND profile, $D_{CJ} > 0$, the Chapman-Jouguet velocity and for every wave speed $D > D_{CJ}$ there is a unique ZND profile moving with that wave speed. The parameter $f = (D^2/D_{CJ}^2)$ measures the degree of overdrive of the detonation and satisfies $f \geq 1$. In studying the instability of a fixed ZND profile, a natural intrinsic length scale is the half-reaction length scale, the distance required for half the reactant to be depleted in the ZND wave. We always use the half-reaction length $L_{1/2}$ of the appropriate ZND wave as the unit length scale throughout the paper. We also normalized the sound speed in the unburnt medium to be $\sqrt{\gamma}$. With this normalization, the specification of $L_{1/2}$ as a length unit entirely determines the time scale and the normalization of K in the reaction rate w.

Linear Stability Theory

The theory of linearized stability for ZND waves was pioneered by Erpenbeck.[19] Since the perturbed precursor shock front is also an unknown, this linearized theory of stability for ZND waves involves a complex free surface problem for variable coefficient systems of linear hyperbolic equations. Recently, Lee and Stewart[22] have developed an efficient shooting method for the numerical solution of the problem in one-space dimension to compute a solution that simultaneously satisfies the reactive Euler equations linearized around the ZND solution, along with linearized Rankine-Hugoniot jump conditions at the perturbed shock and a boundness condition at infinity. Their approach was successfully implemented in the "galloping" detonation study by the authors [12] as a preliminary step to a more complete non-linear asymptotic theory which was compared with the direct simulation results. The numerical method for computing linearized instability is easily extended here to two-space dimensions by performing a Fourier transform in the direction transverse to the shock main direction of propagation, introducing a transverse wave number k. Using the overdrive ratio and the wave number as bifurcation parameters, we now have the capability of computing continuous neutral stability curves (see for example, cases 1 and 2 in next section). As a by-product of the shooting calculation, the spatial structure of the eigenmodes are computed and can be used both as initial data perturbations and for comparison of the direct simulation solutions. One of the typical features of those eigenmodes is that the perturbation in the reaction progress variable Z is confined to a very narrow region behind the shock, whereas the other components of the flow have a much larger spatial extent (see Fig. 5, case 2 in next section). This feature was an important ingredient in our earlier analysis explaining the "galloping" instability in a single-space dimension.[12,14]

Numerical Method

Here we present a new numerical method to solve the reactive Euler equations in two-space dimensions. The global procedure is a very natural fractional step scheme which involves two ingredients per time step.

In the first fractional step, the hydrodynamics part of the problem is solved: the first three equations in Eq.(1) are advanced with the reactant mass fraction advected as a passive scalar. In the second fractional step, the species equation is advanced explicitly by integrating the ordinary differential equation for the mass fraction given the temperature field from the previous fractional step.

The hydrodynamics solver in the first step combines a higher order Godunov method with conservative front tracking and adaptive mesh refinement. This new solver is designed to meet following requirements: 1) sharp and robust representation of the detonation front; 2) appropriate representation of all relevant length scales of the problem, including the small length scales associated with the stiff chemistry near the front; and 3) accurate representation of the smooth regions of the flow and accurate, stable capturing of all the other discontinuities behind the leading front

The higher-order extension of Godunov's method implemented here is second-order accurate in space and time, and captures shock waves and other discontinuities with minimal numerical overshoot and dissipation. The method was first introduced by Woodward and Colella,[24] the version used here is a two-dimensional unsplit version as described by Colella.[25]

This scheme is the basic solver for a simplified version of Berger and Colella's adaptive mesh refinement procedure.[26] To concentrate the computational effort in the region near the front with stiff chemistry, we superimpose on our basic uniform grid a rectangular patch of refined grid in the neighborhood of the leading shock. In the present approach, the width of the fine grid is fixed and the fine grid position is regularly updated along with the entire computational domain to follow the leading shock.

We also implement here the conservative front tracking procedure developed by Chern and Colella.[27] The leading front is represented as a polygonal curve moving through the finite difference mesh and treated as an internal boundary whose motion, along with the fluxes across it, is computed explicitly using Rankine-Hugoniot jump conditions. This procedure avoids the averaging process across the discontinuity as well as the additional numerical viscosity necessary to a stable capture of the shock: the present approach is much more reliable when attempting to compute the physical instability of the discontinuity.

The tracking procedure is fully conservative. This allows us to capture accurately the other discontinuities behind the tracked leading front. In particular, triple-point configurations are computed, where the Mach stem and the incident shock are tracked as parts of the leading front, while the reflected shock and the contact discontinuity, intersecting the leading front from behind, are captured by the finite difference scheme. We will also show that this procedure allows us to compute front instabilities with transverse amplitude too small (few percent of the half-reaction length) to be effectively represented by a finite difference grid but that can generate significant perturbations of the flow in the wake for distances of several half-reaction lengths. A detailed description and documentation for the numerical code is given in Ref.13.

In the present calculation, periodic boundary conditions are imposed in the direction transverse to the front main direction of propagation. The initial data consist of the ZND profile to which we add some small perturbations based on the linear stability calculations.

Numerical Results

We will now describe a series of numerical experiments performed using the new high resolution method. The parameters for the two first cases correspond to problems very close to the linear neutral stability curve.

Case 1 In the first case, γ is set to 1.2, a typical value for gaseous detonation. The other parameters are activation energy $E^+ = 50$ and heat release

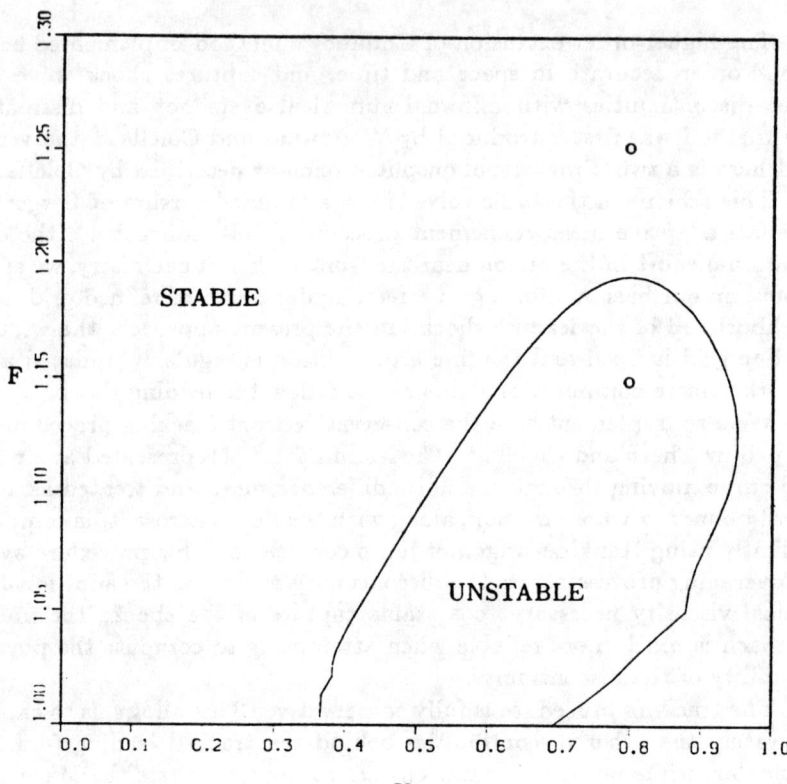

Fig. 1 Case 1, neutral linear stability curve: transition overdrive ratio f as a function of the transverse wave number k.

parameter $q_o = 0.3$ (those parameters were selected earlier by Erpenbeck.[28]) Using the shooting method, the neutral stability curve is computed with the overdrive ratio f and the transverse wave number k as parameters (Fig. 1). For the particular chemical parameters selected here, there exists a critical overdrive, at $f_{cr} = 1.19$ (corresponding to wave number $k_{cr} = 0.8$) above which the detonation is stable at all wave numbers. Calculations are performed on a channel of transverse width $L_y = 7.85$; this width is selected so that the only one unstable wave number geometrically compatible with channel is the critical number $k_{cr} = 0.8$. The calculations are performed both in the stable part of the parameter domain, $f = 1.25$, and in the unstable part, $f = 1.15$. The initial data consist of the ZND profiles corresponding to the two overdrive ratios, superimposed with perturbations corresponding to the eigenmodes at the critical overdrive $f_{cr} = 1.19$. The amplitude of the perturbation corresponds to a front perturbation disturbance of 1% of a half-reaction length. In Fig. 2, we show the evolution of that front perturbation location for both the theoretically stable and unstable overdrives. The exact front location and shape are tracked con-

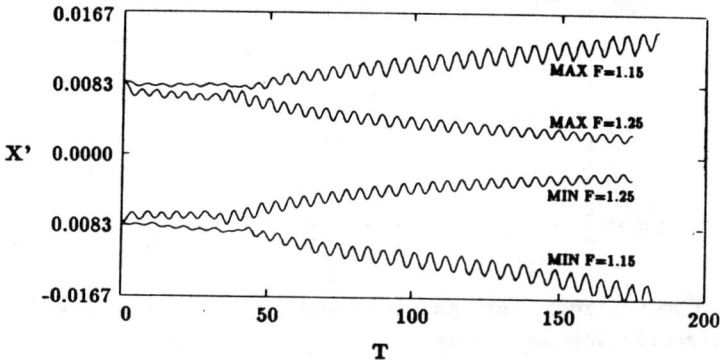

Fig. 2 Case 1, minimum and maximum amplitude of the front position perturbation as a function of time (direct simulation); the growth for the case $f = 1.25$ and the decay for $f = 1.15$ match qualitatively the theoretical prediction.

tinuously as a polygonal curve, so it is easy to compute the maximum and minimum departures in the x direction of that front location with respect to an average location. The numerical results indeed display a growth in the perturbation for $f = 1.25$ and a decrease for $f = 1.15$.

Case 2

In this case, γ is set equal to 3, which is a common approximation to model condensed phase detonations using a polytropic gas law. Again, the activation energy is $E^+ = 50$ and the heat release is chosen small enough, $q_o = 0.125$, so that there exists a critical overdrive (our calculations with the linear stability code have shown that, for higher heat releases, there exists a finite band of unstable wave numbers at very high overdrive ratios that might persist even at infinite overdrive). For the above parameters, the transition overdrive is $f_{cr} = 2.55$ at wave number $k_{cr} = 1.6$. We perform the calculation at the slightly unstable overdrive $f = 2.5$ where the complex eigenvalue for the growth rate is $\alpha = 0.001408 + $ i 1.8517 (in units of the inverse of the half-reaction time). Again the width of the channel is selected to be compatible only with the most unstable wave number. The flow was initialized with the ZND profile and the eigenmode at $f = 2.5$.

In Fig. 3, we show the front perturbation amplitude as a function of time. As predicted, the amplitude grows: after an initial adjustment delay, an exponential growth is observed and then saturation to a maximum amplitude of 1% of the half-reaction length (or 10% of the finest cell size used here, ten times the initial amplitude). Although such saturation cannot be predicted by a strictly linear theory, Majda and Roytburd[15] have developed recently a non-linear asymptotic theory that can describe this type of mode self non-linear interaction, resulting in a Landau-Stuart

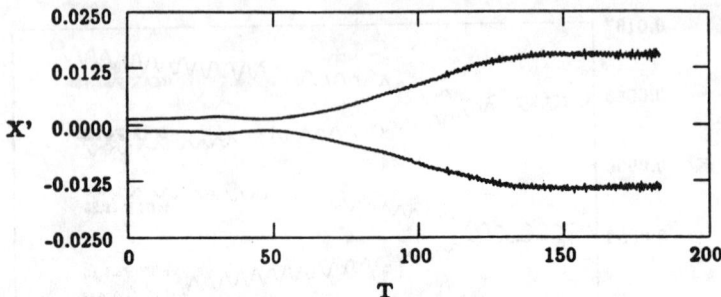

Fig. 3 Case 2, minimum and maximum amplitude of the front position perturbation as a function of time (direct simulation): a slow exponential growth as predicted by the linear stability theory followed by the saturation predicted by the nonlinear extension are observed.

equation to describe the evolution of the perturbation amplitude. Their method has already been used successfully in predicting the same type of saturation behavior in the direct simulations for the case in one-space dimension discussed earlier.[12]

In Fig. 4, we show successive representations of the front as computed explicitly by the tracking procedure (the front actually computed is reproduced three times). In this picture, and in similar pictures for most cases to be described below, the scales for the vertical dimension and for the perturbation amplitude are exaggerated compared to the distance between successive profiles to make the perturbation visible. The front retains a smooth sinusoidal shape, while it is moving down and to the right. In Fig. 5, spatial structure for the linear eigenmodes for pressure, specific volume and reactant mass fraction are displayed on the left, while the corresponding direct numerical simulation results are displayed on the right

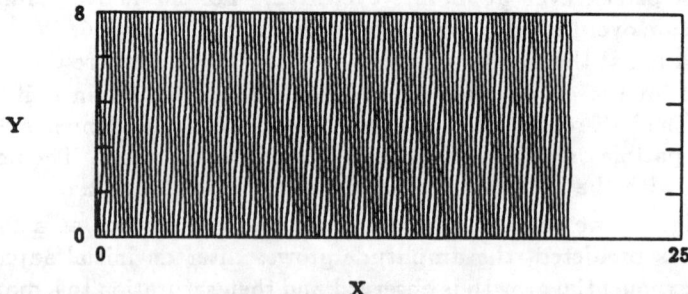

Fig. 4 Case 2, front track (polygonal representation of the tracked front as a function of time - the front actually computed is reproduced twice in the vertical direction by taking advantage of the periodic boundary conditions); the front retains a smooth shape as it is traveling to the right, with the transverse perturbation traveling down.

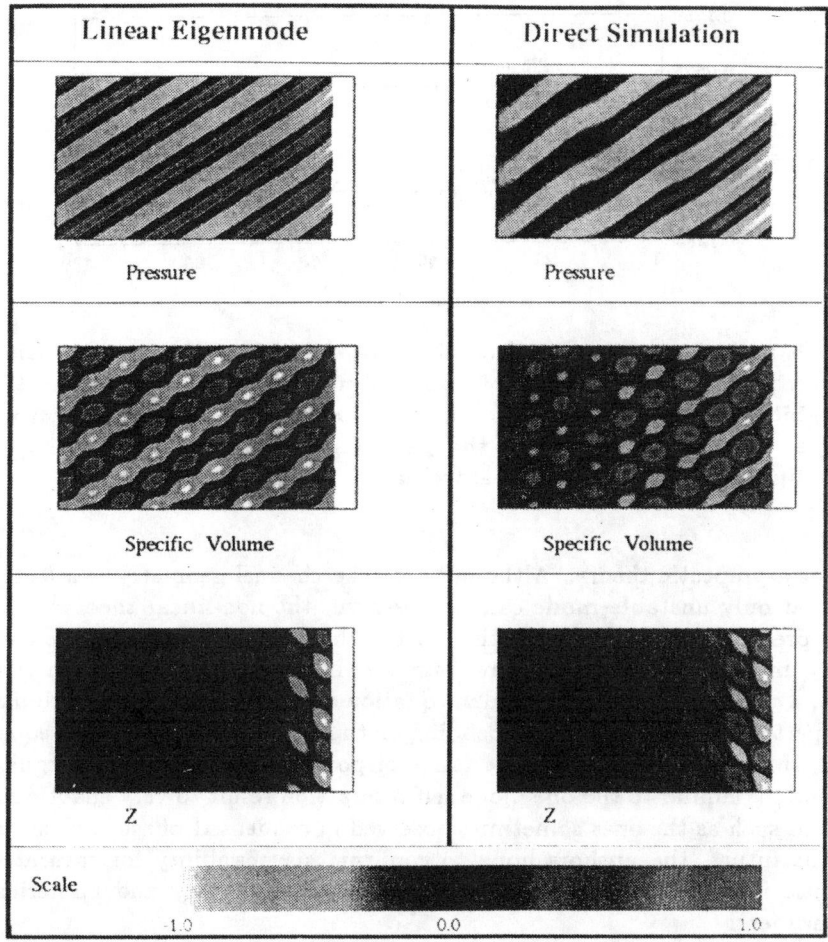

Fig. 5 Case 2, comparison between linear stability eigenmodes (left) and direct simulation results (right); the steady ZND profile is subtracted from the numerical solution to show the small perturbation structure.

after having subtracted the theoretical ZND profile (again reproduced three times vertically). The pictures cover three channel widths (12 half-reaction lengths) and 20 half-reaction lengths in the longitudinal direction, the domain was covered with a mesh refinement of eight points per $L_{1/2}$. The agreement between the prediction and the direct simulation is excellent, especially if one takes into account the extremely small amplitude of the perturbations (1% of $L_{1/2}$ for the front location), resulting in some numerical dissipation as seen in the pictures for the pressure and vertical velocity perturbations. This agreement between the eigenmode spatial structure and the direct simulation perturbation solution is predicted by the non-

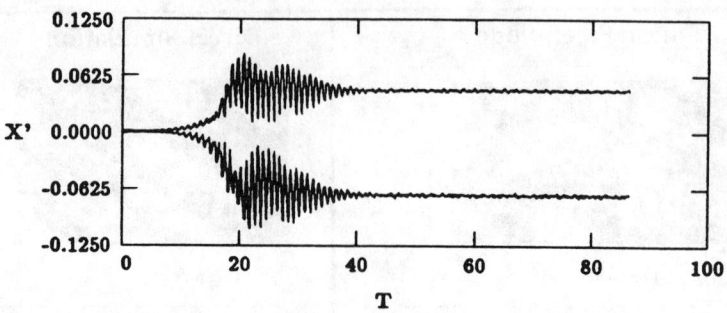

Fig. 6 Case 3, minimum and maximum amplitude of the front position perturbation as a function of time from the direct simulation; the instability is excited with some random perturbation in the initial front shape. As the amplitude of the perturbation grows, it locks into a traveling wave with constant extrema.

linear asymptotic theory. Although here the channel geometry is selected so that only unstable mode can be observed, the non-linear theory[15] can also predict multimode interactions if the channel width was, on the contrary, much larger than the wave numbers in the unstable band. In that case, a complex Guinsburg-Landau equation describes the time evolution of the perturbation amplitude; depending on the exact coefficients of the equation, this could result in a wide range of possible behaviors, from regular pulsations similar to the one mode self-interaction result to very chaotic solutions such as the ones sometimes observed in condensed phase explosives. In the future, the authors hope to confirm this possibility for saturated chaotic instabilities through a combination of the theory and numerical experiments.

The two tests above confirm the high accuracy of the method. The kind of transition test performed here in two-space dimensions could not be passed by most regular shock capturing methods even in one-space dimension. It is obvious from the saturation amplitude of the front perturbation that capturing methods would require an incredibly large number of points in the shock neighborhood to be able to capture this type of instability.

Case 3

This case has the same physical parameters as in case 2, but the overdrive ratio is reduced to $f = 1.2$, away from the neutral stability curve into the unstable parameter domain. The initial data are the ZND profile. The only perturbation applied in this case is a "random" perturbation in the front shock location. In Fig. 6, the history of that perturbation is shown. The perturbation grows very fast to an amplitude of one fine mesh size or $1/8\ L_{1/2}$. The shock tracking scheme allows us to resolve such details in the leading front. First, large oscillations are observed, but later on the perturbations lock into a single-headed mode, with the maximum and

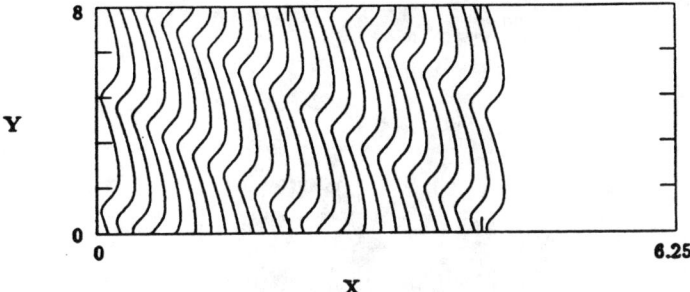

Fig. 7 Case 3, front track: the traveling transverse wave is clearly seen, but no kink that would indicate the existence of a transverse shock wave is observed.

minimum amplitude constant in time. This is confirmed in Fig. 7, where the front track is shown as function of time. The front remains similar to itself while traveling down and to the right. In Fig. 8, the temperature map of the flow at one time is shown, with darker regions corresponding to higher temperatures (solution reproduced for three channel widths). A strong transverse wave is observed corresponding to the strong gradient region in the front shock shape. This type of structure is very similar to the marginal "spinning" detonations sometimes observed in round tubes, even though the heat release in the present case is apparently much too small to generate discontinuities and sustain a full fledged triple-point configuration.

Case 4

This case corresponds to parameters selected earlier by Erpenbeck.[28] Now we have $\gamma = 1.2$, the activation energy is set to $E^+ = 20$ and the heat release is $q_o = 3$. The width of the channel, $L_y = 5.7 L_{1/2}$ corresponds to wave number $k = 1.1$. The flow is initialized using the corresponding linear unstable mode, with eigenvalue $\alpha = 0.138+$ i 1.29. In Fig. 9, the front amplitude history initially displays the expected growth of the transverse traveling wave, but then saturates into a cellular mode, with two Mach triple-points. The maximum and minimum amplitude showing large oscillations as the cells go through a complete cycle. The transition from a traveling transverse wave to a cellular pattern is clearly visible in Fig. 10, where the time track of the front is shown. Fig. 11 shows a grey shade map for the temperature field, at a time just before the collision of two reflected waves. A similar picture for vorticity would show equally straight and regular lines and the vorticity amplitude is very small.

Case 5

In this case, $\gamma = 1.2$, the activation energy is $E^+ = 10$ and the heat release is $q_o = 50$. The overdrive ratio is $f = 1.2$ and the channel width is $L_y = 10$. Those parameters clearly correspond to a unstable problem;

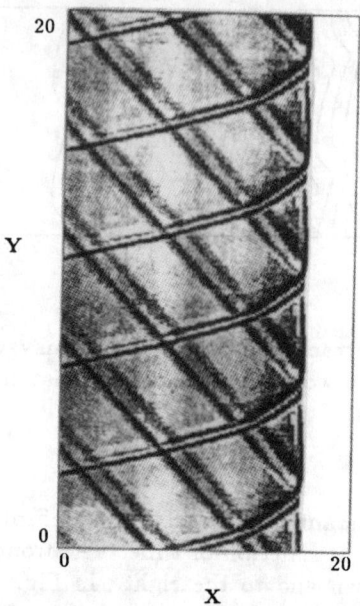

Fig. 8 Case 3, temperature map at one particular time (the actual computation is reproduced five times by taking advantage of the periodic boundary conditions). In this temperature map and the ones to follow, lower temperatures are represented by darker shades, also discontinuities such as shocks and contact discontinuities appear as black contour lines (using a special graphic procedure).

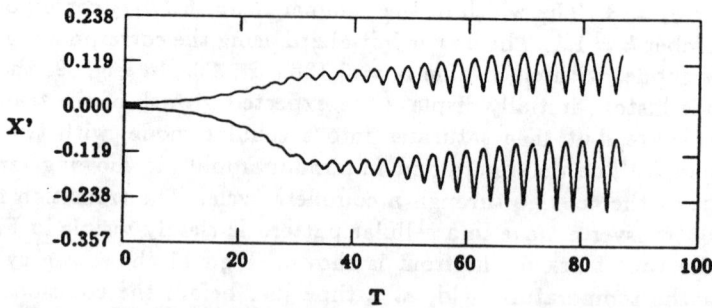

Fig. 9 Case 4, minimum and maximum amplitude of the front position perturbation as a function of time from the direct simulation. The instability is excited using the linear stability traveling mode (no oscillations in min and max); as the perturbation grows, it bifurcates to a symmetric cellular mode (oscillations in min and max corresponding to the different stages of the cellular pattern).

UNSTABLE DETONATIONS IN TWO-SPACE DIMENSIONS 57

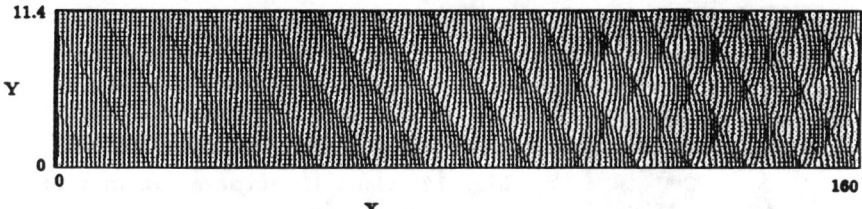

Fig. 10 Case 4, front track: bifurcation from a traveling mode to a cellular mode.

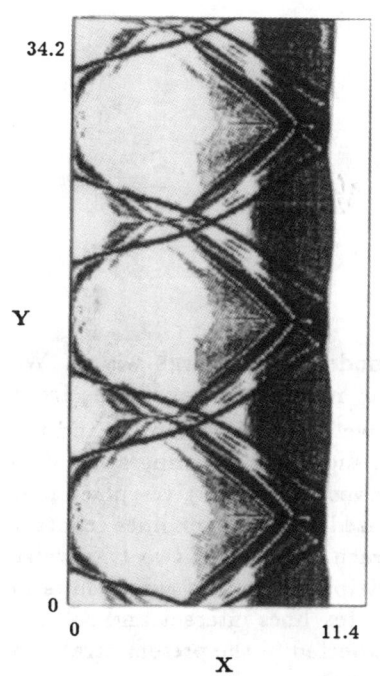

Fig. 11 Case 4, temperature map at one particular time (the actual computation is reproduced three times). Pair of triple points are about to collide as the transverse shocks are moving into the lower temperature region (unreacted gas-darker shades).

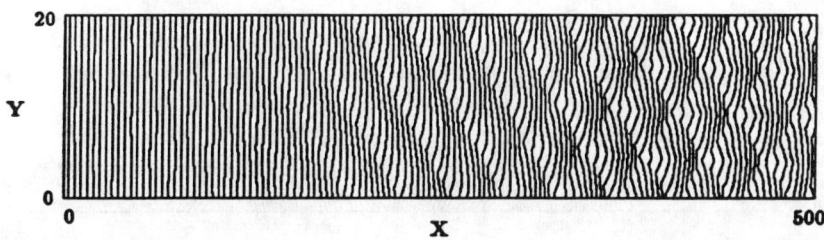

Fig. 12 Case 5, front track: bifurcation from a traveling mode to a cellular mode.

Fig. 13 Case 5, temperature map at one particular time (the actual computation is reproduced three times). Vortex lines generated at the collision of triple points interact, roll up and generate pairs of large eddies.

in fact, there are at least three unstable modes for the given width. We excite the flow with a small perturbation of mode 1 and again the front track in Fig. 12 shows the growth of the traveling transverse wave and the bifurcation to a regular cellular pattern. In Fig. 13, the temperature field is shown. The very regular structure observed in case 4 gives place now to a more complex, yet still regularly organized flow. Vortex lines (contact discontinuities) are generated in pairs at each collision of two transverse waves; the fairly large chemical energy contribution in this case results in bigger amplitude perturbations. Pairs of vortex lines interact and roll up. Regular plumes are observed, as might be expected in the present stratified flow.

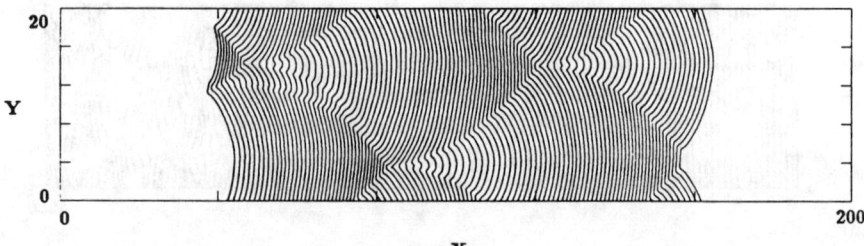

Fig. 14 Case 6, partial front track: excellent agreement with experimental data[29] is observed.

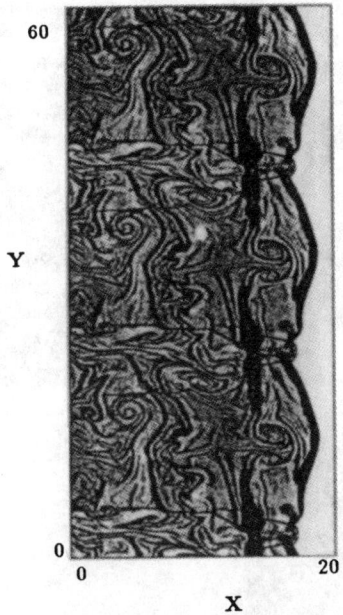

Fig. 15 Case 6, temperature map at one particular time (the actual computation is reproduced three times). The regular pattern from case 5 has given place to a fully turbulent flow as the problem has become more unstable.

Case 6

If the activation energy is now increased to $E^+ = 50$, the stability theory predicts even stronger instabilities. This flow is actually even unstable in one-space dimension and early times of the direct simulation in two-space dimensions are dominated by this longitudinal instability. However, the perturbation quickly becomes fully two dimensional and again a cellular structure is observed. In Fig. 14, we display the numerical front track and notice the surprising good qualitative agreement in the shape of the cell boundaries with earlier experimental results,[29] even though no effort was made to match the complex chemistry of the gas used in the experiment. Fig. 15 shows the temperature map: the regular structure observed in case 5 has now degenerated into a fully turbulent flow. This numerical simulation strongly supports Lee's conjecture regarding the role of turbulence in detonation flows[23]; in this particular case, the front instability itself generates a great amount of turbulence. In Fig. 16, we present the density, reactant mass fraction Z and pressure maps, along with the wave structure deduced from those maps. The letters in the wave diagram correspond to the ones used by Voitsekhovskii et al.[30] The complex Mach structure observed in the present calculation is in remarkable qualitative

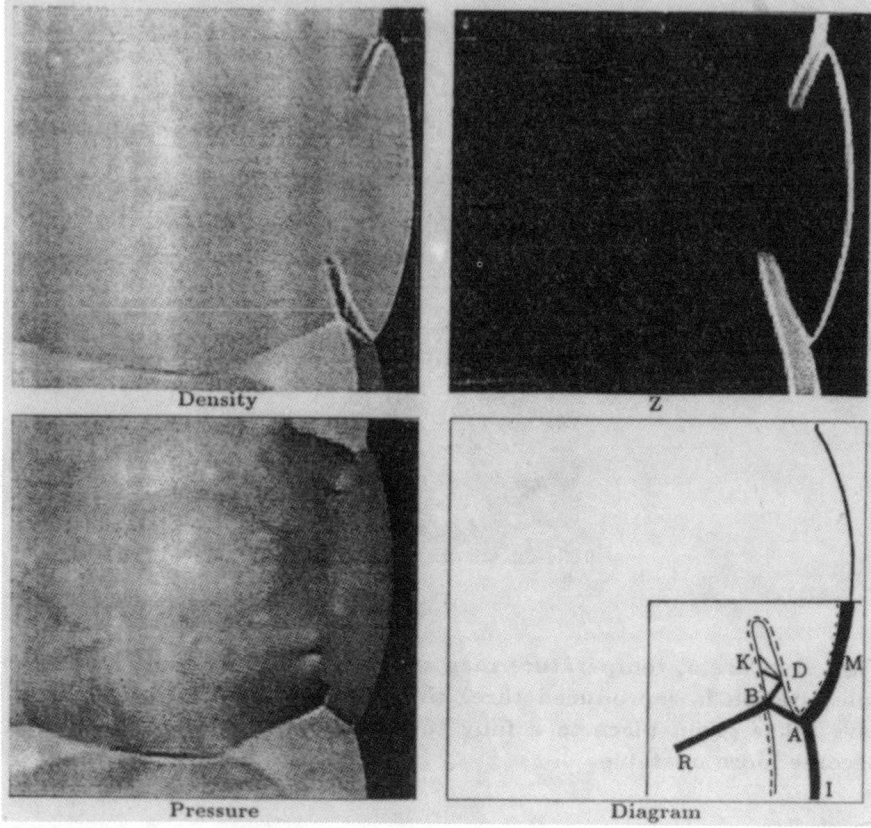

Fig. 16 Case 6, pressure, density, and reactant mass fraction maps at one particular time along with the wave diagram deduced from the flowfield. The detailed multiple Mach stem structure in the wake of the leading front is in surprising agreement with experimental results.[30]

agreement with the experimental results: even though the shock structure in the wake of the front is captured by the finite difference scheme and not explicitly tracked, the sharpness of the computed structure benefits greatly from the accurate computation of the tracked leading front: to our knowledge, such a multiple Mach stem structure has never been observed before in direct numerical simulations.

Conclusion

A new high resolution scheme is developed to solve the reactive Euler equations in two-space dimensions. By combining a higher-order Godunov's method, explicit front tracking and adaptive mesh refinement, we are able to compute accurately a wide variety of instability regimes, ranging from pulsations on the length scale of the half-reaction length to large amplitude cellular structures. The accuracy of the large-scale numerical simulations is confirmed by comparison with theoretical and experimental results. In particular, new regimes of instability involving traveling pulsating modes without Mach stem structures have been found in regimes near the stability boundaries; these types of instabilities are qualitatively like those observed in various condensed phase experiments.[3] In the regimes of strong instability, the numerical calculations indicate that there is the full complexity of two-dimensional turbulence in the wake regime following the unstable detonation wave; these results support the ideas of Lee[23] that unstable detonations are a significant source of turbulence. A more complete and systematic study of both the numerical results and the physical phenomena presented here as well as comparison with theories for detonation instability will be reported elsewhere in the near future.[16]

Acknowledgments

The research of both authors was partially supported by grants A.R.O. DAAL03-89-K-0013 and Office of Naval Research, N0014-89-J-1044. The computations reported in this paper were performed on a CRAY YMP at the Pittsburgh Supercomputer Center through National Science Foundation. grant DMS900001P. The authors want to thank V.Roytburd of the Rensselaer Polytechnic Institute for helpful conversations and Ph.Colella of the University of California at Berkeley for generous advice in the implementation of the numerical method.

References

[1] Oppenheim, A. K. and Soloukhin, R. I., "Experiments in Gasdynamics of Explosions," *Annual Review of Fluid Mechanics*, Vol. 5, 1973, pp. 31-58.
[2] Lee, J. H. S. and Moen, I. O.,"The Mechanism of Transition from Deflagration to Detonation in Vapor Cloud Explosion," *Progress in Energy and Combustion Science*, Vol. 6, 1980, pp. 359-389.
[3] Fickett, W. and Davis, W. C., *Detonation*, University of California Press, Berkeley, 1979.
[4] Taki, S. and Fujiwara, T., "Numerical Analysis of Two-Dimensional Nonsteady Detonations," *AIAA Journal*, Vol. 16, 1973, pp. 73.
[5] Taki, S. and Fujiwara, T., "Numerical Simulation of Triple Shock Behavior of Gaseous Detonations," *Eighteenth Symposium (International) on Combustion*, The Combustion Institute, 1981, pp. 1641-1649.
[6] Oran, E. S., Young, T. R., Boris, J. P., Picone, J. M., and Edwards, D. H., "A Study of Detonation Structure: The Formation of Unreacted Gas Pockets," *Nineteenth Symposium (International) on Combustion*, The Combustion Institute, 1982, pp. 573-582.

[7] Kailasanath, K., Oran, E. S., Boris, J. P., and Young, T. R., "Determination of Detonation Cell Size and the Role of Transverse Waves in Two-Dimensional Detonations," *Combustion and Flame*, Vol. 61, 1985, pp. 199-209.

[8] Guirguis, R., Oran, E. S., and Kailasanath, K., "Numerical Simulation of the Cellular Structure of Detonations in Liquid Nitromethane- Regularity of the Cell Structure," *Combustion and Flame*, Vol. 65, 1986, pp. 339-365.

[9] Boris, J. P., and Oran, E. S., *Numerical Simulation of Reactive Flow*, Elsevier, New York, 1987.

[10] Majda, A. and Roytburd, V., "Numerical Modeling of the Initiation of Reacting Shock Waves," *Computational Fluid Mechanics and Reacting Gas Flows*, edited by B. Engquist, M. Luskin, and A. Majda, the IMA Volumes in Mathematics and its Applications, Vol. 12, 1988, pp. 195-217.

[11] Colella, P., Majda, A. and Roytburd, V., "Theoretical and Numerical Structure for Reacting Shock Waves," *SIAM J.Sci.Stat.Comput.*, Vol. 7, 1986, pp. 1059-1080.

[12] Bourlioux, A., Majda, A. and Roytburd, V., "Theoretical and Numerical Structure for Unstable One-Dimensional Detonations," *SIAM Journal of Applied Mathematics*, Vol. 51, 1991, pp. 303-343.

[13] Bourlioux, A., Ph.D. Thesis, Princeton University, Princeton, NJ 1991.

[14] Bourlioux, A., Majda, A. and Roytburd, V., "Nonlinear Development of Low Frequency One-Dimensional Instabilities for Reacting Shock Waves," *Dynamic Issues in Combustion*, edited by Fife P., Linan, A., and Williams F., IMA Volumes in Mathematics and its Applications, Vol. 35, 1991, pp. 63-83, Springer-Verlag, New York.

[15] Majda, A. and Roytburd, V., "Low Frequency Instabilities of Reacting Shock Waves," *Studies in Appled Mathematics*, Vol. 87, pp. 135-174, 1992.

[16] Bourlioux, A. and Majda, A., "Theoretical and Numerical Structure of Two-Dimensional Unstable Detonations," *Combustion and Flame*, Vol. 90, pp. 211-229, 1992.

[17] Majda, A., "A Qualitative Model for Dynamic Combustion," *SIAM Journal of Applied Mathematics*, Vol. 41, pp. 70-93, 1981.

[18] Majda, A. and Rosales, R., "Weakly Non-Linear Detonation Waves," *SIAM Journal of Applied Mathematics*, Vol. 43, 1983, pp. 1086-1118.

[19] Erpenbeck, J. J., "Stability of Steady-State Equilibrium Detonations," *Physics of Fluids*, Vol. 5, 1962, pp.604-614.

[20] Fickett, W. and Wood, W. W., "Flow Calculations for Pulsating One-Dimensional Detonations," *Physics of Fluids*, Vol. 9, 1966, pp.903-916.

[21] Moen, I. O., Ward, J. W., Rude, G. M., and Thibault, P. A., "Detonation Length Scales for Fuel-Air Explosives," AIAA *Progress in Astronautics and Aeronautics*, Vol. 94, 1984, pp.55-79.

[22] Lee, H. I., and Stewart, D. S., "Calculation of Linear Detonation Instability: One-Dimensional Instability of Plane Detonations," *Journal of Fluid Mechanics*, Vol. 216, 1990, pp.103-132.

[23] Lee, J. H. S., "On the Universal Role of Turbulence in the Propagation of Deflagrations and Detonations," edited by B. Engquist, M. Luskin, and A. Majda, The IMA Volumes in Mathematics and its Applications, Vol. 12, 1988, pp.169-193.

[24] Colella, P. and Woodward, P., "The Piecewise Parabolic Method (PPM) for Gas-Dynamical Simulations," *Journal of Computational Physisc*, Vol. 54, 1984, pp. 174-201.

[25] Colella, P., "Multidimensional Upwind Methods for Hyperbolic Conservation Laws," *Journal of Computational Physics*, Vol. 87, 1990, pp. 171-200.

[26] Berger, M. and Colella, P., "Adaptive Mesh Refinement for Shock Hydrodynamics," *Journal of Computational Physics*, Vol. 82, 1989, pp. 64-84.

[27] Chern, I. and Colella, P., "A Conservative Front Tracking Method for Hyperbolic Conservation Laws," Lawrence Livermore National Laboratory, Livermore, California, Preprint UCRL-97200, 1987.

[28] Erpenbeck, J. J.,"Nonlinear Theory of Two-Dimensional Detonations," *Physics of Fluids*, Vol. 13, 1970, pp. 2007-2026.

[29] Strehlow, R. A., "Gas Phase Detonations: Recent Developments," *Combustion and Flame*, Vol. 12, 1968, pp.81-101.

[30] Voitsekhovskii, B. V., Mitrofanov, V. V., and Topchian, M. E., "Structure of the Detonation Front in Gases," *Fizika Goreniya Vzryva*, Vol. 5, 1969, pp.385-395.

Simulation of Cellular Structure in a Detonation Wave

M. H. Lefebvre*
Royal Military Academy, Brussels, Belgium
E. S. Oran† and K. Kailasanath‡
Naval Research Laboratory, Washington, DC 20375
and
P. J. Van Tiggelen§
Université Catholique de Louvain, Louvain-la-Neuve, Belgium

Abstract

The cell structure of a detonation wave in a hydrogen-oxygen-argon mixture is studied computationally using a two-dimensional, time-dependent solution of the compressible reactive-flow equations. A phenomenological model is used to represent the conversion of reactants to products and energy release. The particular case studied here resolves the structure and dynamics of one detonation cell in great detail. The computations show that the interaction between colliding transverse waves first occurs at about three-quarters of the cell length in a region of fully reacted material. The simulations also show the presence of a small unburned gas pocket in the last 10% of the cell length. Quantitative comparisons to experimental data show general agreement in the computed and measured pressure field and, in particular, the high-pressure spike at the end of the cell confirms the reinitiation model. The computation also correctly predicts the decay of the leading shock velocity along the axis of the cell. The comparisons also show that an improved chemical and thermophysical model is required for better quantitative agreement.

This paper is delcared a work of the U.S. Government and is not subject to copyright protection in the United States
 *Lecturer, Department of Chemistry
 †Senior Scientist for Reactive Flow Physics
 ‡Head, Center for Reactive Flows and Dynamical Systems, Laboratory for Computational Physics and Fluid Dynamic
 §Professor, Laboratoire de Physico-Chimie del la Combustion

Nomenclature

D	=	detonation velocity, m/s
E	=	density of total energy, J/m^3
L	=	cell length, mm
M	=	Mach number
N	=	total number of moles, mol/kg
P	=	static pressure, Pa
R	=	ideal gas constant, J/mol K
T	=	temperature, K
t_{char}	=	characteristic time, μs
t_{ind}	=	induction time
U	=	specific internal energy, J/kg
ΔU_r^o	=	specific reaction energy at 0 K, J/kg
v	=	scalar velocity, m/s
$\mathbf{v}$	=	vectorial velocity
x	=	spatial variable along the propagating axis
γ	=	isentropic expansion coefficient
λ	=	cell width, mm
ρ	=	density, kg/m^3
τ	=	reaction progress parameter

Introduction

The structure of detonation waves in gases has been extensively studied and physical models have been established. The unsteady character of the leading shock throughout a detonation cell has been recognized by all investigators. With his study on the detonation limit, Soloukhin showed that transverse shock waves behind the leading shock are crucial in sustaining detonation propagation in gases.[1] Two-dimensional detonation propagation has been numerically simulated[2] and subjects such as the development of cellular structure[3] and the irregularity in the structure of detonation cells[4] has been investigated numerically. A review of the numerical studies of multidimensional detonations can be found in Oran et al.[5]

The original macroscopic detonation-cell investigations of the average velocity of the detonation wave and soot patterns formed on the walls of detonation tubes have now been extended to finer measurements inside the detonation cell itself.[1-7] A characteristic time t_{char}, defined by Dormal et al.[8] as the time period for a detonation cell to be completed, is equal to the ratio of the cell length L over the mean detonation velocity D. This characteristic time has been well correlated with the induction time computed along the centerline of a cell. A detailed study of the evolution of induction times inside the detonation cell has been carried out.[9]

The purpose of this paper is to simulate the detailed structure of a detonation cell in a mixture of $H_2/O_2/Ar$ (2:1:7) at 50 Torr and 298 K in order to obtain a self-sustained wave and then use the simulation to investigate and to interpret some previously obtained experimental data. In particular, we focus on the pressure profiles near the end of the cell, the interaction of the transverse waves, the velocity profile along its centerline, and the variation of the induction time inside the detonation cell.

Model and Method for Numerical Analysis

The numerical calculation of gas-phase detonations solves the compressible, time-dependent, conservation equations for total mass density ρ, momentum density $\rho\mathbf{v}$, and energy E. The time-dependent compressible gasdynamic equations are as follows:

$$\frac{\partial \rho}{\partial t} = -\nabla \cdot \rho \mathbf{v} \qquad (1)$$

$$\frac{\partial \rho \mathbf{v}}{\partial t} = -\nabla \cdot (\rho \mathbf{v} \mathbf{v}) - \nabla P \qquad (2)$$

$$\frac{\partial E}{\partial t} = -\nabla \cdot (E\mathbf{v}) - \nabla \cdot (\mathbf{v}P) \qquad (3)$$

The energy E represents the total energy and is the sum of the kinetic and internal energy.

We use a two-step reaction model to simulate the chemical reactions and energy release. The first step is a thermoneutral induction period during which there is no energy release. During this step, the reaction taking place is

$$H_2 + O_2 + \text{diluent} \longrightarrow \text{radicals}, \quad \Delta U = 0$$

The second step corresponds to the energy release reaction and can be written symbolically as

$$H_2 + O_2 + \text{radicals} \longrightarrow \text{product}, \quad \Delta U = \Delta U_r^o$$

with appropriate conversion of internal energy to sensible heat. A progress parameter τ tracks the evolution of the reactions. The parameter τ is initially zero and is allowed to increase to unity as the induction time passes. We have used a table of induction times which were obtained by integrating a detailed set of reaction rates.[10] The induction time t_{ind}, is calculated for the current conditions of temperature and pressure. The progress parameter τ is calculated so that

$$\frac{\partial \tau}{\partial t} = \frac{1}{t_{\text{ind}}[T(t), P(t)]} \qquad (4)$$

When the reaction occurs in a fluid in motion, the time derivatives in Eq. (4) denote a substantial derivative which follows the flow and the equation become

$$\frac{\partial \rho \tau}{\partial t} + \nabla \cdot \rho \tau \mathbf{v} = \frac{\rho}{t_{\text{ind}}} \tag{5}$$

valid for τ going from 0 to 1. During the induction period, the parameter τ is convected in a Lagrangian sense and yields the time history of the temperature and pressure seen by a reacting fluid element. When τ reaches unity, the induction period ends and the energy release is initiated. During the energy-release process, τ varies linearly from 1 to 2, that is,

$$\frac{\partial \tau}{\partial t} = \frac{1}{5.0 \times 10^{-6}} \tag{6}$$

which says that the time elapsed for the full reaction is constant and equal to 5 μs. The total released energy $\Delta U_r^o{}_{\max}$ is assumed equal to -820 kJ/kg, and varies linearly with τ,

$$\frac{\mathrm{d}\Delta U_r^o / \Delta U_r^o{}_{\max}}{\mathrm{d}t} = \frac{1}{5.0 \times 10^{-6}} \ \mathrm{s}^{-1}$$

We can also write Eq. (6) as

$$\frac{\partial \rho \tau}{\partial t} + \nabla \cdot \rho \tau \mathbf{v} = \frac{\rho}{5 \times 10^{-6}} \tag{7}$$

which is valid for τ going from 1 to 2. Equations (5) and (7) are now in conservative form and can be solved by the same method used to solve Eqs. (1–3). The gas is also assumed to be ideal. The relation between specific internal energy U and temperature is then

$$U = \frac{NR}{\gamma - 1} T + \Delta U_r^o \tag{8}$$

Thus the pressure can be deduced from

$$P = (\gamma - 1)\left(E - \rho \Delta U_r^o - \frac{1}{2}\rho v^2 \right) \tag{9}$$

We assume that both molecular mass $(1/N)$ and γ remain unchanged during the progress of the reaction and are equal to their initial values.

The equations presented above, (1–3), (5), and (7), do not have any explicit term representing physical viscosity, that is, that the flow is assumed to be inviscid. Other diffusive transport processes such as thermal conduction and molecular diffusion are also not included. These are reasonable assumptions for the high-speed flow discussed in this paper. However, near the reaction front itself, the diffusive transport processes may become more important due to the high temperatures and the large gradients in temperature and species concentrations. The numerical resolution required to resolve these gradients adequately is beyond the scope of this investigation.

The simulation is based on solutions of the time-dependent Euler equations (1–3), (5), and (7) using the flux-corrected transport algorithm.[11] This is an explicit, fourth-order accurate, nonlinear finite-difference method that reproduces discontinuities with small dispersion errors and minimal numerical diffusion. The fluid dynamics and chemistry terms in the conservation equations are solved separately and then coupled by using timestep-splitting methods. The boundaries of the computational domain are solid walls, represented by free-slip boundary-conditions. The computational cell spacing is held fixed in time and its typical size ($\Delta x \times \Delta y$) is 0.015×0.030 cm^2. The timestep $\Delta t=$ is variable. This ensures a resolution in the reaction zone of about 12 cells. Typical values for Δt are about 0.05 μs. A calculation is run with 850×156 computational cells, and is carried out until the detonation reaches the boundary. This allows the computation of about one and a half detonation cells. A complete calculation for a wave going from cell 150 to 790, i.e., for the detonation wave to move 96 mm, takes about 4000 s of CPU-time on the NRL CRAY X-MP/24 computer.

Results and Comparison With Experiments

The numerical simulation was initiated by placing a square pocket of hot, unburned gas, centered on the axis, behind a planar steady detonation. When the pocket reacts, it sends out pressure disturbances which interact with the detonation front. Two transverse waves are formed and after a few collisions with the channel walls, the transverse waves establish the detonation front structure. After 4500 timesteps (about 200 μs), a row of three similar detonation cells was recorded and there is a stable self-sustained detonation wave.

Unless noted differently, all experimental data reported in this paper are based on series of experiments performed by Libouton[12] and Dormal.[8,9] The experiments were carried out in a rectangular 32×92 mm^2 tube, with the same mixture ($H_2:O_2:Ar/2:1:7$) and at the same pressure (50 Torr) as the computational simulation. In the experiments, the pressure is measured by piezoelectric gauges, the local induction time by the emission intensity of OH radical, and the local shock velocities by platinum gauges located on the channel walls.

Some General Observations

From previous calculations,[4] we know that choosing the right channel width is important for obtaining a complete detonation cell. If the channel width is less than what could be called its "natural cell size," the computed cell appears flattened. By performing series of calculations in which the channel width was varied, we found that, for the chemical model described above and a channel 46.8 mm wide, the locus of the triple points forms a complete, single shaped detonation cell. The triple points reflect without any time or space gap on the boundaries. The evolution of the pair of

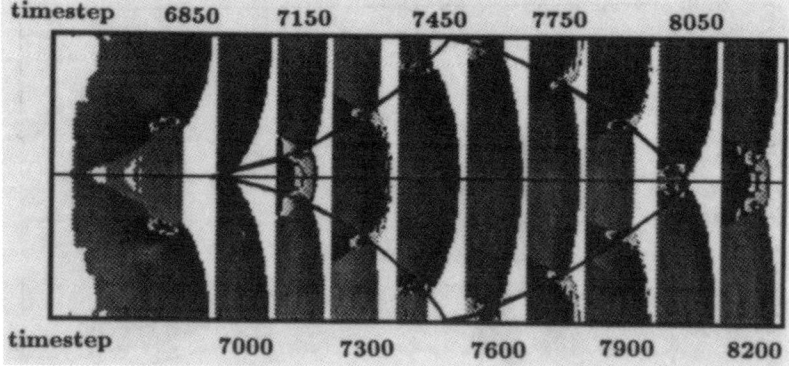

Fig. 1 Composite of pressure contours; the path of the triple points are added to show their movement. The detonation is propagating in a 50-Torr stoichiometric H_2/O_2 mixture diluted with 70% argon.

triple points in our current calculation is exhibited in Fig. 1 (steps 6850–8200). From the figure, we estimate the cell length L to be about 77 mm. The computed ratio of cell-width to cell-length (λ/L) is about 0.61. Experimental data give a cell width (λ) and length of about 92 and 170 mm, respectively, i.e., a value of λ/L equal to 0.54 so that the numerical and the experimental lengths differ by about a factor of two. Because of the larger value of the computed ratio λ/L, the general appearance of the structure exhibited in Fig. 1 is slightly more compact than the experimental one.

The average computed velocity of the wave along the entire structure in Fig. 1 is 1623 m/s. The experimental and Chapman-Jouguet detonation velocities are 1475 and 1619 m/s, respectively. The difference between the computed and experimental velocity is consistent with the larger experimental cell size. Now we consider some of the physical properties of the cell in more detail, and use this information to determine the role of the transverse waves in supporting the detonation.

Local Detonation Velocities

The detonation-velocity profile on the cell centerline is computed by following the pressure jump throughout the numerical grid. Figure 2 shows a smoothed curve of the experimental data (solid line) and computed detonation velocities (triangles). From Fig. 2, we learn that the wave velocity just after the collision of the triple points, that is, in the early part of the structure, is about 1.48 times the average value of 1623 m/s. The shock speed then decreases gradually throughout the cell and reaches its lowest value at the end of the cell. This lowest velocity reached is 0.76 times the mean velocity. We do not notice any increase of the wave velocity when the system approaches the head of the cell. Although such an increase has been reported previously,[6] it was never observed in our experiments, and

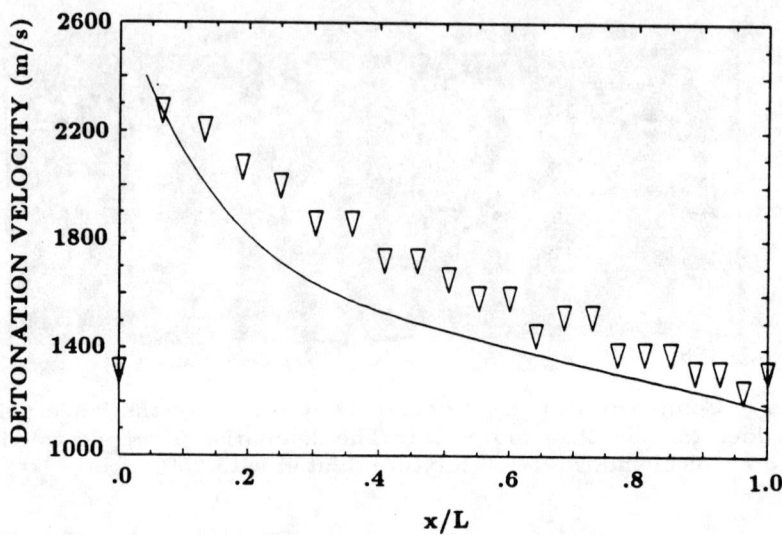

Fig. 2 Evolution of the experimental[8] and computed shock velocity inside the detonation cell: the experimental data are smoothed (solid line); triangles are computed velocities (this work).

in this calculation, the calculated wave velocity exhibits a monotonic decay in the initial region of the cell.

Pressure Profiles

The evolution of the pressure jumps at the cell centerline is consistent with the decaying shock velocity. The pressure spikes observed in the beginning of the cell are very high ($P/P_o = 52$ at $x/L = 0.10$) and decrease rapidly throughout the structure as shown in Fig. 3. In the second half of the cell, the leading shock jump is relatively constant (about $P/P_o = 20$ at $x/L = 0.70$). The pressure ratios can be calculated by the equation for a steady shock,

$$\frac{P}{P_o} = \frac{2\gamma}{(\gamma + 1)} M^2 - \frac{(\gamma - 1)}{(\gamma + 1)} \tag{10}$$

where M is the Mach number of the incident shock. Using the computed shock velocities in Eq. (10), we found

$$P/P_o = 48 \quad \text{at} \quad x/L = 0.10$$

and

$$P/P_o = 19 \quad \text{at} \quad x/L = 0.70$$

which are in good agreement with the computed pressure spikes. To better understand the pressure profile along the centerline of the cell, we must look

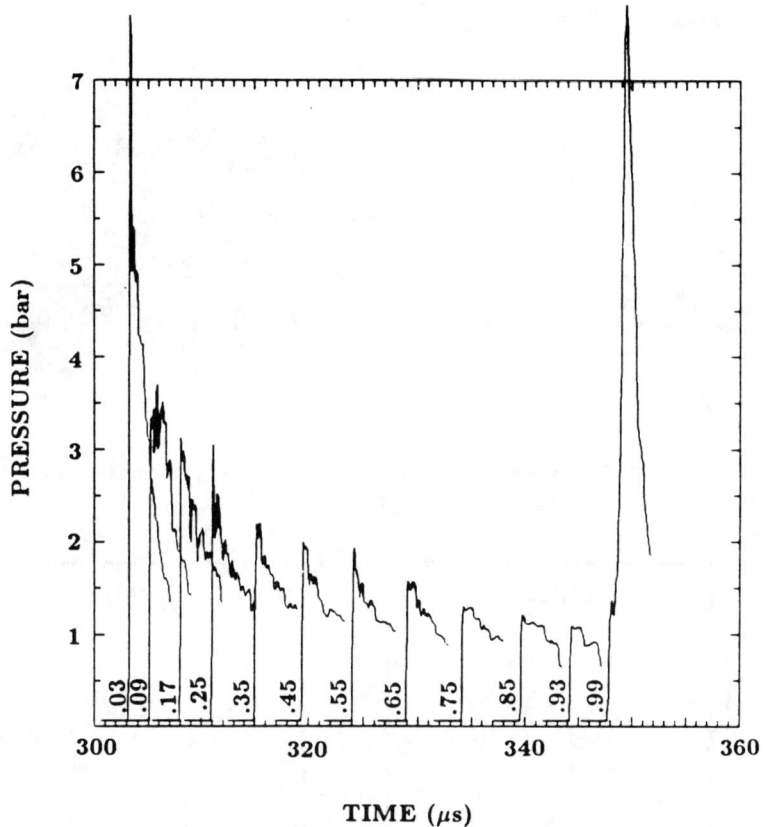

Fig. 3 Computed pressure rises along the centerline of the detonation cell: the labels refer to the location of the probe inside the cell (x/L).

at the movement and the curvature of the transverse waves (TWs) as they approach the end of the cell, as shown in Fig. 4. The portions of the TWs very close to the triple point are almost parallel to the centerline. From step 7700 (leading shock at $x/L = 0.73$), the tails of the TWs become more and more convergent. At timestep 7900, the tails of the TWs collide weakly with each other on the centerline, considerably behind the leading shock front. Since they are quite flat at this time, this weak second shock collision on the centerline occurs almost simultaneously along the line from $x/L = 0.25$ to 0.69 (see black dashes in Fig. 4). At this time, the leading shock has already reached $x/L = 0.93$. The incident shock is also considerably weakened and from $x/L = 0.75$, the collision of TWs produces a pressure rise greater than that of the leading shock. Experimental observations report this second pressure rise at about $x/L = 0.70$. The time between the arrival of the leading shock and the strong collision of the TWs (the "pressure delay time") decreases continuously as the cell develops and vanishes as the cell structure is completed, as shown in Fig. 5. Experimental pressure delay

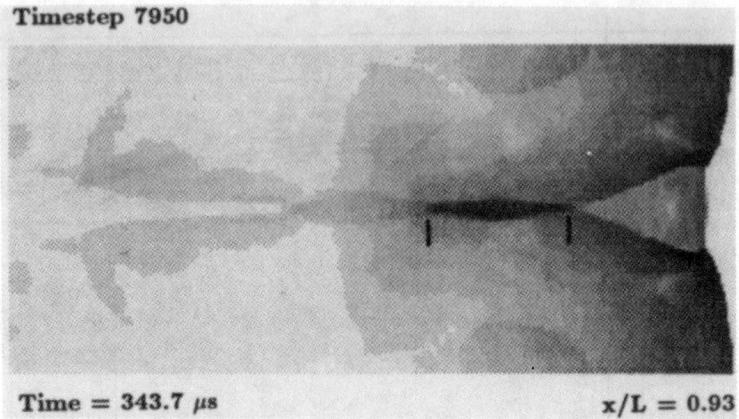

Fig. 4 Isobars around the detonation front; the x/L values refer to the location of the leading shock (timesteps 7950).

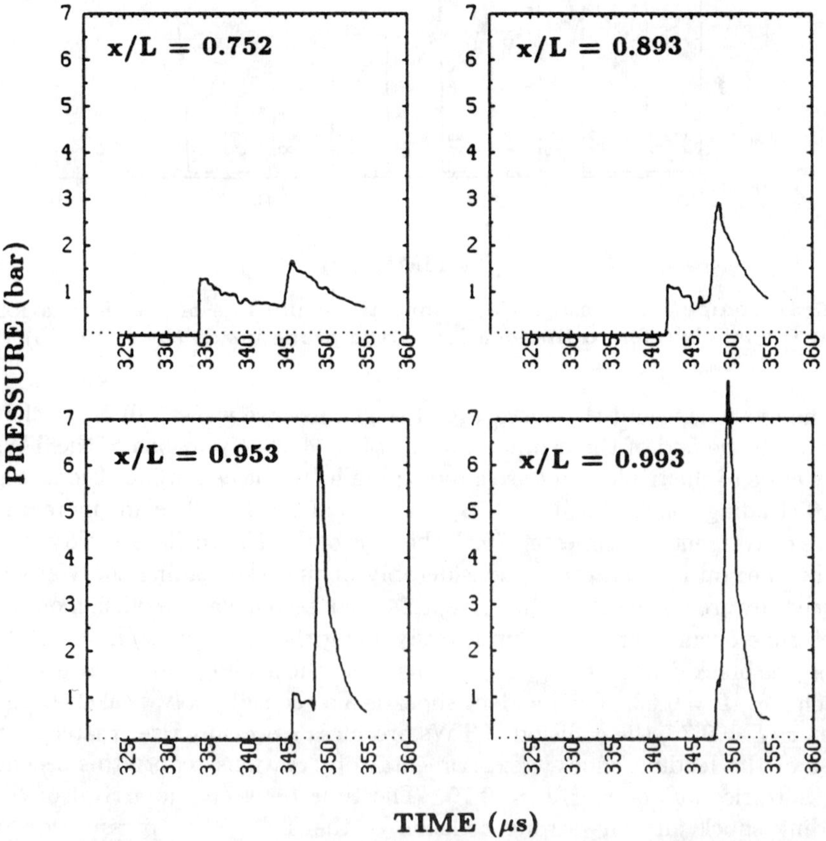

Fig. 5 Pressure profiles computationally recorded with probes located on the centerline of the cell showing the interaction of the TWs. From $x/L = 0.75$, the second pressure rise becomes stronger and stronger.

Table 1 Experimental and computed pressure delay time, μs

x/L	< 0.70	0.84	0.88	0.92	0.97
Experiment[7]	–	10.5	8.0	5.5	2.7
Calculation	–	7.52	5.75	3.9	1.8

times along a cell centerline are summarized and compared with the present calculations in Table 1.

We again observed that the numerical results correctly describe the global trend of the phenomenon. The difference in scale is consistent with the reduced size of the computed cell structure and the larger computed detonation velocity.

Reaction Zone–Induction Time

As the triple points approach the head of the detonation structure, the reaction zone falls farther and farther behind the leading shock front. Table 2 gives an overview of the computed induction zone along the cell centerline, that is, the distance between the incident shock and the beginning of the region where heat release occurs.

A comparison between the calculated and experimental induction time makes little sense since the detonation velocities are different. A more meaningful quantitative comparison should be the nondimensional induction time Θ defined as[8]

$$\Theta = \frac{t_{ind}}{t_{char}} \quad \text{with} \quad t_{char} = \frac{L}{D_{average}}$$

The work of Dormal et al.[8] shows that on the centerline of the cell, the parameter Θ evolves in the same way, independent of the dilution or the initial pressure. Experiments in many different mixtures show a real consistency in the parameter Θ. For our computational cell, t_{char} is equal to 45.9 μs. As it may be seen in Fig. 6, there are notable differences between the experimental and calculated Θ.

Table 2 Space between the shock and the end of the induction zone

x/L^a	0.10	0.25	0.50	0.75	0.93
Distance, mm	0.30	0.45	1.20	2.85	4.35

[a] Refers to the location of the incident shock.

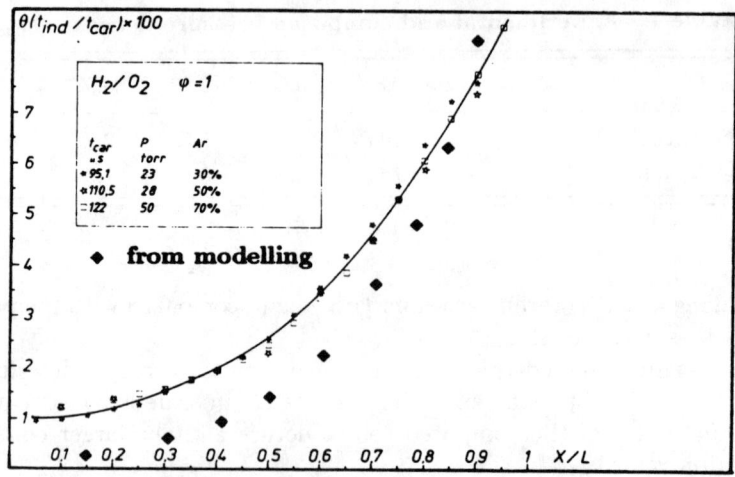

Fig. 6 Experimental[8] and calculated ratio t_{ind}/t_{char}.

A small unburned gas zone is cut off by the TWs at $x/L = 0.99$ and this zone is about 9% of the entire length of the cell. This trapped and more slowly reacting gas explodes nearly homogenously when the incident shock reaches the head of the cell. The chronology of the TW's interaction and the chemical process occurring at late times behind the shock is shown unambiguously in Fig. 7. The temperature profiles recorded before $x/L = 0.89$ show three phases: the first temperature rise corresponds to the arrival of incident shock, the second one to the exothermic reaction, and the third one to the collision of TWs. The maximum reaction temperature is high (about 4000 K at $x/L = 0.89$). In particular, the pressure rise due to the interaction of the TWs takes place in completely reacted gases and thus does not induce any new exothermic chemical processes. For $x/L > 0.91$, the second temperature jump disappears; the collision of TWs triggers the chemical reaction instantaneously. The maximum reaction temperature drops progressively and reaches only 3000 K at the end of the cell. The reaction spreads out from $x/L = 0.91$ to the apex almost simultaneously, as can be seen by superimposing the profiles of Fig. 7. These observations can be related to the measurements of OH emission[1] which indicate a recrudescence of intensity beyond $x/L = 0.80$.

Discussion

In general, the simulation compares well qualitatively with the experimental data[8-12] and also helps explain some of the experimental observations. The global decay of the shock velocity is appropriately described but the relatively slow decay in the beginning of the cell is not consistent with the experimental records which exhibit an exponential decay. This steep decrease is also consistent with the theoretical argument in the work of Van Tiggelen.[13] In the case we have simulated, a strong interaction be-

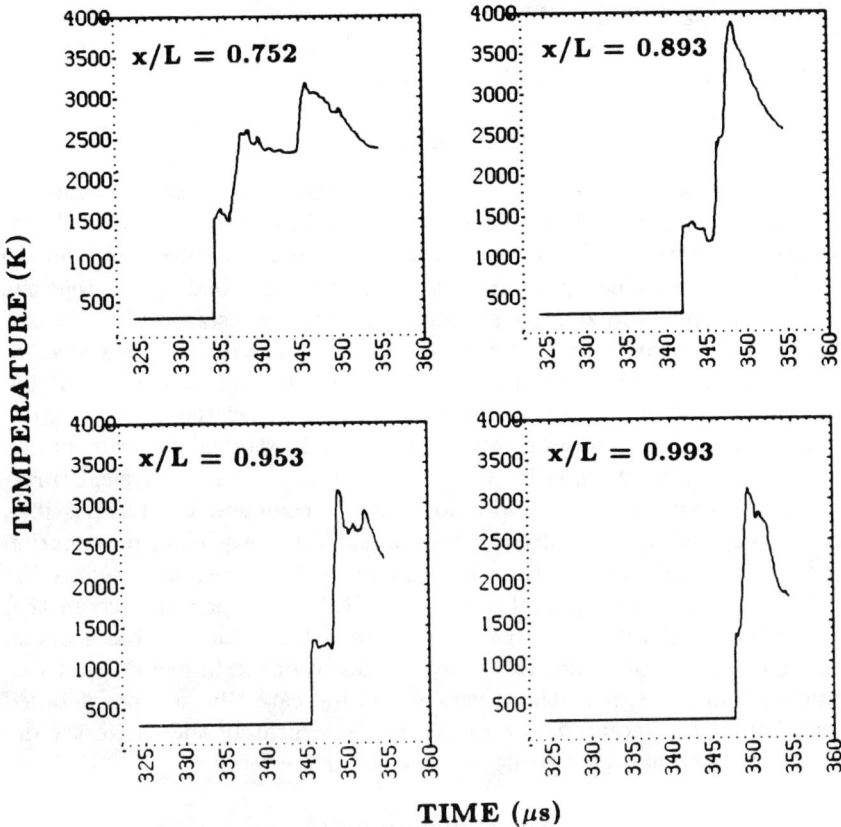

Fig. 7 Temperature profiles at several locations on the axis of the detonation cell.

tween both TWs occurs at 75% of the total length of the cell, a result in fairly good agreement with experimental observations. It is important to notice that, according to our calculations, up until 90% of the cell length, the TWs collide in burned gases; hence, in this portion of the cell, they can not cause the cell reinitiation process. An unburned region extending over the last 10% of the cell length is formed and explodes when the colliding TWs interact with the pocket. However, the OH-emission records show an intense reaction region spreading out on the last 15% of the cell. The way the calculated and experimental induction times differ from each other is consistent with the differences in calculated and experimental shock velocities, but we might expect the reduced induction times (t_{ind}/t_{char}) to agree better. There are a number of factors that might help explain the relatively small size of the cellular structure and the disagreement between the calculated and experimental values of the ratio λ/L. Moreover, multidimensional effects such as boundary layers, and three-dimensional reflected shocks (slapping waves) perturb the conditions on the centerline. However, the errors in the chemical and the thermochemical models we have used may be more important than these factors.

Conculsions

In this paper, we attempt to provide deeper theoretical understanding of the mechanics leading to the establishment of the sustained cellular detonation structure. We have used a nonlinear monotone algorithm to solve the reactive Euler equations in two dimensions. Making sure that the induction and reaction zones are always numerically resolved by at least a few grid cells, we are able to describe accurately the dynamical processes in a detonation cell and, in particular, the reacceleration process occurring when the triple points collide and a new cell is generated. The mixture we have considered has already been extensively studied experimentally. Comparisons between experimental and numerical data of physical (pressure, velocity) and chemical (induction time) parameters are thus possible. Because the fundamental physical mechanisms in detonation propagation are convection and energy release from chemical reactions, and because the overall behavior of the physical parameters (velocities, pressure, geometry) agrees rather well with the experiments, we believe that the basic model and its solution is adequate. However, the deviations in induction-time calculations from the experimental values well indicate that a more detailed chemical model is required to describe more accurately the chemistry dependent phenomena. This is the subject of current work.

Acknowledgments

This work was sponsored by the Naval Research Laboratory through the Office of Naval Research and by the Department of Defence of Belgium.

References

[1] Soloukhin, R.I., "Shock Waves and Detonations in Gases," *Mono Book Corp.*, Baltimore, MD, 1966, pp. 138–162.

[2] Taki, S., and Fujiwara, T., "Numerical Analysis of Two-Dimensional Nonsteady Detonations," Vol. 16, Progress in Aeronautics and Astronautics, AIAA, NY, pp. 73–77.

[3] Kailasanath, K., Oran, E.S., Boris, J.P., and Young, T.R., Determination of Detonation Cell Size and the Role of Transverse Waves in Two-Dimensional Detonation," *Combustion and Flame*, Vol. 61, 1985, pp. 199–209.

[4] Guirguis, R.H., Oran, E.S., and Kailasanath, K., "Numerical Simulation of the Cellular Structure of Detonation in Liquid Nitromethane," *Combustion and Flame*, Vol. 65, 1986, pp. 339–365.

[5] Oran, E.S., Kailasanath, K., and Guirguis, R.H., "Numerical Simulations of the Development and Structure of Detonations," Vol. 114, *Progress in Aeronautics and Astronautics*, AIAA, NY, 1988, pp. 155–168.

[6] Takai, R., Yoneda, K., and Hikita, T., "Study of Detonation Wave Structure," *15th Symposium (International) on Combustion*, 1974, pp. 69–78.

[7] Libouton, J.C., Dormal, M., and Van Tiggelen, P.J., "Reinitiation Process at the End of the Detonation Cell," *Gasdynamics of Detonations and Explosions*, Vol. 75, *Progress in Astronautics and Aeronautics*, AIAA, NY, 1981, pp. 358–369.

[8] Dormal, M., Libouton, J.C., and Van Tiggelen, P.J., "Etude Expérimentale des Paramètres à L'intérieur d'une Maille de D'etonation," *Explosifs*, Vol. 36, 1983, pp. 76–94.

[9] Dormal, M., Libouton, J.C., and Van Tiggelen, P.J., "Evolution of Induction Time in Detonation Cell," *Acta Astronautica*, Vol. 6, 1979, pp. 875–884.

[10] Oran, E.S., Young, T.R., and Boris, J.P., "Weak and Strong Ignition. I. Numerical Simulations of Shock Tube Experiments," *Combustion and Flame*, Vol. 48, 1982, pp. 135–148.

[11] Boris, J.P., and Book, D.L., Vol. 16, *Methods in Computational Physics*, Academic Press, New York, 1976, pp. 85–129.

[12] Libouton, J.C. and Van Tiggelen, P.J., "Influence de la Composition du Mélange Gazeux sur la Structure des Ondes de Détonation," *Acta Astronautica*, Vol. 3, 1976, pp. 759–769.

[13] Van Tiggelen, P.J., and Libouton, J.C., "Evolution des Variables Chimiques et Physiques à L'intérieur d'une Maille de Détonation," *Annales Phyiques Françaises*, Vol. 14, 1989, pp. 649–660.

Mach Reflection of Detonation Waves

J. Meltzer,* J. E. Shepherd,† R. Akbar,* and A. Sabet*
Rensselaer Polytechnic Institute, Troy, New York 12180

Abstract

The diffraction of a nominally planar gaseous detonation at a wedge was investigated to determine the critical wedge angle for transition from regular to Mach reflection. Experiments were conducted in a square 83-mm cross-section detonation tube using stoichiometric mixtures of hydrogen-oxygen at 0.2 bars. Experimental results for the triple-point trajectory angle produced during Mach reflection were obtained using the smoke foil technique and are compared with analytic calculations made using three-shock theory and the oblique detonation polars. Measurements of the cell size behind the overdriven Mach stem are also reported. Both analytic and experimental results are compared with work from previous investigations to address apparent discrepancies in the existing literature.

Introduction

The motivation for this study is to resolve discrepancies in previous investigations regarding Mach reflection of a detonation wave. Whereas the diffraction of a shock wave by a wedge in a nonreactive gas has been the topic of considerable investigation, the reactive (detonation) case has received only limited attention in studies by Ong,[1] Gvozdeva and Predvoditeleva,[2] Manzhalei and Subbotin,[3] Gavrilenko et al.,[4] Gavrilenko and Prokhorov [5,6] and Edwards et al.[7] In each of these investigations, the detonation is given similar treatment as a nonreactive shock and standard[8] inviscid two- and three-shock theories applied to predict the triple-point trajectory and the critical angle for transition from regular to Mach reflection. Although detonations can be simplistically analyzed as shock waves followed by a thin chemical reaction zone, there are notable differences between the dynamics of nonreactive shocks and detonations.

Copyright © 1991 by the American Institute of Aeronautics and Astronautics, Inc. All rights reserved.
* Graduate Student. Department of Aeronautical Engineering.
† Associate Professor. Department of Mechanical Engineering.

Foremost among these are the three-dimensional transverse instabilities which give rise to the cellular structure of the detonation front.[9] Furthermore, while the simplest model of a detonation, the Chapman-Jouguet (CJ) model, assumes that the reactions takes place instantly across a reactive shock, a detonation is known[10] to involve a coupling between an incident nonreactive shock followed by a chemical reaction zone. Additionally, expansion in the products leads to a Taylor wave, an expansion wave which follows the incident detonation. In all the analyses the transverse waves, reaction zone structure, and the Taylor wave are neglected. To reduce the complexity of the problem, this investigation, like those previous, also neglects these influences in the theoretical treatment.

In this investigation, experiments over a range of wedge angles were performed in a square cross-section detonation tube using stoichiometric hydrogen-oxygen mixtures at 0.2 bars. The results are compared to computations based on the three-shock model using realistic thermochemistry to predict the trajectory and transition angles. Both the experimental results, obtained using the smoke foil technique, and the calculated predictions are compared to the results of the previous studies to provide clarification.

Although a detonation wave is known[9] to posses a complex three-dimensional structure, in a one-dimensional sense, the Zel'dovich, von Neumann, Döring (ZND) model of a detonation idealizes the wave as a strong shock followed by a chemical reaction zone.[10] As such, the diffraction of a detonation would be expected to exhibit similar modes of behavior as a nonreactive shock as long as the reaction zone is thin compared to any other length scale present in the problem. For the case of Mach reflection, this implies that this simple treatment is valid when the Mach stem is much larger than the reaction zone thickness. The diffraction of nonreactive shocks has been investigated in depth in studies by von Neumann,[11] Law and Glass,[12] Ben-Dor and Glass,[13] Henderson,[14] and Hornung and Taylor.[15]

When a planar shock wave encounters a wedge, the incident shock is reflected by the wedge surface and the induced flow behind it is deflected by the wedge corner. Experimental investigations using interferometry and Schlieren techniques have shown that at least four types of reflections are possible depending on the wedge angle, Mach number of the incident wave, and initial conditions of the test gas. The reflections may be categorized into two principal modes, regular and Mach reflection. The latter of these cases may be further subdivided into single, complex, and double Mach reflections. All three types of Mach reflections share the similar features of a triple point with an incident shock and Mach stem, with the distinction arising from the the shape of the reflected wave. In our experimental investigation performed for the detonation case, no optical methods were available to determine the distinction in the type of Mach reflection. Therefore, the discussion of transition between modes of reflection presented here will be limited to the case of transition from regular to Mach reflection and the specific type of Mach reflection is not known.

A second noteworthy feature of the diffraction of a detonation by a wedge is that it also affords the opportunity to investigate the stability of overdriven detonations. The Mach stem produced during Mach reflection is an overdriven detonation, that is, a detonation which propagates with a velocity higher than the Chapman-Jouguet velocity. The Chapman-Jouguet velocity

(U_{CJ}) is the calculated detonation wave speed that results in sonic flow in the products with respect to the detonation front. Freely propagating detonations usually travel at wave velocities (U) which are 95-98% of U_{CJ}. As the degree of overdrive ($U/U_{CJ} = M/M_{CJ}$) increases, the transverse wave spacing (cell size λ) decreases until the stability limit, at which point the transverse instabilities are damped out. In this investigation the stability limit is determined from smoke foil measurements of the cell size following the Mach stem.

The problem of the diffraction of a nominally planar detonation was first investigated by Ong[1] in his doctoral thesis at Michigan in 1955. Ong's investigation had two objectives: the first to predict the shape of the reflected wave generated during Mach reflection, the second to predict the critical angle for transition from regular to Mach reflection. Calculations of the shape of the reflected wave were made by applying perturbation techniques to the equations of motion in the psuedosteady frame. The critical wedge angle was determined by applying two-shock theory with constant specific heats (perfect gas) assumption to the case of regular reflection. Three-shock theory was applied to the case of Mach reflection to determine the trajectory of the triple point. To check his predictions, Ong performed experimental wedge studies in a rectangular detonation tube using an equimolar hydrogen-oxygen mixture at an initial pressure of 20 psia. Initiation of the detonation was through the deflagration to detonation transition (DDT) mechanism. Results, obtained experimentally from Schlieren photographs, showed the diffraction to be a single Mach reflection with the reflected wave being nearly circular. Whereas comparison is made with the reflected wave and trajectory angle, the experimental transition angle is not reported. Calculations performed by Ong may be in question due to a possible error in the values used for the initial conditions. In DDT, the pressure of the mixture ahead of the flame is increased by compression. The detonation is initiated at a pressure higher than the initial pressure, resulting in a change in the actual CJ conditions from those based on the original properties of the mixture.

The first published investigations were conducted by Russian researchers in the mid to late 1960s. Gvozdeva and Predvoditeleva[2] used Schlieren photography to investigate diffraction in methane-oxygen mixtures at 1 atm. Double- and triple-shock theories were used to determine the transition and trajectory angles. Calculations included realistic chemistry and were performed for both the frozen and equilibrium cases. Gvozdeva reports that the calculations did not compare well with the experiments, predicting regular reflection for some cases in which Mach reflection was observed experimentally. Also, Schlieren photographs for some cases indicate the reflected wave to be kinked, suggesting a complex or double Mach reflection process.

In the late 1970s Gavrilenko et. al.[4] studied the diffraction of a detonation wave as a method of producing overdriven detonations. Initial experiments were performed in a rectangular detonation tube with wedge inserts used to narrow the cross section. Results obtained from streak photographs were reported for the degree of overdrive as a function of wedge angle. Calculations were also performed using three-shock theory to predict the critical wedge angle. Gavrilenko reports that these calculations showed little dependence on mixture composition and initial pressure. A calculated critical wedge angle of 34 ± 0.4

deg. is reported for mixtures of $2H_2 + O_2$, $H_2 + O_2$, $4H_2 + O_2$, $C_2H_2 + 2.5O_2$, C_2H_2 + air (stoichiometric), $CO + 2O_2$, $CH_4 + 2O_2$, and CH_4 + air (stoichiometric) at pressures from 0.1 to 1 atm. Dilution with argon was predicted to increase the angle by approximately 3 deg.

Gavrilenko and Prokhorov[5,6] conducted additional experiments to determine the triple-point trajectory and the critical angle using single wedges in a rectangular detonation tube with results obtained by smoke foils, and Schlieren and streak photographs. Stoichiometric mixtures of hydrogen-oxygen and acetylene-oxygen at initial pressures ranging from 0.05 to 1 atm were tested and the critical angle for both mixtures found to be 40 ± 1 deg. Gavrilenko and Prokhorov do not report the exact pressures or cell sizes for which this was determined and make no mention of the sensitivity of the angle to either of these parameters. The maximum degree of overdrive is reported to be 1.3 and occurs at 38 deg, close to the critical angle. Gavrilenko states that the triple point motion is not self-similar, and that a unique correlation exists between the dimensions of the Mach stem and the detonation cell size of the incident wave. However, little detail is provided about this conclusion.

The most recent work, published by Edwards et al.[7] in 1984, had the objective of determining the effect of cell size on the diffraction process. Experiments were performed in a rectangular detonation tube with a wedge insert using stoichiometric mixtures of hydrogen-oxygen and acetylene-oxygen with different degrees of argon dilution. A range of incident cell sizes, 3-8 mm, were investigated by varying the mixture composition and initial pressure. Schlieren photographs and smoke foils were used to determine the trajectory of the triple point, critical wedge angle, and change in cell size as a function of overdrive. Barthel's[16] acoustic model for predicting the transverse wave spacing is used to estimate cell sizes behind the overdriven Mach stem and comparison is made with values measured from smoke foils. Barthel's model compares well up to an overdrive of 1.2, however, beyond this the model overpredicts the wave spacing. While the triple-point trajectory is shown to decrease with decreasing cell size, it is unclear whether this effect is due to the three-dimensional transverse wave structure or the change in initial pressure and Mach number of the incident wave.

In a later publication Nettleton[17] compares the experimental data from Edwards et al.[7] with predictions from Whitham's method for the diffraction of a shock at a compression corner and standard two- and three-shock theories. Whitham's method is an area-Mach number relationship developed for nonreactive shocks. It does not account for any interactions of pressure disturbances, generated behind the front with the diffracting wave.[18] Application of this method to a detonation using the equilibrium CJ model, is being studied at Rensselaer and will be presented in a later paper. As applied to the case of a compression corner, Whitham's method requires the existence of a Mach stem and is known[18] to give inexact predictions for the critical angle. Nettleton reports the two-shock calculation for $2H_2 + O_2$ + Ar mixture yields a value of 65 deg, compared to the experimental value of 48 ± 2 deg. This is in disagreement with Gavrilenko and Prokhorov[5,6] who report a calculated value of 34 ± 0.4 deg and a measured value of 40 ± 1 deg.

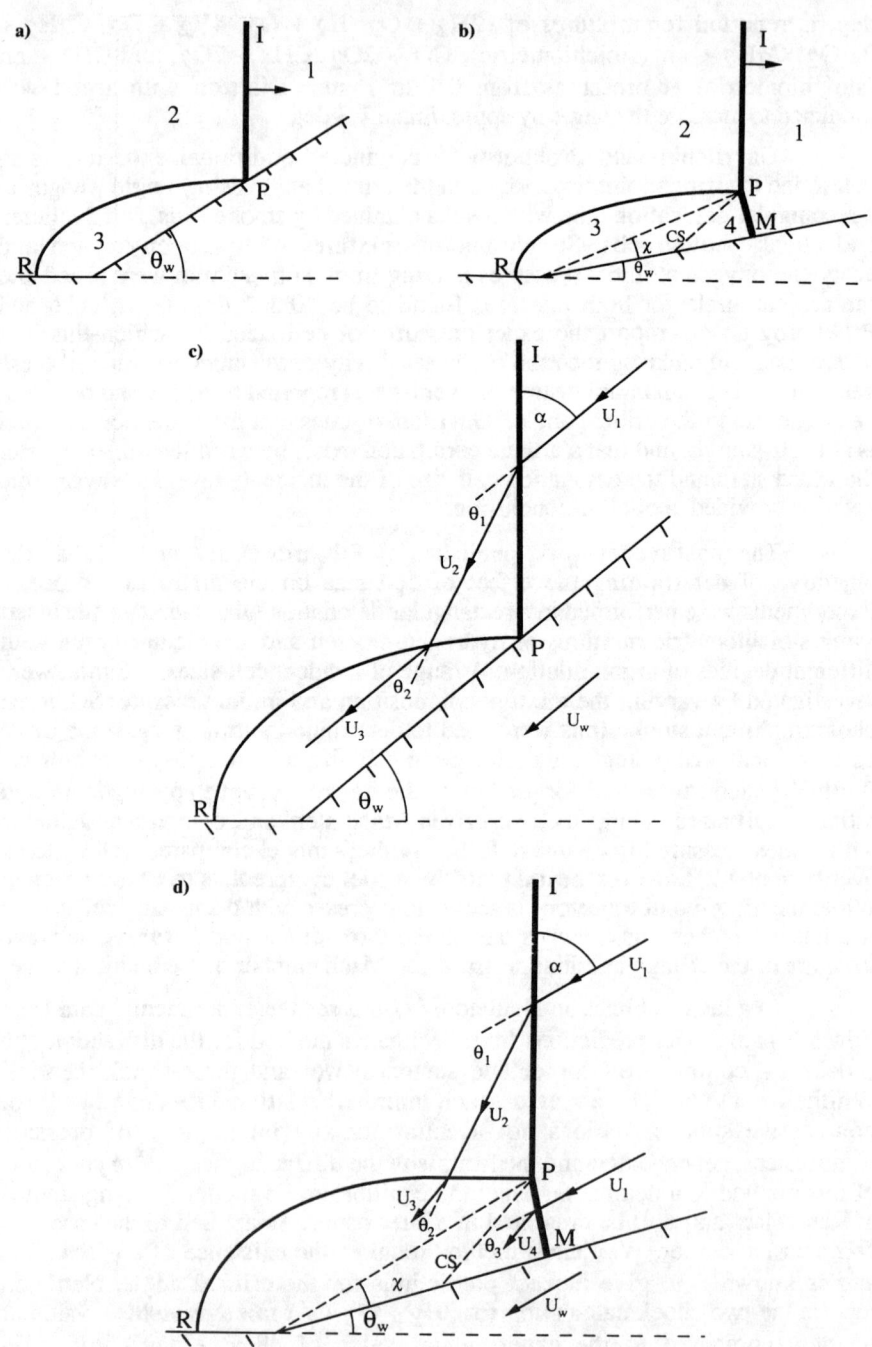

Fig. 1 Reflection of a shock wave incident on a wedge viewed in the lab frame: a) Regular reflection; b) Single Mach reflection; c) regular and d) Mach reflections viewed in the psuedosteady frame.

Theory

The standard inviscid analysis of shock reflection considers the process as psuedosteady when viewed in a reference frame fixed with respect to the reflection or triple point. Figure 1 shows the cases of regular and Mach reflection as viewed in the laboratory and psuedosteady frames. The process is termed psuedosteady since the reflecting surface is not at rest with respect to the reflection or triple point, P in Fig. 1. In the case of regular reflection, the wall moves with the velocity of the incident flow U_1, while in Mach reflection, the wall not only translates but also appears to recede from the triple point with velocity U_w. In this frame, the standard[8] steady oblique shock relations in the form of a shock polar may be used to analyze the problem.

In regular reflection viewed in the psuedosteady frame, Fig. 1c, the initial flow U_1 is parallel to the wedge making an angle α with the incident shock I. After the shock, the flow is deflected by an amount θ_1. For an inviscid flow, the wedge establishes a boundary condition that the flow must be parallel to the surface. From this it is evident that the reflected wave R must be of sufficient strength to return the flow parallel to the wall, that is, $\theta_1 + \theta_2 = 0$. This situation is clearly illustrated by constructing the shock polar in the pressure-deflection (P-θ) plane. As shown in Fig. 2a, the flow returns to zero deflection when the reflected polar crosses the pressure axis.

From both analytic and experimental treatments of nonreactive shocks, it is known that for fixed initial conditions, as the wedge angle θ_w is decreased, the mode of reflection undergoes a transition from regular to Mach reflection. Mach reflection, as shown in Figs. 1b and 1d, is characterized by the appearance of a third shock, the Mach stem M, which is required to achieve the needed flow deflection. The position of the triple point grows linearly with time in a self-similar manner, along a straight trajectory at an angle χ to the wedge surface. A contact surface CS separates the gas processed by the incident and reflected shocks, regions 2 and 3, from that processed by the Mach stem, region 4. In the psuedosteady frame, Fig. 1d, the incoming flow is parallel to the triple-point trajectory at an angle α with the incident wave. As in regular reflection, it is deflected by an amount θ_1 by the incident wave and θ_2 by the reflected wave. However, in the case of Mach reflection, the deflection following the reflected wave is insufficient to return the flow to the original flow direction so that a third shock, the Mach stem, is needed to produce a deflection θ_3. The three-

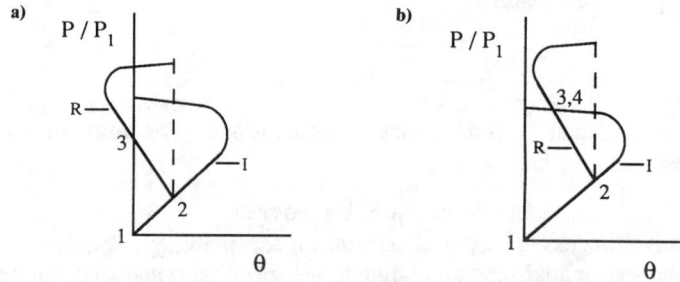

Fig. 2 Idealized shock polars illustrating a) regular and b) Mach reflections.

shock configuration is determined by matching the flow deflection angle and the pressure at the slipstream. From the matching condition across the slipstream $\theta_1 + \theta_2 = \theta_3$, that is the total deflection across the incident and reflected waves must be the same as the deflection across the Mach stem, and $P_3 = P_4$. The graphical solution to this matching process is given by finding the intersection of the polars for the incident and reflected shocks as shown in Fig. 2b.

Once the resulting wave configuration has been determined, the triple-point trajectory angle χ may be computed from geometric relations assuming that the Mach stem is straight and normal to the wall. Law and Glass[12] used this assumption in a three-shock analysis for nonreactive shocks in air and compared calculations for trajectory angle with experiments at a fixed Mach number. They concluded that the three-shock model predicted $\chi(\theta_w)$ accurately away from the regular-Mach reflection transition region, however, within 5 deg of the transition point, the experimental values of χ begin to substantially deviate from the three-shock predictions.

Using three-shock theory and the oblique detonation polars, theoretical predictions for the trajectory and critical angles were computed. A summary of the method developed by Sabet is presented here with a more detailed description given in Refs. 19 and 20. STANJAN,[21] a chemical equilibrium computer code, was run to determine the CJ conditions for the incident detonation. Using overdriven velocities, a modified version of STANJAN is used along with the Rankine-Hugoniot relations to construct the equilibrium detonation adiabat. A polynomial fit is obtained to describe the relationship $w_2=f(w_1)$, where w is the normal component of velocity and the subscripts denote the respective upstream and downstream states. This relation is used to numerically solve the oblique shock relations at a specified angle of incidence α. From this, the overdriven detonation polar is constructed in the P-θ plane. The reflected polar is generated in a similar fashion by computing equilibrium shock adiabats for gas initially at CJ conditions behind the incident (reactive) wave. The oblique detonation polar for $\alpha = 60$ deg is shown in Fig. 3. The polars start at the same point, the CJ point corresponding to state 2 in Fig. 1d. The intersection point corresponds to the matching condition across the slipstream.

Knowing the pressure P_4, in the region behind the Mach stem, the detonation adiabat may be used to find an estimate of the normal velocities across the Mach stem, w_{1m} and w_4. The trajectory angle χ is then computed solving the quadratic equation

$$(\tan \chi)^2 - \frac{(1 - \frac{w_4}{w_{1m}})}{\tan \theta_3} \tan\chi + \frac{w_4}{w_{1m}} = 0 \qquad (1)$$

which can be obtained from geometric considerations. The value of w_{1m} may be improved using the computed χ,

$$w_{1m} = U_1 \cos\chi \qquad (2)$$

and an iterative process is used to determine an acceptable χ. The final step is to determine the wedge angle corresponding to the specified α and calculated χ,

$$\theta_w = \frac{\pi}{2} - \alpha - \chi \qquad (3)$$

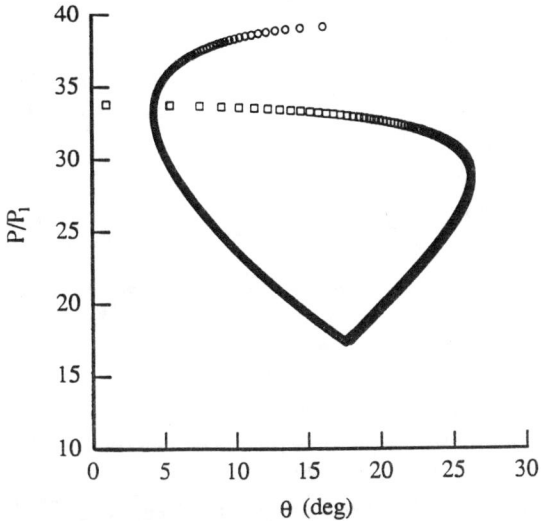

Fig. 3 Oblique detonation polar for $2H_2 + O_2$ mixture at 0.2 bar with $\alpha = 60$ deg.

To check the validity of our method, some nonreactive shock cases were computed and the results compared with previous experimental (Henderson[14]) and theoretical results (Ben-Dor and Glass[13]). As shown in Fig. 4, our results compared extremely well with both sets of results.

Experimental

Experiments[19] were performed in a detonation tube 4.88 m in length using wooden wedges varying from 10 to 50 deg in 5-deg increments. The tube had a 83-mm-square cross-section with 6.35-mm radius rounded corners. The wedges, which were mounted to the end flange of the tube, were fitted with a 3.2 mm-thick aluminum face milled to a sharp leading edge so as to rest flat on the bottom of the tube. Smoke foils were placed along the side of the wedge to record the triple-point trajectory and the cell structure behind the incident wave and Mach stem. Foils were cut from 0.635 mm-thick aluminum sheet and prepared over an open kerosene lantern. It was determined that the best traces, i.e., with the most detail, were obtained when the foils were lightly sooted for approximately 10 min.

Stoichiometric mixtures of hydrogen-oxygen at an initial pressure of approximately 0.2 bars were used throughout the experiments. Initiation of the detonation was achieved through use of an exploding wire system. Confirmation was obtained by comparing the measured wave velocity and pressure with known CJ conditions. Two PCB model 113A21 piezoelectric pressure transducers were used to obtain pressure traces and to time the speed of the incident detonation. For the test mixture, the calculated[21] CJ pressure was 3.63 bar and the velocity was 2750 m/s, yielding a Mach number equal to 5.1. The measured wave velocity was consistently 95-98% of the calculated CJ velocity.

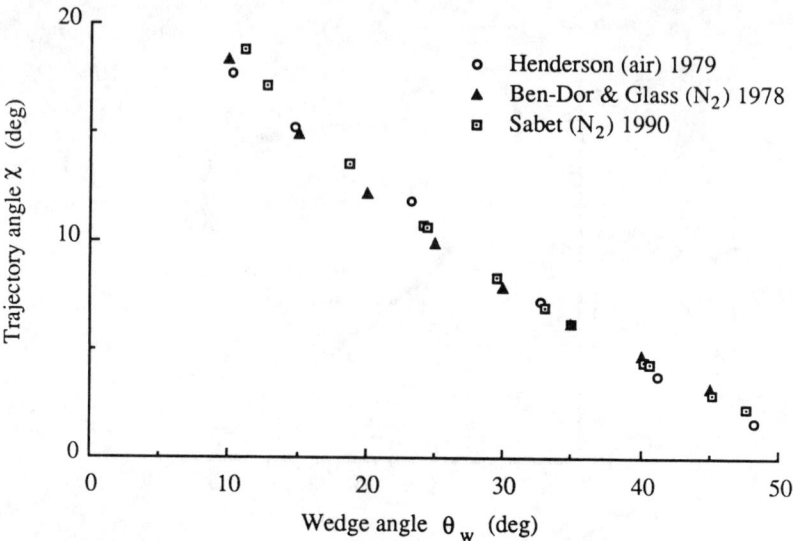

Fig. 4 Comparison of calculated trajectory angle for nonreactive shocks; Sabet method is compared with results from Henderson[14] and Ben-Dor and Glass.[13]

Results

Results for the triple-point trajectory angle and cell sizes were determined from smoke foils. The regularity of the cell structure behind the incident wave was moderate, with the average cell size ranging from 7-8 mm, which agrees well with published data.[22] Interpretation of the foils was hindered by a loss in quality due to the irregular interference of the slapping waves. Identification of the triple-point trajectory was made by noting the change in cell size between the incident wave and Mach stem on the side foil. It was also observed that cells behind the Mach stem were aligned parallel to the wedge, while those behind the incident wave were aligned along the tube axis. Measurements of the cell size behind the overdriven Mach stem were taken from foils placed on the wedge surface. Direct measurements were made by viewing the foils under a microscope.

An example of a smoke foil obtained for a 30-deg wedge is shown in Fig. 5, while the results of the triple-point trajectory and the overdriven cell size are summarized in Table 1. For the 10 and 15-deg wedges, the degree of overdrive is slight and the average ratio of the overdriven cell size to the incident cell size, λ_M/λ_{CJ}, ranged from 0.58 to 0.82. The limited reduction in cell size led to difficulty in measuring the trajectory.

The clearest results for the triple point trajectory were obtained for the 20-, 25-, and 30-deg cases. For these cases, the average ratio of the cell sizes, λ_M/λ_{CJ}, ranged from 0.26 to 0.48 with the cells behind the Mach stem being large enough to see clearly by eye, yet having sufficient reduction to be easily distinguishable from the incident cells. The higher overdrive at these angles greatly increases the regularity of the overdriven cell structure as seen on the front foils.

Table 1 Summary of experimental results for trajectory angle and overdriven cell size.

Wedge angle θ_W (deg)	Trajectory angle χ (deg)	Overdriven cell size λ_M (mm)	Cell size ratio, S λ_M/λ_{CJ}
10	17 ± 3	5.1 - 7.1	0.68 - 0.95
15	15 ± 3	3.0 - 5.6	0.41 - 0.75
20	12 ± 2	2.5 - 4.6	0.34 - 0.61
25	7 ± 1	2.7 - 3.2	0.36 - 0.42
30	6 ± 1	1.8 - 2.0	0.24 - 0.27
35	3 ± 1	1.3 - 1.5	0.17 - 0.20
40	2.5 ± 0.5	0.5 - 0.8	0.07 - 0.10
45	*	*	*

* indicates regular reflection

At 35 deg the trajectory angle is small, 3 ± 1 deg, and the cell structure behind the Mach stem on the side foil is no longer visible by the naked eye. At 40 deg the Mach stem has become marginally stable. The front foil reveals only scattered patches of cells that appear, die out, and reappear. Viewing the region under the trajectory on the side foil reveals the same evidence, intermittent patches of cells with the structure dying out. Where cells existed, λ_M/λ_{CJ} was 0.09. At 45 and 50 deg there is no evidence of a triple-point trajectory or of overdriven cells on the front foil. From this, it was concluded that the critical angle for transition from Mach to regular reflection is between 40 and 45 deg.

Figure 6 shows a plot of the trajectory angle χ as a function of wedge angle. The large error bars for the 10- and 15-deg cases reflect the difficulty in determining the correct angle for these cases. For the remaining cases the trajectory angle was distinguishable to ± 2 deg. The zero angle data point at 45 deg indicates regular reflection.

The ratio of the overdriven to incident cell size as a function of wedge angle is plotted in Fig. 7. The data points represent the average overdriven cell sizes normalized by the average incident cell size, 7.5 mm. The range of the error bars correspond to the range of overdriven cell sizes measured on the front foil. The small error bars for the 30-, 35-, and 40-deg, wedges indicate the high regularity of the overdriven cell structure. The 40-deg angle was taken as the stability limit. The stability limit does not represent a condition of zero cell size, rather the condition at which the overdrive is sufficient to cause damping of the natural transverse disturbances.

Discussion

The three-shock method described previously was used to compute triple-point trajectory angles for stoichiometric hydrogen-oxygen mixtures at 0, 30 and 70% dilution with argon. The results, shown in Fig. 8, demonstrate that the predicted trajectory is not very sensitive to the degree of dilution. The critical angle was determined as the extrapolated wedge angle at which $\chi = 0$. From the figure it was calculated to be approximately 34 deg, in excellent agreement with

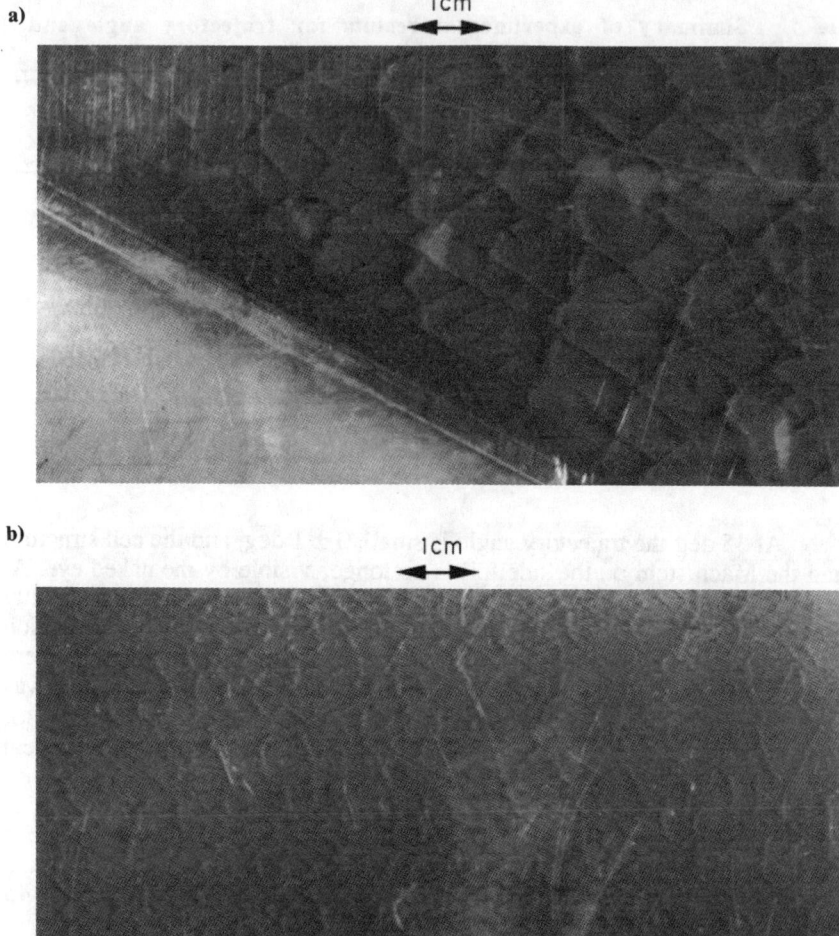

Fig. 5 Smoke foils for 30-deg wedge: a) side foil; b) front foil.

the value determined by Gavrilenko.[4] Nettleton[17] reports calculated critical angles of 50 deg for argon-diluted $2C_2H_2 + 5O_2$ mixtures and 65 deg for argon-diluted $2H_2 + O_2$ mixtures. These results were obtained with nonreactive shock dynamics.[18] A more appropriate reactive flow version of Whitham's method must be carried out before the validity of this technique can be judged. Such a comparison is being worked on by the present authors.

In Fig. 6 the experimentally determined values for χ are compared with computed values for the 0% dilution case. the experimental critical angle is between 40 and 45 deg for the case studied in the present investigation. Our three-shock analysis consistently underpredicts the triple point trajectory χ, and

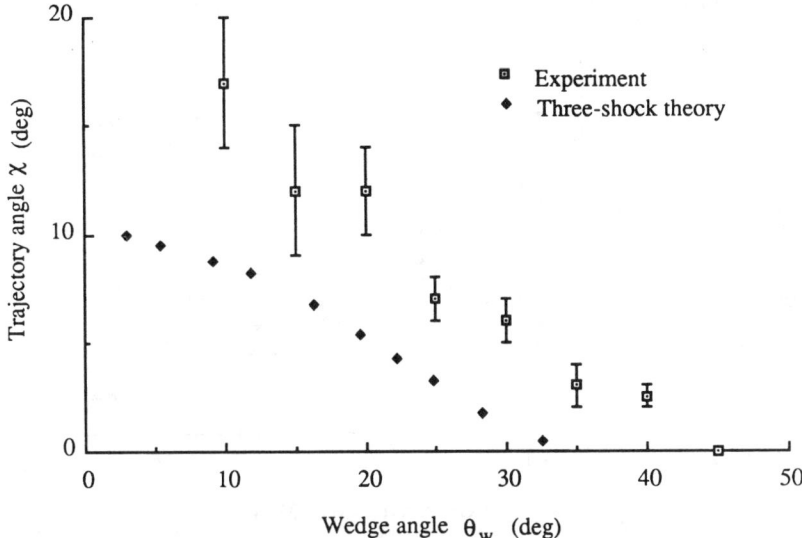

Fig. 6 Comparison of theoretical and experimental trajectory angles for detonations in $2H_2 + O_2$ mixtures at 0.2 bar.

the critical angle, by 6-10 deg. Our experimental results are similar to those of Gvozdeva[2] and Gavrilenko and Prokhorov.[6] They determined a critical angle of about 35 deg for stochiometric CH_4-O_2 and 40 deg for stochiometric H_2-O_2 and C_2H_2-O_2 mixtures. It has been reported[23] that similar detonation diffraction phenomena depend strongly on cellular regularity. In addition, we also anticipate the cell width to be an important parameter. We would only expect a self-similar diffraction process if the cell width is much smaller than the Mach stem. A limited exploration of these issues was performed in the study by Edwards et al.[7], who tested three mixtures with various cell widths. The critical wedge angle $\theta_{w,crit}$ was found to be different for all three mixtures. It is interesting that the three mixtures correspond to three different levels of cellular structure irregularity. These are, in order of increasing regularity: $C_2H_2 + O_2$, $\theta_{w,crit} = 33$ deg; $2H_2 + O_2 + Ar$, $\theta_{w,crit} = 45$ deg and $2H_2 + O_2 + 70\%Ar$, $\theta_{w,crit} = 48$ deg. Further tests are needed to clarify this issue. In addition, a systematic study of the triple point trajectories as a function of cell width would also be useful.

The comparison shows that for a detonation the observed Mach reflection regime extends into the theoretical region of regular reflection. This effect, which is the same as reported by Gvozdeva,[2] is opposite of the effect observed in the diffraction of nonreactive shocks. In the nonreactive shock case, regular reflection has been observed to occur in the theoretical Mach reflection region. This has been attributed to viscous effects by Hornung and Taylor.[15] Hornung and Taylor's argument is as follows. For regular reflection in the psuedosteady frame, the velocity of the flow following the reflected wave will be less than the velocity of the wall, which moves at the same speed as the incident

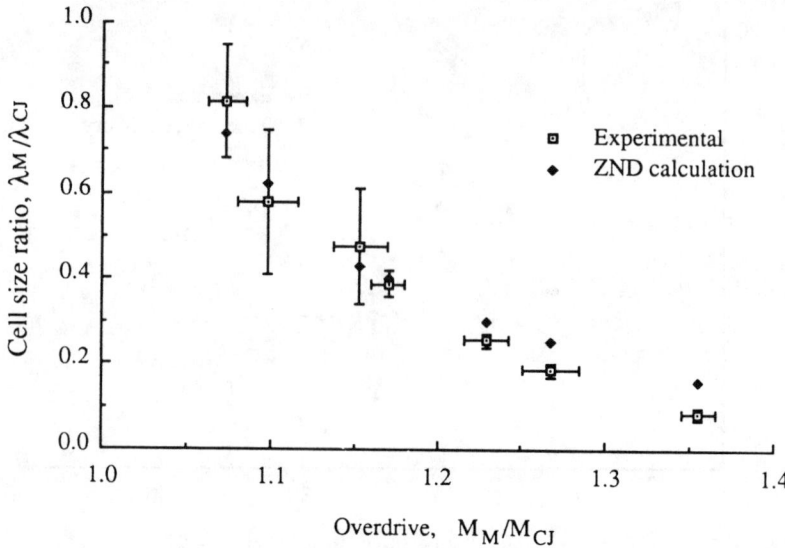

Fig. 7 Comparison of measured overdriven cell size ratio with ZND calculations.

flow. Considering the flow as viscous, this leads to a boundary layer with a negative displacement thickness, resulting in an apparent flow into the wall. The deflection condition for the viscous case requires that the flow following the reflected wave must be turned parallel to the effective wall. As a result, the deflection of the flow after the reflected wave is less than in the inviscid case by an amount ε and the deflection condition becomes $\theta_1 + \theta_2 = \varepsilon$. The maximum incident flow deflection which can be returned by the reflected wave occurs at a lower wedge angle than in the inviscid case. That is, the critical wedge angle, θ_w, is lower in the viscous case than in the inviscid case. Hornung and Taylor[15] performed a careful experimental study of the viscous effect and concluded that the critical angle for Mach reflection could be decreased by as much as 7 deg. The flow behind the detonation should be expected to experience the same type of viscous effects as in the nonreactive shock case. This would imply that there should be a lowering of the critical wedge angle, up to 7 deg from the calculated inviscid value, extending regular reflection into the Mach reflection regime. This is opposite of the experimental observations, in which the critical wedge angle is observed to be 6-10 deg higher than the predicted value.

Whereas no definite conclusion has been obtained, we propose that the discrepancy between the three-shock theory and the experimental data is due to the three-dimensional instability wave structure of the detonation. The three-shock model is a one-dimensional treatment and does not account for interaction of the transverse structure with the diffracting wave. The finite thickness of the reaction zone could be another reason for the failure of the three-shock analysis to correctly predict the triple-point trajectory and critical angle. In the present experiments, the reaction zone length behind the undisturbed detonation is calculated to be 0.24 mm and is about 0.05 mm behind the Mach stem at an overdrive of $U/U_{CJ} = 1.3$, near the critical angle. For wedge angles less than 30 deg, the Mach stem is greater than 10 mm high at the end of the wedge, suggesting that the reaction zone effects should be small for $\theta_w < 30$ deg. However, if the nominal cell size of 7.5 mm is used as the effective reaction zone length, then it is not possible to conclude that effects of the detonation structure are negligible. Experiments using a systematic variation of the incident wave cell size are needed to investigate this issue further. One experiment which might give further insight would be to test $2H_2 + O_2 + 7Ar$ mixtures at high initial pressures, which have very regular and small amplitude instability waves.[22]

Further investigation including Schlieren photography is also needed to determine whether the detonation experiences single, complex, or double Mach reflection. Although the idealized three-shock theory does not account for any type other than single Mach reflection, based on evidence from nonreactive shocks it is not felt that the possible presence of complex or double reflections can account for the discrepancy observed in the detonation case. Ben-Dor and Glass[13] compared experimental results for the trajectory angle as a function of wedge angle and Mach number for shocks in N_2 with predictions calculated using a three-shock theory and polar method developed by Law and Glass.[12] The comparison showed good agreement to ± 2 deg over the range $20 < \theta_w < 40$ deg and $1 < M < 8$. This is despite the fact that for a fixed Mach number, for example M = 5 as in our detonation case, the shock was observed experimentally

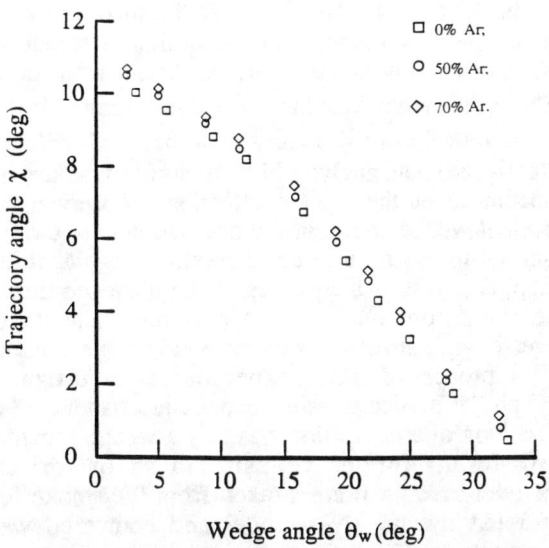

Fig. 8 Theoretical trajectory angles for CJ detonation in $2H_2 + O_2$ mixtures at 0.15 bar diluted with argon: □ 0% Ar; ○ 50% Ar; ◇ 70% Ar.

to undergo single Mach reflection for $\theta_w < 14$ deg, complex Mach reflection for $14 < \theta_w < 25$ deg, and double Mach reflection for $25 < \theta_w < 48$ deg.

Another shortcoming of our analysis to predict χ may be the assumption that the Mach stem is straight and normal to the wall. As noted by Hornung,[24] experimental evidence for the Mach reflection of shocks has shown that the slipstream is curled toward the Mach stem. This is due to a wall pressure gradient which is formed from the difference in stagnation pressures for flows on either side of the slipstream striking the wedge. This wall pressure gradient can also result in a strong wall jet which may interact with the Mach stem leading to curvature.

Analytic predictions for the overdriven cell size ratio as a function of overdrive were made using a computer code[25] which uses the ZND model to resolve the reaction zone structure. A comparison with experimental data, shown in Fig. 7, indicates that at lower overdrives the results are close but that the calculated values slightly overpredict values at higher angles. Without direct measurements available, the overdriven velocities at each angle were determined from geometric relations using the experimental values for χ. The uncertainty in the χ data results in the displayed uncertainties in overdrive. The effect of the overdrive on the cell size has been reported previously by various authors.[26-29] In particular, Desbordes et al.[27-29] overdrive the detonation by an abrupt transition from a stochiometric mixture to a lean mixture. The highest overdrive reported by them[29] is D/D_{CJ} of about 1.3 (oxygen-acetylene mixture).

Conclusion

The critical angle for transition from regular to Mach reflection for a stoichiometric hydrogen-oxygen mixture at 0.2 bar was determined experimentally to be 40-45 deg. This is above the theoretical value of 34 deg computed using three-shock theory and the oblique detonation polars. The experimental results of this investigation are consistent with results reported by Gavrilenko et al.[4-6] and are bracketed by the values reported by Edwards et al.[7] Our three-shock theoretical analysis agrees with that of Gavrilenko et al.[4-6] and disagrees with the approximate analysis of Nettleton.[17] Unlike experiments with nonreactive shocks in which the regular reflection is observed to extend below the theoretical critical wedge angle, in the detonation case the region of Mach reflection is observed to extend into the theoretical regular reflection regime. This is most probably due to inadequacies of the idealized three-shock theory when applied to the detonation case. In particular, the three-dimensional detonation transverse wave structure is expected to have a substantial influence on the diffraction process. Further experimental investigation, including Schlieren photography, is needed to gain a better understanding of the problem.

The diffraction of a detonation wave by a wedge provides an effective way of obtaining results for the cell size in an overdriven detonation. Measurements of overdriven cell sizes taken from the smoke foils are in line with values predicted by the ZND model and compared well with other experimental values reported in previous investigations. From our experiments the stability limit was found to occur at a wedge angle of about 40 deg, at a calculated overdrive of approximately $M/M_{CJ} = 1.36$.

Acknowledgments

Funding for this research was provided by a National Science Foundation Engineering Research Equipment Grant MSM - 8806189. Partial support was also provided by Lawrence Livermore National Laboratories under Contract B055778. We would like to thank J. H. S. Lee, R. Knystautas and D. Desbordes for their useful comments.

References

[1] Ong, R. S., "On the Interaction of a Chapman-Jouget Detonation Wave with a Wedge," PhD. Thesis, University of Michigan, 1955.

[2] Gvozdeva, L. G., and Predvoditeleva, O. A., "Triple Configurations of Detonation Waves in Gases," Fizika Goreniya i Vzryva, Vol. 5, No. 4, 1969, pp. 451-461.

[3] Manzhalei, V. J., and Subbotin, V. A., "Stability of an Overcompressed Detonation," Combustion, Explosion, and Shock Waves, Vol. 12, No. 6, 1977, pp. 819-825.

[4] Gavrilenko, T. P., Nikolaev, Y. A., and Topchiyan, M. R., "Supercompressed Detonation Waves," Combustion, Explosion, and Shock Waves, Vol. 15, No. 5, 1980, pp. 659-692.

[5] Gavrilenko, T. P., and Prokhorov E. S., "Compressed Detonation Wave in a Real Gas," Combustion, Explosion, and Shock Waves, Vol 17, No. 6, 1982, pp. 689-692.

[6] Gavrilenko, T. P., and Prokhorov E. S., "Overdriven Gaseous Detonations," Shock Waves, Explosions and Detonations: Progress in Astronautics and Aeronautics, edited by M. Summerfield, Vol. 87, AIAA, New York, 1983, pp. 244-250.

[7] Edwards, D. H., Walker, J. R., and Nettleton, M. A., "On the Propagation of Detonation Waves Along Wedges," Archivum Combustion, Vol. 4, No. 3, 1984, pp. 197-209.

[8] Thompson, P. A., Compressible Fluid Dynamics, McGraw-Hill, New York, 1972.

[9] Lee, J. H. S., "Dynamic Parameters of Gaseous Detonations," Annual Review of Fluid Mechanics, Vol. 16, 1984, pp. 311-336.

[10] Fickett, W., and Davis, W. C., Detonation, University of California Press, Berkley, CA, 1979.

[11] von Neumann, J., Collected Works, Pergamon, Oxford, England, 1963.

[12] Law, C. K., and Glass, I. I., "Diffraction of Strong Shock Waves by a Sharp Compressive Corner," C. A. S. I. Transactions, Vol. 4, 1971, pp. 2-12.

[13] Ben-Dor, G., and Glass, I. I., "Domains and Boundaries of Non Stationary Oblique Shock Wave Reflexions. Part 1. Diatomic Gas," Journal of Fluid Mechanics, Vol. 92, 1979, pp. 459-496.

[14] Henderson, L. F., "On the Whitham Theory of Shock Wave Diffraction at a Convex Corner," Journal of Fluid Mechanics, Vol . 99, 1980, pp. 801-811.

[15] Hornung, H., and Taylor, J. R., "Transition from Regular to Mach Reflection of Shock Waves. Part 1. The Effect of Viscosity in the Psuedosteady Case," Journal of Fluid Mechanics, Vol. 123, 1982, pp. 143-153.

[16] Barthel, H. O., "Predicted Spacings in Hydrogen-Oxygen-Argon Detonations," The Physics of Fluids, Vol. 17, No. 8, 1974, pp. 1547-1553.

[17] Nettleton, M. A., Gaseous Detonations: Their Nature and Control, Chapman and Hall, New York, 1987.

[18] Whitham, G. B., "A New Approach to the Problem of Shock Dynamics. Part 1. Two Dimensional Problems," Journal of Fluid Mechanics, Vol. 2, 1957 pp. 145-171.

[19] Meltzer, J. S., "The Diffraction of a Detonation by a Wedge," Master's Thesis, Rensselaer Polytechnic Institute, Troy, NY, Dec. 1990.

[20] Sabet, A. I., "Investigation of Equilibrium and Chemical Kinetic Behavior of Detonation Waves in Hydrogen-Air Mixtures and an Evaluation of the Oblique Detonation Wave as the Combustor for a Scramjet," Master's Thesis, Rensselaer Polytechnic Institute, Troy, NY, May 1990.

[21] Reynolds, W. C., "The Element Potential Method for Chemical Equilibrium Analysis: Implementation in the Interactive Program STANJAN; Version 3," Department of Mechanical Engineering, Stanford Univ., Palo Alto, CA, Jan. 1986.

[22] Strehlow, R. A., and Engel, C. D., "Transverse Waves in Detonations II. Structure and Spacing in H_2-O_2, C_2H_2-O_2, C_2H_4-O_2 and CH_4-O_2 Systems," AIAA Journal, Vol. 17, No. 3, 1969, pp. 492-496.

[23] Shepherd, J. E., Moen, I. O., Murray, S.B., and Thibault, P. A., "Analyses of the Cellular Structure of Detonation," 21st Symposium [International] on Combustion, The Combustion Institute, Pittsburgh, PA, 1986, pp. 1649-1658.

[24] Hornung, H., "Regular and Mach Reflections of Shock Waves," Annual Review of Fluid Mechanics, Vol. 18, 1986, pp. 35-38.

[25] Shepherd, J. E., "Chemical Kinetics of Hydrogen-Air Diluent Detonations," Dynamics of Explosions: Progress in Astronautics and Aeronautics edited by Martin Summerfield, Vol. 106, AIAA, New York, 1986, pp. 263-293.

[26] Vasiliev, A. A., and Nikolaev, Y., "Theoretical Model of a Detonation Cell," Acta Astronautica, Vol. 5, Pergamon Press, New York, 1978, pp. 983-996.

[27] Desbordes, D., and Vachon, M., " Critical Diameter of Diffraction for Strong Plane Detonations," Dynamics of Explosions: Progress in Astronautics and Aeronautics, edited by M. Summerfield, Vol. 106, AIAA, New York, 1986, pp. 131-143.

[28] Desbordes, D., "Transmission of Overdriven Plane Detonations: Critical Diameter as a Function of Cell Regularity and Size," Dynamics of Explosions: Progress in Astronautics and Aeronautics, edited by M. Summerfield, Vol. 114, AIAA, New York, 1988, pp. 170-

[29] Desbordes, D., and Lannoy, A., "Effects of a Negative Step of Fuel Concentration on Critical Diameter of Diffraction of a Detonation," Dynamics of Detonations and Explosions: Detonations: Progress in Astronautics and Aeronautics, edited by A.R. Seebass, Vol. 133, AIAA, New York, 1990, pp. 170-186.

Formation and Propagation of Photochemical Detonations in Hydrogen-Chlorine Mixtures

Norihiko Yoshikawa*
Toyohashi University of Technology, Toyohashi, Japan
and
John H. Lee†
McGill University, Montreal, Quebec, Canada

Abstract

This paper reports the physical and chemical mechanisms of the detonation initiation occurring in photosensitive hydrogen-chlorine mixtures irradiated by continuous monochromatic light beams. Detailed simulations of the one-dimensional reactive flow fields for different radiation intensities are used for the identification of the three different regimes of the development of pressure waves. For the first regime, typical transition from deflagration to detonation is observed below a certain theshold level of the radiation intensity. The second is the direct initiation regime in which a reacting shock wave very rapidly grows via an effective shock wave amplification mechanism due to the coherent energy release during the propagation in preconditioned reacting media. The detonation wave structures at the steady propagation phase are similar to those of the usual Chapman-Jouguet waves in which the shock heating of the von Neumann spike plays a dominant role for triggering the detonative chemical reactions. Further increase of the radiation intensity in the third regime rather suppresses the formation of the steep pressure gradient of shock wave. The effect of the photochemical enhancement by the radiation source exceeds the gasdynamic compression effect for triggering the chemical reactions. Combustion waves with relatively thick reaction zone and lower pressure peak are found at the quasisteady propagation phase.

Copyright © 1992 by the American Institute of Aeronautics and Astronautics, Inc. All rights reserved.
  * Associate Professor, Department of Energy Engineering.
  † Professor, Department of Mechanical Engineering.

Introduction

The initiation phenomena of gaseous detonation occur under various conditions. The experimental studies that have been most extensively done are those of the blast initiations and the deflagration-to-detonation transitions. In addition to these initiation modes, a uv radiation to photosensitized explosive mixture is an effective means for rapid detonation initiations, although only a few studies of the phenomena have been done so far. Since the pioneer work of Norrish[1] and his co-workers in the flash photolysis in 1950s, the photochemical method has been recognized as a good tool for studying the spectroscopies of the intermediate free radicals in combustion chemistry. Thrush[2] first reported that a detonation wave propagates in a long reaction tube due to an inhomogeneity of flash irradiation. Wardsworth[3] was the first to intentionally use the flash photolysis for initiating detonation. The initiation process has been first observed by a direct Schlieren photography of the phenomena, see Lee et al.[4] The experiments have shown that a rapid detonation initiation is due to an effective amplification of reacting shock wave propagation in the preconditioned, photochemically reacting media. The initiation mechanism has been referred to as shock wave amplification through coherent energy release (SWACER). An experimental study for trying a possiblity of the mechanism operative in initiating detonation without any strong external sources[5] has demonstarted direct initiations. The initiation occurs in a gradient field of temperature and free radicals developed by a mixing between a hot jet of burned gas and the surrounding fresh gas. The observations suggested that the SWACER mechanism is responsible for rapid initiation. The role of the SWACER mechanism has been discussed with a more generalized view.[6]

In spite of all the research efforts mentioned above, the shock amplification mechanism has not been elucidated to a satisfactory extent. In particular, quantitative basis is in a poor situation. Although some numerical analyses have been tried,[7] the computations have not been sufficiently done because of the limitation of the computer efficiencies in the middle of 1970s. The present numerical analysis of a photochemical detonation initiation is a first step of the theoretical re-examinations of the SWCAER mechanism with more reliable computer analyses.

Theoretical Model

The photodissociation process of chlorine molecules by monochromatic light beam is derived by the Beer-Lambert photon-absorption law.

$$I = I_0 \exp(-\varepsilon \int_0^x [Cl_2]dx) \qquad (1)$$

where I, W/cm², is the radiation intensity at distance x; I_0 radiation intensity at the window; ε absorption cross section, cm²/mol; [Cl₂] concentration of chlorine molecules, mol/cm³; x distance from the window, cm. We assumed that a monochromatic light beam of 330 nm continuously irradiates a photosensitized gas mixture through a window with a constant radiation intensity. The wavelength is close to the peak of the Cl_2 absorption band at room temperature. The corresponding value of ε is 1.52×10^5 (Ref. 8). The bond energy of Cl_2 is 498.9 nm. The difference between irradiation energy and the bond energy is used to heat up the gas mixture. Therefore, the terminology photothermal may be more appropriately used than photochemical when the energy difference is large. About one third of the photon energy is converted into thermal energy for the gas mixture. The photon energy absorbed at distance x is equal to $-\partial I/\partial x$ as given by Eq. (2):

$$\frac{\partial I}{\partial x} = I \varepsilon [Cl_2] \tag{2}$$

The quantity $I \varepsilon [Cl_2]/\rho$ is the absorbed photon energy per unit mass of the gas mixture. Here, ρ denotes density of gas mixture. This term is the external energy source in the energy conservation equation. The production rate of chlorine atoms via photodissociation W_{Ph} is given by Eq. (3)

$$W_{Ph} = 2 I \varepsilon Y_c [Cl_2]/E_{Ph} \tag{3}$$

where Y_c is the quantum yield; E_{Ph} photon energy, erg/mol. The values of E_{Ph} and ε/E_{Ph} are 0.363×10^{13} erg/mol and 4.19×10^{-8} cm²/erg, respectively. The quantum yield Y_c is estimated about 0.3-0.4 at room temperature and increases largely with increasing temparature. Accurate data for Y_c were not available. In this study, $Y_c=1$, viz., 100% dissociation, was assumed, because the photothermal effect enhances temperature rise in the early stage of the reaction.

The photodissociation of Cl_2 molecules into Cl atoms rapidly initiates the chain reactions of $H_2+Cl \rightarrow HCl+H$ and $H+Cl_2 \rightarrow HCl+Cl$. These chain-propagation reactions are exothermic and thus effective to regenerate the chain carriers of H and Cl atoms. The elementary reaction rate coefficients of H_2-Cl_2 system are obtained in the literature.[9] The thermal data of chemical species are approximated by fourth-order polynomials of temperature using the data of JANAF Thermochemical Tables.[10]

Computational Methods

The one-dimensional reactive Euler equations are numerically solved. The mass Lagrangian coordinate system is advantageous over the Euler coordinate system for the following reasons:

1) it avoids unnecessary numerical diffusion between mesh points; 2) it provides fine resolution near shock waves with smaller number of mesh points. The computation procedures are split into two parts: 1) photochemical constant-volume combustion and 2) non-reacting hydrodynamic processes. The change of chemical species concentrations and temperature in the constant-volume combustion are computed by an implicit single-step method of Lomax and Bailey.[11] After a systematic comparison between some existing single-step schemes, we chose this scheme because of its stability and accuracy. The hydrodynamics are solved by a combination of second-order two-step explicit MacCormack scheme[12] and FCT scheme of Boris and Book.[13] The MacCormack scheme is favorable because it does not need any interpolations of mesh points that might have introduced errors in the reaction computations. One cycle of the splitting method comprises the following three parts: 1) constant-volume combustion with time step Δt; 2) hydrodynamic process with time step $2\Delta t$; 3) constant-volume combustion with time step Δt. In a typical computation, 2000 mesh points were used for covering flow length of 40 cm. The Lagrangian spatial mesh size, which is equal to $\rho_0 \Delta x$, was set with $\Delta x = 0.2$ mm so that the most exothermic part of reaction zone was resolved with 30-40 mesh points. The time step was controlled by choosing a smaller characteristic time between the hydrodynamic and reaction times. The reaction time was determined by an empirical time-step criterion of Fujii.[14] The FACOM VP-200 supercomputer at Nagoya University was used for a vectorized computer code. In a typical computation, 40,000 cycle iterations were necessary to get sufficient data for analyzing the time evolution of a flowfield.

Results and Discussion

The stoichiometric H_2-Cl_2 mixtures at 300 K and 100 Torr are the initial conditions for all the computations. Temperature and pressure are denoted by T and p, respectively. The quantities of the initial conditions are indicated with the subscripts $_0$. The sound velocity c_0 is 304.9 m/s. The characteristic e holding length [$I/I_0 = 1/e$ in Eq. (1)] is about 2.5 cm for the gas mixture. The radiation intensity I_0 was changed for studying its effects on detonation initiation. Characteristic properties of the Chapman-Jouguet detonation in the gas mixture are given with the subscript $_{CJ}$ as follows: detonation velocity $D_{CJ} = 1656$ m/s; detonation Mach number $M_{CJ} = 5.43$; $p_{CJ}/p_0 = 19.2$; $\rho_{CJ}/\rho_0 = 1.82$; $T_{CJ}/T_0 = 9.68$. The von-Neumann spike conditions are given with the subscript $_{VN}$ as follows: $p_{VN}/p_0 = 34.4$; $\rho_{VN}/\rho_0 = 5.95$; $T_{VN}/T_0 = 5.78$. The parameters of the constant-volume explosion are given with the subscripts $_{CV}$ as follows: $p_{CV}/p_0 = 10.8$; $T_{CV}/T_0 = 6.35$.

Figure 1 shows the time evolution of pressure profiles for a transition from deflagration to detonation. The arrow marks of 10% HCl mole fraction in the figure indicate the locations of reaction fronts behind which dominant heat release process occurs. The profiles 1-5 show a development of a deflagration wave. The reaction front generates a shock wave 2-5 mm ahead. The shock heating effect becomes more contributive than the external irradiation for enhancing the chemical reactions. A rapid growth of the shock wave occurring in the profiles 5-7 is unstable. After the momentary decoupling of the reaction front from the shock wave in the profile 7, the wave interaction is re-established, and so a transition occurs. The rapid pressure rise in the profile 7 exceeds the von Neumann spike. On the basis of the profiles 8 and 9, the mean velocity of the detonation wave is about 1600 m/s. This velocity is quite comparable with the usual Chapman-Jouguet velocity. A computation for I_0 = 0.5 kW/cm^2 gave a more salient behavior: a shock wave of p/p_0 = 63 was observed. Figure 2 also gives a transition case. The transition instability, that appears in the time history of the pressure waves, becomes smaller. The profiles 7-9 show a detonation propagation with a certain instability.

The direct initiation occurs by a radiation intensity about 2 kW/cm^2 given in Fig. 3. The coupling between the reaction front and the shock wave is established at an early stage of pressure wave development. The profiles 1-3 show that the gas mixture in front of the shock wave is already reacting with a small pressure rise. The energy release from the preconditioned media is coherently added to the propagating shock wave. Therefore, the SWACER mechanism is responsible for the efficient shock wave amplification. The profiles 6-9 show a steady propagation of deto-

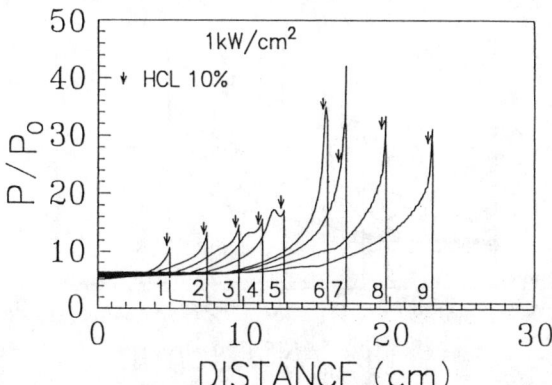

Fig. 1 Detonation transition with I_0 = 1 kW/cm^2 radiation, time (μs) of profiles 1-9: 1, 114; 2, 139; 3, 160; 4 174; 5, 187; 6, 210; 7, 217; 8, 233; 9, 253.

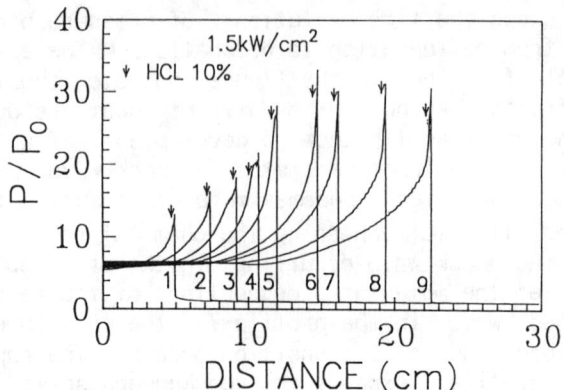

Fig. 2 Detonation transition with $I_0 = 1.5$ kW/cm² radiation, time (μs) of profiles 1-9: 1, 96.7; 2, 119; 3, 134; 4 146; 5, 156; 6, 172; 7, 182; 8, 202; 9, 221.

nation wave with a small fluctuation. The shock pressure is about 10% lower than the von Neumann spike pressure of the usual Chapman-Jouguet wave. The effects of external radiation on triggering the chemical reactions become much smaller after the detonation wave is established. The shock heating is more dominant in the steady propagation of the detonation wave. Figure 4 shows a case with a radiation intensity near the upper critical threshold for direct initiation. The SWACER mechanism is also responsible for the pressure wave development in this case. The difference from the case of Fig. 3 concerns the structure of the detonation wave in steady propagation as shown in the profiles

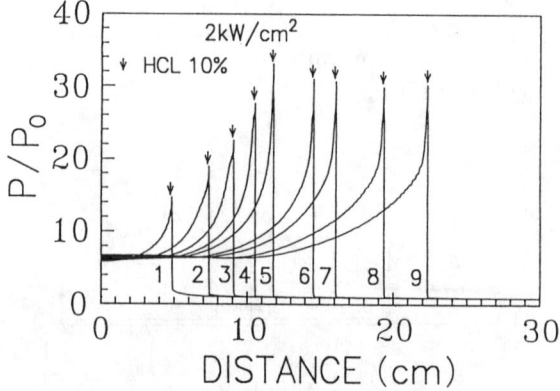

Fig. 3 Direct detonation initiation with $I_0 = 2$ kW/cm² radiation, time (μs) of profiles 1-9: 1, 86.1; 2, 107; 3, 120; 4 131; 5, 139; 6, 156; 7, 166; 8, 187; 9, 206.

DETONATIONS IN HYDROGEN-CHLORINE MIXTURES

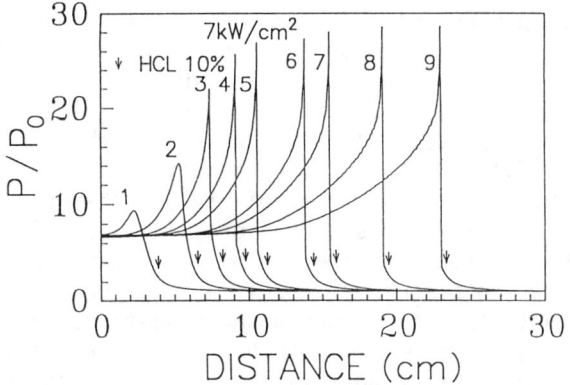

Fig. 4 Direct detonation initiation with $I_0 = 7$ kW/cm^2 radiation, time (μs) of profiles 1-9: 1, 38.9; 2, 59.4; 3, 72.2; 4, 83.4; 5, 92.7; 6, 113; 7, 124; 8, 147; 9, 171.

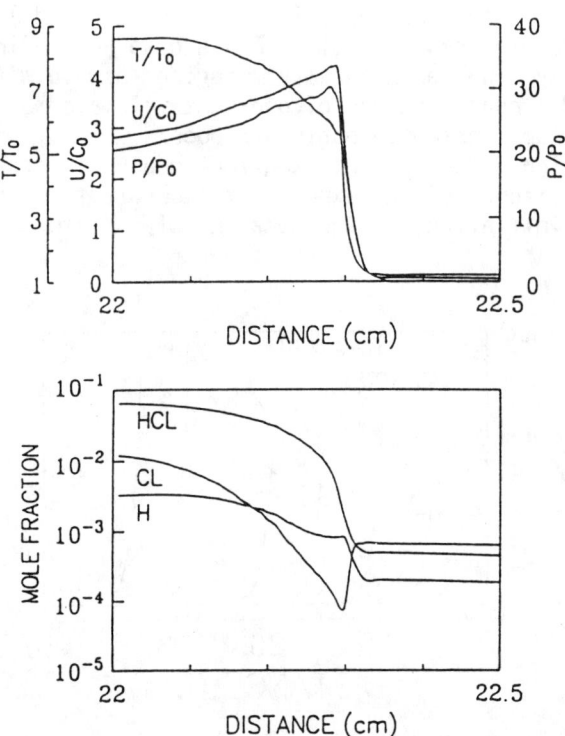

Fig. 5 Structure of photochemical detonation wave. (Close-up view of profile 9 in Fig. 4.)

6-9. A closeup view at the time of the profile 9 is given in Fig. 5. A very intense radiation penetrates the shock wave. The excess radiation further enhances the photodissociation and forms a precursor reaction zone. In this case, the shock heating is not the only process for triggering the chemical reactions. The change of detonation structure results in the shock pressure that is slightly lower than that of Fig. 3. The mean velocity is about 1600 m/s on the basis of the profiles 7 and 9.

With further increasing the radiation intensity, as shown in Figs. 6 and 7, the precursor reaction zone becomes larger and more contributing. This phenomenon, however, does not enhance the SWACER mechanism any more, rather it suppresses the wave amplification. Compression waves with thick reaction zones appear. The peak pressure of the wave is much lower than that of the Chapman-Jouguet wave. Although the pressure waves in both figures are growing even at the last profiles, the rate of pressure rise is low, and so the location of shock wave formation is far from the window. With increasing radiation power the wave peak pressure decreases. But the wave velocity contrarily increases; on the basis of the locations of the reaction fronts of the profiles 9 and 11, the mean velocities of Figs. 6 and 7 are about 1600 and 1900 m/s, respectively. For a case of infinite radiation intensity, we can naturally imagine a wave with infinite velocity and a peak pressure of constant-volume combustion, viz., the whole volume instantaneously explodes.

In conclusion, the present numerical analysis provided some clear explanations of the experimental observations of the photochemical initiation in the same H_2-Cl_2 mixtures[4] at least

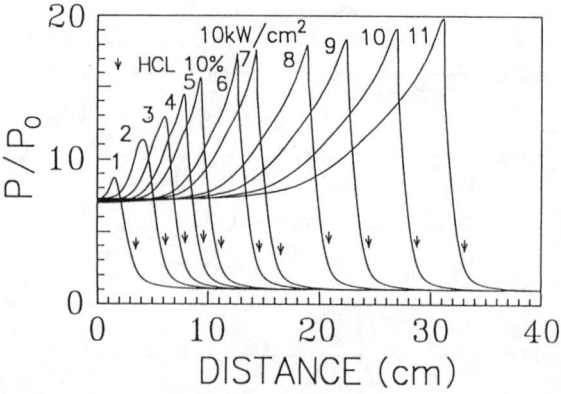

Fig. 6 Pressure wave with I_0 = 10 kW/cm² radiation, time (μs) of profiles 1-11: 1, 29.4; 2, 46.7; 3, 58.0; 4, 68.4; 5, 77.6; 6, 98.3; 7, 109; 8, 138; 9, 159; 10, 187; 11, 213.

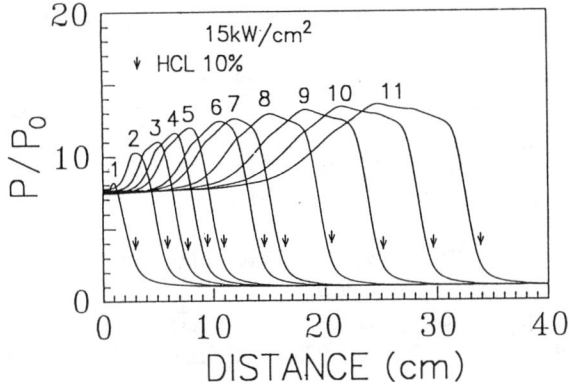

Fig. 7 Pressure wave with I_0 = 15 kW/cm^2 radiation, time (μs) of profiles 1-11: 1, 21.3; 2, 35.3; 3, 45.1; 4, 53.7; 5, 61.6; 6, 79.5; 7, 89.1; 8, 112; 9, 134; 10, 157; 11, 179.

qualitatively. The results demonstrated that detailed computer simulations are viable tools for analyzing the wave amplification processes. However, more computations of different types of preconditioned flowfields, such as temperature gradient fields, are necessary for establishing generalized criteria for the SWACER mechanism to be operative.

References

[1] Norrish, R. G. W., "The Study of Combustion by Photochemical Methods," Proceedings of the 10th (International) Symposium on Combustion, Plenary Lecture, 1965, pp. 1-18.

[2] Thrush, B. A., "The Homogeneity of Explosions Initiated by Flash Photolysis," Proceedings of the Royal Society of London, A233, 1956, pp. 439-460.

[3] Wadsworth, J., "Use of Flash Photolysis to Initiate Detonation in Gaseous Mixtures," Nature, Vol. 190, 1961, pp. 623-624.

[4] Lee, J. H., Knystautas, R., and Yoshikawa, N., "Photochemical Initiation of Gaseous Detonations," Acta Astronautica, Vol. 5, 1978, pp. 971-982.

[5] Knystautas, R., Lee, J. H., Moen, I., and Wagner, H. Gg., "Direct Initiation of Spherical Detonation by a Hot Turbulent Gas Jet," Proceedings of the 17th Symposium (International) on Combustion, 1979, pp. 1235-1245.

[6] Lee, J. H. S., and Moen, I. O., "The Mechanism of Transition from Deflagration to Detonation in Vapor Cloud Explosions," Progress in Energy and Combustion Science, Vol. 6, No. 1, 1980, pp. 359-389.

[7] Yoshikawa, N., "Coherent Shock Wave Amplification in Photochemical Initiation of Gaseous Detonations," Ph.D. Dissertation McGill University, 1980.

[8] Calvert, J. G., and Pitts, J. N., Jr., Photochemistry, Wiley, 1966, pp. 176-185.

[9]Cohen, N., Jacobs, T. A., Emanuel, G., and Wilkins, R. L., "Chemical Kinetics of Hydrogen Halide Lasers. 1. the H_2-Cl_2 System," International Journal of Chemical Kinetics, Vol. 1, 1969, pp. 551-569.

[10]Stull, D. R., and, Prophet, H., JANAF Thermochemical Tables, NSRDS-NBS 37, 2nd ed., 1971.

[11]Lomax, H., and Bailey, H. E., "A Critical Analysis of Various Numerical Integration Methods for Computing the Flow of a Gas in Chemical Nonequilibrium," NASA TN D-4109, 1967.

[12]MacCormack, R. W., "The Effect of Viscosity in Hypervelocity Impact Cratering," AIAA Paper 69-354, 1969.

[13]Boris, J. P., and Book, D. L., "Flux-Corrected Transport I. SHASTA, a Fluid Transport Algorithm that Works," Journal of Computational Physics, Vol. 11, 1973, pp. 38-69.

[14]Fujii, H., Nakano, M., Tamura, M., and Yoshikawa, N., "Application of Lomax-Bailey Implicit Scheme to Reactive Flows," Proceedings of the International Symposium on Computational Fluid Dynamics (Nagoya), 1990, pp. 887-892.

Mechanism of Unstable Detonation Front Origin

A. N. Dremin*
Russian Academy of Sciences, Moscow, Russia

Introduction

The mechanism of unstable detonation front origin is still a real problem.[1,2] Shchelkin was the first to introduce the idea of detonation kinetic instability.[3] In essence, later most investigators developed Shchelkin's approach to the problem. In this approach, the stationary detonation complex stability has been investigated, the complex consisting of the shock front and the chemical reaction behind it. The complex was advanced in 1940 by Rosing and Chariton.[4] According to Rosing and Chariton an explosive does not suffer any chemical change within the detonation wave shock front; it is only compressed within the front and its decomposition takes place behind the front at a shock-compressed state. The chemical reaction zone is characterized by high pressure and is called a "chemical spike."

It is evident from the following experimental findings that Schelkin's approach to the detonation stability problem is not universal in favour of the statement. First, it has been found that the normal Chapman-Jouguet (CJ) detonations of physically homogeneous explosives always originate from strong detonation. It is obvious that if the initiating shock pressure is higher than the CJ pressure the strong detonation will appear in the beginning. As for the case of the initiating shock pressure lower than the CJ pressure, it is the so-called shock-to-detonation transition process and it has been investigated in detail.[5-7] In particular, it has been found that under the shock effect a layer of shock-compressed explosive (SCE) appears and the SCE detonation arises first at the piston-SCE interface. The SCE detonation pressure exceeds considerably the pressure of the CJ

Copyright © 1992 by the Institute of Chemical Physics (Chernogolovka). All rights reserved.
*Head, High Dynamic Pressure Department, Institute of Chemical Physics.

detonation of the explosive at normal state. Therefore, it results in the strong detonation appearance just as it overtakes the initiating shock wave front. So, the CJ detonation appears from the strong detonation at any way of it initiation. Second, it turned out that the question as to why one explosive has a stable CJ detonation and another has not settled just during the transition process. Third, it has been revealed that the question of whether the CJ detonation will be stable or not is governed by the chemical reaction breakdown phenomenon.[5,8,9] It has been shown that the CJ detonation is stable if the breakdown does not take place during the transition process. On the contrary, if the breakdown occurs the detonation will be unstable.

Explosive Shock Decomposition Regularities and Breakdown Phenomenon

What is the breakdown phenomenon? As known there is a rarefaction wave behind the shock wave front in most explosion processes involved in detonics. Therefore, the explosive temperature behind the shock front is governed by two factors: 1) the explosive decomposition tends to self-heating and 2) the adiabatic cooling in the rarefaction wave tends to decrease the explosive temperature. If this cooling effect compensates the self-heating of the explosive the decomposition reaction will cease. This is the very case of the breakdown phenomenon. On the other hand, if the cooling effect does not compensate the self-heating of the explosive, the decomposition reaction will accelerate and in some induction time will assume an explosion character.

It follows from the foregoing that the correlation between the two rates – the rate of adiabatic cooling and the rate of initial heat evolution just behind the shock wave front – governs the breakdown phenomenon display. Unfortunately, at present there is little known about condensed explosives detonation decomposition mechanism. It is still the least investigated in the theory of detonation. Most investigators try to elucidate how the high parameters of the explosive shock-compressed state (in particular pressure and density) affect the chemistry of the explosive molecules. But how the state forms itself, that is, the process of explosive shock compression is not taken into account. A fundamental new concept of the problem has been developed in the Institute.[10] According to the concept the activation and nonequilibrium destruction of some part of explosive molecules take place inside the detonation wave shock front ($\simeq 10^{-10} s$). The process results in the origin of some active particles (radicals, ions and so on). The particles behave as if they had been injected into the compressed and heated explosive. Naturally, they influence explosive subsequent decomposition. As a rule the activation energy of the active particles interaction with explosive molecules is low (5–15 kcal/mole),[11] that is, the interaction proceeds with speed.

The number of explosive molecules decomposed inside the detonation wave shock front is not known for the present. Obviously it is some function of the wave intensity. Therefore, not only the explosive shock compression temperature but also the number of particles increases with the wave intensity increase. Both of these factors (the temperature and the amount of the increase) result in the increase of the initial rate of heat evolution just behind the detonation wave shock front. It follows from the just proposed notion of the chemical reaction breakdown phenomenon that the heat evolution rate increase diminishes the phenomenon probability.

Breakdown Phenomenon and Unstable Detonation Front Structure

The question arises – how does the breakdown phenomenon govern the detonation wave instability origin as well as the unstable front structure? It goes without saying that the CJ detonation wave will have a stable front if the explosive decomposition reaction breakdown phenomenon does not occur during the process of the strong detonation transition to the normal one. Detonation waves with a stable front have been observed experimentally in some liquid explosives.[5] Accordingly, the stable detonation front is more probable in the case of power explosives. This statement is based on the fact that the initial rate of heat evolution just behind the shock front inside the chemical spike of power explosives CJ detonation waves is larger than that of the weak explosives. Therefore, the reaction breakdown phenomenon is of low probability in the case of power explosives CJ detonation. Nevertheless, weak explosives as a rule have unstable detonation fronts. In the case of weak explosives the reaction breakdown occurs easily during the process of strong detonation transition to the CJ detonation. After the breakdown has occurred the continuing process will resemble the above-mentioned shock-to-detonation transition process.[5,7] As in the process a layer of SCE originates in the wave front. The wave parameters (pressure and pressure gradients) change in such a way that the SCE adiabatic cooling governed by the rarefaction wave behind the shock front becomes unable to compensate the SCE self-heating due to its decomposition; the decomposition accelerates and, finally, it results in the SCE explosion origin at the detonation products-SCE interface. (It should be mentioned that for a strong but short initiating shock wave the reaction may not originate again after the first breakdown. However this case is of no interest for the present discussion.)

The instability of the detonation front is a kinetic property of some explosives and is inevitable; that is, the SCE decomposition reaction breakdown and reinitiation of the explosions in the front will occur repeatedly if the breakdown has once occurred. Really, the SCE explosion appears at the detonation products-SCE interface and gives rise to the detonation in the SCE. The detonation pressure exceeds considerably that of the initial explosive CJ detonation. For instance, in nitromethane the pressure in the SCE layer is approximately equal to the CJ pressure (130 kbars);[5,12] if the detonation wave front were stable, the chemical spike pressure would be equal to 225 kbars but the SCE detonation pressure is approximately equal to 350 kbar. Therefore, the strong detonation will appear again when the SCE detonation overcomes the shock wave front.

It should be mentioned that for some casual reasons the reaction breakdowns and the subsequent SCE explosions in reality do not take place simultaneously over the whole detonation front, but occur at randomly distributed individual sites. The SCE detonation from each site spreads hemispherically in the SCE. Each individual detonation overcomes the shock front and causes a very strong divergent detonation in the initial undisturbed explosive. The rarefaction rate behind these detonation fronts is larger in comparison with that behind the plane front and, therefore, the reaction breaks down easily again. Each breakdown results in the appearance of some islet of a new SCE and in each increasing islet a new SCE detonation can originate in the induction time corresponding to the SCE state, and so on. In the process of the transition from the initial strong detonation to the normal one, the pressure in each new SCE islet is somewhat smaller than that in the previous one, and finally when the CJ detonation is reached, the pressure in the SCE islet appears to be approximately equal to the CJ pressure.[5] This implies that the normal detonation front pulsations start from the pressure level which equals to the CJ pressure.

It should be emphasized that the breakdown phenomenon in the hemispherical strong detonations will also occur even in the case when the detonation resulting CJ pressure is maintained constant. The scale of the detonation front pulsations in this case will correspond to that of the ideal detonation.

After the reaction breaks down in some strong divergent local detonation the SCE detonation spreads over the SCE layer like a ring. Figure 1 shows half sections of four pictures (a, b, c, d) that represent the ring evolution in time and space in sequence. In the figure the vertical arrows show the detonation wave direction; A is the point of SCE detonation origin, B is the point of chemical reaction breakdown, M is the triple-point. There are some triple-shock configurations in the crosssection on of every ring, with layers (1, 2, 3) of the SCE building up between them. The triple-shock configurations consist of the straight wave (S); the principal element of the configuration, the transverse wave (T) representing the SCE detonation; and the oblique wave (O) which is, as the matter of fact, the strong detonation in the undisturbed explosive. a) initial SCE layer 1, A_1 the SCE_1 detonation origin; b) principal triple-shock configuration just before SCE_2

detonation origin (point A_2); c) principal 1 and following 2 triple-shock configurations just before SCE_3 detonation origin (point A_3); d) principal 1, second 2 and third 3 triple-shock configurations just before a new principle triple-shock configuration appearance (point A_4 is analogous to point A_1 and is the point of a new SCE_1 detonation origin); e) flow structure at dark zones appearance phenomenon in liquid explosives detonation according to Refs.[11,12]; D_{cj} detonation wave front velocity, v velocity of chemical reaction breakdown in the detonation wave front.

The farther away from the triple-point M_1, the lower the pressure is in the front of the oblique strong detonation wave O_1. B_1 is the point where this detonation is eliminated indicating that the explosive decomposition reaction ceased due to the breakdown phenomenon. However, experiments have shown that the reaction can start again (even many times) at some distance from the triple-point M_1.[12] The matter is that the SCE pressure just after the breakdown has occurred is considerably larger than the CJ pressure. During the process of the ring increase the pressure decreases. However, rarefactions originating due to the pressure decrease can not overcome the point B_1. Obviously, the flows structure in this case is somewhat analogous to that of the structure at the so-called dark zones origin process at liquid explosives detonation[13] (see Figs. 1b and 1e). In both cases the SCE flow velocity U is larger than the sound velocity C. For this reason the time for each consecutive part of the explosive, which enters the point B, to reach the point A is longer than for previous one. When the time is equal to the SCE explosion induction period the explosion takes place and it results in the SCE-detonation origin. According to available experimental data[5,14] the detonations are observed to move mostly toward the point M_1 of the principal triple-shock configurations and their velocities are equal to the transverse detonation velocity of the principal configurations. As a consequence many new triple-shock configurations can appear behind the triple-point M_1; the induction time of all new explosions in sequence increasing up to the value corresponding to that of the explosion at the point A_1.

Thus, different-order explosions and, correspondingly, different-order triple-shock configurations can arise in the highly unstable detonation front. In outward appearance they originate chaotically in time and space, and therefore the unstable detonation front structure seems to be extremely complicated. However, as the matter of fact, there is an inner law in this apparent chaos: It is the first order explosions (FOEs) (point A_1 in Fig. 1) that determine the magnitude of the largest size of the detonation front pulsations. Moreover, the existence of the detonation wave with the unstable front is governed by the FOEs; if the explosions do not occur, the detonation wave will disappear. The FOEs are responsible also for regulating triple-shock configuration number in the strong detonation wave front during its transition to the normal detonation regime as well as in the front of the spherical detonation wave. Indeed, it follows from the preceding that the reaction breakdown leads to the SCE islet appearance at each local diverging strong detonation; the islet size increases during the ring detonation's progress over the detonation wave front. At the process the islets combine all together with the SCE continuous layer origin. However, since the FOEs originate at the beginning chaotically, the SCE continuous layer is not a constant thickness. Therefore new FOEs arise again nonsimultaneously. Obviously, the SCE ring detonations corresponding to each new FOE spread over the detonation wave front surface and consume adjacent parts of the SCE layer with the result that the FOEs have no time to occur in the parts. Just for this reason the FOE quantity in the detonation fronts decreases during the process of the strong detonations transition to the CJ detonation. An average size of the CJ detonation wave front inhomogeneity (unsmoothness) is also settled by this mechanism.

The presented considerations are based mostly on the analysis of the available experimental data for liquid explosives.[5,14] However, it is obvious

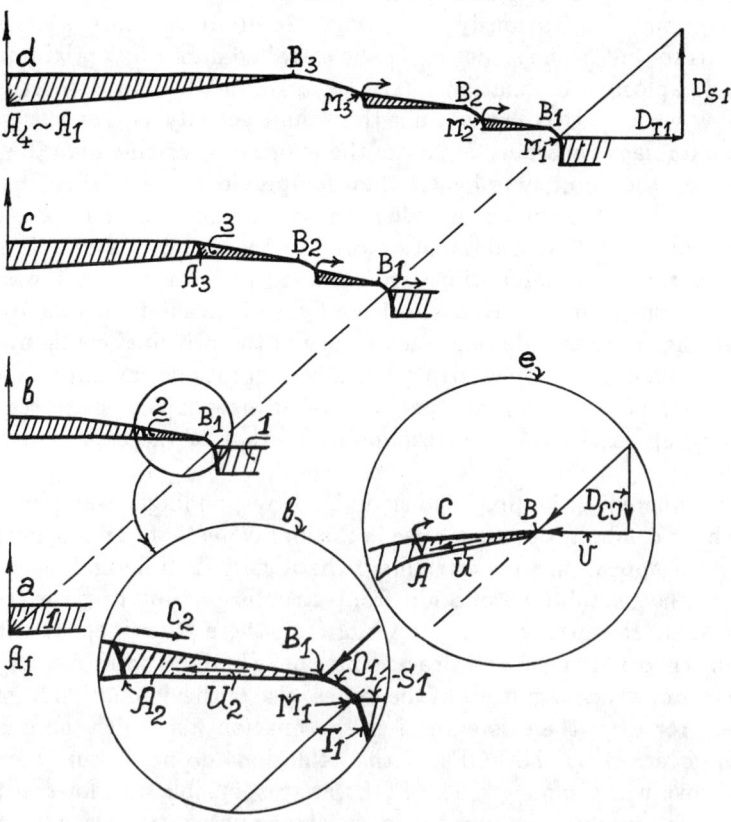

Fig. 1 Qualitative diagram of SCE zones (1,2,3) appearance and triple-shock configurations motion over unstable detonation wave front.

that the instability origin nature for gaseous detonation is analogous to that for liquids. The reaction breakdown phenomena during the process of strong detonations transition to normal detonation are the main factor for the detonation instability origin and for the detonation front structure in both cases. But outwardly there is some difference between the two. In particular, unlike the unstable detonation front in liquids, no new reaction reinitiation arises behind the B_1 point except for the ring center A_1 in gases at the CJ detonation (see Fig. 1b). On the other hand, one can observe the oblique wave instability at spinning detonation in gases.[15]

References

[1] Taki, S., and Fajiwara, T., "Numerical Analysis of Two-Dimensional Nonsteady Detonations," *AIAA Journal*, Vol. 16, *No* 1, 1978, pp. 73–77.

[2] Tarver, C., and Calef, D., "The Detonation of High Explosives," *Energy and Technology Review, High Explosives*. LLNL, Jan.-Feb., 1988, pp. 1–8.

[3] Schelkin, K. I., "Two Kinds of Unstable Burning," *Zurnal Experimentalnoi i teoretitsheskoi Fiziki*, Vol. 36, 1959, pp. 600–606.

[4] Rosing, V. O., and Chariton, Yu. B., "Explosive Detonations at Small Charge Diameter," *Doklady Akademii Nauk SSSR*, Vol. 26, *No* 4, 1940, pp. 360–361.

[5] Dremin, A. N., Savrov, S. D., Trofimov, V. S., and Shvedov, K. K., "Detonation waves in condensed media," Moscow, Nauka, 1970; *Translated from Russian by Foreign Technology Division, Wright-Patterson Air Force Base*, OH, Aug. 1972.

[6] Campbell, A. W., Devis, W. L., and Travis, J. R., "Shock Initiation of Detonation in Liquid Explosives," *Physics of Fluids*, Vol. 4, *No* 3/4, 1961, pp. 498–510.

[7] Chaiken, R. F., "Comments on Hypervelocity Wave Phenomena in Condensed Explosives," *Journal Chemical Physics*, Vol. 33, 1960, pp. 760–768.

[8] Dremin, A. N., "Critical Phenomena in the Detonation of Liquid Explosives," *12th Symposium (International) on Combustion*, Poitiers, France, 1968, pp. 691–699.

[9] Trofimov, V. S., and Veretennicov, V. A., "On condensed explosives detonation front structure," *Chemical Physics of Combustion and Explosion Processes. Detonation*, Chernogolovka, Nauka, 1977, pp. 9–11.

[10] Dremin, A. N., Klimenko, V. Yn., Davidova, O. N., Zoludeva, T.A., "Multiprocess Detonation Model," *Proceedings of the Ninth Symposium (International) on Detonation*, Portland, USA, 1989, pp. 287–291.

[11] Kondratiev, V. N., "Gas-Phase Reactions Rate Constant," Moscow, Nauka, 1971.

[12] Dremin, A. N., "Pulsating Detonation Front," *Fizika Gorenia i Vzriva*, *No* 4, 1983, pp. 159–169.

[13] Dremin, A. N., and Trofimov, V. S., "The Calculation of Liquid Explosives Detonation Failure Diameter," *Zurnal Prikladnoi Mekhaniki i Technitsheskoi Fiziki*, No 1, 1964. pp. 126–131.

[14] Persson, P. A., and Persson, G.,"High Resolution Photography of Transverse Wave Effects in the Detonation of Condensed Explosives," *Proceedings of the 6th Symposium (International) on Detonation*, Coronado, USA 1976, pp. 414–425.

[15] Schelkin, K. I., and Troshin, Ya. K., "Gasdynamic of Combustion," Moscow, 1963.

Numerical Modeling of Galloping Detonation

S. M. Aksamentov,* V. I. Manzhaley,† and V. V. Mitrofanov‡
Lavrentyev Institute of Hydrodynamics, Novosibirsk, Russia

Abstract

The problem of gaseous detonation propagation through a channel has been solved numerically in a one-dimensional nonstationary statement taking into account heat transfer and gas retardation in a boundary layer. A mathematical model comprises a flame front propagation through a mixture behind a leading shock wave and mixture burnout after the induction period. The product state is defined with the use of approximate chemical equilibrium equations. For the $CH_4 + 2O_2$ mixture with the initial data corresponding to those in Ul'yanitsky's experiments of 1981, the calculated data are in satisfactory agreement with the experimental results. Depending on the initiation conditions, the wave either decays or is in a self-sustained regime with longitudinal pulsations. In the absence of losses on the walls, the pulsation pitch is of the order of the cell length in a multiheaded structure of a real wave in a wide tube. Because of the introduction of momentum, heat losses, and, in particular, gas removal from the wave region caused by the boundary layer, the pulsation pitch increases by

---

Copyright © 1992 by the American Institute of Aeronautics and Astronautics, Inc. All right reserved.
* Associate Research Scientist
† Associate Professor
‡ Professor

dozens of times. A spontaneous development of the secondary, i.e., behind the shock front detonation, after the primary one has decayed is reproduced in the calculations. Below a certain pressure value or channel diameter size, a galloping detonation is transformed into a stationary or weakly pulsating low-velocity detonation, in which the shock wave is sustained by fast flame propagation according to a conductive-convective mechanism.

Introduction and Experimental Data

The galloping detonation was first described by Mooradian and Gordon.[1] The name was suggested by R. E. Duff in the discussion concerning the work by Manson et al.[2] The most detailed information on the wave structure was obtained by Saint-Cloud et al.[3] for the C_3H_8 + 5 O_2 + 10 N_2 mixture in a 10 x 20-mm tube with a rectangular cross section, then by Ul'yanitsky[4] for the CH_4 + $2O_2$ mixture in the channel with an internal diameter d = 12.8 mm. The concentration limits of galloping detonation of the H_2, CH_4, C_2H_4, and C_3H_8 with air mixtures in the tube with d = 70 mm were determined by Borisov et al.[5] A "gallop" is observed in gaseous mixtures with irregular cellular structures of a detonation front far from the limits. The latter is related to a high value of E_a/RT_s, which was shown for a number of mixtures by Manzhaley.[6] A gallop is characterized by considerable longitudinal pulsations with pitches of more than 100 d. When reaching the detonation limit due to the decrease in either initial pressure p_o or concentration of a missing component or tube diameter, the cellular front structure is successively transformed into a spinning and then a galloping structure. A relative deficiency of the wave velocity $\varepsilon = (D-D_o)/D_o$ thereby increases (D_o is the ideal detonation velocity without losses for the same mixture and p_o). The previously mentioned authors obtained the values of $\varepsilon \leqslant 0.2$. For some mixtures, after the galloping detonation regime, there may exist a steady-state regime with low velocity D = 0.45 -0.6 D_o; it was discovered recently by Manzhaley[7,8] for narrow channels (d < 3 mm). The assumption concerning the possibilities of an analogous detonation propagation regime in rough-walled tubes at D = 500 m/s was previously made by Zel'dovich.[9] So far, however, the existence range of a low-velocity regime has been studied insufficiently. In the argon-diluted mixtures that have regular cells a gallop has not been discovered yet, and a spinning detonation regime is assumed to be limited.

According to the experimental data, each period of galloping detonation pulsations consists of the following stages (phases): 1) a jumplike transformation of weakened wave into an overdriven detonation wave with a fine-cellular structure, 2) a monotonic wave weakening to D_o and its transformation into a spinning wave sustained during 2-3 rotations of a transverse wave, 3) a transverse wave decay and burning zone separation from the shock as the velocity continues to decrease, and 4) the end of the velocity decrease and transition to a slow wave acceleration with some decrease of the distance between the shock and the burning zone. The larst stage is terminated by the formation of a secondary fine-cellular detonation wave in compressed gas, which overtakes the leading shock. Then the cycle is repeated.

Ul'yanitsky[4] has developed a simple theoretical model of galloping detonation. It is based on an approximate law of leading shock propagation assuming that the chemical reaction behind the shock in the second phase is instantaneous, and in the third phase it comes to a complete end. When the ignition delay is over in front of the interface of the combustion products, shock compressed gas is burnt out instantaneously and completely. The moment of time when the above-mentioned transition takes place is found from the given value of mean wave velocity. Momentum and heat transfer to the walls are not taken into account. The model satisfactorily describes the main features of the process except for front acceleration and secondary detonation wave development.

In our experiments with the $C_2H_2 + 2.5\ O_2$ mixture in the 1-mm-diam tubes, a galloping regime was observed inside the pressure limits 7.2 kPa $< p_o <$ 9.2 kPa. The gallop pitch ranged from 0.32 to 0.36 m at a velocity deficiency $\varepsilon = 0.3 \mp 0.1$. Under higher initial pressure, a spinning regime took place, and under lower pressure a low-velocity detonation (LVD) was observed. As initial pressure and velocity increased, a transition from LVD to a galloping regime occurred due to adiabatic self-ignition inside the LVD-wave structure, between a shock front and flame. There a new wave was first a multiheaded and then a spinning decaying one, and finaly it was the wave with a LVD-structure. After a minimum velocity of about 800 m/s it was accelerated to its expected stationary value of about 1400 m/s, which is impossible to achieve due to self-ignition. In the narrow tubes used, a relative LVD phase duration turned out to be higher than in the tubes of about 10 mm in diameter and larger. This duration accounted for 60-85% as compared to the gallop pitch.

Reddy et al.[10] succeeded in obtaining a galloplike regime by a two-dimensional numerical detonation modeling in a channel with arectangular cross section with one gas-permeable wall. Heat transfer and friction were not taken into account; a chemical reaction occurred inside the volume according to the kinetics used previously by Levin and Markov.[11]

The principal aim of this study is to elucidate the role of the flame front and the boundary layer at the wall. The authors used a more economical one-dimensional numerical model of detonation in a tube to investigate in more detail the effects of mixture properties, losses on the walls and combustion peculiarities on the existence region, and parameters of a galloping detonation wave. The value of the efficient flame velocity in shock compressed gas, is which took into account (approximately) two-dimensional effects of boundary-layer formation on the tube walls, was introduced.

Theoretical Model

The model of flow at galloping detonation in a tube adopted by the authors is based on the following assumptions:

1) Transverse flow oscillations are insignificant, and the medium motion may be described by a one-dimensional equations for perfect gas with momentum and heat transfer from the channel walls as follows:

$$\frac{\partial \rho}{\partial t} + \frac{\partial (\rho u)}{\partial x} = 0$$

$$\frac{\partial (\rho u)}{\partial t} + \frac{\partial (\rho u^2 + p)}{\partial x} = -\frac{4\tau}{d}$$

$$\frac{\partial(\rho e)}{\partial t} + \frac{\partial(\rho u e)}{\partial x} - p\frac{d\rho}{dt} = \frac{4(\tau u - q)}{d}$$

$$\frac{d}{dt} = \frac{\partial}{\partial t} + u\frac{\partial}{\partial x}, \quad p = \frac{\rho RT}{\mu} \tag{1}$$

All of the designations have been universally adopted.

2) The shear stress τ and heat flux to the walls, q, are defined by the well-known expressions for a steady-state flow in tubes using the Reynolds analogy:

$$\tau = \lambda\rho u|u|/8, \quad q = \lambda\rho u[H-H(T_0)]/8 \tag{2}$$

where

$$\lambda = 6/Re \quad \text{if} \quad Re < 12$$
$$\lambda = 0.316/Re^{0.25} \quad \text{if} \quad 1200 < Re < 10^5$$
$$\lambda = 0.021 \quad \text{if} \quad Re > 10^5$$
$$H = e + p/\rho, \quad Re = \rho u d/\eta$$

3) An immediate chemical equilibrium in combustion products is achieved. It is described by the approximate equations, obtained by Nikolayev and Fomin[12]:

$$F(\rho,T,\mu) = \frac{\rho}{\mu}\left(1 - \frac{\mu}{\mu_{max}}\right)^2 - K\cdot\left(\frac{\mu}{\mu_{min}} - 1\right)\cdot\exp(-E/RT) = 0$$

$$e(T,\mu) = \frac{RT}{\mu(\gamma_2 - 1)} + E/\mu - h \tag{3}$$

where K, μ_{max}, μ_{min}, h, and E are constants, defined for a given mixture using exact calculations of the equilibrium at one of the points. Before exothermal transformations, internal energy was defined by

$$e = e_0 + \int_{T_0}^{T} c_v dT \tag{4}$$

where the dependence of c_v on T was approximated on the basis of standard thermodynamics table data.

4) There are two alternative combustion mechanisms: a) volumetric exothermal reaction in gas particles after the induction period, and b) the combustion in a plane small-thickness flame propagating relative to the particles with some velocity U before the end of the induction period. At the combustion front, it is appropriate to choose a mechanism that provides a higher velocity. In principle, a simultaneous combustion is possible by the two different mechanisms at different points.

5) A residual portion of induction period is defined by

$$Y = 1 - \int_{t_s}^{t} \frac{dt}{\tau} \qquad (5a)$$

where t_s is the moment of time when a shock passes through a particle; and τ the induction time behind a rectangular-profile shock wave dependent on T, p, and mixture composition. The exothermal reaction starts at $Y = 0$ and is described by any generalized kinetic equation known for the mixture being used

$$\frac{d\beta}{dt} = f(\beta,\rho,T) \qquad (5b)$$

or it may be assumed to be instantaneous. Here β is the parameter related to energy release.

6) Two-dimensional peculiarities of a real flow behind the shock wave significantly affect the flame propagation. Because of gas retardation in the boundary layer, the flow velocity in the tube center is higher than a mean one by some value Δu which is to be added to the flame velocity. Assuming the flow between the jump and flame to be quasistationary and using the results obtained by Mirels[13] for a turbulent boundary layer, we get

$$U = u_n + \frac{D}{W}\left(\frac{d}{l_m}\right)^{0.8} \cdot \left(\frac{x_s - x}{d}\right)^{0.8} \qquad (6)$$

where

$$\frac{l_m}{d} = \left(\frac{1}{4\beta_1}\right)^{1.25} \cdot \left(\frac{a_{st}\rho_{st}d}{\eta_{st}} \cdot \frac{p_0}{p_{st}} \cdot M_s\right)^{0.25} \cdot \frac{p_0}{p_s} \cdot \frac{W}{W-1}$$

$$\beta_1 = [8.6 + (1.47\gamma^2 - 4.91\gamma + 3.1)M_s] \cdot 10^{-2}$$

and where the first term in Eq. (6) is normal flame velocity dependent on T as

$$u_n = B + B_1 \cdot (T/T_{st})^{\nu} \qquad (7)$$

B, B_1, and ν are constants, D the shock velocity, a the sound speed, $M_s = D/a_0$, $W = \rho_s/\rho_0$, and $l_m = (x_s - x)_{max}$ is the distance from the shock front at which gas out of the boundary layer reaches velocity D; subscript st denotes the parameters under standard conditions, and $\gamma = C_p/C_v$. The expression for the value β_1 approximates the calculated data obtained by Mirels[13] for $M_s = 2$-6, $\gamma = 1.3$-1.67. A boundary layer is turbulent when $u l_m \rho (W-1)/\eta > 5 \cdot 10^6$. The latter condition is fulfilled in the cases considered below.

To initiate detonation at $t = 0$ and x in the range $[0, l_0]$, the pressure $p_1 = (200$-$400) p_0$ was prescribed, where mixture density was invariable. The discontinuity decay at $x = l_0$ and the following flow were calculated. The latter was determined on the basis of the finite difference analogy of the system of Eqs. (1-6), with the addition of explicit expressions for $\tau(p,T)$ and exothermal reaction velocity, where the boundary conditions were $u = 0$ at $x = 0$. A numerical Lagrangian method with artificial viscosity, developed by Yagafarov and Zaguskin,[14] based on the implicit difference scheme of first-order accuracy was used. The Courant number was equal to 0.5. The computational network consisted of 1000 cells, half of which were in a wave head, where the network was uniform in Lagrangian coordinates and covered approximately one pitch of the galloping detonation. As the distance from the front increased, the network cell size also increased, as well according to a special algorithm.

Calculations were performed for the $CH_4 + 2O_2$, $2H_2 + O_2$, and $2H_2 + O_2 + 7A$ mixtures.

For the first mixture, the expression for ignition delay in Eq. (5a) was taken from Borisov et al.[15] as follows:

$$\tau = \frac{\xi_{met}^{0.3} \xi_{ox}^{-1} p^{-0.7}}{2.5 \cdot 10^{10} \exp(-25000/T) + 2.5 \cdot 10^4 \exp(-10000/T)} \quad (8)$$

where ξ_{met}, ξ_o are the molar fractions of CH_4 and O_2. The exothermal reaction velocity was described by the decrease of fraction β of unreacted matter, following the equation

$$\frac{d\beta}{dt} = -1.22 \cdot 10^{14} \beta \left(\frac{p}{T}\right)^{0.75} \exp(-16000/T) \quad (9)$$

Here, as in Eq.(8), the pressure unit is $[p] = 10^5$ Pa, $[T] = K$, and $[t] = s$. Equation (9) was obtained from the data contained in the monograph by Schetinkov[16] for oxidation of CO limiting a final stage of methane combustion. Inside the reaction zone, Eqs.(3) were used, where the energy equation was added by the term βh. The value h was found using the continuity conditions for variables ρ, p, T, μ and e at the transition from the induction zone to the reaction zone. For both mixtures containing hydrogen, the expression for induction time was taken from the paper by Levin and Markov[11] and Reddy et al.[10]:

$$\tau = k_1 \cdot \rho^{-1} \exp(9800/T) \qquad (10)$$

where $k_1 = 3.33 \cdot 10^{-12}$ s·g/cm^3. The expression for burnout kinetics was taken from the same paper,[10] but it did not take into account inverse reaction, since the equilibrium was provided by Eq. (3) as follows:

$$\frac{d\beta}{dt} = - k_2 \cdot p^2 \cdot \beta^2 \exp(-2000)/T) \qquad (11a)$$

where $k_2 = 1.5 \cdot 10^{-5}$ Pa^{-2} s^{-1}. In the other variant, the kinetics suggested by Nikolayev[17] were used:

$$\frac{d\beta}{dt} = 4 \cdot K_+ \cdot \rho \cdot F(\mu, \rho, T) \qquad (11b)$$

from whence the first part of Eqs.(3) follows after equilibrium has been reached. The constants in Eqs.(3) and (11b) for the mixtures under consideration were obtained by Nikolayev and Fomin.[12] In Eq.(7) the following values of constants for normal flame velocity in the CH$_4$ +O$_2$ mixture were chosen from the work of Lewis and Von Elbe[18] and Ref.19: B = 0.975 m/s, B_1 = 2.275 m/s, and v = 1.7. They were experimentally verified over the temperature range of 273-1273 K. For both hydrogen-oxygen mixtures, these constants were assumed to be equal to 0, 9 m/s, and 1.75 respectively. However, for the mixtures containing argon, the necessary experimental data were not found.

In some calculations, the models were used where flame propagation was not taken into account.

Calculation and Discussion

Figure 1 displays the calculated wave velocity in the $CH_4 + 2O_2$ mixture along the x coordinate of the tube. The initial data $p_0 = 7 \cdot 10^3$ Pa, $T_0 = 293$ K, and $d = 12.7$ mm correspond to the experimental conditions used by Ul'yanitsky[4] under which the galloping detonation regime is the most stable. As is seen, the wave pulsations with the pitch $L_{cal} = 1.9 \mp 0.2$ m are repeated. The experimental value $L_{exp} = 2.2$ m $\mp$ 0.2 m. The calculated value of mean velocity is $\overline{D}_{cal} = 1830$ m/s; the experimental value is $\overline{D}_{exp} = 1850$ m/s, and an ideal detonatio velocity without losses ($\lambda = 0$) is $D_0 = 2222$ m/s. The values of velocity maxima and the dynamics of variation in D are close to the experimental ones. The most significant difference is observed in the values of velocity minima, $D_{min\ cal} \cong 1500$ m/s, and $D_{min\ exp} = 1250 \mp 150$ m/s.

High-frequency velocity oscillations at the closing stage of the cycle are seen on the calculated curve. Their period is of the order of the time of the sonic wave travel between the shock front and the flame

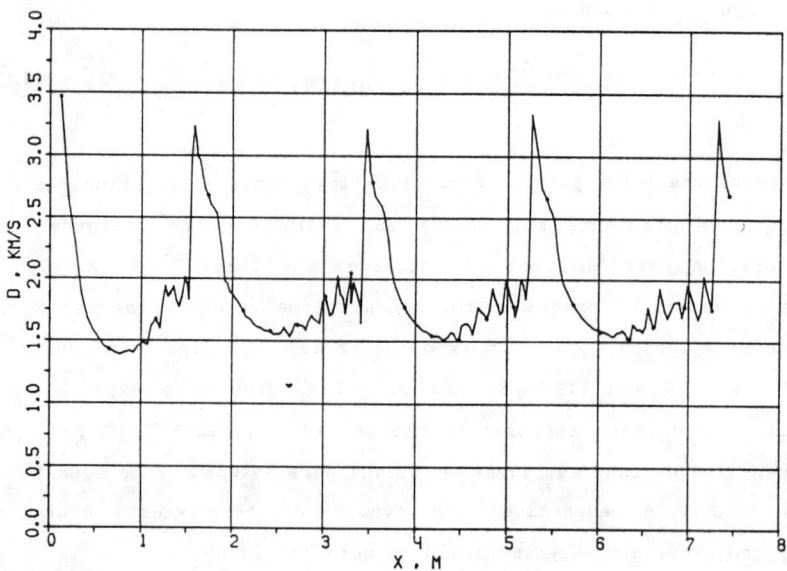

Fig. 1 Calculated D-x dependence for the galloping detonation regime in $CH_4 + 2O_2$ mixture; $p = 0.07 \cdot 10^5$ Pa, $d = 12.7$ mm.

front. In the minimum velocity region and farther, almost up to the end of the cycle, the mixture burnout occurs in the flame at $Y > 0$. The initial calculation data, presented in Fig. 2, are the same as in Fig. 1, but the exothermal reaction is assumed to be instantaneous after the induction period. There is no significant difference in $D(x)$; the gallop pitch is the same. Only the oscillatory process regularity decrease is observed here. The lower plot in Fig. 2 illustrates the trajectory of the shock and flame fronts corresponding to the last period of gallop on the upper plot. As is seen, the distance between these two fronts over most of the second half of the period is approximately constant. High-frequency oscillations are not noticeable here as they are on the streak-records of a real process. The wave structure at this stage corresponds to low-velocity detonation (LVD) without self-ignition, when the shock front is sustained by combustion on the flame front; the induction period is still continuing $(Y > 0)$. A general view of the lower picture is rather close to the streak-camera records of the real process.

The induction period comes to an end in front of the flame near the cycle end, where the transition to self-ignition causes a quick formation of a detonation wave in the compressed gas. The last process, described by Aksamentov et al.,[20] is illustrated in Fig. 3. Here a "spontaneous" mechanism of deflagration-to-detonation transition (DDT), first described by Zel'dovich et al.,[21] or SWAGER-mechanism, as it was called by Knustaustas et al.,[22] manifests itself.

Figure 4 illustrates variation in $D(x)$ as pressure decreases for the same mixture. The data in the upper plot are the same as in Fig. 1. The pressure decrease results in the fundamental oscillation amplitude decrease (the middle plot), there DTD inside the induction zone disappears. Then only high-frequency oscillations are observed (the lower plot), and the wave structure corresponds to LVD.

Presented in Fig. 5 are the calculated results for the $2H_2 + O_2$ mixture, $d = 12.7$ mm. The initial pressures are $p_0 = 0.56, 0.28, 0.14, 0.07,$ and $0.03 \cdot 10^5$ Pa, for a, b, c, d, and e plots, respectively. The plot f shows the variation of the relative deficiency of mean velocity ε vs pressure.

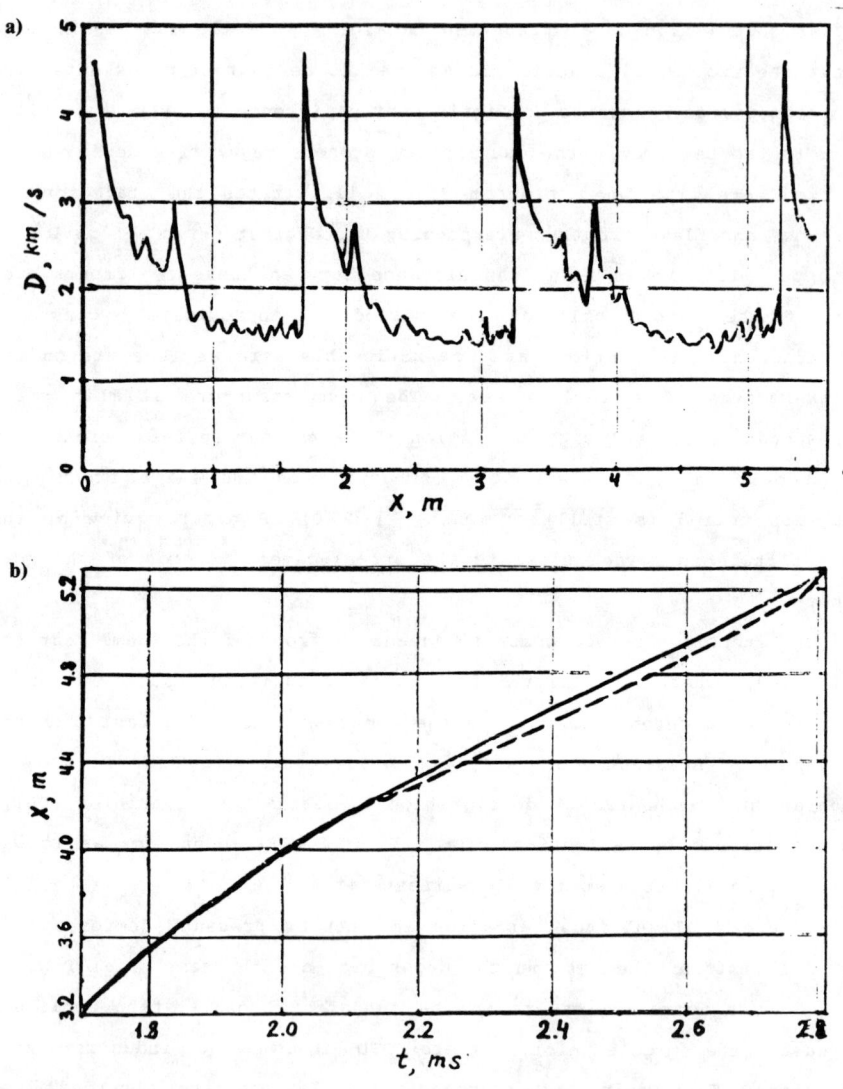

Fig. 2 The **galloping** detonation regime in $CH_4 + 2O_2$ mixture in the case of instantaneous exotermal reaction for $v = 0$: a) $D-t$ dependence; b) the $x-t$ graph for the leading shock front (solid line) and the flame front (dashed line), this corresponds to the last period (bettween the pointers) on the upper curve.

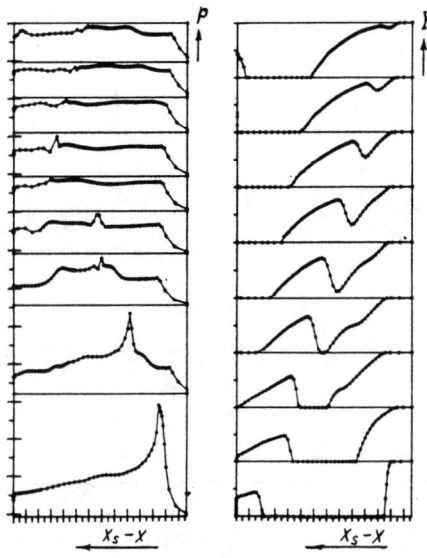

Fig. 3 DDT at the end of the galloping detonation; pressure profiles at sequential time moments.

Under higher pressure, the wave pulsations, described earlier by Levin and Markov,[11] are represented. They are a one-dimensional analogy of the real cellular detonation structure first described by Voitsekhovsky et al.[23] The pulsation pitch is several times as high as the experimental value of a longitudinal cell size under the same pressure. The combustion zone at the end of pulsations follows the shock at a distance less than the tube diameter, and boundary-layer effects do not manifest themselves. If the flame motion mechanism is not taken into account in the calculations ($U = 0$), the results are not changed. The influence of friction and heat transfer is also small ($\varepsilon \cong 0.02$).

At a two-fold decrease in p_0, the oscillations become very irregular. At a further decrease in p_0, the transition to a rather regular galloping regime with the pitch $L \cong 3$ m and then to LVD with insignificant oscillations in velocity takes place. It should be noted that periodical LVD oscillations were also considered in our experiments in narrow tubes, mentioned in the first section.

The change of the calculated mean pulsation pitch L, with the tube diameter d decrease from 127 mm to 6 mm at $p_0 = \text{const} = 7 \cdot 10^3$ Pa, is shown

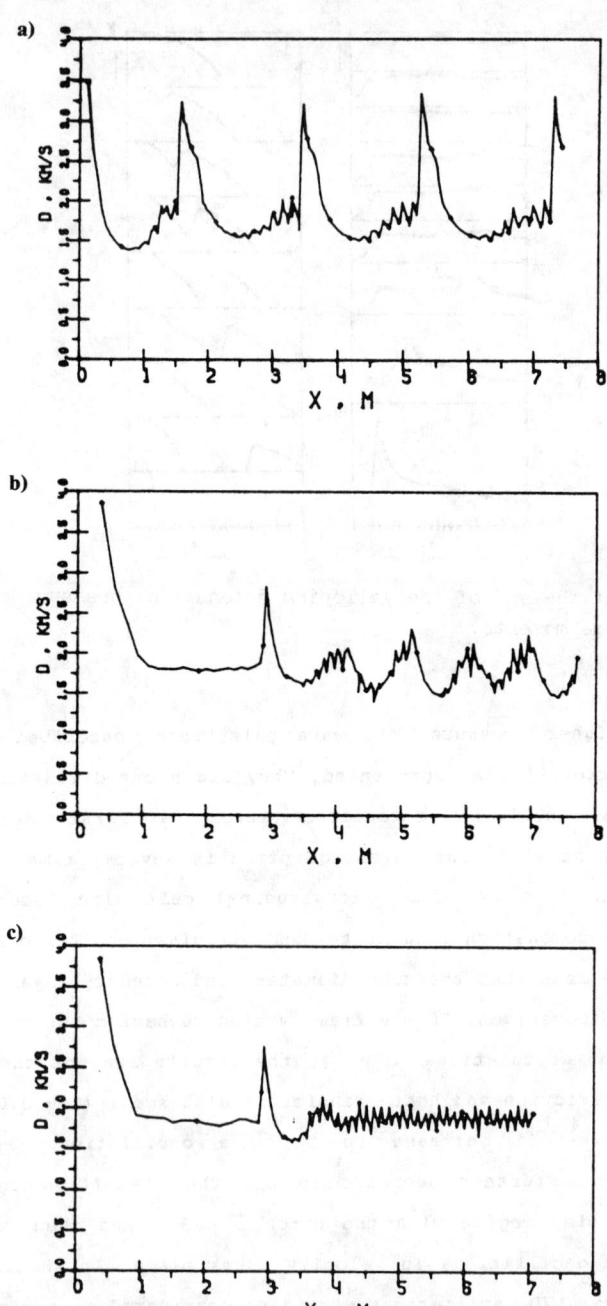

Fig. 4 Transition from the gallop to LVD in $CH_4 + 2O_2$ mixture with the pressure drop p: a) $7 \cdot 10^3$ Pa, b) $1.5 \cdot 10^3$ Pa, and c) $0.75 \cdot 10^3$ Pa; $d = 12.7$ mm.

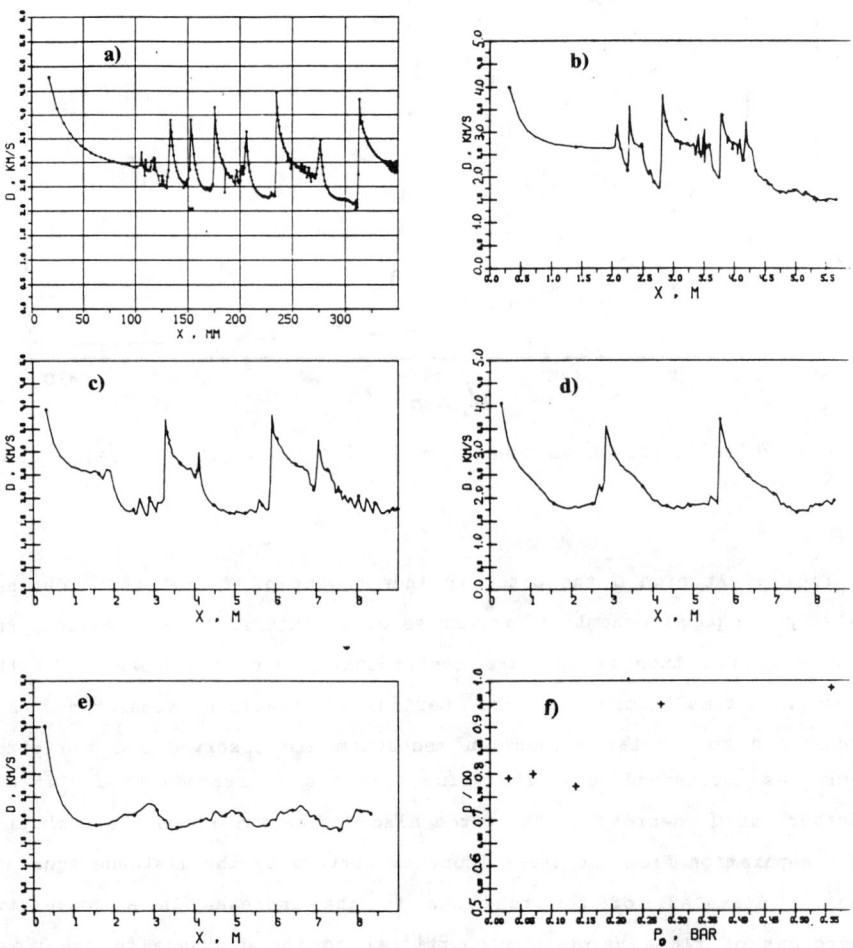

Fig. 5 Wave front velocity along the x coordinate in $2H_2 + O_2$ mixture under varying pressure p: a) $0.56 \cdot 10^5$ Pa, b) $0.28 \cdot 10^5$ Pa, c) $0.14 \cdot 10^5$ Pa, d) $0.67 \cdot 10^5$ Pa, e) $0.03 \cdot 10^5$ Pa, and f) mean velocity deficiency change, $\varepsilon = (D_o - D)/D_o$, with pressure.

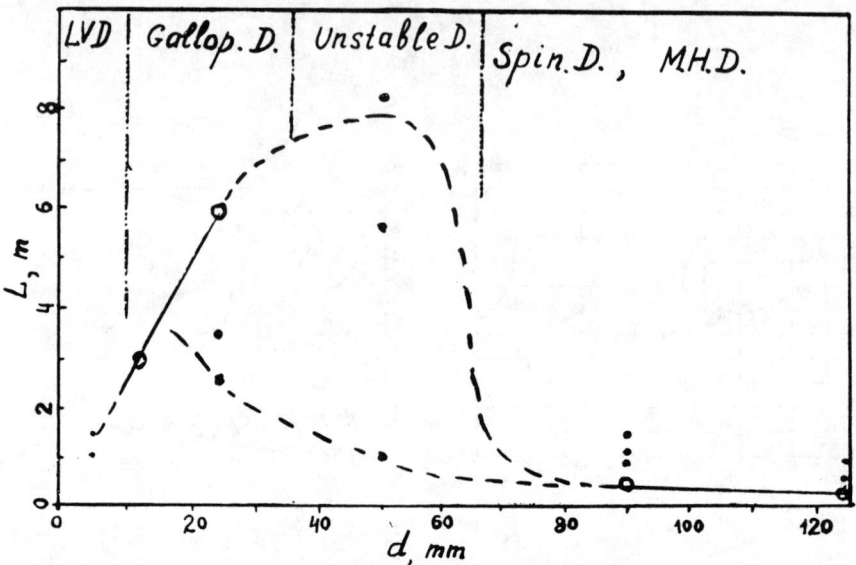

Fig. 6 Pulsation pitch vs tube diameter d for $p = 0.07 \cdot 10^5$ Pa, mixture $2H_2 + O_2$.

in Fig. 6. At high d the pitch is independent of d, and it is defined only by the physicochemical properties of a mixture. As d decreases, the losses on the tube walls are contributive, and $\overline{D}$ decreases. As the induction time increases, the periodical ignition separations and transition to frontal combustion mechanism are observed and the pitch increases correspondingly. The value $L \simeq 6$ m is reached at $d = 25$ mm. Further, as d decreases, the pitch also decreases, since the combution zone separation from the shock front is limited by the distance equal to several diameters of the tube due to the increase in a convective component of flame velocity proportional to the distance to the power 0.8. At $d = 6$ mm the solution is provided by LVD.

The presence of the galloping regime in the $2H_2 + O_2 + 7Ar$ mixture (Fig. 7) was not confirmed by the numerical modeling, in accordance with the experiments and the approximate model developed by Ul'yanitsky.[4] The calculations show either regular oscillations corresponding to the cellular structure (the upper plot) or the LVD regime (the lower plot). In intermediate cases (the middle plot) there are rear irregular pulsations.

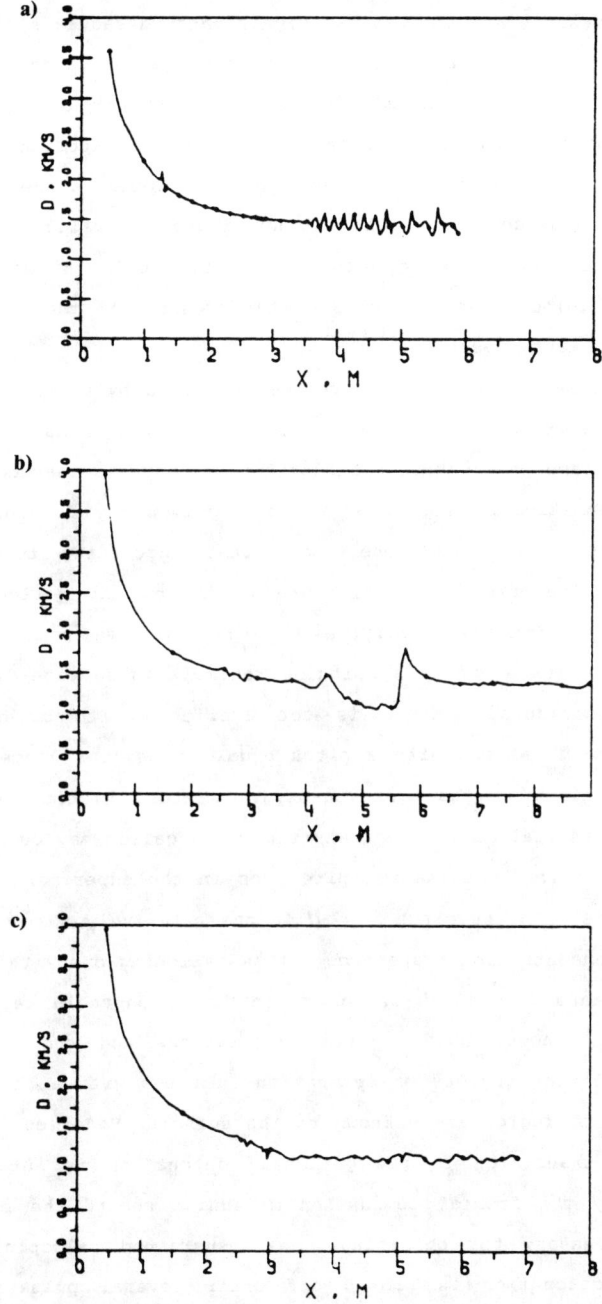

Fig. 7 Detonation calculation in $2H_2 + O_2 + 7A$ mixture for the varying pressure p: a) $0.06 \cdot 10^7$ Pa, b) $0.04 \cdot 10^7$ Pa, c) $0.02 \cdot 10^7$ Pa; tube diameter $d = 12.7$ mm.

This model, however, has a significant disadvantage in that detonation is unlimited, since flame front propagation is not restricted by any conditions, and the mixture is always burnt out completely. Actually, a part of the mixture near the walls does not burn, and at critical pressures, the flame is impossible to propagate due to losses. Because of this disadvantage, the LVD speed and the value of D for a galloping detonation turned out to be overestimated. In addition, the dependence of velocity deficiency ε on the pressure in the low-pressure region is also misrepresented.

For the same reason, the LVD regime, obtained by calculations, may actually be unavailable in some mixtures if, at the decrease of p_0 or d, the limit of flame propagation behind the shock wave is reached earlier. Thus the computational range over which a gallop may develop is much wider than an experimental one. Note that, according to the data obtained by Ul'yanitsky,[4] the pressure range for the gallop is very narrow, and even with small additives of argon this regime is impossible (detonation is broken off in a spinning regime). In our one-dimensional numerical experiments, a spin is not available. It corresponds to longitudinal oscillations with a pitch equal to several diameters of a tube, and a gallop in the hydrogen-oxygen mixture is observed over a wide range of initial data. The transition to a gallop may be identified by an abrupt change in pulsation pitch, as in the experiment where the spinning regime with the pitch $L \simeq 3d$ in the relatively narrow range of pressure or concentration transforms into a galloping one with the pitch $L > 100\ d$. The intermediate region demonstrates unstable regimes with abrupt pitch changes, both in the calculations and the experiment. Regular oscillation regimes of detonation in tubes with a pitch inside the intermediate region are unknown to the authors. Under our numerical modeling the transition to the galloping detonation is automatically accompanied by the frontal combustion mechanism behind the shock. The latter is necessary for obtaining a self-sustained galloping regime: without it, detonation is broken off after several pulsations with different, usually increasing, pitches.

The presented calculation results indicate the necessity of more accurate experimental research of galloping and low-velocity detonation regimes. By making use of this model, it is possible to measure gaseous flame velocities in the high-temperature regions close to autoignition temperatures. At the same time, the results of theoretical calculations of flame velocity can be tested, taking into account in particular autoturbulization effects, which were ignored above.

Conclusion

The following were concluded from the described galloping detonation study:

1) A one-dimensional model of detonation in perfect inviscid gas without taking into account flame front propagation does not allow obtaining the galloping detonation parameters in agreement with the experimental results: a) the pulsation pitch turns out to be several times lower, b) the pulsations themselves are extremely irregular, and c) with an actual value of shear coefficient, the detonation breaks off.

2) Introduction into the model of an efficient flame front velocity, taking into account velocity variation on the tube axis under the action of the atwall viscous boundary layer, allows the parameters of an galloping regime close to experimental ones to be obtained by the calculations.

3) The galloping detonation wave structure at the closing stage of stabilization and low-speed growth is analogous to stationary low-velocity detonation. The shock front is thereby sustained by the flame at a distance of several tube diameters from the shock. The flow regime at this stage is unsteady: when the ignition delays in gas particles before the flame has come ended , its slow acceleration is replaced by a very quick transition to detonation of compressed nonuniformly heated gas by the spontaneous mechanism.

References

1. Mooradian, A. J., and Gordon, W. E., "Gaseous Detonation - Initiation of Detonation," *Journal of Chemical Physics, Vol. 19*, 1951, pp. 1166-1172.

2. Manson, N., Brochet, C., Brossard, J., and Puiol, Y., "Vibration Phenomena and Instability of Self-Sustained Detonations in Gases," *Ninth Symposium (International) on Combustion*, The Combustion Institute, Pittsburgh, PA, 1963, pp. 461-469.

3. Saint-Cloud, J. P., Guerraud, C., Brochet, C., and Manson, N., "Quelques Particularities des Detonations Tres Instables dans les Melanges Gazeux," *Astronautica Acta*, Vol. 17, 1972, pp. 487-498.

4. Ulyanitsky, V. Yu. "Study of Galloping Regime of Gaseous Detonation," *The Physics of Combustion and Explosion (in Russian)*, Vol. 17, No. 1, 1981, pp. 118-124.

5. Borisov, A. A., Gelfand, B. E., Loban, S. A., Mailkov, A. E., and Chomik, S. V., "Study of the detonation limits of fuel-air mixtures in smooth and rough tubes," *Chemical Physics (in Russian)*, Vol. 1, No. 6, 1982, pp. 848-853.

6. Manzhaley, V. I., "Fine Structure of the Head Front of Gaseous Detonation," *The Physics of Combustion and Explosion (in Russian)*, Vol 13, No. 3, 1977, pp. 470-472.

7. Manzhaley, V. I., "Structure of near-limit gas detonation," *International Conference Dedicated to the Centennial of the Paper by V. A. Michelson of the Combustion Theory*, Moscow, Russia, 1990.

8. Manzhaley, V. I., "Gas Detonation Regimes In Capillaries," *13th International Colloquium on Dynamics of Explosions and Reactive Systems, Abstracts and Information*, Nagoya International Center, Nagoya, Japan, 1991, p. 49.

9. Zel'dovich, Ya. B., *Theory of Deflagration and Detonation in Gases*, USSR Academy of Sciences, Moscow-Leningrad, 1944.

10. Reddy, K. V., Fujiwara, T., and Lee, J. H., "Role of the transverse waves in a detonation wave - a stady based on propagation in a porous wall chamber," Memoirs of the Faculty of Engineering, Research Rept. Vol. 40, No. 1, Nagoya University, Nagoya, Yapan, 1988.

11. Levin, V. A., and Markov, V. V., "Detonation Occurrence at Concentrated Energy Supply," *The Physics of Combustion and Explosion (in Russian),*" Vol 11, No. 4, 1975, p. 623.

12. Nikolaev, Y. A., and Fomin, P. A., "On Calculation of Equilibrium Flows of Chemically Reactive Gases," *The Physics of Combustion and Explosion (in Russian)*, Vol. 18, No. 1, 1982, pp. 66-72.

13. Mirels, H., "Shock Tube Test Time Limitation Due to Turbulent - Wall Boundary Layer," *AIAA Journal*, Vol. 2, No. 1, 1964, pp. 114-126.

14. Yagafarov, F. G., and Zaguskin, V. A., (1984) "An economical numerical method for the equations of gas dynamic and elastic-plastisity," Central Institute of Technical Information, No. DR-453, (in Russian).

15. Borisov, A. A., Zamanskii, V. M., Lisjanskii, V. V., Scatchkov, G. I., "Estimation of Detonation Initiation Critical Energy of Gaseous Systems Using Ignition Delays," *Chemical Physics (in Russian)*, Vol. 12, No. 5, 1986, pp.1683-1689.

16. Schetinkov, E. C., *Physics of Gas Combustion*, Moscow, 1965.

17. Nikolaev, Y. A., "Model of Kinetic of Chemical Reactions at a High Temperatures," *The Physics of Combustion and Explosion (in Russian)*, Vol. 14, No. 4, 1978, pp. 73-86.

18. Lewis, B., and Elbe, G., *Combustion, Flames and Explosions of Gases*, Academic Press, New York, 1961.

19. Barnett, H. C., and Hibbard, R. R. (eds.), "Basic Consideration in the Combustion of Hydrocarbon Fuels With Air," Lewis Flight Propulsion Lab. Rept. 1300., Cleveland, OH, 1957.

20. Aksamentov, S. M., Matsukov, D. I., Mitrofanov, V. V., "On Mechanism of the Secondary Blast Waves Origin behind the One-Dimensional Detonation Wave in Gas," *The Physics of Combustion and Explosion (in Russian)*, Vol. 26, No 5, 1990, pp. 135-136.

21. Zel'dovich, Ya. B., Librovich, V. B., Makhviladze, G. M., and Sivashinsky, G. I., "On the Development of Detonation in a Non-Uniformly Preheated Gas," *Astronautika Acta*, Vol. 15, 1970, pp. 313-322.

22. Knystautas, R., Lee J. H., Moen, I., and Wagner H. G., "Direct Initiation of Spherical Detonation by a Hot Turbulent Gas Jet," *17th Symposium (International) on Combustion*, The Combustion Institute, Pittsburg, PA, 1978, pp. 1235-1245.

23. Voitsekhovskii, B. V., Kotov, B. E., Mitrofanov, V. V., and Topchian, M. E., "Optical Investigations of Transverce Detonation Waves," *News of Siberian Division USSA Academy of Sciences (in Russian)*, No. 9, 1958, pp. 74-80.

Experimental Study of the Fine Structure in Spin Detonations

Z. W. Huang* and P. J. Van Tiggelen†
Université Catholique de Louvain, Louvain-la-Neuve, Belgium

Abstract

The spinning detonations characterize the behavior of the detonation phenomena close to the limit in round tubes. The description of the spin detonation has been investigated extensively in the past, particularly in regard to the shock wave configuration. However, by a closer inspection of the spin detonation, a multiple-waves structure behind the transverse shock is noticed, we call this the fine cell structure. The purpose of this paper is to study the relationship between the size of this fine cell structure and initial conditions (pressure, composition, and nature of the fuel). A round aluminum tube with a diameter of 90mm is used in the study. Mean detonation velocity is measured by ionization gauges, and the wave structure is recorded on soot deposited on metal plates inserted in the tube. Among different gaseous mixtures tested in the experiments, the occurrence of the fine structure is found in $2H_2 + O_2$ and $C_2H_2 + 7.5O_2 + 34Ar$ mixtures. The initial pressure ranges in which the spin detonation exists for the two mixtures are 4-15 Torr and 20-40

Copyright © 1992 by the American Institute of Aeronautics and Astronautics, Inc. All rights reserved.
*Research Engineer; on leave from University of Science and Technology of China, Hefei.
†Professor, Laboratoire de Physico-Chimie de La Combustion.

Torr, respectively. It has been shown for both systems that the cell size of the fine structure increases obviously when the initial pressure decreases within the pressure range of spin detonation. Quantitative research shows an approximate linear relation between the logarithm of the fine cell size and the logarithm of the initial pressure.

Introduction

The phenomenon of spin detonation was discovered by Campbell and Woodhead[1] in 1926. Detailed experimental studies conducted by them and by Bone et al.[2] in 1935 confirmed that near the detonation limit, there appears a brightly luminous region on the detonation head which rotates around the axis of the round tube in which the detonation propagates.

After the discovery, various attempts to explain spin detonations have been made by different authors. A flow scheme with a transverse wave was suggested by Voytsekhovskiy[3] on the basis of many experiments in 1957 and the proposed structure was examined comprehensively with the advanced experiments by Voytsekhovskiy et al.,[4] and by Schott.[5] A general pattern of the wave structure confirmed by them is shown in Fig. 1. A leading shock (A_1AA_2) appears in front of the detonation wave (see Fig. 1). The chemical induction zone length behind the leading shock is different from point to point because of the variable strength of the leading shock. The position of the reaction zone (RZ) is depicted by the dotted line on Fig. 1. A transverse detonation front BC is formed, therefore, and propagates along the wall as a spiral line. The triple points A, B, and C in Fig. 1 can leave clear traces on the sooted tube wall, so the soot technique is often used for the study of the wave structure.

According to the measurements and to the two-dimensional analysis by Ref.4 and by Nikolaev et al.[6] the segment BC is an overdriven detonation spreading in the compressed gas in front of it. It was also noticed in some specific mixtures,[5,7-9] that BC has its own internal structure called, in this paper, the fine structure. But, to the best of our knowledge, there is not as yet a quantitative systematic study of the problem. Some preliminary experimental results concerning the fine cell size are presented in this paper as an approach to the study on the BC segment.

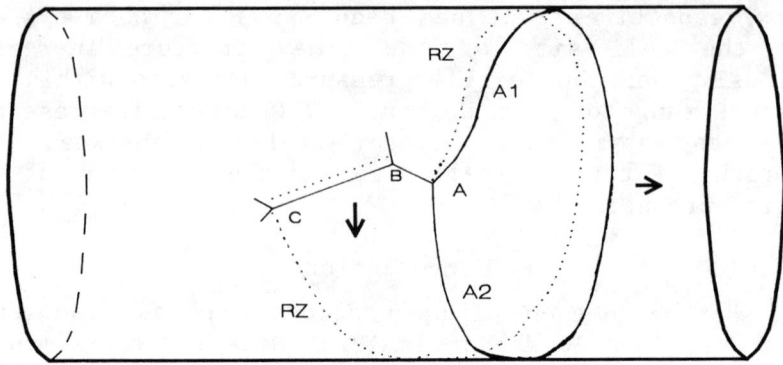

Fig. 1a Structure of the front of a spin detonation in a round tube : ── shock, ━━ tube wall; A, B, C are triple points and RZ is the reaction zone.

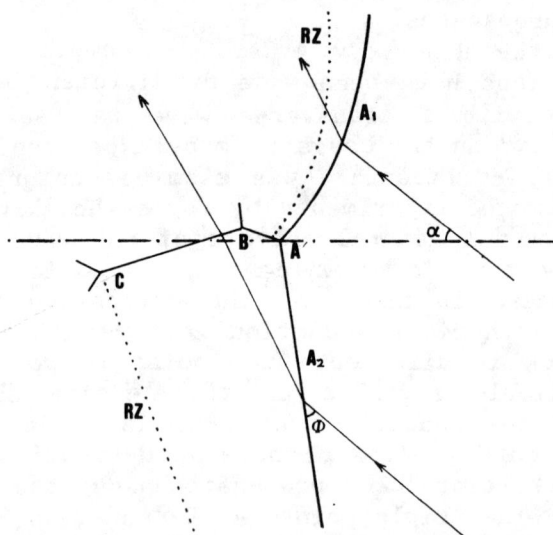

Fig. 1b Details of the transverse wave configuration of a spin detonation :─··─···─ tube axis; ──▶── flow line.

Experimental Set Up

A round tube with an internal diameter of 90 mm is used in the experiments. The total length of the tube is 6 m, including a working section 4 m long and a driving section 2 m. A flange with an orifice (30 mm in diameter) is located between the two sections. A mylar diaphragm covers the hole allowing the use of mixtures with different initial pressure and initial composition in both sections. The mixtures are prepared by employing a dynamic procedure to fill up the tube. Gas flow of each component of the mixture is monitored through sonic nozzles.

The mixture used in the driving section is normally a stoichiometric hydrogen-oxygen mixture without any diluent. To initiate the detonation, a deflagration in the driving gas is, at first, realized by either a weak electric spark, or a hot-wire set up. Then, the deflagration accelerates itself gradually during propagation, and transits abruptly to a detonation before arriving at the end flange of the driving section. This is achieved with the help of obstacles located inside the tube. Breaking the diaphragm and passing through the orifice, the detonation coming from the driving section directly ignites a detonation in the working section filled with the mixture to be studied. Normally, the initial detonation in the working tube is an overdriven one, and has to decay to a stable regime before reaching the test section.

The average detonation velocity is measured by six ionization gauges which are located in the test section 300 mm apart. The rise time of the gauges is less than 2 μs. Platinum gauges were also used to measure the local slope angle of the leading shock. Their rise time is less than 1 μs.

Soot technique is adopted to record the wave structure as printed on the internal wall of the tube. A sooted steel sheet (75 cm long) with a thickness of 0.1 mm is rolled and inserted into the working section near the end part of the detonation tube. Cellular patterns can be found after the detonation has swept over the sooted plate. Figure 2 shows a typical soot record of the spin detonation with fine structure. The picture corresponds to the detonation run in a stoichiometric $CH_4/H_2/O_2$ mixture at 4 Torr with a ratio
$\beta = H_2/(H_2 + CH_4) = 0.8$. The upper part of Fig. 2a is the overall soot trace which is sketched in the lower part. Figure 2b is an enlargement of the central track according to the frame in Fig. 2a.

Experimental Results and Discussion

Before presenting the experimental results, some definitions of the parameters on spin detonation are given in Fig. 3 which shows schematically the trace

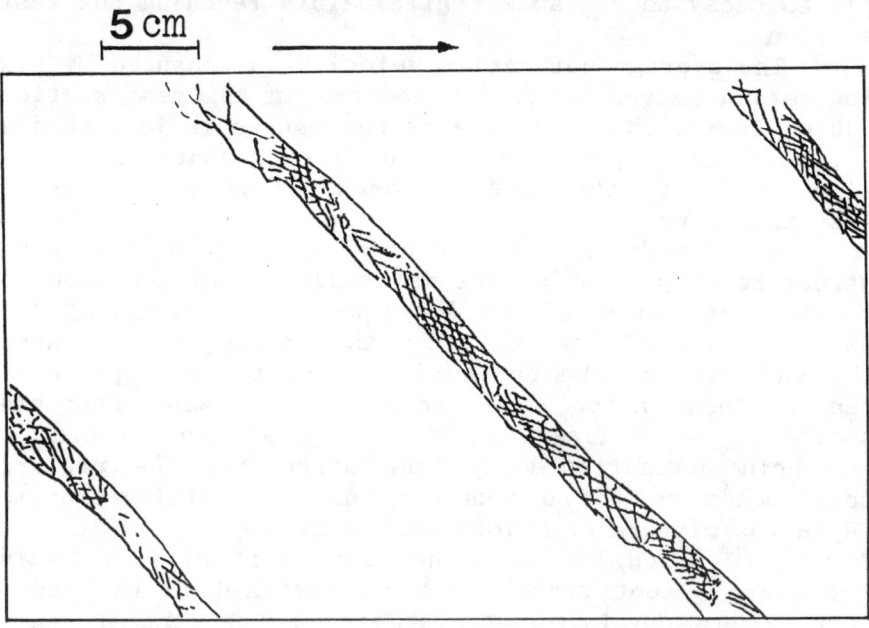

Fig. 2a Soot trace records of a spin detonation in a stoichiometric $CH_4/H_2/O_2$ mixture at an initial pressure with P_o = 4 Torr and ß = 0.8: general view of the spin, sketch in lower part; scale as stated of the drawing and direction of propagation according to the arrow.

FINE STRUCTURE IN SPIN DETONATIONS

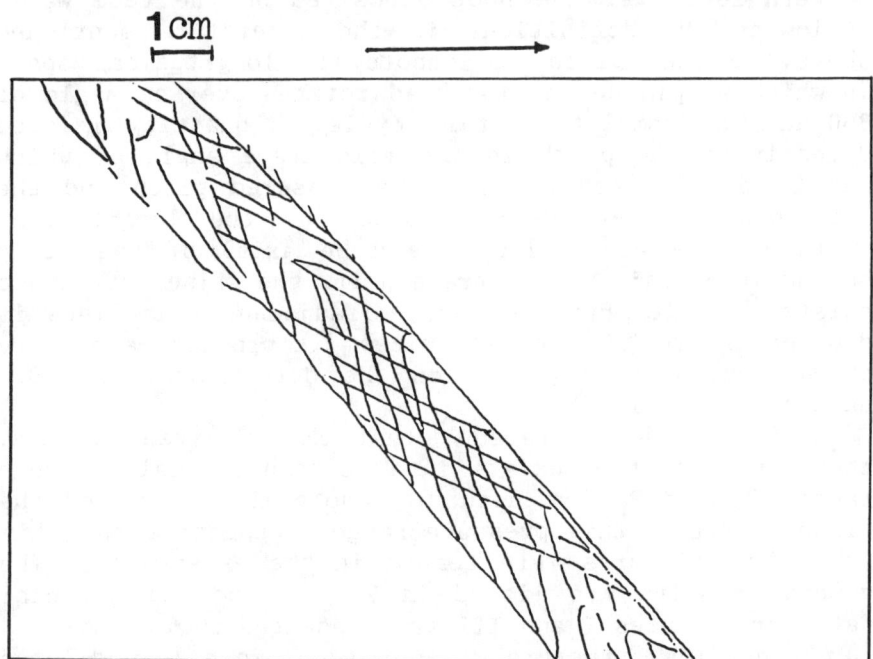

Fig. 2b Soot trace records of a spin detonation in a stoichiometric $CH_4/H_2/O_2$ mixture at an initial pressure with $P_0 = 4$ Torr and $ß = 0.8$: enlargement according to the frame, sketch in lower part; scale as stated of the drawing and direction of propagation according to the arrow.

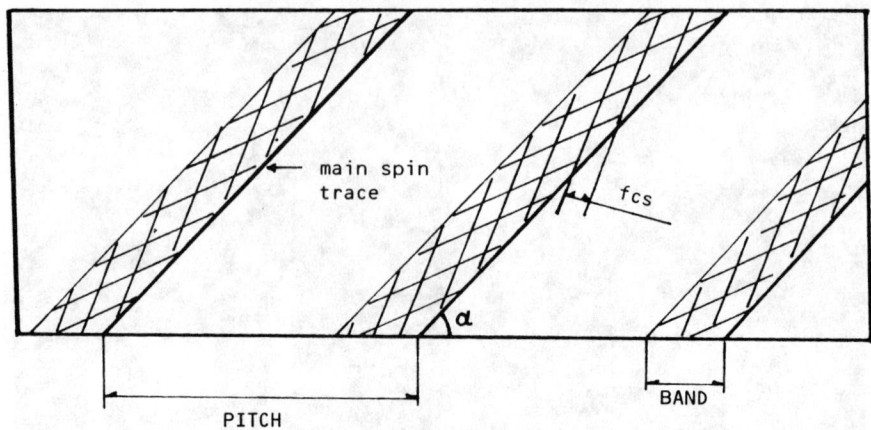

Fig. 3 Definition of parameters characterizing the structure of the spin.

pattern left over the soot deposited on the tube wall. Following the definitions in the literature mentioned above, we use "pitch" to denote the longitudinal space in which a spin detonation head rotates over an angle of 360 deg to complete a full cycle. A quantity related directly to the pitch is the main trace angle α, which can be deduced easily from the measured pitch and the tube diameter. We define "band" as the longitudinal width (as measured along the tube axis) of the strip behind the main spin trace. If the fine structure exists, the distance between two adjacent fine lines is denoted as "fcs" (fine cell size). It provides a measure of the cellular structure on the detonation front BC, marked on Fig. 1.

It was demonstrated[10] that for a given mixture, the spin detonation exists in a certain initial pressure range. P_l and P_h are used to denote the lower and the higher limit of this pressure range, respectively.

Five mixtures were tested in the experiments. The general results are listed in Tab. 1 and some special data for mixtures I and III are presented in Tab. 2.

The fine structure was not found in a lean diluted hydrogen-oxygen mixture ($H_2 + O_2 + 4.67Ar$) and in a stoichiometric (1.8 CO + O_2 + 0.2 H_2 + 3 Ar) mixture. With methane-oxygen mixtures there are fine cells inside the band, and the width of the band changes drastically even during a single cycle of the spin, exhibiting very unsteady features. We also observed in our experiments

Table 1 Investigated mixtures and the average values of spin pressure range, pitch, band width and pitch angle

No	MIXTURE	P_1, Torr	P_h, Torr	Pitch, mm	Band, mm	α, deg
I	$2H_2 + O_2$	8	17	260 ± 10	30 ± 5	47.4
II	$H_2 + O_2 + 4.67Ar$	21	35	265 ± 5	40 ± 5	46.8
III	$C_2H_2 + 7.5O_2 + 34Ar$	20	40	250 ± 10	45 ± 5	48.5
IV	$1.8CO + O_2 + 0.2H_2 + 3Ar$	26	50	280 ± 5	45 ± 5	45.3
V	$CH_4 + 2O_2$	17	25	265 ± 10	--	46.3

Table 2 Spin detonation velocities (experimental and cj values) in H_2-O_2 mixture (I) and in C_2H_2-O_2-Ar mixture (II)

Mixture I ϕ = 31 deg			
P_0, Torr	D_{exp}, m/s	α, deg	D_{cj}, m/s
8	2281	47.9	2582
10	2362	46.3	2601
12	2520	47.4	2616
15	2343	46.9	2628
17	2459	47.4	2634
Mixture III ϕ = 30 deg			
P_0, Torr	D_{exp}, m/s	α, deg	D_{cj}, m/s
20	1179	48.5	1401
25	1190	48.8	1409
30	1198	48.5	1415
35	1232	48.5	1421
40	1246	49.2	1426

with mixtures containing methane some irregular structure attached on the other side of the main spin trace. This corresponds to instabilities related to the segment AA_1 of the leading shock. This behavior also was mentioned by Manzhalei and Mitrofanov.[9]

With $2H_2 + O_2$ and $C_2H_2 + 7.5O_2 + 34Ar$, the regular fine structures of spin detonation were observed behind the main spin trace. The dependences of the fcs with respect to the initial pressure for those two last mixtures are presented in Figs. 4 and 5. It can be seen

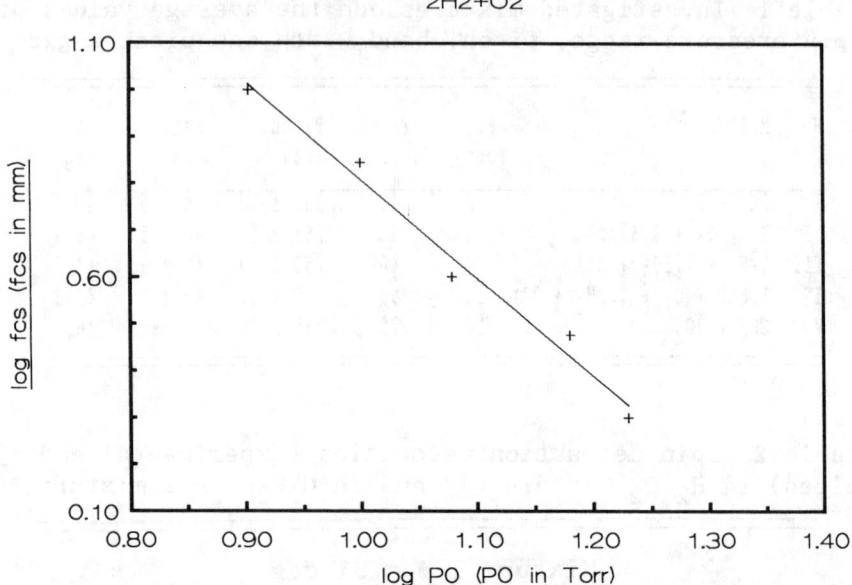

Fig. 4 Pressure dependence of the fine cell structure of spin detonations for stoichiometric undiluted hydrogen-oxygen mixture: initial pressure P_o in Torr.

from the data that the logarithm of fine cell size obeys a linear relationship when plotted vs the logarithm of the initial pressure P_o.

If we denote the temperature and the pressure behind the shock AA_2 (in Fig. 1) as T' and P', we get, according to shock relations, the following equation :

$$\frac{T'}{T_o} = \frac{[\gamma M^2 - (\gamma-1)/2][(\gamma-1)M^2/2 +1]}{[(\gamma+1)/2]^2 M^2} \quad (1)$$

$$\frac{P'}{P_o} = \frac{2\gamma M^2 - (\gamma- 1)}{\gamma + 1} \quad (2)$$

where M is the normal Mach number of the shock, and γ the specific heat ratio of the gas. It can be seen from Fig. 1b that the incident velocity of fluid particle to shock AA_2 is $D/\cos \alpha$, because its component along the direction of the tube axis is D. Considering, at same time, that the incident angle of fluid particle

FINE STRUCTURE IN SPIN DETONATIONS

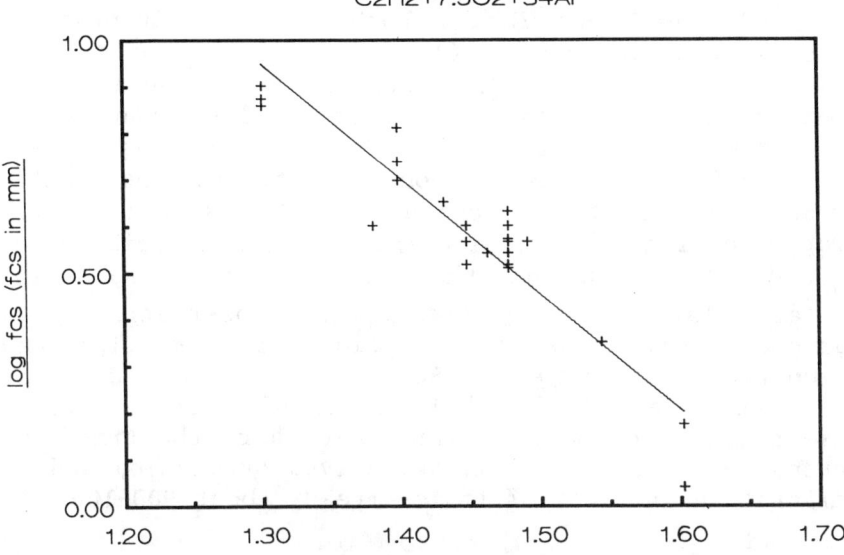

Fig. 5 Pressure dependence of the fine cell structure of spin detonations for lean argon diluted acetylene-oxygen mixture (equivalence ratio = 0.33 and percentage diluent = 80%), initial pressure P_0 in Torr.

to AA_2 is equal to ϕ, we can write the normal incident velocity of fluid particle to AA_2 as $D \sin \phi / \cos \alpha$. Thus, the normal Mach number M can be obtained by the following relation :

$$M = \frac{D \sin \phi}{C_0 \cos \alpha} \quad (3)$$

Here D is the experimental average detonation velocity and C_0 the speed of sound of the fresh gas. The value of ϕ is measured by platinum gauges in experiments. To within the accuracy ± 2 deg we have not found any variation of ϕ when changing the initial pressure in the spin detonation range, ϕ = 30 deg. Therefore, we consider it a constant. It can be seen from Tab. 2 that α can also be treated as a constant. Therefore, the change of the normal Mach number M is mainly from the variation of the average detonation velocity D.

From the data in Tab. 2, it comes out that the variation of D, when the initial pressure changes from

P_l to P_h, is about 7% for both mixtures. Substituting this value into formulas (1) and (2), we found that the values of P'/P_0 and T'/T_0 change for about 9%; so, if we consider M as a constant, the error is less than 10%.

Now let us treat M as a constant and substitute it into Eq.(2). We see at once that, in the initial pressure range of spin detonation, P' is approximately proportional to P_0. Considering the relationship between P_0 and fcs, we conclude that log (P') will also be in a linear relation to log (fcs). This observation is in agreement with the one obtained for multiple-head detonations as studied by Refs. 4 and 11 and discussed by Libouton et al.[12,13] in terms of a characteristic time related to the cell size. But, here the transverse detonation front BC is an overdriven detonation and the temperature in front of it is already about 900-1000 K.

Acknowledgments

The authors acknowledge the financial support of the Fonds de la Recherche Fondamentale et Collective (F.R.F.C.) under Contract 2.9006.88. Huang Z.W. is also indebted for a doctoral fellowship from Nobel Explosifs Belgium.

References

[1] Campbell, C., and Woodhead, D.W., "The Ignition of Gases by an Explosive Wave." Pt. I, Carbon Monoxide and Hydrogen Mixtures, Journal of the Chemical Society, Vol. 130, 1927, pp. 1572-1578.

[2] Bone, W.A., Fraser, R.P., and Wheeler, W.H., "A Photographic Investigation of Flame Movements in Gaseous Explosive." Philosophical Transactions of the Royal Society of London, Vol. A 235, 1935, pp. 29-78.

[3] Voytsekhovskiy, B.V., "Structure of Spin Detonation." Doklady Akademii Nauk SSSR, Vol. 114, No. 4, 1957, pp. 717-731.

[4] Voytsekhovskiy, B.V., Mitrofanov, V.V., and Topchiyan, M. Y., "Structure of Detonation Front in Gases," Novosibirsk : IZD-VO Sibirsk. Otdel. Akademii Nauk SSSR, 1963.

[5] Schott, G.L., "Observation of the Structure of Spinning Detonation." Physics of Fluids, Vol. 8, 1965, pp. 850-865.

[6] Nikolaev, Y.A., Topchiyan, M.E., and Ul'yanitskii, V.Y., "Experimental and Theoretical Study of the Spin Detonation Configuration." *Fizika Gorenia i Vzryva*, Vol. 14, No. 6, 1978, pp. 106-109.

[7] Denisov, Y.N., and Troshin, Y.K., "Pulsating and Spinning Detonation of Gaseous Mixtures in Tubes." *Doklady Akademii Nauk SSSR*, Vol. 125, 1959, pp.110-113.

[8] Denisov, Y.N., and Troshin, Y.K., "The Fine Structure of Spinning Detonation." *Combustion and Flame*, Vol. 16, 1971, pp. 141-145.

[9] Manzhalei, V.I., and Mitrofanov, V.V., "The Stability of Detonation Shock Waves with a Spinning Configuration." *Fizika Gorenia i Vzryva*, Vol. 9, No. 5, 1973, pp. 703-710.

[10] Gordon, W.E., Mooradian, A.J., and Harper, S.A., "Limit and Spin Effects in Hydrogen-Oxygen Detonations." *Seventh Symposium (International) on Combustion*, Butterworths, London, 1959, pp. 752-759.

[11] Strehlow, R.A., and Engel, C.D., "Transverse Waves in Detonation." *AIAA Journal*, Vol. 7, No. 3, 1969, pp. 492-496.

[12] Libouton, J.C., Dormal, M., and Van Tiggelen, P.J., "The Role of Chemical Kinetics on Structure of Detonation Waves." *Fifteenth Symposium (International) on Combustion*. The Combustion Institute, Pittsburgh, PA, 1975, pp. 79-86.

[13] Libouton, J.C., and Van Tiggelen, P.J., "Influence of the Composition of the Gaseous Mixture on the Structure of Detonation Waves," *Acta Astronautica*, Vol. 3, No. 9-10, 1976, pp. 759-769

Influence of Fluorocarbons on H_2O_2Ar Detonation: Experiments and Modeling

M. H. Lefebvre,* E. Nzeyimana,† and P. J. Van Tiggelen‡
Université Catholique de Louvain, Louvain-la-Neuve, Belgium

Abstract

The influence of CF_4 and CF_3H on gaseous detonations propagating in $H_2/O_2/Ar$ mixtures has been investigated. Detonations velocities and cell sizes have been recorded and measured. The reference mixture was a stoichiometric H_2/O_2 mixture diluted with 50% argon at an initial pressure of 200 Torr. From experimental data, an unexpected behavior of detonation velocity appears when the diluent is modified from pure argon to mixtures containing variable amounts of fluorocarbons. The Chapman-Jouguet model fails to explain the observed behavior. An explanation based on a detailed chemical kinetic mechanism of the heat release is suggested which allows most of the experimental results to be accounted for.

Introduction

Although halocarbons are typical flame inhibitors,[1] their influence on gaseous detonations is quite complex. They can be viewed as inhibitors in carbon monoxide-hydrogen-oxygen mixtures[2] and in some acetylene-oxygen mixtures.[3] On the other hand, CF_4 definitely acts as a promotor in hydrogen-oxygen-argon mixtures as demonstrated by Nzeyimana and Van Tiggelen.[4] From those experimental data, a slight increase of the detonation velocity is noticeable when argon diluent is replaced gradually by variable amounts of CF_4. The regularity of the cell structure is vanishing and the CJ calculation does not allow an explanation of the observed data. D_{CJ} remains, indeed, constant up to 17% of CF_4 in the total initial mixture, contrasting with the increase of the experimental detonation velocities D_{exp}.

Copyright © 1992 by the American Institute of Aeronautics and Astronautics, Inc. All rights reserved.
*Senior Research Engineer, Royal Military Academy, Belgium.
†Research Chemist, Laboratoire de Physico-Chimie de la Combustion
‡Professor, Laboratoire de Physico-Chimie de la Combustion.

The purpose of this paper is to confirm the unexpected behavior of CF_4 using different tube geometries and to extend the study to another fluorocarbon compound, namely, CF_3H. An original chemical kinetic mechanism is developed that indicates the role of the induction time, the pulse time (the time elapsed during the heat release), and the specific energy release.

Apparatus, Experimental Data, and CJ Model

The apparatus used in the present study has been described previously by Libouton et al.[2] and Vandermeiren et al.[5] Two detonation tubes of different cross sections have been used: 1) a rectangular one of 9.2x3.2 cm^2 and 2) one with a square cross section of 3.2x3.2 cm^2.

The mixing was achieved during the filling of the tube and the composition of the gas flow was controlled by employing shocked flow conditions through nozzles for each individual gas.

The experimental conditions for the investigated $H_2/O_2/Ar/CF_4$ or CF_3H systems are as follows: stoichiometric $H_2/O_2/Ar$ mixtures ($\phi = 1$) with a constant dilution ($\alpha = 0.5$), at the initial pressure of 200 Torr. The amount of CF_4 or CF_3H in the fresh gases is varied in such a way that the either Ar-CF_4 or CF_3H amounts remain constant, i.e., equal to 50% of the total content ot the mixture. However, it must be kept in mind that the added CF_3H quantities modify the overall equivalence ratio ϕ of the mixture, as stated in Table 1. Detonation cells size has been recorded from the traces left over glass plates covered with soot and located 4 m behind the diaphragm. The detonation velocity was measured by ionization gauges.

The Tables 1a and 1b give the data for the experiments run in the tube with a rectangular cross section: experimental detonation velocities D_{exp}, experimental Mach numbers M_{exp}, cell lengths L; as well as some computed parameters: CJ detonation velocities D_{CJ}, CJ Mach numbers M_{CJ} and the heat released at CJ condition Q_{CJ}. Values are computed by taking into account the following species in the burnt gases: H, H_2, C(g), O, OH, H_2O, F, HF, CO, CF, O_2, Ar, CO_2, CF_2, CF_2O, CF_3, CF_3H or CF_4. Calculations with solid carbon have also been made using the code Fortran SIN provided by Mader,[6] but it did not influence the final results, the amount of solid carbon remained quite negligible at this pressure range.

Quite unexpectedly, the experimental study exhibits a slight increase of the detonation velocity when we add a few percents of CF_4. Furthermore, for mixtures with either 20% or 25% CF_4 added, the experimental detonation velocity D_{exp} is larger than D_{CJ} (Fig. 1), irrespective of the tube geometry.

For mixtures containing CF_3H (Fig. 2), the general trend seems to be more in agreement with the CJ model. But, the limits of detonation are reached at 14% in the square tube and at 28% in the rectangular one, in any case, well before the large decrease of D_{CJ} occuring at 33% of CF_3H in the total mixture. Close to the limit we observe a drastic increase of the cell size (Fig. 3). When the mixtures contain 5% or more of fluorocarbon, the regularity of the cell

Table 1a CF_4 experimental data and CJ values as measured in 9.2x3.2 cm² tube

%	D_{exp}, m/s	M_{exp}	L, mm	D_{CJ}, m/s	M_{CJ}	Q_{CJ}, kJ/kg
0	1830	4.84	10.7	1874	4.95	1741
5	1859	5.23	11.1	1875	5.28	1780
7.5	1861	5.39	10.9	1875	5.43	1797
10	1854	5.52	11.6	1874	5.58	1811
15	1828	5.72	14.3	1872	5.86	1835
20	1798	5.70	14.6	1697	5.56	1502
25	1604	5.48	48.5	1574	5.38	1483
30	--	--	--	1488	5.29	1529
35	--	--	--	1414	5.21	1567
40	--	--	--	1346	5.13	1590

Table 1b CF_3H experimental data and CJ values as measured in 9.2 x 3.2 cm² tube

%	ϕ	D_{exp}, m/s	M_{exp}	L, mm	D_{CJ}, m/s	M_{CJ}	Q_{CJ}, kJ/kg
0	1.00	1830	4.84	10.6	1874	4.95	1741
2.5	1.07	1863	5.03	9.4	1895	5.12	1808
5	1.17	1885	5.22	7.7	1913	5.28	1872
7.5	1.22	1898	5.34	7.6	1930	5.43	1928
10	1.26	1922	5.51	6.2	1945	5.58	1979
15	1.45	1928	5.73	10.9	1970	5.85	2064
20	1.56	1924	5.92	13.1	1987	6.11	2199
25	1.71	1925	6.11	27.6	1996	6.33	2148
28	1.84	1924	6.21	--	1996	6.44	2144
30	1.85	--	--	--	1993	6.51	2132
35	2.04	--	--	--	1916	6.43	1996
40	2.15	--	--	--	1736	5.98	1974

structure can be classified as "poor", i.e., slightly irregular.[7] We can already conclude that the CJ approach does not describe the effects of CF_4 and CF_3H added to the original mixture over the whole composition range and, therefore, we will suggest an explanation based on a chemical kinetic point of view.

Description of the Kinetic Model and Results

The chemical kinetics scheme used here consists of about 16 elementary reactions relating 14 species: H_2, O_2, OH, H_2O, O, H, Ar, CF_4 (or CF_3H),

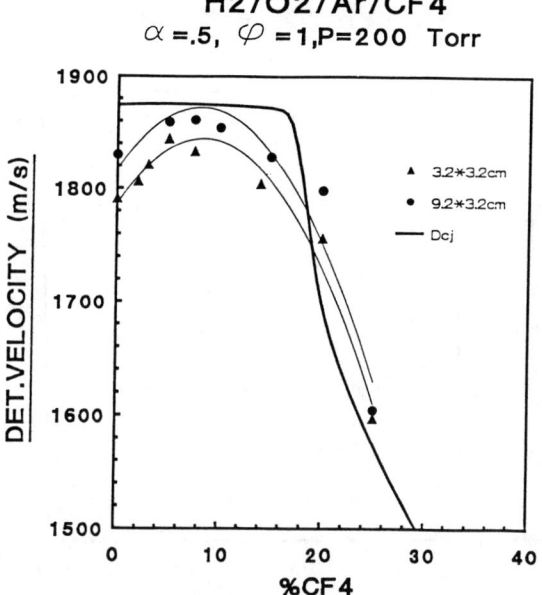

Fig. 1 Comparison of values of D_{CJ} and experimental velocities in both tube geometries for $H_2/O_2/Ar/CF_4$ mixtures vs the CF_4 percentage.

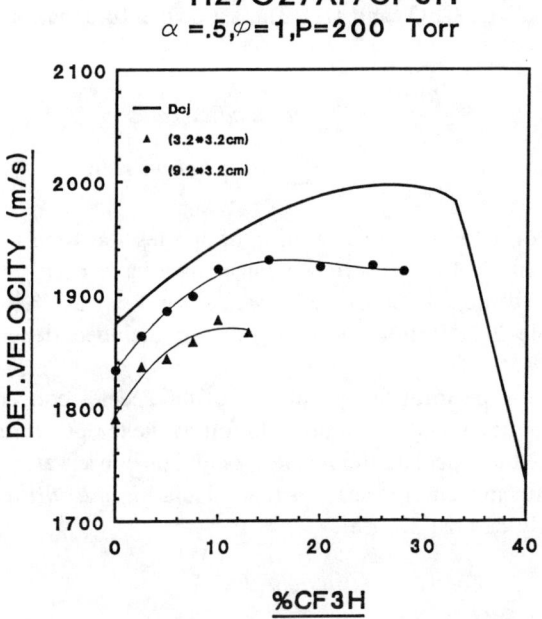

Fig. 2 Comparison of values of D_{CJ} and experimental velocities in both tube geometries for $H_2/O_2/Ar/CF_3H$ mixtures vs the CF_3H percentage.

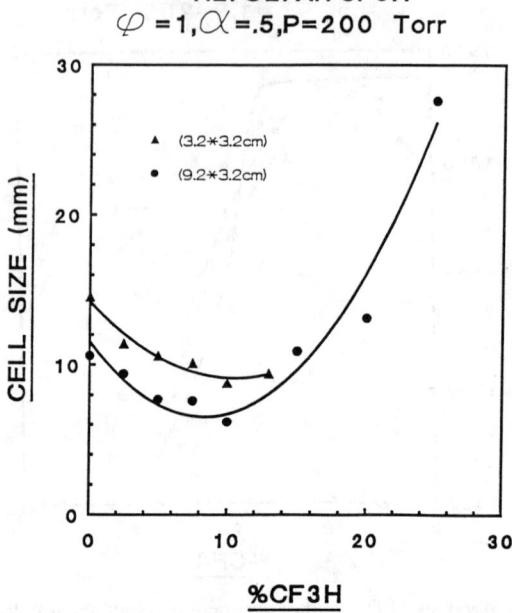

Fig. 3 Cell length vs CF_3H percentage at variable tube geometries in $H_2/O_2/Ar/CF_3H$.

CF_3, CO, CO_2, CF_2O, CFO, and HF. The set of the 14 equations for reaction rates is written as

$$\frac{d[i(t)]}{dt} = \sum \text{production rates} - \sum \text{consumption rates} \quad (1)$$

where [i(t)] in mol/m³ is the concentration of species i at time t. The reaction scheme is given in Table 2. Thermochemical data have been taken from the JANAF tables[8] and the rate constants expressed as $k_i = AT^m \exp[E/RT]$ are collected in Table 2. Reactions 12a or 12b are included depending on the mixture studied.

The code used to perform the calculation of the species concentrations also solves the conservation equations in order to follow the temperature profile. The temperature is used to calculate the reaction rate. The conservation equation for mass, momentum, and energy used in the calculation are written for a one-dimensional steady detonation, i.e.,

$$\rho u = \rho_1 u_1 \quad (2)$$

$$P + \rho u^2 = P_1 + \rho_1 u_1^2 \quad (3)$$

Table 2 Kinetical mechanism

No.	Reaction	E kcal/mol	A mol⁻¹cm⁻³s⁻¹	m	Ref
1	$H_2 + O_2 \rightarrow 2\ OH$	40	1.00E+11		9
2	$H + O_2 \rightarrow OH + O$	16.8	2.24E+14		9
3	$OH + O \rightarrow O_2 + H$	0	1.13E+13		9
4	$O + H_2 \rightarrow OH + H$	9.45	1.74E+13		9
5	$OH + H \rightarrow O + H_2$	7.3	7.33E+12		9
6	$OH + H_2 \rightarrow H_2O + H$	5.15	2.19E+13		9
7	$H_2O + H \rightarrow OH + H_2$	20.10	8.41E+13		9
8	$H_2O + O \rightarrow OH + OH$	18.0	5.75E+13		9
9	$OH + OH \rightarrow H_2O + O$	0.78	5.75E+12		9
10	$H + OH + M \rightarrow H_2O + M$	0	1.17E+17		9
11	$H_2O + M \rightarrow H + OH + M$	105.0	2.20E+16		9
12a	$CF_3H + H \rightarrow CF_3 + H_2$	12.7	2.40E+14		10
12b	$CF_4 + H \rightarrow CF_3 + HF$	42.0	2.00E+14		11
13	$CF_3 + OH \rightarrow CF_2O + HF$	0	1.00E+13		a
14	$CF_2O + H \rightarrow CFO + HF$	25.0	1.00E+14		a
15	$CFO + H \rightarrow CO + HF$	0	1.00E+13		a
16	$CO + OH \rightarrow CO_2 + H$	-0.77	1.29E+07	1.3	12

[a] estimated

$$\frac{u_1^2 - u_2^2}{2} = \sum n_i \int_{T_1}^{T} C_{pi} dT - Q \tag{4}$$

The ideal gas law is assumed so that

$$\frac{P}{\rho} = \sum n_i\ RT \tag{5}$$

The quantities ρ, u, P, T, C_{pi}, and Q are the specific mass (kg/m³), gas velocity with respect to the shock, pressure, temperature, heat capacity (J/mol K) of species i, and specific heat release, respectively. The n_i are the specific composition (mol/kg). The subscript 1 refers to the initial condition. Note that the heat content is calculated as a quantity depending on temperature and composition. The heat capacities C_{pi} are fit to an eight-degree polynomial by the method of least squares from the JANAF data.[8] The initial velocity u_1 is equal to the experimental detonation velocity D_{exp}. The subroutines of the computation program are provided in Fig. 4. The computation stops when the calculated Mach number is equal to 1.0 indicating that a pseudo-Chapman-Jouguet plane is reached. At that time, indeed, a sonic condition is attained, but the composition has not yet reached the full equilibrium, meaning that the CJ condition is not fulfilled.

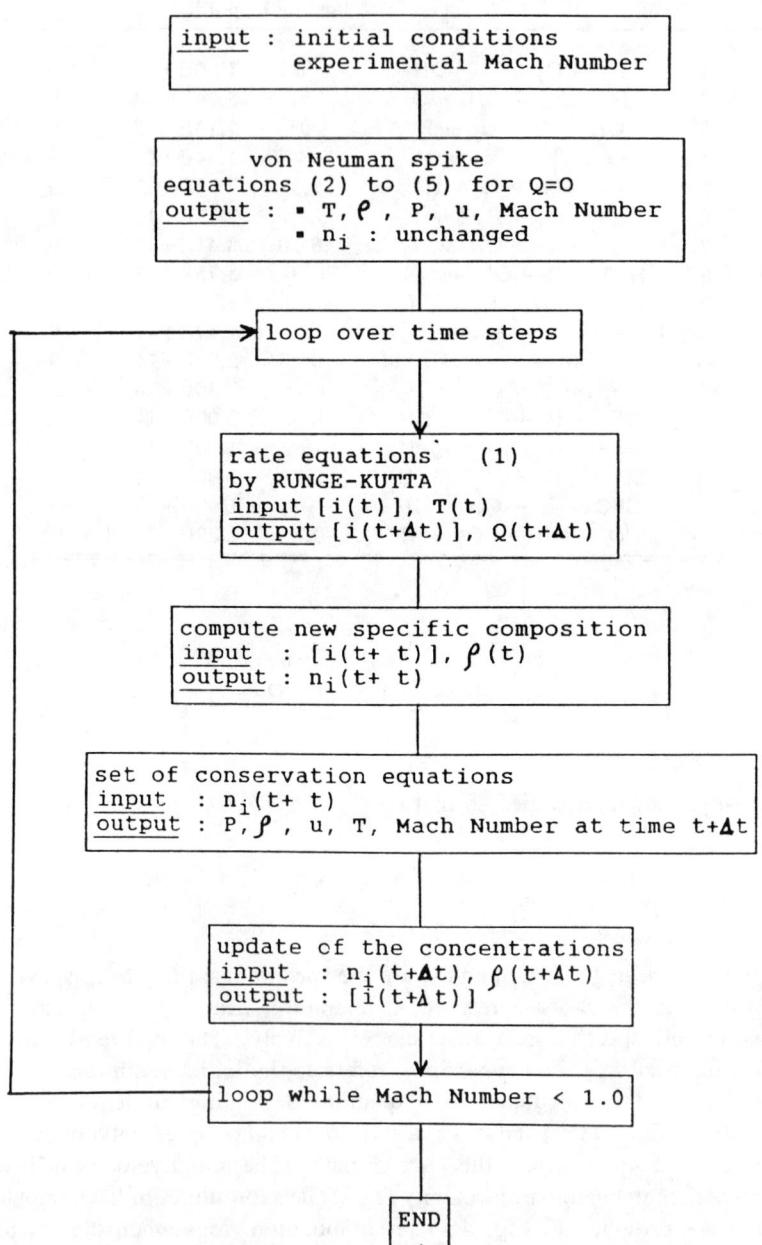

Fig. 4 Flow chart of the computation program.

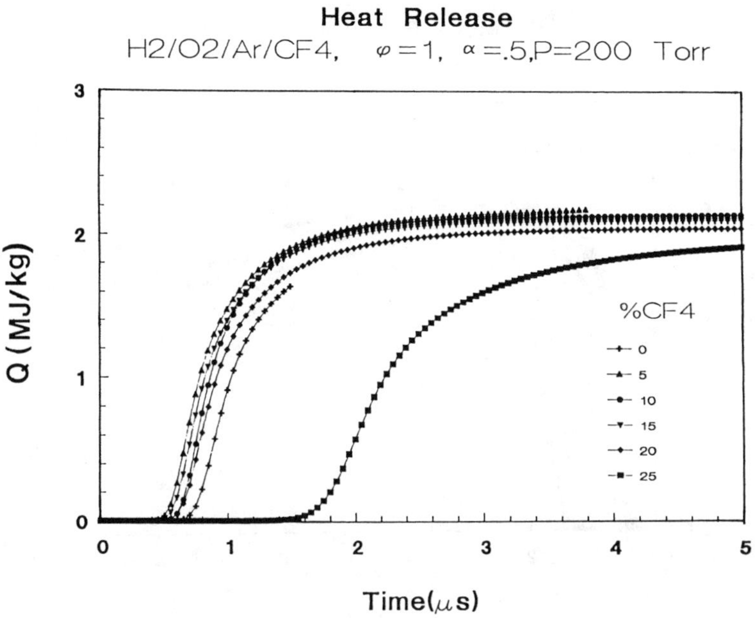

INFLUENCE OF FLUOROCARBONS 151

Heat Release
H2/O2/Ar/CF4, $\varphi=1$, $\alpha=.5$, P=200 Torr

Fig. 5 Heat release curves for CF_4 mixtures vs time: + 0%, ▲ 5%, ● 10%, ▼ 15%, ◆ 20%, ■ 25%.

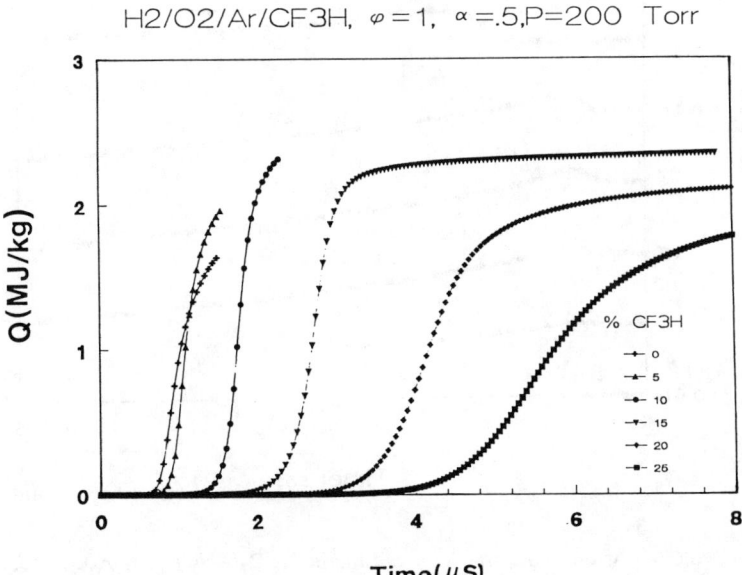

Fig. 6 Heat release curves for CF_3H mixtures vs time: + 0%, ▲ 5%, ● 10%, ▼ 15%, ◆ 20%, ■ 25%.

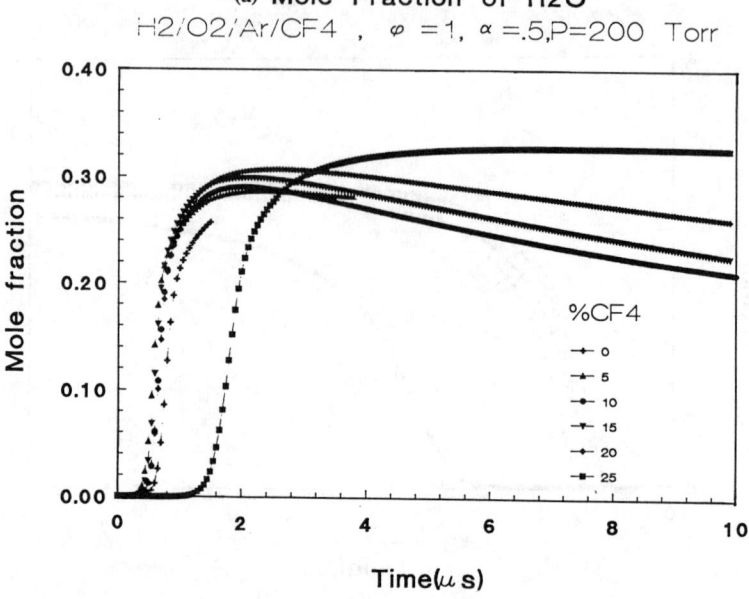

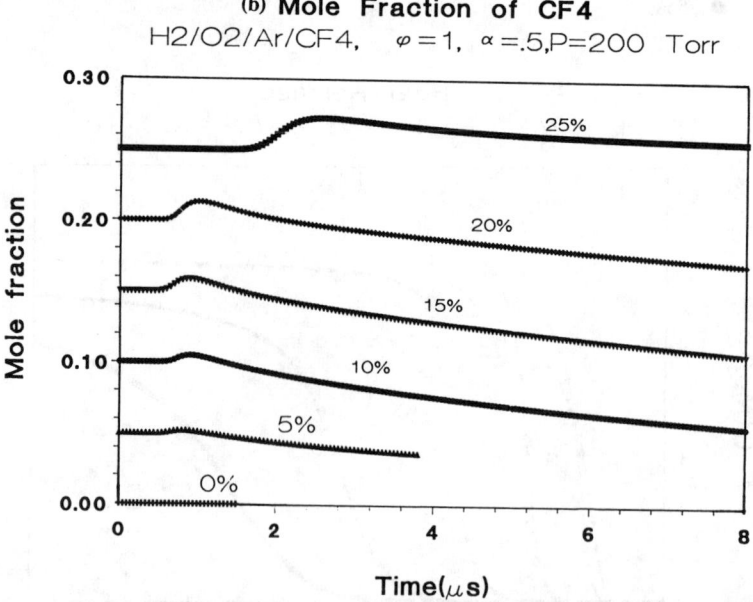

Fig. 7 Evolution of the mole fractions vs time in $H_2/O_2/Ar/CF_4$: a) H_2O; b) CF_4; c) HF; d) CF_2O.

INFLUENCE OF FLUOROCARBONS 153

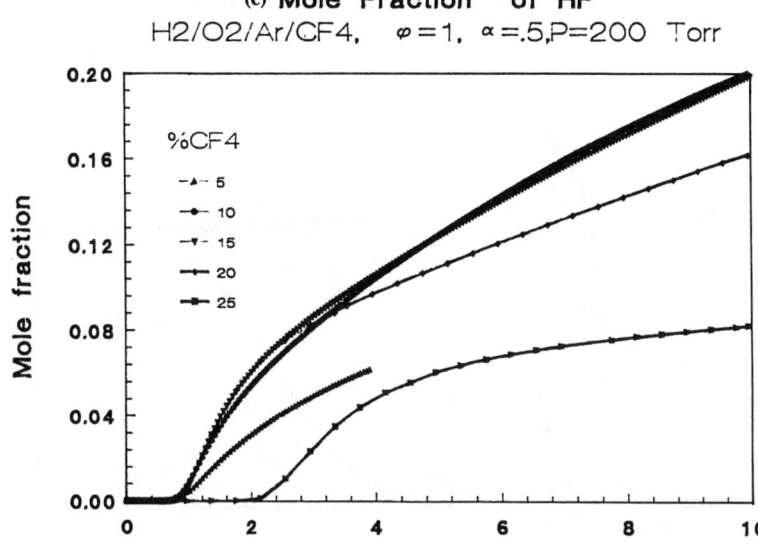

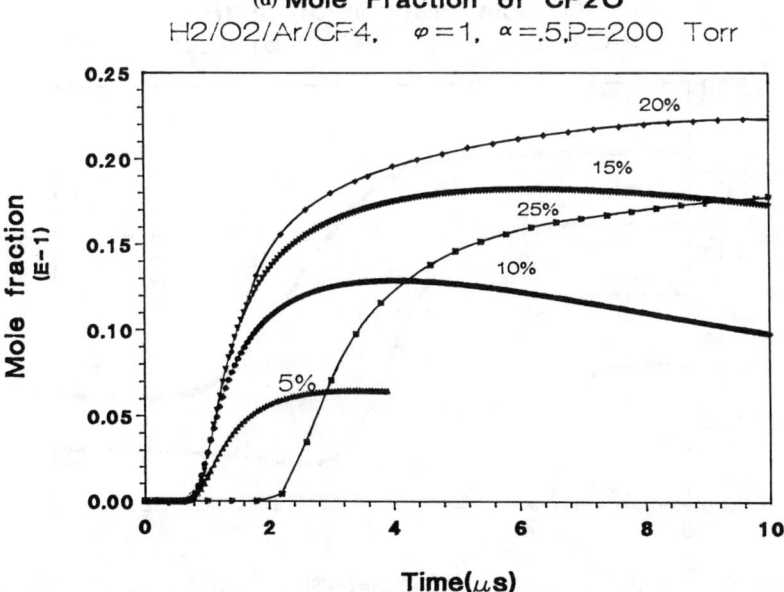

Fig. 7 (continued) Evolution of the mole fractions vs time in $H_2/O_2/Ar/CF_4$: a) H_2O; b) CF_4; c) HF; d) CF_2O.

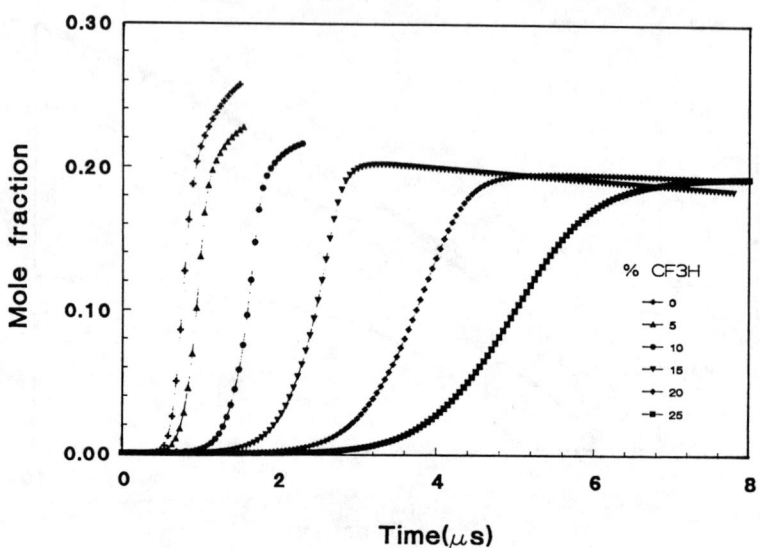

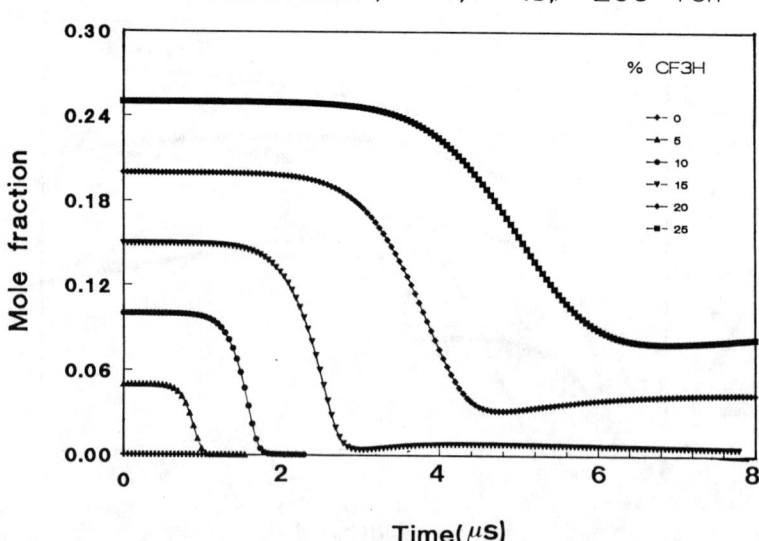

Fig. 8 Evolution of the mole fractions vs time in $H_2/O_2/Ar/CF_3H$: a) H_2O; b) CF_3H; c) HF; d) CF_2O.

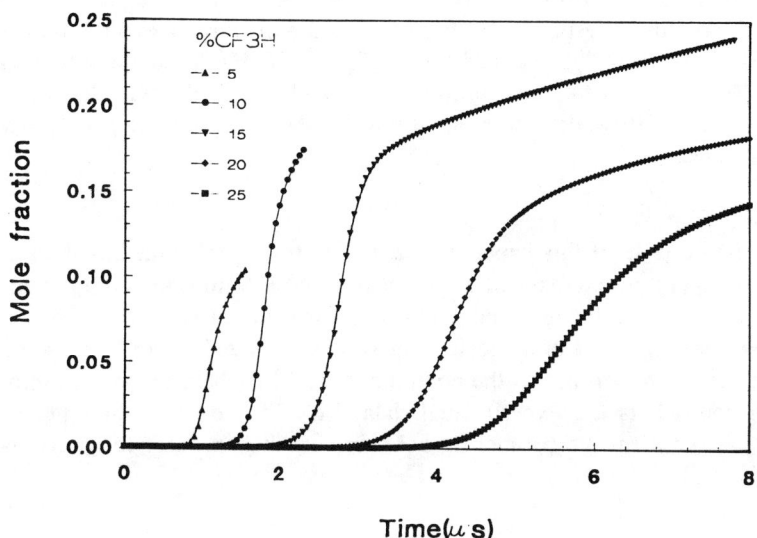

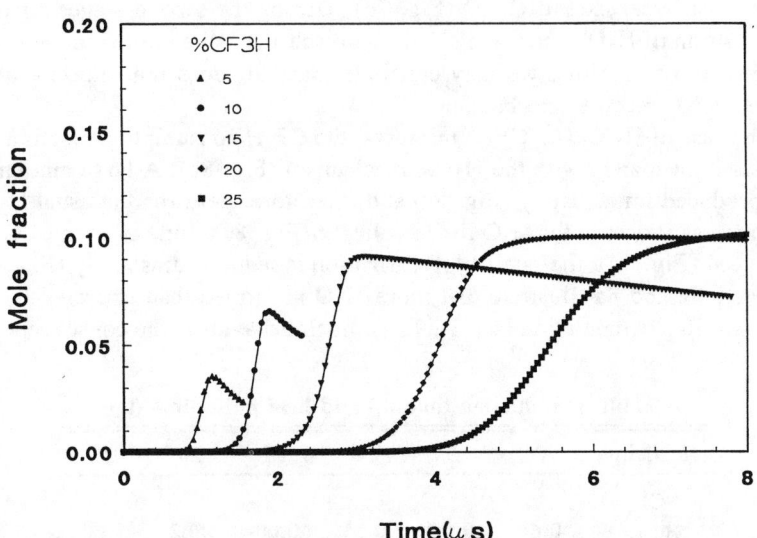

Fig. 8 (continued) Evolution of the mole fractions vs time in $H_2/O_2/Ar/CF_3H$: a) H_2O; b) CF_3H; c) HF; d) CF_2O.

The performed calculations allow the evolution of the physical parameters (heat release, P, T) to be followed as well as the composition of each individual species. In Figs. 5 and 6 the curves of heat release vs time are plotted for CF_4 and CF_3H mixtures, respectively. Figures 7 and 8 exhibit the evolution of some mole fractions, namely, the additive (CF_4 or CF_3H), a stable intermediate product (CF_2O), and the final products (HF and H_2O). We will discuss three characteristics of this calculations successively: the induction time, composition, and heat release.

Induction Time

For the purpose of this paper, the induction time t_i is defined as the time at which a temperature increase of 10 K above its Von Neumann value is noticed. Table 3 shows the calculated induction time for some mixtures.

The inhibiting effect of a species can be characterized by an increase of the induction time. According to the computations, CF_3H behaves as an inhibitor, but this contradicts the experimental data. For CF_4, on the other hand, the calculations confirm the promoting behavior of CF_4, at least for percentages lower than 20%.

Composition

For the $H_2/O_2/Ar/CF_4$ mixtures, we observed a two-stage reaction. First, the H_2O_2 mechanism produced H_2O (Fig. 7a) just as if CF_4 is not present. This first stage is followed by a slow decomposition of CF_4 (Fig. 7b) producing HF (Fig. 7c) and, to a lesser extent, CF_2O (Fig. 7d). During the second stage, a partial decomposition of H_2O occurs (Fig. 7a). To reach the full equilibrium, we need more than 0.5 ms! Thus, we may conclude that CF_4 does not interfere at all with the H_2/O_2 reaction mechanism.

In the case of $H_2/O_2/Ar/CF_3H$ mixtures, the CF_3H consumption reaction 12a competes immediately with the H_2/O_2 mechanism (Fig. 8b). A large amount of HF is produced immediately (Fig. 8c) and, therefore, perturbs the usual H_2/O_2 mechanism as it occurs for H_2O time evolution (Fig. 8a). But, as soon as CF_2O is produced (Fig. 8d), the rate of HF formation is reduced drastically (Fig. 8c).

Figures 7a and 8a illustrate that more H_2O is formed than expected by the full equilibrium computation (see Table 4). Such behavior is the consequence of

Table 3 Induction time (t_i) and heat pulse time (t_p)

% additive	0	5	10	15	20	25
CF_4						
t_i, µs	0.64	0.55	0.55	0.60	0.62	1.60
t_p, µs	0.37	0.32	0.31	0.31	0.40	0.62
CF_3H						
t_i, µs	0.64	0.83	1.25	1.93	2.15	3.65
t_p, µs	0.37	0.42	0.46	0.48	0.70	0.75

Table 4 Mole fraction and heat release at sonic condition

Mixture: % Halogen:	$H_2/O_2/Ar$ 0		$H_2/O_2/Ar/CF_3H$ 10		$H_2/O_2/Ar/CF_4$ 10	
Species	Kin	CJ	Kin	CJ	Kin	CJ
H_2	6.4	6.7	7.1	5.2	1.4	1.6
O_2	2.0	2.2	0.2	0.8	1.3	2.4
Ar	55.0	55.3	40.8	37.0	40.5	37.0
CF_4	--	--	--	--	3.5	E-13
CF_3H	--	--	0.3	E-12	--	--
H	4.0	3.3	1.0	2.6	0.5	1.2
F	--	--	--	0.1	--	0.2
O	1.9	1.5	0.1	0.8	0.4	1.1
OH	5.0	5.0	1.4	2.7	2.4	2.4
H_2O	25.8	26.0	22.2	13.9	18.4	9.0
CF_3	--	--	0.3	E-12	0.02	E-11
CF_2	--	--	--	E-08	--	E-08
CF	--	--	--	E-07	--	E-08
CF_2O	--	--	5.2	E-06	0.7	E-05
CFO	--	--	1.3	E-05	0.2	E-05
HF	--	--	16.9	27.7	24.9	36.0
CO	--	--	2.8	6.7	2.5	5.0
CO_2	--	--	0.2	2.6	3.2	4.1
Q, kJ/kg	1656	1741	2339	1979	2176	1811
D, m/s	1830	1874	1922	1945	1854	1875

the slow reaction 12b for CF_4 mixtures and of the slow reaction 14 for CF_3H mixtures (Table 2).

Heat Release

As exhibited in Figs. 5 and 6, the heat release curves present an interesting shape. When the sonic condition is satisfied, the amount of heat released is much larger if either CF_3H, or CF_4 is added than if argon is the sole diluent. After the step rise of the heat release and for large concentrations of additive (CF_3H or CF_4), the heat release curves level off, due to the slow consumption of either CF_4 or CF_2O in the case of CF_3H. Even more important to notice is that the heat pulse time t_p as characterized in Fig. 9 is shorter for mixtures containing small amounts of CF_4 (see Table 3 and Fig. 10). The heat pulse time is defined as the time elapsed between the end of the induction zone and the second inflection of the power pulse (dQ/dt).

The kinetic calculations indicate that the behavior of mixtures containing fluorocarbons compounds is complicated; an increase of the induction time can be compensated for by a large and fast energy release. Moreover, the

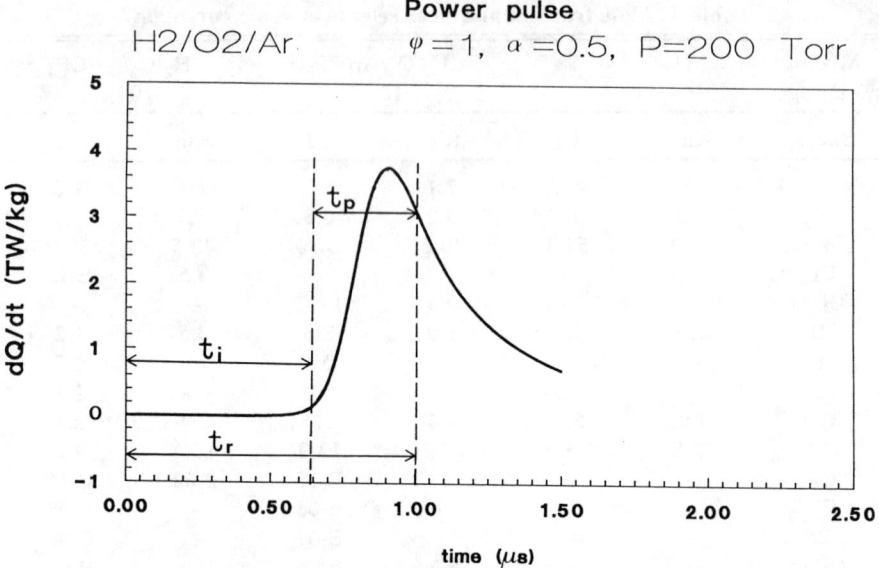

Fig. 9 Evolution of the power pulse in $H_2/O_2/Ar$, t_i=induction time, t_p=heat release time, t_r=reaction time; all times are in microsecond.

composition observed at a time scale consistent with any detonation process is far from the thermodynamic equilibrium.

Comparison of the Kinetic Approach with the CJ Model

The description and the results of the chemical kinetic model show that it is possible to reach the sonic condition by two different approaches: 1) the classical CJ calculation which assumes the equilibrium composition and 2) a time-dependent method employing the simple kinetic model listed on Table 2. Table 4 summarizes the species mole fractions and the heat release calculated for three mixtures: 2 $H_2/O_2/3$ Ar, 2 $H_2/O_2/2.4$ Ar/0.6 CF_4, and 2 $H_2/O_2/2.4$ Ar/0.6 CF_3H. The heading "kin" refers to the results gained from the kinetic code and the heading "CJ" reports the classical Chapman-Jouguet model. The comparison is quite meaningful. Let us describe each mixture separately.

$H_2/O_2/Ar$

For this system, the agreement between both calculations is remarkably good. The fact that the heat release Q at the end of the kinetic model is slightly lower than the CJ value is due to the smaller experimental shock velocity used for the kinetic computation.

Table 5 Comparison between CF_4 and CF_3H additive

	CF_4 mixture	CF_3H mixture
Induction time	Favorable	Unfavorable
Heat pulse time	Favorable	Unfavorable
Heat release, Q	Very favorable	Very favorable

$H_2/O_2/Ar/CF_3H$

A significant difference is noticeable in the mole fractions and the heat release Q. The composition obtained by the kinetic model is far from the equilibrium, in particular for H_2O, CF_2O, and HF. But for other species the agreement with the equilibrium calculation is better. The most important consequence is the discrepancy between the heat releases Q. Almost 20% more heat is generated if we consider the kinetic model.

$H_2/O_2/Ar/CF_4$

The first step of CF_4 attack (reaction 12b) impedes the complete CF_4 consumption as described previously. Once again, we notice a 20% increase of the heat release in the case of the kinetic model.

From those numerical data, it is obvious that the assumption of the chemical equilibrium is not guaranteed for mixture including fluorocarbon compounds. Thus, the Chapman-Jouguet model fails to describe detonation velocity in such

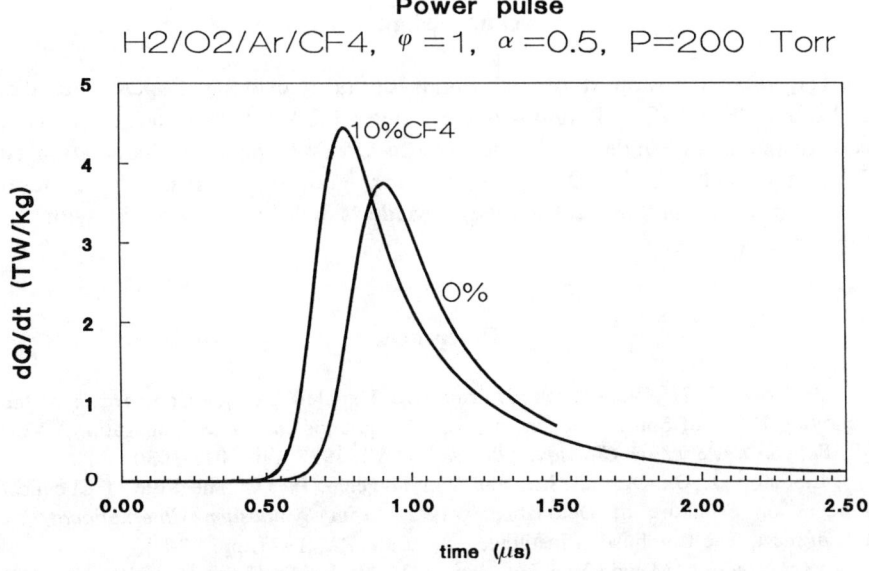

Fig. 10 Comparison of the power pulse in $H_2/O_2/Ar/CF_4$ mixtures containing 0% and 10% of CF_4, respectively.

mixtures, as has been already observed experimentally.[4] The substantial heat release appearing in the chemical kinetic model can be a likely explanation for the increase of the experimental detonation velocities in mixtures containing either CF_4 or CF_3H.

Conclusions

The unexpected behavior of the detonation velocities for mixtures containing variable amounts of CF_4 and CF_3H has been established experimentally in various conditions of confinement. A promoting influence of both additives is definitely noticeable up to 10%. Such an influence contradicts the prediction made by the classical Chapman-Jouguet model for CF_4, even if unusual species are included in the equilibrium computation. The regular character of the cell structure is lost when more than 5% of fluorocarbon is added. An original chemical kinetic scheme has been suggested to explain the values of D_{exp} larger than D_{CJ} for CF_4 mixtures. Three parameters (induction time, heat pulse time and heat release) have been considered in the discussion of the kinetic model and their influence are summarized in Table 5. The positive influence of the heat release Q is obvious. However, such an influence is counteracted by the unfavorable effects of the induction time and the heat pulse time for CF_3H mixtures. In the case of CF_4 mixtures, the combination of three positive influences allows justification of a promoting effect on the detonation velocity larger than the one expected from a mere CJ calculation.

Acknowledgments

The financial support of the Fonds de la Recherche Fondamentale et Collective (F.R.F.C.), Belgium under Contract 2.9006.88 is acknowledged. Nzeyimana is also indebted to the Agence Générale pour la Coopération au Développement (A.G.C.D.), Belgium for granting a research doctoral fellowship. The authors acknowledge Mader's helpful cooperation with the Fortran SIN Code.

References

[1]da Cruz, F. N., Vandooren, J., and Van Tiggelen, P. J., "Comparison of the Inhibiting Effect of Some Halogenomethane Compounds on Flame Propagation," Vol. 97, *Bulletin des Sociétés Chimiques Belges*, 11-12, 1988, pp. 1011-1030.

[2]Libouton, J. C., Dormal, M., and Van Tiggelen, P. J., "The Role of Chemical Kinetics on Structure of Detonation Waves," *15th Symposium (International) on Combustion*, The Combustion Institute, Pittsburgh, PA, 1975, pp. 79-86.

[3]Vandermeiren, M. and Van Tiggelen, P. J., "Role of an Inhibitor on the Onset of Gas Detonations in Acetylene Mixtures," Vol. 114, *Progress in Astronautics and Aeronautics*, AIAA, NY, 1987, pp. 186-201.

[4]Nzeyiama, E. and Van Tiggelen, P. J., "Influence of Tetrafluoromethane on Hydrogen-Oxygen-Argon Detonation," Vol. 133, *Progress in Astronautics and Aeronautics*, AIAA, NY, 1991, pp. 77-88.

[5] Vandermeiren, M. and Van Tiggelen, P. J., "Cellular Structure of Detonation in Acetylen-Oxygen Mixtures," Vol. 94, *Progress in Astronautics and Aeronautics*, AIAA,NY, 1985, pp. 104-111.

[6]Mader, C., *Numerical Modeling of Detonation*, Appendix A, University of California Press, 1979.

[7]Libouton, J. C., Jacques, A., and Van Tiggelen, P. J., "Kinetic Structure and Sustainance of Detonation Waves," *Actes du Colloque Inernational Berthelot-Vieille-Mallard- Le Chatelier*, Vol II, 1981, pp. 437-442.

[8]JANAF Thermochemical Tables, U.S. Department of Commerce, National Bureau of Standards, Washington DC, 1985.

[9]Baulch, D. L., Drysdale, D. D., and Lloyd, A.C., *High Temperature Reaction Rate Data, Vols. 2 and 3: Critical Evaluation of Rate Data for Homogeneous Gas-Phase Reactions of Interest in High-Temperature Systems*, Dept of Physical Chemistry, The University Leeds 2, England, 1968.

[10]Vandooren, J., da Cruz, F. N., and Van Tiggelen, P. J., "The Inhibiting Effect of CF_3H on the Structure of a Stoichiometric $H_2/CO/O_2/Ar$ Flame," *22th Symposium (International) on Combustion*, The Combustion Institute, Pittsburgh, PA, 1988, pp. 1587-1595.

[11]Kondratiev, V.N., "Rate Constants of Gas-Phase Reactions," U.S. Department of Commerce, National Bureau of Standards, Washington DC, 1972.

[12]Baulch, D. L., Drysdale, D. D., Duxbury, D. D. and Grant, S.J., *Evaluated Kinetic Data for High Temperature Reactions, Vol. 3: Homogeneous Gas-Phase Reactions of O_2-O_3 System, and of CO-O_2-H_2 System, and of Sulphur Containing Species*, Butterworths, London, 1976.

Oxidation of Gaseous Unsymmetrical Dimethylhydrazine at High Temperatures and Detonation of UDMH/O_2 Mixtures

Said Abid,* Gabrielle Dupré,† and Claude Paillard‡
National Center of Scientific Research and University, Orléans, France

Abstract

The oxidation mechanism of gaseous unsymmetrical dimethylhydrazine (UDMH) behind a shock wave consists of a rapid decomposition of UDMH followed by the oxidation process after an ignition delay. In the present study, the ignition delays of UDMH decomposition products with oxygen are measured behind a reflected shock wave, in a large range of temperature (850-1550 K), pressure (31-212 kPa) and concentration: $[O_2]/[UDMH]$ = 1-32.3 and Ar molar fraction = 0-0.95. A simple relationship between ignition delays, shock conditions and species molar fractions is obtained in the case of argon-diluted mixtures, providing a value for the activation energy. For non-diluted UDMH/O_2 mixtures, the ignition delay is found to be strongly dependent on $[O_2]/[UDMH]$ ratio. If the shock conditions are such that the ignition delay of the UDMH oxidation reaction is significantly reduced, a detonation wave can be initiated behind the incident shock. The same $[O_2]/[UDMH]$ ratios as for ignition delay measurement, with 75 mol% Ar or less, are used. The critical temperatures and pressures for the onset of a detonation wave are first determined, then the conditions for the propagation of a self-sustained detonation are found. The two diagnostics are, first, velocity measurement via a series of piezoelectric pressure transducers, and, second, detonation cell size determination using the classical smoked foil method. From the ignition delay data, an

---

Copyright @ 1991 by the American Institute of Aeronautics and Astronautics, Inc. All rights reserved.
*Graduate Student, Laboratory of Combustion and Reactive Systems, C.N.R.S.
†Chargée de Recherche, Laboratory of Combustion and Reactive Systems, C.N.R.S.
‡Professor, Department of Chemical Kinetics, University of Orléans.

induction distance for the explosive reaction is estimated, according to the definition of the model of Zeldovich, von Neumann and Döring. The factor of proportionality found between induction distance and cell size is found to be inaccurate and strictly valid for diluted mixtures.

Introduction

There has been a constant interest in hydrazine and its methyl derivatives for rocket propellant use. Among them, 1,1 Dimethylhydrazine, $(CH_3)_2N_2H_2$, commonly called unsymmetrical dimethylhydrazine (UDMH), is used as a fuel component generally mixed with hydrazine and oxidizers. A major drawback of using UDMH in aerospace applications is its instability. It is an endothermic compound: the enthalpy of formation at 298 K is $+51.63$ kJ mol^{-1} for liquid UDMH and $+84.3$ kJ mol^{-1} for the ideal gas.[1] It is hypergolic with dinitrogen tetroxide and can react violently with oxygen. The variation of the enthalpy of complete oxidation reaction for gaseous UDMH at 298 K

$$(CH_3)_2NNH_2(g) + 4O_2(g) \longrightarrow 2CO_2(g) + N_2(g) + 4H_2O(g)$$

is -1974 kJ mol^{-1}.

As a result of the aerospace interest, research work involving UDMH attracted considerable attention in the 1960s. Several studies dealing with thermal decomposition or combustion in static or flow reactors,[2-6] pyrolysis, or oxidation behind shock waves[7,8] have been published. As shown in our previous paper,[8] the addition of small quantities of oxygen up to 12 mol% has very little influence on the half-life of UDMH but greatly increases the reaction rate at higher concentration. The only data available up to now on ignition delays[7,9] has revealed that the oxidation mechanism consists of a rapid decomposition of UDMH followed by the oxidation process after an ignition delay depending on temperature, pressure, and concentration. Concerning UDMH detonation, very little data are reported in the literature, except two values of measured and calculated velocities of $UDMH/O_2/N_2$ mixtures, given by Just.[7]

The aim of the present investigation is, first, to express the ignition delays of UDMH oxidation in terms of temperature and species concentration, and to compare them with the half-times of the thermal decomposition. Second, our investigation consists of finding the conditions for the initiation of a detonation in the $UDMH/O_2$ mixtures and to investigate their sensitivity to detonation. From the explosion safety point of view, the question of ignition delays and detonability is of great importance for aerospace purpose.

Experimental Details

Gaseous unsymmetrical dimethylhydrazine (UDMH) is stored at a vapor pressure corresponding to room temperature in a Pyrex vessel and is eventually mixed with oxygen and argon, according to the partial pressure law. The purity of the mixture is checked by mass spectrometry.

Two shock tubes are used, one for the determination of ignition delays, the other for the detonation study. The test mixtures are introduced into the low-pressure section at ambient temperature T_1 and total pressure P_1, and helium is introduced into the driver section at a pressure P_4. The characteristics of the shock tubes are as follows:

1) The low-pressure section of the first tube, 50 mm i.d. and 6.44 m long, is made of Pyrex glass. The initiation of the shock wave is produced by a sudden rupture of the diaphragm pierced by a four-bladed steel knife propelled by a pneumatic system. A pair of CaF_2 optical windows, mounted flush with the inside wall of the test section, is located at about 10 mm from the tube end. A monochromator and a photomultiplier are used to follow UDMH absorption and emission at a 220-nm wavelength.

2) The second shock tube, 38 mm i.d., made entirely of stainless steel, is able to function at pressures close to 1 MPa and at a higher shock strength, via a two-diaphragm system (Fig. 1). This system consists of a dismounting intermediate tube section connected to the driver and driven sections with two "terphane" diaphragms of a thickness appropriate to the required bursting pressure. A 220-mm-long dismounting element of the driven section can receive a smoked terphane foil for the detonation structure study.

Both shock tubes are equipped with a series of piezoelectric pressure transducers for the measurement of incident shock or detonation velocity and determination of shock parameters.

Results

Ignition Delay of the Oxidation of Unsymmetrical Dimethylhydrazine

The ignition delays of the oxidation reaction of $UDMH/O_2$ mixtures, diluted or not diluted with argon, have been measured behind a reflected shock wave in a large range of conditions, using the emission signal at 220 nm. The emission at this wavelength is very intense and is due to the explosive oxidation of the decomposition products. A simultaneous emission occurs at 600 nm as well as a strong pressure rise due to the reaction exothermicity. At 220 nm, there is also a strong absorption of UDMH that permits following the kinetics of its decomposition at rather

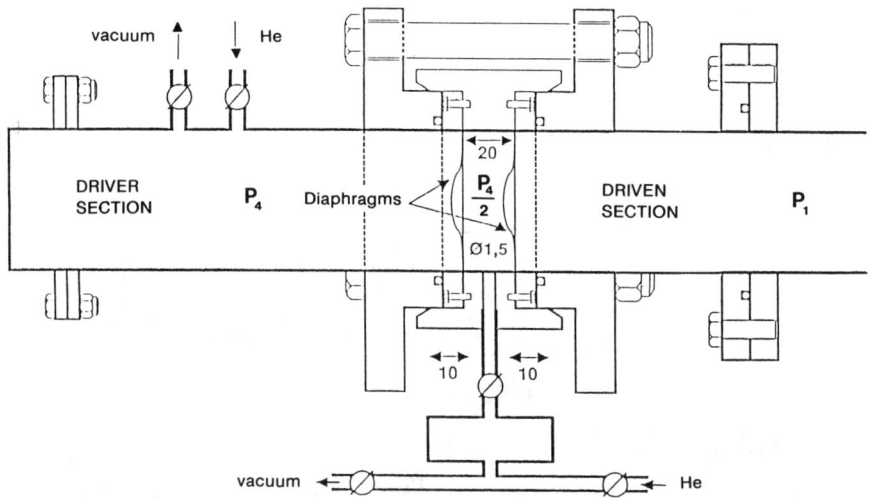

Fig. 1   Schematic of the two-diaphragm system on the 38-mm-i.d. shock tube.

low concentration with an optical path limited to 50 mm. At this wavelength, the absorption due to the reaction products is low, but their contribution has been taken into account for the analysis of the absorption signal. It allows the simultaneous following, at 220 nm, of UDMH decomposition using the absorption signal and, of its oxydation using the emission signal.

A first set of experiments consisted of studying argon-diluted mixtures, xUDMH/yO$_2$/zAr, where x, y, and z are the molar fractions of UDMH, O$_2$, and Ar, respectively, with x = 0.01-0.058, y = 0.029-0.275, and z = 0.691-0.95. The range of shock temperatures and pressures is T = 980-1555 K and P = 31-108 kPa.

A typical record is shown in Fig. 2. Simultaneous with the pressure signal given by a piezoelectric transducer located at the same distance (10 mm) as the optical windows upstream the tube end, it shows, first, according to Just,[7] the UDMH absorption signal that grows at the shock arrival, then diminishes as UDMH begins to decompose, and second, a very important emission signal coming from the species formed during the oxidation of the decomposition products. The time interval between the reflected shock front and the emission signal is defined as the ignition delay of the oxidation reaction. Thus, it is clear that the oxidation reaction is a two-phase mechanism, consisting first of a rapid decomposition of UDMH, and then of the oxidation of the decomposition products.

Figure 3 shows the logarithmic variation of ignition delay with inverse reflected shock temperature for three different xUDMH/yO$_2$/zAr mixtures. The data of Just[7] (curves 1 and 2), obtained

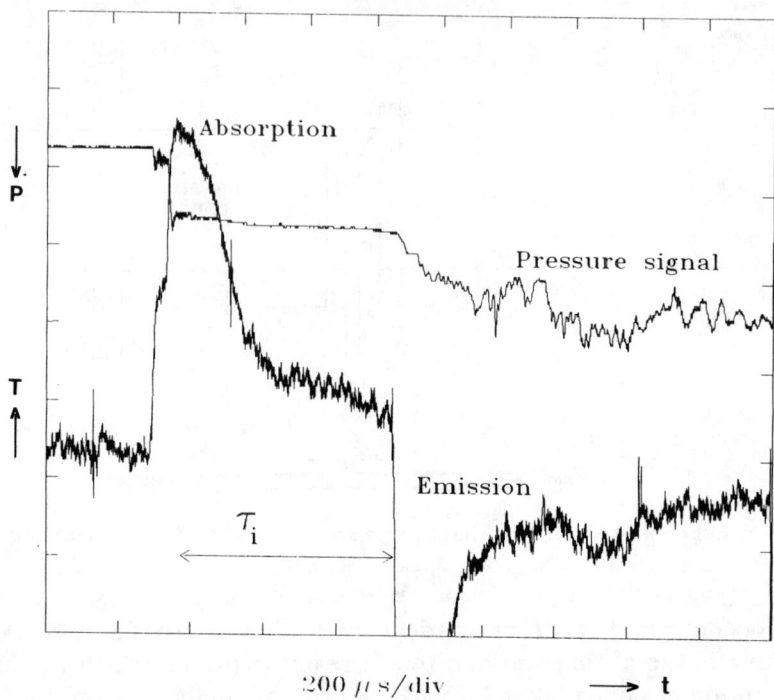

Fig 2 Simultaneous record of the pressure signal, the UDMH absorption signal at 220 nm and the emission signal at the same wavelength corresponding to the oxidation of UDMH decomposition products: mixture: 0.0345 UDMH/0.275 O_2/0.6905 Ar ; reflected shock conditions: T_5 = 1039 K ; P_5 = 45.7 kPa.

in a similar range of temperatures (900-1400 K), for a larger range of pressure extended toward higher pressures (50-200 kPa), and for larger Ar molar fractions (0.955-0.975) are also included.

The analysis of the data of 118 experiments leads to the expression of the oxidation ignition delay τ_i as follows:

$$\tau_i = C\,[UDMH]^m\,[O_2]^n\,[Ar]^p \exp(E/RT)$$

where E is the apparent activation energy, R the perfect gas constant, T the reflected shock temperature, m, n, and p exponents, and C a constant. Results presented in Fig. 4 represent the function:

$$Y = \ln \tau_i - (m \ln [UDMH] + n \ln [O_2] + p \ln [Ar])$$

vs inverse reflected shock temperature. The constants C and E/R are deduced from the best linear correlation obtained for particular values of m, n, and p. With a correlation factor better than 0.98, one finds: m = 0.06, n = -1.44, p = 0.94, E ≃ 108 kJ mol^{-1}, and C = 1.0 10^{-9} (mol m^{-3})$^{0.44}$ s^{-1}.

The oxidation ignition delay of UDMH diluted with argon measured behind a reflected shock wave, independent of wall effect, can be expressed in terms of temperature, pressure, and molar fractions as

$$\tau_i = 1.0 \; 10^{-9} \; [x]^{0.06} \; [y]^{-1.44} \; [z]^{0.94} \; [P/RT]^{-0.44} \exp(12960/T) \quad (1)$$

where τ_i, P, and T are given in s, kPa, and K respectively, and R is equal to 8.314 J mol^{-1} K^{-1}.

The relationship (1) is valid in the range of pressure (31-108 kPa), temperature (980-1555 K), and argon dilution (69-95 mol %). The standard deviation of the ratio, of measured τ_i and τ_i calculated from the above relationship, is 24%. A quasiconstant apparent activation energy can be deduced, of the order of 108 kJ mol^{-1}, for the different xUDMH/yO$_2$/zAr mixtures studied.

The variation of the ignition delay vs inverse temperature, given in Fig. 3, shows a trend for the activation energy of the oxidation reaction to

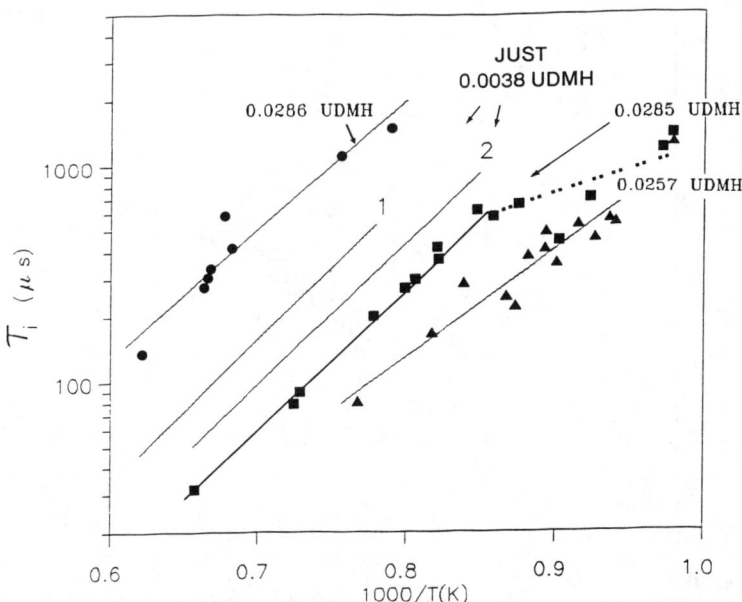

Fig. 3 Logarithmic variation of the ignition delay of the oxidation reaction of UDMH decomposition products vs inverse reflected shock temperature, for different xUDMH/yO$_2$/zAr mixtures: the curves 1 and 2 correspond to the data of Just[7]: 1) x = 0.00376, y = 0.0211, z = 0.97514 ; 2) x = 0.00363, y = 0.0414, z = 0.95497.

diminish as the mixture becomes poor, in accordance with Just's results[7] obtained in much more diluted mixtures. For the stoichiometric mixture, it seems that the activation energy decreases with temperature. In fact, the relationship (1) shows that, for given molar fractions, the ignition delay varies with P/T ratio. The data corresponding to a given mixture have been obtained in a large range of pressure over temperature ratios. This explains a certain scattering in the delay ignition values observed in Fig. 3. Moreover, the less diluted the mixture, the more important the role of reaction exothermicity on ignition delays. For nondiluted $UDMH/O_2$ mixtures, it is no more possible to get a simple formulation of the ignition delay, as shown in the second series of experiments.

Three different mixtures $xUDMH/yO_2$ have been examined, corresponding to three sets of shock temperature and pressure ranges:

x = 0.15	850 < T(K) < 1037	54 < P(kPa) < 111
x = 0.08	952 < T(K) < 1056	120 < P(kPa) < 163
x = 0.03	907 < T(K) < 1040	171 < P(kPa) < 212

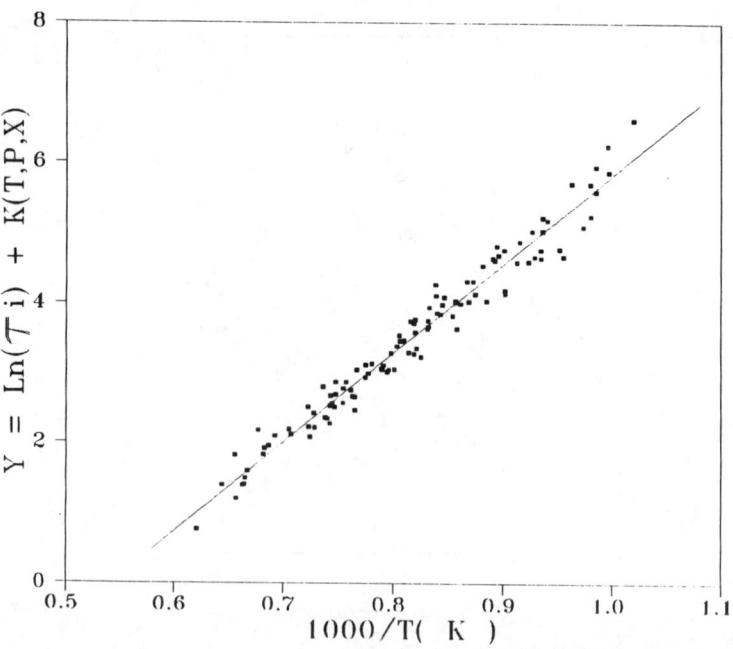

Fig. 4 Variation of the expression Y vs inverse reflected shock temperature for $xUDMH/yO_2/zAr$ mixtures with x = 0.01-0,058: $Y = \ln \tau_i - (m \ln [DMHA] + n \ln [O_2] + p \ln [Ar])$.

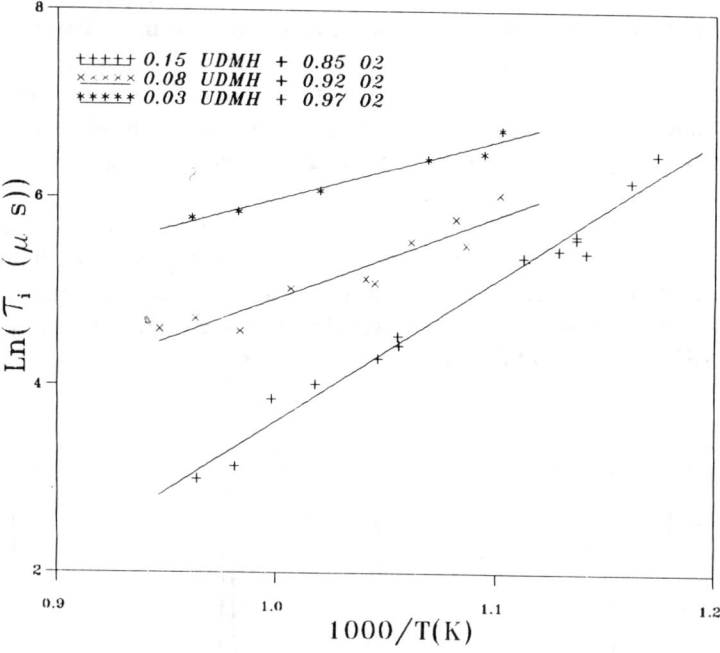

Fig. 5 Logarithmic variation of the ignition delay of the oxidation reaction of UDMH decomposition products vs inverse reflected shock temperature for nondiluted xUDMH/yO$_2$ mixtures.

Figure 5 gives the logarithmic values of the oxidation ignition delay τ_i for the considered mixtures vs inverse reflected shock temperature. Contrary to the case of largely diluted mixtures, for which τ_i could be expressed by a unique and simple relationship, no similar expression could be obtained for nondiluted UDMH/O$_2$ mixtures. The slopes of the linear curves $\tau_i = f(1/T)$ vary greatly with the mixture concentration.

Sensitivity to Detonation of UDMH/O$_2$ (+Ar) Mixtures

The shock temperature and pressure can be chosen so as to reduce significantly the ignition delay of the UDMH oxidation reaction and to induce a coupling between the shock wave and the reaction zone. The conditions for the onset of a detonation wave behind an incident shock have been first examined, then the propagation of a stable self-sustained detonation have been found. Several mixtures have been investigated: lean UDMH/8O$_2$ and stoichiometric UDMH/4O$_2$ mixtures diluted with 0, 50, and 75 mol% Ar. Runs were made at ambient initial temperature T_1 over the initial pressure range P_1 = 0.67-10 kPa and initial pressure of helium P_4 in the driver tube between 200 and 700 kPa.

The explosive mixture is submitted to a temperature T_2 and a pressure P_2 behind the incident shock that can be calculated from initial pressure, temperature and molar fractions, and from the experimental incident shock velocity. T_2 and P_2 values depend on shock strength which is approximately proportional to the initial pressure ratio P_4/P_1 across the diaphragm.

A detonation can be formed if the initial pressure P_1 of a given mixture is above a limit value, or if the ratio P_4/P_1 is high enough to initiate a coupling between the shock and the reaction front. Figure 6 shows the variation of the experimental wave velocity V_S vs P_4 for UDMH/$8O_2$ mixture at three initial pressures P_1.

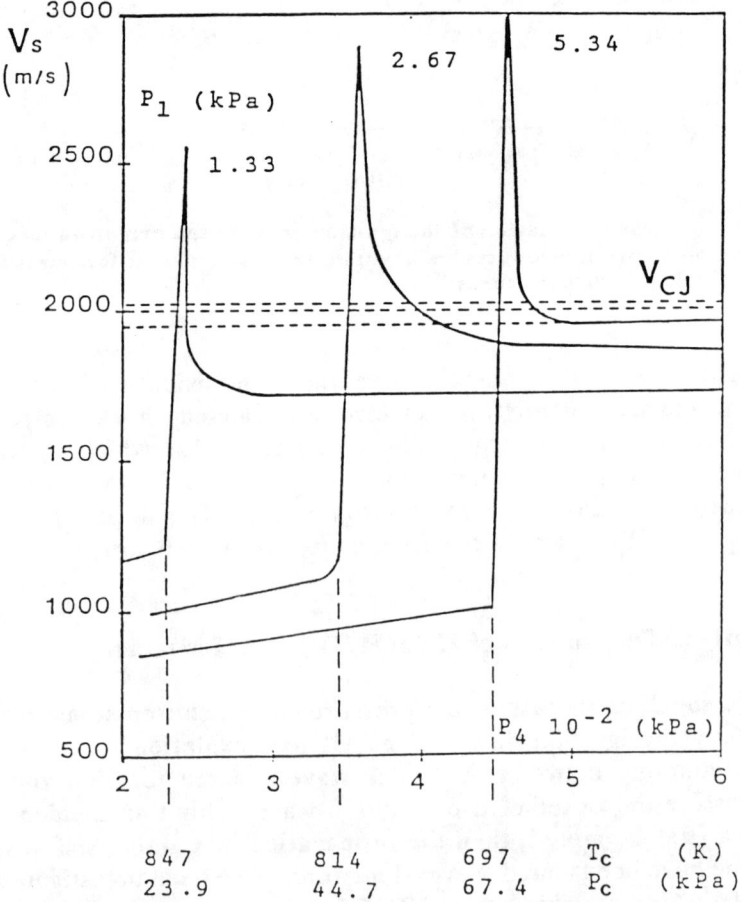

Fig. 6 Evolution of the wave velocity V_S vs driver gas pressure P_4, for three initial pressures P_1 of UDMH/$8O_2$ mixture ; the corresponding critical temperatures T_c and pressures P_c for the transition to detonation are indicated.

Table 1 Critical pressures P_c and temperatures T_c for the transition to detonation, for $UDMH/4O_2$ and $UDMH/8O_2$ mixtures diluted with Ar, at various initial pressures P_1; z is Ar molar fraction.

Mixture	z	P_1, kPa	P_c, kPa	T_c, K
$UDMH/4O_2$	0	1.32	27.1	806
		2.63	45.7	736
	0.5	2.63	36.6	805
		3.95	51.6	777
		5.26	71.7	796
	0.75	3.95	52.2	926
$UDMH/8O_2$	0	1.32	23.6	847
		2.63	44.1	814
		5.26	66.4	697
	0.75	5.26	66.9	970

For low P_4 values, the wave velocity increases smoothly. At a critical value corresponding to the onset of coupling, the wave velocity suddenly increases, varying between the incident shock velocity value V_I ($\simeq$ 1000 m/s) and an overdriven detonation velocity value, much higher than Chapman-Jouguet velocity V_{CJ}. The "coupling" temperature T_c and pressure P_c, associated with the coupling P_{4c} value, can easily be calculated. Their values are given in Fig. 6. If a pressure P_4 higher than P_{4c} is used, the wave velocity tends to become more stable as P_4 increases and, finally, a stable self-sustained detonation of a quasiconstant velocity, lower than V_{CJ}, is observed along the tube. For a given mixture, an increase of initial pressure P_1 leads to an increase of the coupling pressure P_{4c}, corresponding to an increase of critical pressure and a decrease of critical temperature. The experimental detonation velocity is closer to CJ value for higher P_1. If we compare the stoichiometric and lean mixtures, equally diluted with argon at the same initial pressure, the onset of detonation occurs at a lower P_4 value (consequently, at higher T_c and lower P_c) for the lean mixture than for the stoichiometric one. A similar behavior is observed in the case of a given $xUDMH/yO_2/zAr$ mixture at identical P_1 but diluted with increasing argon concentrations. Table 1 summarizes the critical temperature and pressure conditions for the onset of a detonation in $xUDMH/yO_2/zAr$ mixtures.

The experimental velocities V_D of the stable self-sustained detonation have been compared to the theoretical Chapman-Jouguet velocities V_{CJ}. The Chapman-Jouguet characteristics were computed

Table 2 Experimental detonation velocities V_D measured by Just[7] and theoretical Chapman-Jouguet detonation velocities V_{CJ} calculated by Just[7] and in the present work, using Stanjan code,[10] for two different $xUDMH/yO_2/zN_2$ mixtures; x, y, and z being the molar fractions.

P_1, kPa	x	y	z	V_D, m/s	V_{CJ}, m/s Ref.7	V_{CJ}, m/s Ref.10
5.34	0.153	0.77	0.077	2150	2130	2169
5.60	0.220	0.67	0.110	2320	2310	2365

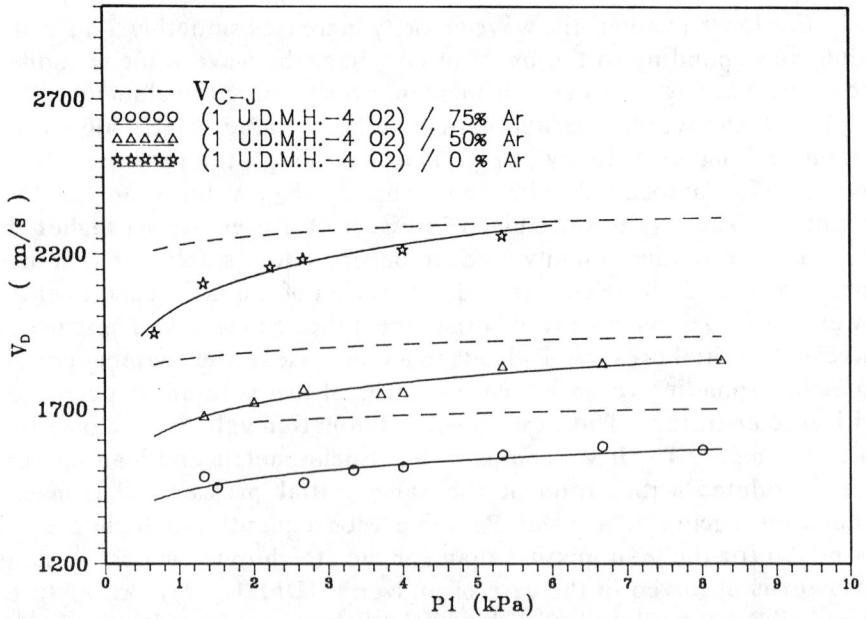

Fig. 7 Evolution of the experimental detonation velocity V_D vs initial pressure P_1 of the stoichiometric $UDMH/4O_2/zAr$ mixtures, differently diluted with argon; comparaison with theoretical Chapman-Jouguet velocity V_{CJ}.

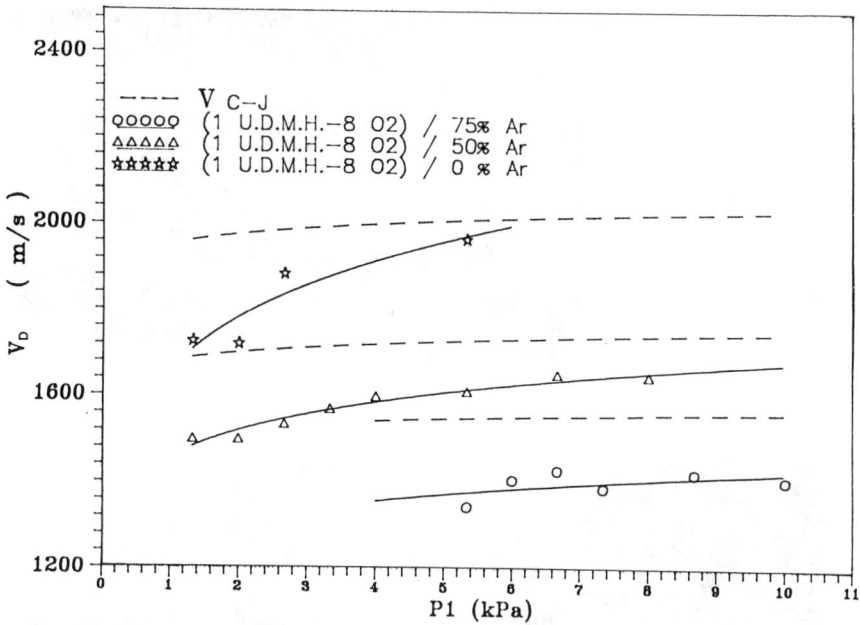

Fig. 8 Evolution of the experimental detonation velocity V_D vs initial pressure P_1 of the lean UDMH/$8O_2$/zAr mixtures, differently diluted with argon; comparaison with theoretical Chapman-Jouguet velocity V_{CJ}.

using the STANJAN code,[10] taking into account a value for the enthalpy of formation of gaseous UDMH of 84.3 kJ mol^{-1} and the presence of 20 species at thermodynamic equilibrium, nine of them (H_2O, CO, N_2, OH, H, H_2, O_2, CO_2 and NO) with a molar fraction larger than 10^{-3}. No data about UDMH/O_2 detonation exists in the literature, except the detonation velocities measured and calculated by Just[7] for two UDMH/O_2/N_2 mixtures (Table 2). Just's values of the CJ velocities are smaller than both the experimental one and our own CJ calculations. The discrepancy may arise from a different but unspecified value chosen by Just for UDMH enthalpy of formation.

The influence of initial pressure and mixture composition has been studied. Figures 7 and 8 show the variation of V_D with initial pressure P_1 for stoichiometric and lean mixtures, diluted with increasing argon concentrations. The velocity deficit $\Delta V/V_{CJ} = (V_{CJ} - V_D)/V_{CJ}$, significantly high at low pressure, is reduced as P_1 increases. However, V_D remains lower than V_{CJ} in the case of a self-sustained detonation. Such a case occurs for P_4/P_1 ratios larger than the critical one corresponding to transition to detonation but smaller than a value leading to overdriven or unstable detonation.

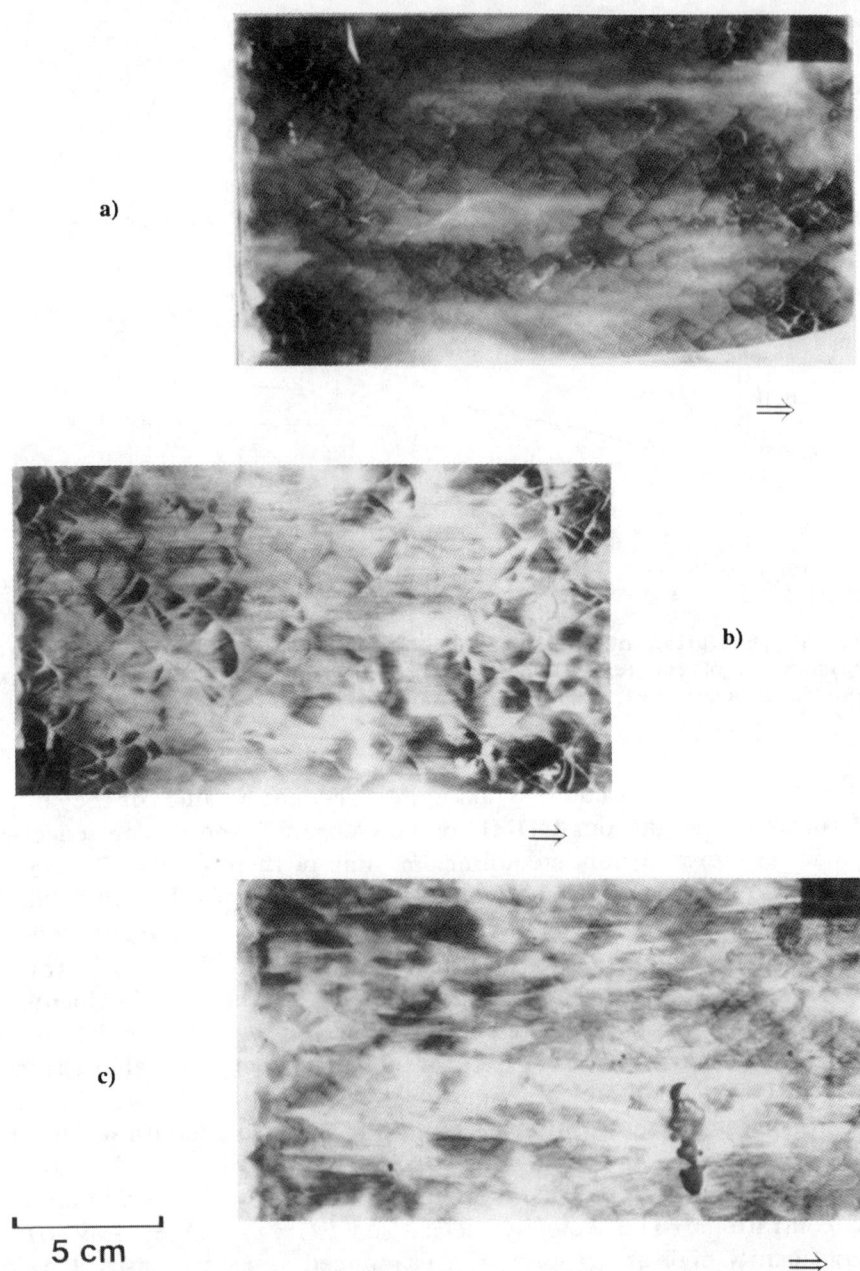

Fig. 9 Series of smoked foil records showing the tridimensional structure of the detonation wave for different UDMH/4O$_2$/zAr mixtures at different initial pressures: a) z = 0; P$_1$ = 4.0 kPa; b) z = 5; P$_1$ = 4.0 kPa; c) z = 5; P$_1$ = 8.3 kPa; d) z = 15; P$_1$ = 4.0 kPa; e) z = 15; P$_a$ = 2.7 kPa; f) z = 15; P$_1$ = 1.3 kPa.

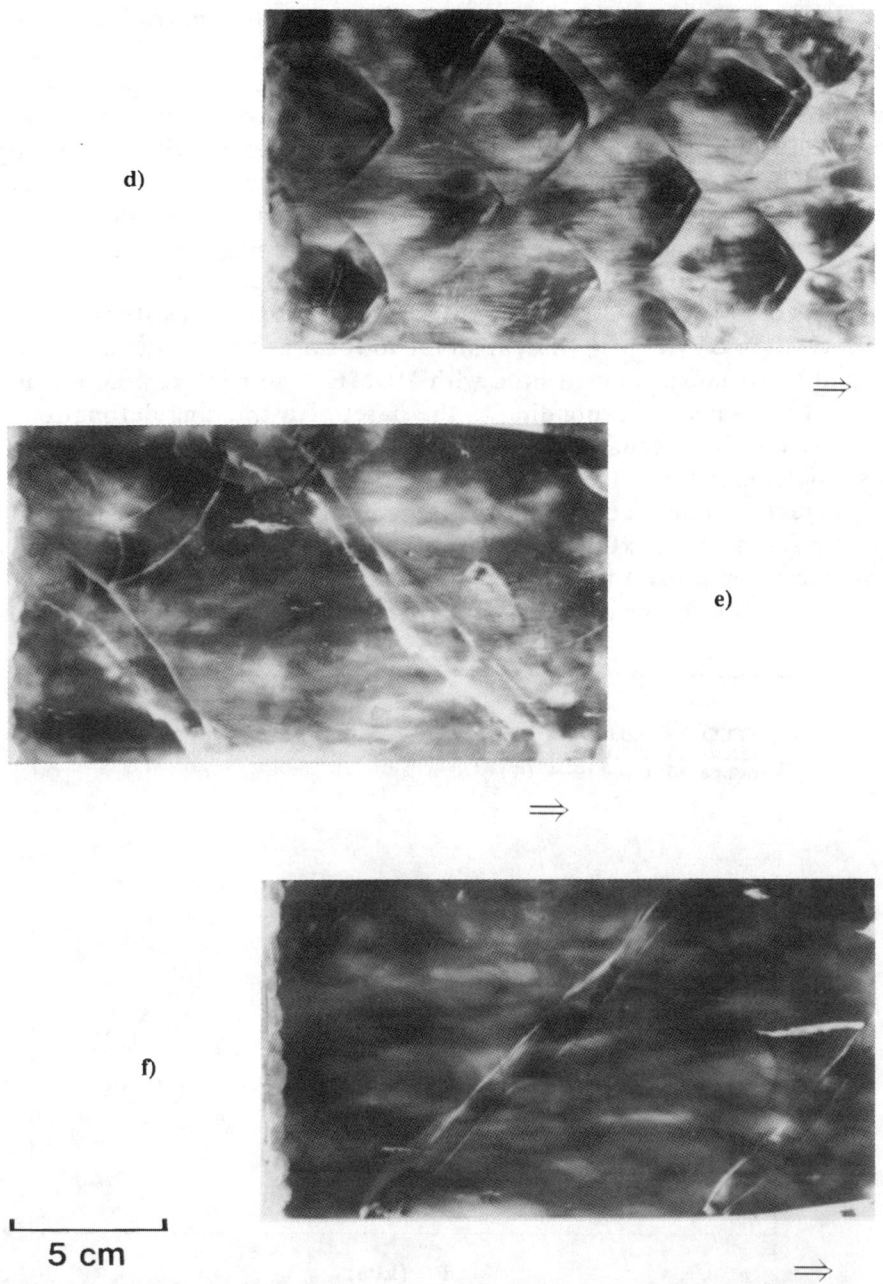

Fig. 9 (continued) Series of smoked foil records showing the tridimensional structure of the detonation wave for different UDMH/$4O_2$/zAr mixtures at different initial pressures: a) $z = 0$; $P_1 = 4.0$ kPa; b) $z = 5$; $P_1 = 4.0$ kPa; c) $z = 5$; $P_1 = 8.3$ kPa; d) $z = 15$; $P_1 = 4.0$ kPa; e) $z = 15$; $P_a = 2.7$ kPa; f) $z = 15$; $P_1 = 1.3$ kPa.

In the case of a stable self-sustained detonation, the tridimensional structure is recorded on a smoked foil located on the test section. Figures 9a-f show some photographs of these structures corresponding to stoichiometric UDMH/$4O_2$ mixtures for various dilutions and initial pressures. For a given mixture, the cell size λ increases with decreasing pressure (Fig. 9d-f). When reaching a certain pressure, λ becomes equal to the tube perimeter πd (Fig. 9e), then for an even smaller pressure, a spinning detonation is observed (Fig. 9f). A spinning detonation takes place when the velocity deficit $\Delta V/V_{CJ}$ becomes important. Table 3 gives the values of $\Delta V/V_{CJ}$, λ, and λ/d, for the different mixtures studied. Spin is observed for $\Delta V/V_{CJ} > 10\%$, at all the lower pressures as the mixture is less diluted or more concentrated with UDMH. The mixture composition or initial pressure corresponding to the onset of a spinning detonation is considered as the detonability limit.[11] Such a limit is shown in Table 3 for the studied mixtures.

Figure 10 and Table 3 show the evolution of cell size with initial pressure for mixtures xUDMH/yO_2 diluted or nondiluted with argon. A linear variation of log λ with log P_1 is shown as well as the increase of cell size with Ar dilution or with [O_2]/[UDMH] ratio, other parameters being

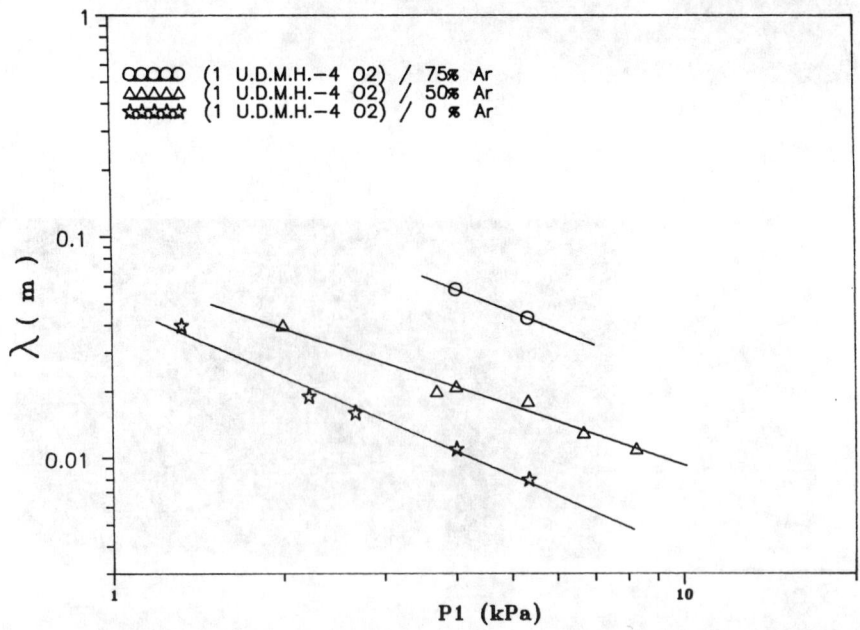

Fig. 10 Logarithmic variation of the detonation cell size λ vs initial pressure P_1 of UDMH/$4O_2$ mixtures, differently diluted with argon.

Table 3 Detonation velocity deficits $\Delta V/V_{CJ}$, cell size λ, and λ/d ratios for UDMH/4O$_2$ and UDMH/8O$_2$ mixtures diluted with argon, at various initial pressures P_1: tube diameter is d = 38 mm ; z is Ar molar fraction.

Mixture	z	P_1, kPa	$\Delta V/V_{CJ}$ %	λ, mm	λ/d
UDMH + 4 O$_2$	0	0.67	11.70	SPIN	>π
		1.33	6.02	40	1.05
		2.22	3.57	19	0.50
		2.67	3.52	16	0.42
		4.00	3.23	11	0.29
		5.34	1.87	08	0.21
	0.5	1.33	10.27	SPIN	>π
		2.00	9.34	40	1.05
		3.70	8.16	20	0.53
		4.00	8.23	21	0.55
		5.33	4.27	18	0.47
		6.67	4.19	13	0.34
		8.25	3.92	11	0.29
	0.75	2.67	12.29	SPIN	>π
		4.00	10.53	58	1.53
		5.33	8.22	43	1.13
UDMH + 8 O$_2$	0	1.33	11.72	SPIN	>π
		2.00	12.90	52	1.37
		2.67	5.09	30	0.79
		5.33	1.69	12	0.32
	0.5	2.67	11.82	SPIN	>π
		4.02	7.09	31	0.82
		5.33	6.94	27	0.71
		6.67	4.90	23	0.61
		8.03	5.40	19	0.51
	0.75	5.33	12.13	SPIN	>π
		6.67	8.04	110	2.89
		8.67	8.26	59	1.55
		10.22	10.10	22	0.58

identical. As the detonation cell size is a measure of the detonation sensitivity of an explosive mixture, it allows a comparison of the sensitivity of different mixtures. Figure 11 gives the logarithmic variation of the detonation cell size λ vs log P_1 for the lean and stoichiometric UDMH/O_2 mixtures, compared to the cell size of other stoichiometric fuel/oxygen mixtures (H_2, CH_4, C_2H_2, C_2H_4, C_3H_8, N_2H_4).[12,13] The detonation cell size of UDMH/$4O_2$ and, consequently, its detonation sensitivity, is between C_2H_4 and C_2H_2 stoichiometric values and lower than that of N_2H_4/O_2. Lean UDMH/$8O_2$ mixture has a smaller detonation hazard, similar to C_2H_4/O_2 one.

The ignition delay, in the case of the Zeldovich, von Neumann, and Döring (ZND) detonation model representing the induction period separating the shock front and the reaction zone, can be calculated from the relationship (1). The corresponding induction distance Δ_i is related to τ_i as follows:

$$\Delta_i = \tau_i (V_{CJ} - v)$$

where v is the particle velocity. For a certain number of explosive mixtures such as ClO_2 (Ref.14), N_2H_4 (Ref.13), and fuel/air or oxygen,[12] it is

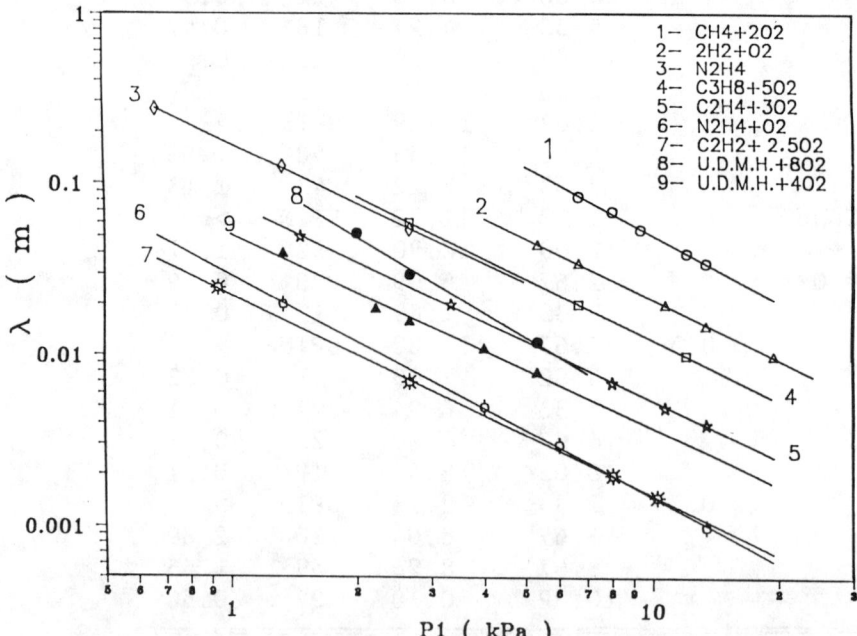

Fig. 11 Logarithmic variation of the detonation cell size λ vs initial pressure P_1 of UDMH/$4O_2$ and UDMH/$8O_2$ mixtures, compared with the cell size of various stoichiometric fuel/O_2 mixtures.

Table 4 Proportionality factor $A = \lambda/\Delta_i$ between cell size and induction distance for different UDMH/oxygen mixtures diluted with argon.

Mixtures	A
UDMH + 8 O_2 + 27 Ar	7 - 37
UDMH + 8 O_2 + 9 Ar	16 - 40
UDMH + 4 O_2 + 15 Ar	22 - 41
UDMH + O_2 + 5 Ar	13 - 16

verified that the detonation cell size is directly proportional to the detonation induction distance as

$$\lambda = A \Delta_i$$

Table 4 gives the values of the proportionality factor A obtained for different $xUDMH/yO_2/zAr$ mixtures. A is found to be included between 7 and 40. This value is much larger than that found for ClO_2 (A ≃ 6),[14] and of the order of that of N_2H_4 (A ≃ 29)[13] or fuel/air or oxygen mixtures [A ≃ 29 (Ref.15) or 22 (Ref.16)]. It is not a constant value: this may be due to the extrapolation of the ignition delay relationship at the upper limit of the validity range of temperature and pressure, or to the inaccuracy of the cell size measurement in the case of irregular structure. It could be also possible that A is not a constant, but a function of mixture composition for example. However, even an approximate value of A is useful for estimating the detonation cell size from induction time measurements, in case of an unreadable structure on the smoked foil, and for predicting the detonation hazard. For nondiluted UDMH/oxygen mixtures, values of A are largely scattered because of the lack of ignition delay measurement in the range of shock temperature and pressure reached for these mixtures.

Conclusion

The present investigation on the oxidation reaction of unsymmetrical dimethylhydrazine behind a shock wave has confirmed the results of Just[7]: it is a two-phase mechanism, consisting of a rapid UDMH decomposition followed by the combustion of the decomposition products after an ignition delay. For argon-diluted UDMH/oxygen mixtures only, a simple relationship between the ignition delay, shock temperature and pressure, and species molar fractions could be found leading to an apparent activation energy of 108 kJ mol^{-1}. The ignition delay measurements, carried out at shock temperatures and pressures close to von Neumann values, provide a means of estimating the induction distance between the shock front and the reaction zone in a detonation wave.

The critical temperatures and pressures for the onset of a detonation wave have been determined and the conditions of propagation of a self-sustained detonation in UDMH/O_2/Ar mixtures established. Detonation velocities and cell size have been carefully measured. It is clearly shown that a spinning detonation occurs if the cell size over tube diameter ratio is of the order or above π or if the velocity deficit is larger than 10%, as shown in a previous study.[17] For diluted mixtures only, an approximate correlation has been found between induction distance deduced from ignition delay measurement and detonation cell size.

Acknowledgment

This work was sponsored by the Ministère de la Défense, Délégation Générale pour l'Armement, Direction des Recherches, Etudes et Techniques, under Contract No 87/157.

References

[1]Marsh, W. R., and Knox, B. P., "Hydrazine fuels," *USAF Propellant Handbooks*, AFRPL-TR-69-149-1, Bell Aerosystems, Vol. I, Buffalo, NY, 1970.

[2]Cordes, H. F., "The Thermal Decomposition of 1,1-Dimethylhydrazine," *Journal of Physical Chemistry*, Vol. 65, No. 9, 1961, pp. 1473-1477.

[3]Eberstein, I. J., and Glassman, I., "The Gas-Phase Decomposition of Hydrazine and its Methyl Derivatives," *10th Symposium (International) on Combustion*, The Combustion Institute, Pittsburgh, PA, 1965, pp. 365-374.

[4]Golden, D. M., Solly, R. K., Gac, N. A., and Benson, S. W., "Very Low-Pressure Pyrolysis. VII Decomposition of Methylhydrazine, 1,1 - Dimethylhydrazine, 1,2 - Dimethylhydrazine and Tetramethylhydrazine. Concerted Deamination and Dehydrogenation of Methylhydrazine," *International Journal of Chemical Kinetics*, Vol. IV, 1972, pp. 433-448.

[5]Gray, P., and Spencer, M., "Combustion of Unsymmetrical Dimethyl Hydrazine : Spontaneous Ignition in Decomposition and Oxidation," *Combustion and Flame*, Vol. 6, No. 4, 1962, pp. 337-345.

[6]Gray, P., and Spencer, M., "Studies of the Combustion of Dimethyl Hydrazine and Related Compounds," *9th Symposium (International) on Combustion*, The Combustion Institute, Pittsburgh, PA, 1963, pp. 148-157.

[7]Just, T., "Experimentelle Untersuchungen zur Kinetik der Reaktion von 1.1 Dimethylhydrazin mit Sauerstoff," Deutsche Forschungs- und Versuchsanstalt für Luft- und Raumfahrt, Institüt für Physikalische Chimie der Verbrennung, DLR FB 70-34, 1970, pp. 1-26.

[8]Gulati, S. K., Abid, S., and Paillard, C. E., "Pyrolyse de la diméthylhydrazine asymétrique derrière une onde de choc réfléchie," *Comptes-Rendus de l'Académie des Sciences Paris*, Vol. 309, Série II, 1989, pp. 1469-1474.

[9]Bhaskaran, K. A., Gupta, M. C., and Just, T., "Shock Tube Study of the Effect of Unsymmetric Dimethyl Hydrazine on the Ignition Characteristics of Hydrogen-Air Mixtures," *Combustion and Flame*, Vol. 21, No. 1, 1973, pp. 45-48.

[10]Stanjan Chemical Equilibrium Solver, Version 3.84 IBM-PC, Stanford University, Stanford, CA, 1987.

[11]Dove, J. E., and Wagner, H. G., "A Photographic Investigation of the Mechanism of Spinning Detonation," *8th Symposium (International) on Combustion*, Williams and Wilkins, Baltimore, MD, 1962, pp. 589-600.

[12]Knystautas, R., Lee, J. H., and Guirao, C., "The Critical Tube Diameter for Detonation Failure in Hydrocarbon-Air Mixtures," *Combustion and Flame*, Vol. 48, No. 1, 1982, pp. 63-83.

[13]Pedley, M. D., Bishop, C. V., Benz, F. J., Bennett, C. A., McClennagan, R. D., Fenton, D. L., Knystautas, R., Lee, J. H., Péraldi, O., Dupré, G., and Shepherd, J. E., "Hydrazine Vapor Detonations," *Dynamics of Explosions*, Progress in Astronautics and Aeronautics, edited by A.L. Kuhl, J.R. Bowen, J.-C. Leyer, and A. Borisov, Vol. 114, AIAA, New York, 1988, pp. 45-63.

[14]Paillard, C., Dupré, G., Al Aiteh, H., and Youssefi, S. "Correlation between Chemical Kinetics and Detonation Structure for Gaseous Explosive Systems", *Dynamics of Detonations and Explosions: Detonations*, Progress in Astronautics and Aeronautics, edited by A.L. Kuhl, J.-C. Leyer, A.A. Borisov, and W.A. Sirignano, Vol. 133, AIAA, New York, 1991, pp. 63-76.

[15]Westbrook, C. K., and Urtiew, P. A., "Chemical Kinetic Prediction of Critical Parameters in Gaseous Detonations," *19th Symposium (International) on Combustion*, The Combustion Institute, Pittsburgh, PA, 1982, pp. 615-623.

[16]Shepherd, J. E., "Chemical Kinetics of Hydrogen-Air-Diluent Detonations," *Dynamics of Explosions*, Progress in Astronautics and Aeronautics, edited by J.R. Bowen, J.-C. Leyer, and R.I. Soloukhin, Vol. 106, AIAA, New York, 1986, pp. 263-293.

[17]Dupré, G., Knystautas, R., and Lee, J. H., "Near-Limit Propagation of Detonation in Tubes," *Dynamics of Explosions*, Progress in Astronautics and Aeronautics, edited by J.R. Bowen, J.-C. Leyer, and R.I. Soloukhin, Vol. 106, AIAA, New York, 1986, pp. 244-259.

Digital Signal Processing Analysis of Soot Foils

J. J. Lee,* D. L. Frost,† J. H. S. Lee,† and R. Knystautas†
McGill University, Montreal, Quebec, Canada

Abstract

The current method of subjectively classifying the irregularity of cellular patterns of soot-foils and estimating the average cell size "by eye" can be improved by using digital image processing techniques. The image of the cellular pattern is first digitized, then analyzed by computer using two methods: the computation of a histogram of the cell sizes in the pattern, and the computation of the autocorrelation function of the pattern. For this study, explosive mixtures known to have varying degrees of irregularity were analyzed. The histogram provided a value of mean cell size close to that estimated by eye, and the variance of the cell size provided a quantitative measure of the level of irregularity in the pattern. The autocorrelation function method provided a spectrum of cell sizes for each soot-foil as well as a qualitative assessment of the irregularity. For regular cellular patterns, the dominant cell size in the spectrum was found to correspond to the

---

Copyright © 1992 by the American Institute of Aeronautics and Astronautics, Inc. All rights reserved.
*Electrical Engineering Department.
†Professor of Mechanical Engineering Department.

cell size estimated by eye. For the more irregular cases, a single dominant cell size was more difficult to discern. However, the smallest value of cell size in the spectrum was found to correspond closely to the cell size estimated by eye.

1. Introduction

The major advances made in the past decade in gaseous detonations revolve around the use of the cell size to characterize the length scale of the reaction zone thickness of a real cellular detonation. This originates from the paper of Edwards et al.[1] in 1979 pointing out the possible universal validity of the $d_c=13\lambda$ correlation between the critical tube diameter and the cell size, a result first noted by Mitrofanov and Soloukhin[2] some fifteen years earlier. Subsequent experimental studies by Knystautas et al.[3] and numerous other investigators have confirmed Edwards' conjecture for a variety of fuel-oxygen and fuel-air mixtures. A simple analytical model for the critical energy for the direct initiation of a detonation was also developed by Lee.[4] The model gave a fairly good prediction of the critical energy based on cell size data. A review of these advances in the early 1980s was given by Lee.[5]

In the mid 1980s, Moen et al.[6] pointed out the failure of the $d_c=13\lambda$ criterion for mixtures having very regular cell patterns. For example, it was found that for C_2H_2-O_2 diluted with 75% Ar, the critical tube diameter could be as high as 26λ as compared with mixtures having more irregular cellular structures where $d_c=13\lambda$ applies. Moen et al.[6] also demonstrated that the velocity deficit in smooth tubes depends on cell regularity. For mixtures with irregular cell patterns, the velocity deficit deviates significantly from the prediction based on the Fay-Dabora model. More recent work by Dupré et al.[7] lends further support to Moen's results in that detonation limits are found to depend on cell regularity as well. Mixtures with very regular cell patterns fail more easily than mixtures with highly irregular cell patterns. It appears that in addition to the cell size λ, there is a need for a parameter capable of quantifying the regularity of the detonation cell pattern in order to make a more meaningful correlation with the dynamic parameters of detonations.

Although the sooted foil is commonly used to study the three-dimensional structure of gaseous detonations, the present methods of analyzing soot-foil records are limited. Experimenters are generally able to obtain an average cell size by estimating the characteristic length scale of the cellular pattern on soot-foils "by eye." Irregularity, however, has only been classified qualitatively, notably by Libouton and Van Tiggelen.[8] Even the estimation of the cell size is not always obvious, especially in the case of very irregular cellular patterns, and requires an experienced eye in order to minimize inconsistencies due to subjectivity.

The first attempt to achieve a more objective analysis of soot-foils by using digital image processing techniques was made by Shepherd et al.[9] and Shepherd and Tieszan.[10] In his work, Shepherd calculated the Fourier spectrum of the digitized image of soot-foils, thus obtaining a spectrum of spatial frequencies. Although he was able to obtain an estimate of the characteristic cell size from the dominant spatial frequency, the full spectrum was difficult to interpret and only qualitative observations pertaining to cell irregularity were made.

The problem of studying irregular or chaotic behavior has been researched in other fields using a number of different techniques. The autocorrelation function (ACF) has been successfully used to study the onset of chaotic behavior in the patterns generated by surface ripple waves in a liquid.[11] The histogram, although commonly used, has not been applied to the study of irregularity in detonation structure, presumably due to the difficulties involved in implementing it. Digital techniques have made it easier to apply standard statistical analysis to soot-foil cellular patterns. It should be noted that in this paper, the irregularity of the cellular pattern is measured in terms of the fluctuations in the line spacings of the pattern even though it can be due to many factors. The application of the histogram and the ACF to the analysis of soot-foil cellular patterns is explored.

2. Soot-Foil Use in the Analysis

For the present study, two sets of soot-foils were analyzed. A first set was obtained from detonation experiments conducted in a large aluminum tube 3.65 m

Table 1 Mixture compositions and initial pressures of the trials from which the foils were obtained

Foil name	Mixture	Initial pressure
sh 20	$(C_2H_2 + 2.5\ O_2) + 0\%A_R$	12 Torr
sh 10	$(C_2H_2 + 2.5\ O_2) + 50\%A_R$	40 Torr
sh 25	$(C_2H_2 + 2.5\ O_2) + 75\%A_R$	100 Torr
Van Tiggelen good	$(H_2 + 0.5\ O_2) + 50\%He$	175 Torr
Van Tiggelen poor	$(C_2H_2 + 1.25\ O_2) + 0\%A_R$	50 Torr
Van Tiggelen very poor	$(C_2H_2 + 5N_2O) + 80\%A_R$	100 Torr

long with an inner diameter of 145 mm. The detonable mixture inside the tube was ignited by a high voltage spark discharge at one extremity. A soot-foil approximately 30 x 30 cm lined the inner wall near the other extremity. After each experiment, a contact print was immediately taken of the soot-foil to have the cleanest possible record of the detonations' cellular pattern. These contact prints were subsequently used in the analysis. The compositions of the gaseous mixtures used in the trials of this experiment (i.e., sh 20, sh 10, and sh 25), are shown in Table 1. The acetylene-oxygen mixtures were diluted with increasing concentrations of argon to cover a wide range of cellular irregularity. The initial pressures of the mixtures were chosen such that the cell size $\lambda \approx d/20$, where d is the tube diameter. This was done to reduce the boundary effects of the tube wall. By minimizing the coupling between the acoustic modes of the tube and oscillatory modes of the detonation, the natural instability of the mixture manifested by the cellular pattern can be studied.

A second set of soot-foils provided by Van Tiggelen[12] were also analyzed. These soot-foils were obtained in a 4 m.

Fig. 1 The cellular pattern on a section of an actual soot-foil from the large tube experiment:
sh 20; ($C_2H_2 + 2.5O_2$) + 0% Ar.

long rectangular channel with inner dimensions 32 x 92 mm. The mixture compositions for these foils are also given in Table 1. These soot-foils were chosen as representative of "good," "poor," and "very poor" regularity in cellular structure according to Van Tiggelen's subjective classification.

3. Digitization of the Soot-Foils

Perhaps one of the most common experimental methods of studying the three-dimensional structure of gaseous detonations is to use soot-foils. These soot-covered sheets inserted inside a detonation tube show the typical diamond-shaped cellular pattern etched on the foil by a passing detonation wave (Fig. 1).

Before the actual analysis of the cellular pattern can be performed, one-dimensional digital signals containing the information of the pattern must be obtained. The conversion of the image of the cellular pattern into digital signals involves several steps; namely, the elimination of noise by hand tracing, the digitization of the image, and the extraction of one-dimensional signals from the digitized image. The entire procedure will be described using, as an example, a typical soot-foil obtained from the large tube experiment [sh 20, ($C_2H_2 + 2.5 O_2$) + 0%Ar].

3.1 Noise Elimination

In estimating the cell size by eye, one measures the characteristic spacing between adjacent lines on the foil. The human eye can easily recognize the lines forming the cellular pattern, effectively ignoring noise in the foil image such as marks caused by dirt swept across the foil, variations in darkness due to nonuniformities in the depositing of the soot layer, and streaks caused by the rapid flow of gas over the soot layer.

The digital signal processing techniques described in this paper require a digitized image of the cellular pattern. This is obtained by scanning gray-scale levels of the contact prints of the soot-foils. The gray scale is simply a numerical representation of the different levels of brightness in an image from white (0) to black (255), and by digitizing the contact print, a two-dimensional array of numbers representing the cellular pattern is obtained. Digital signal processing techniques can then be applied to the digitized image to extract information from it. However, a reasonably low noise signal is necessary to apply these techniques, consequently, the previously mentioned foil image noise must be eliminated. After much experimenting with various digital image enhancing techniques such as high and low pass filters and edge enhancing algorithms, the best image filter was found to still be the human eye. Indeed, when the the cellular pattern was deeply embedded in noise and even partially erased in certain places, only the filtering and pattern recognition ability of the eye was able to distinguish the lines. The simplest and most effective method of obtaining a noise free digital image was to first trace out the lines of the cellular pattern by hand, and then digitize this hand-drawn representation of the soot-foil. The digitized image of the noise-free cellular pattern of the soot-foil of Fig. 1 is shown in Fig. 2.

3.2 Line Profiles

Information on the characteristic cell size and the irregularity of the cellular pattern is contained in the two-dimensional digitized image of the pattern. Although it is possible to analyze this image directly using two-dimensional techniques, it is simpler to consider a one-dimensional representation of the cellular pattern. By taking line profiles of images, one-dimensional signals can be obtained without loss of cell size information.

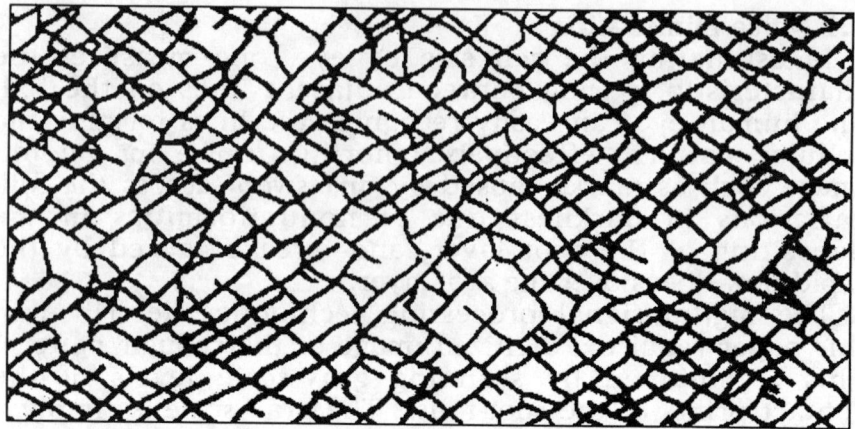

Fig. 2 The digitized image of the hand-traced cellular pattern of the soot-foil shown in Fig. 1.

A line profile is simply a plot of the variation of gray level along a certain line across the digital image. The orientation of the lines on the foil image is chosen to be normal to the direction of propagation of the detonation wave as shown in Fig. 3. In this way, line profiles containing the cell size information are obtained. This is done to be consistent with the physical meaning of the cell size, which is the spacing between the triple-point trajectories across the detonation front. The cell size can be read from the line profile as the spacing between two adjacent spikes (e.g., the typical line profile in Fig. 4), and variations in the spacing between the spikes can be attributed to cell pattern irregularity. For each foil, a set of different line profiles was taken across the cellular pattern to improve the statistical accuracy of the analysis results. For a typical analysis of a foil, 20 line profiles spaced approximately 10 mm apart were taken across the entire digitized image and the analysis results of these lines were averaged together.

The cellular pattern as shown in Fig. 2 is formed by two sets of superposed diagonal lines representing transverse waves in the detonation front propagating in opposite directions. In the case of a detonation tube with a circular cross section, this corresponds to transverse waves propagating in the clockwise and counter-clockwise directions. For a rectangular channel, the transverse waves propagate in the vertical and horizontal axes.

DIGITAL ANALYSIS OF SOOT FOILS 189

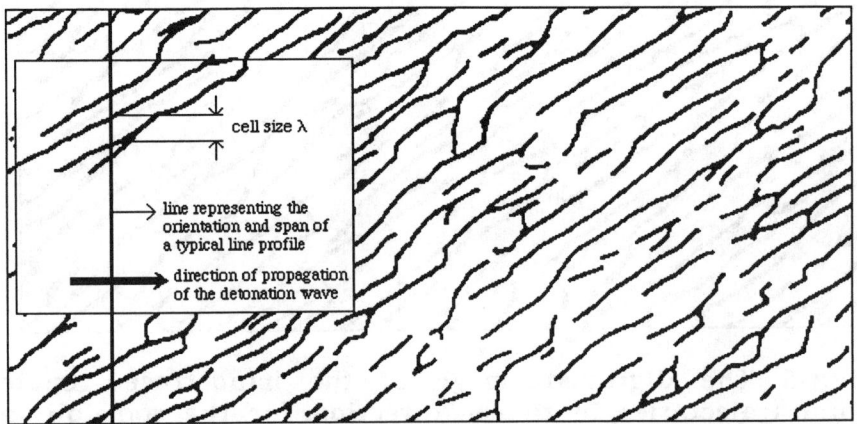

Fig. 3 The digitized image of the hand-traced triple-point trajectories in the θ+ direction for the soot-foil in Fig. 1.

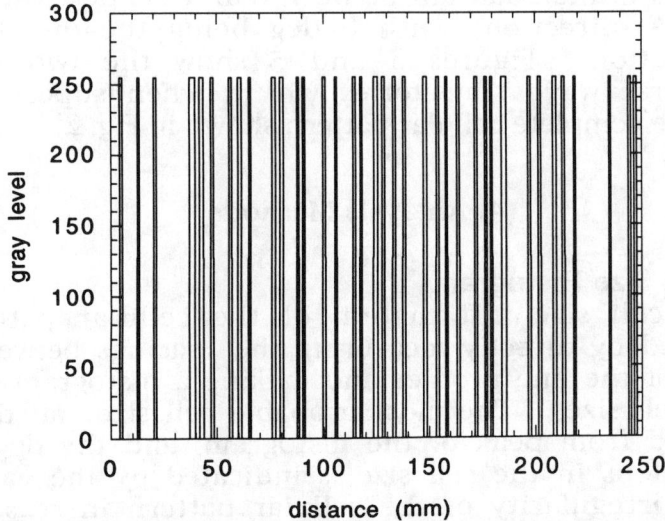

Fig. 4 Plot of a typical line profile from the soot-foil in Fig. 1.

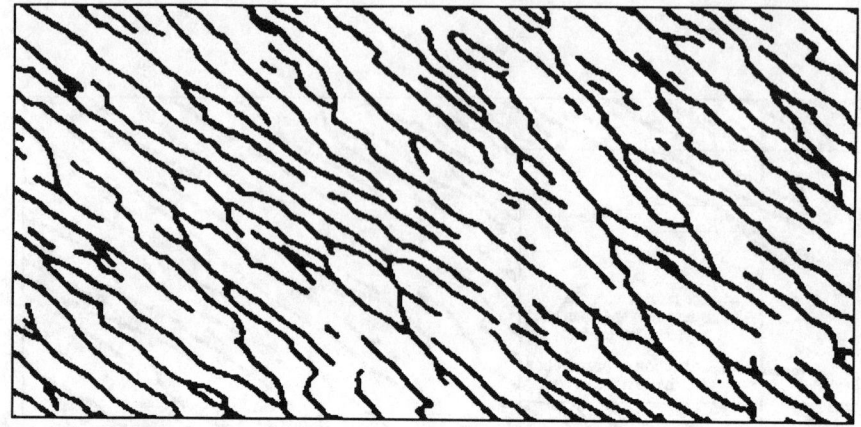

Fig. 5 The digitized image of the hand-traced triple point trajectories in the θ- direction for the soot-foil in Fig. 1.

Consequently, it is best to consider each set of waves separately for the cell size analysis. The soot-foil pattern was traced out in two steps: first the lines in the quadrant between 0 and +90 deg (indicated as the θ+ direction), then the lines in the quadrant between 0 and -90 deg (indicated as the θ- direction), with 0 deg being the direction of propagation. Figures 3 and 5 show the two sets of transverse wave trajectories which, when superimposed form the complete cellular pattern shown in Fig. 2.

4. Analysis Methods

4.1 Cell Size Histogram

The cell size information of the cellular pattern is extracted by directly measuring the spacings between the spikes of the line profiles and making a histogram plot of these cell sizes. The most probable cell size can then be obtained from peak of the histogram, and the degree of fluctuations in the cell size is indicated by the variance. As the irregularity of the cellular pattern increases, the variance is expected to increase as well. To test the reproducibility of the results, the histogram calculation was performed on two soot-foils obtained from two separate trials having identical initial conditions. The resulting variances were found to differ by 18% in the θ+ direction and by 2% in the θ- direction. As previously mentioned, many line profiles were taken across the cellular pattern, so the cell size and variance represent averaged values over the entire foil.

The histograms of the soot-foil in Fig. 1 are shown in Fig.6.

4.2 Autocorrelation Function

When analyzing soot-foils by eye, an experienced experimenter searches for cells within the pattern that are typical in shape and size to the majority of the other cells, i.e., he searches for the largest group of cells that are similar to one another. The ACF performs this mathematically since it shows the level of correlation or similarity within a signal. The ACF represents the level of correlation between a signal and a phase shifted copy of itself. For example, a periodic signal is identical to itself when phase shifted by exactly one period. Hence, the ACF shows a large correlation at shifts equal to integer multiples of the period. The ACF of a periodic signal is also periodic with peaks located at shifts equal to multiples of the period. So if a cellular pattern is perfectly regular, i.e., the spacings of the lines are exactly equal, the the ACF consists of a train of identical peaks separated by shifts exactly equal to the line spacing. If the signal contains random fluctuations, the ACF tends to decay since the signal becomes less correlated. Therefore, the ACF is useful for discerning periodicities or recurring features in a signal, since peaks in the ACF indicate the shifts at which strong self-similarity occurs.

For this analysis, a slightly modified version of the standard autocorrelation function or ACF [13,14] is used. The ACF used was modified in order to eliminate finite length bias, which is an undesirable linear amplitude taper caused by finite signal length.[13,14] The removal of bias is accomplished by calculating the crosscorrelation between the signal and a truncated, zero padded copy of itself.

Consider $x[n]$ to be a discrete signal with an even number of elements M,

$$y[n] = \begin{cases} x[n] & 0 \leq n \leq M/2 \\ 0 & M/2 < n \leq M \end{cases}$$

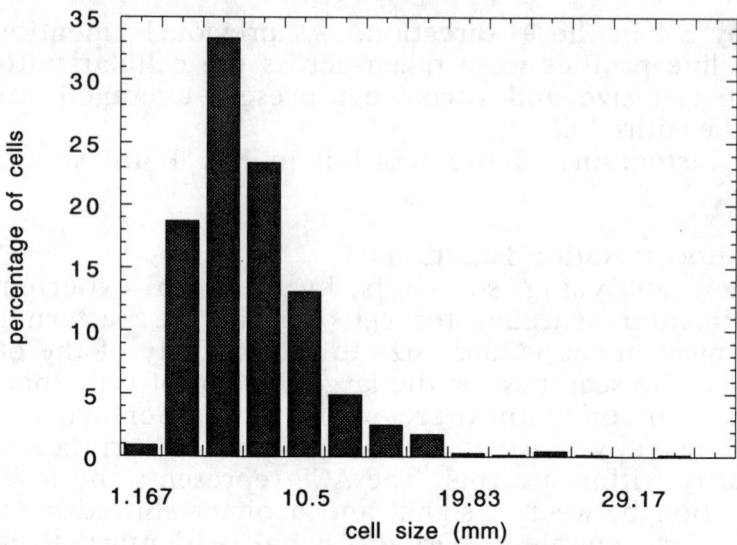

Fig. 6a The histogram of the cell sizes of the triple-point trajectories in the θ+ direction for the soot-foil in Fig. 1: the mean cell size is 7.45 mm and the variance is 13.48 mm.

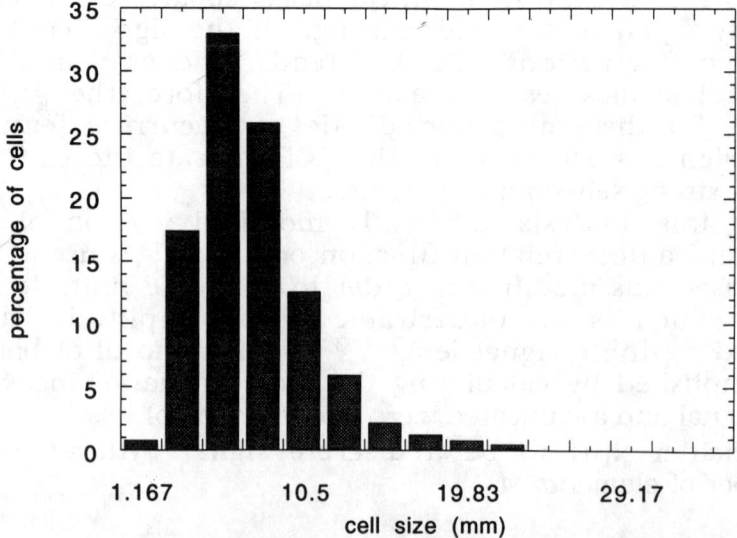

Fig. 6b The histogram of the cell sizes of the triple point trajectories in the θ− direction for the soot-foil in Fig. 1: the mean cell size is 7.40 mm and the variance is 11.61 mm.

then the unbiased autocorrelation function $\phi_{xx}[m]$ of $x[n]$ is given by:

$$\phi_{xx} = c_{xy}[m] \equiv \frac{1}{M} \sum_{n=0}^{M-|m|-1} x[n]y[n+|m|], \quad 0 \leq |m| \leq \frac{M}{2} - 1$$

where $c_{xy}[m]$ is the cross correlation function[13,14] between $x[n]$ and $y[n]$.

The advantages of this scheme is that we obtain an ACF of the signal without finite length bias and with constant variance.[13] There are other ways to correct for bias [13] but they result in a linear taper of the variance.

For the analysis of a soot-foil, the ACF is calculated for several line profiles in the foil image, then the ACFs are averaged together to give an overall ACF for the foil. It should be noted that long signals provide a better approximation of the ACF, so it is necessary to use line profiles spanning as many "cells" as possible.

As an example, it is easier to consider first a soot-foil having a highly regular cellular pattern (e.g. Van Tiggelen's good regularity foil). The ACF of this foil is shown in Fig. 7. Because of small fluctuations in the pattern, the ACF is shown to exhibit a progressive decay in amplitude as the shift increases. It is interesting to note that the shift of the first peak (excluding the peak at zero shift) is found to corresponds closely to the cell size estimated by eye. The first peak occurs at a shift of 14.6 mm for the θ+ waves and 14.07 mm for the θ- waves, which is close to the estimated cell size of 13 mm.

For the more irregular soot soil of Fig.1, the ACF is shown in Fig. 8. In contrast to to Fig. 7, it can be seen that a single dominant cell size is more difficult to discern due to the large fluctuations in amplitude and periodicity of the ACF peaks. However, the first peak, which occurs at a shift of 6.61 mm, still corresponds closely to the value of the cell size measured by eye. However, the ACF also shows other highly correlated spacings in the cellular pattern as indicated, for example, by the large peak at a shift of 33.65 mm. Although the shift of the first peak of the ACF corresponds closely to the cell size estimated by eye, it should be noted that this peak may not be the largest or dominant one. This indicates the possibility of other dominant cell sizes in the pattern. It appears that the eye

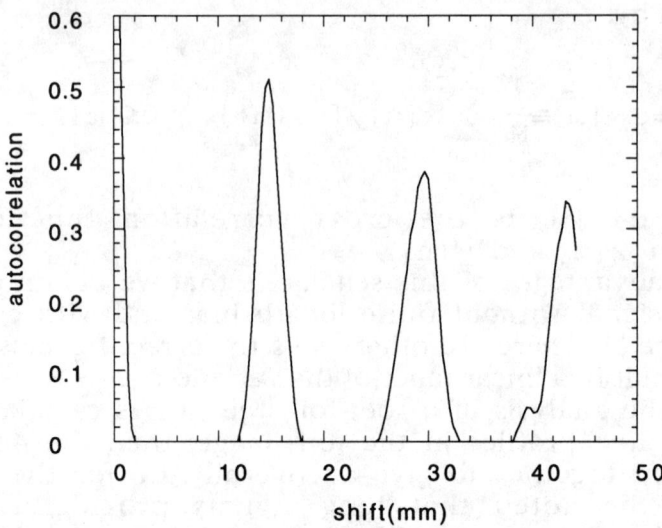

Fig. 7a The average ACF of the triple-point trajectories in the θ+ direction for Van Tiggelen's good regularity foil.

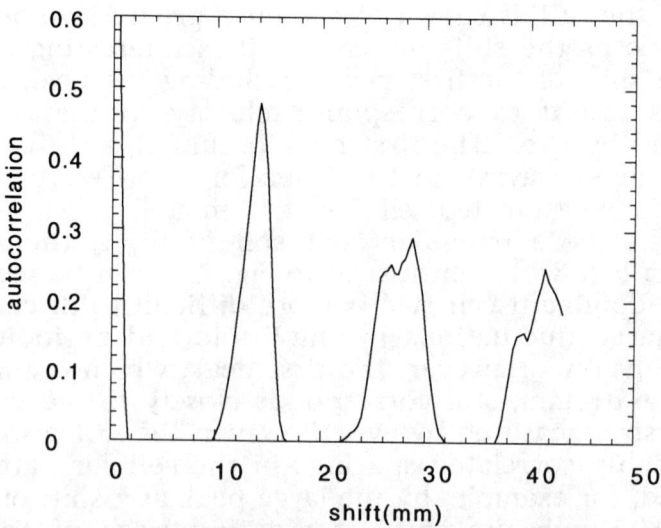

Fig. 7b The average ACF of the triple point trajectories in the θ− direction for Van Tiggelen's good regularity foil.

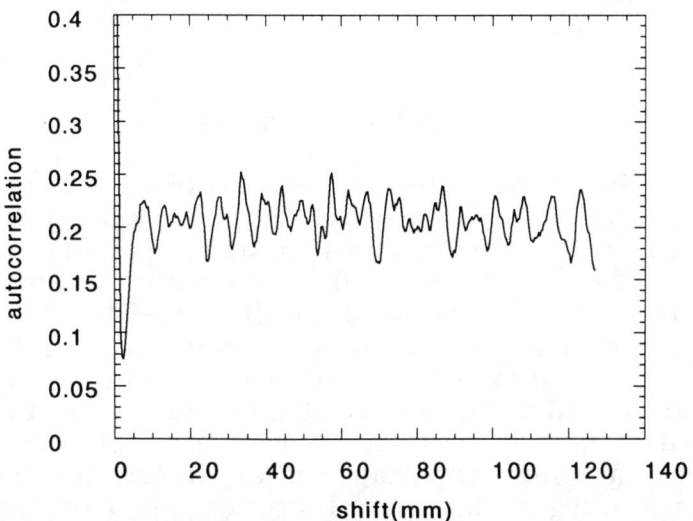

Fig. 8a The average ACF of the triple-point trajectories in the θ+ direction for the soot-foil in Fig. 1:
sh 20; $(C_2\ H_2 + 2.5O_2\ ) + 0\%Ar.$

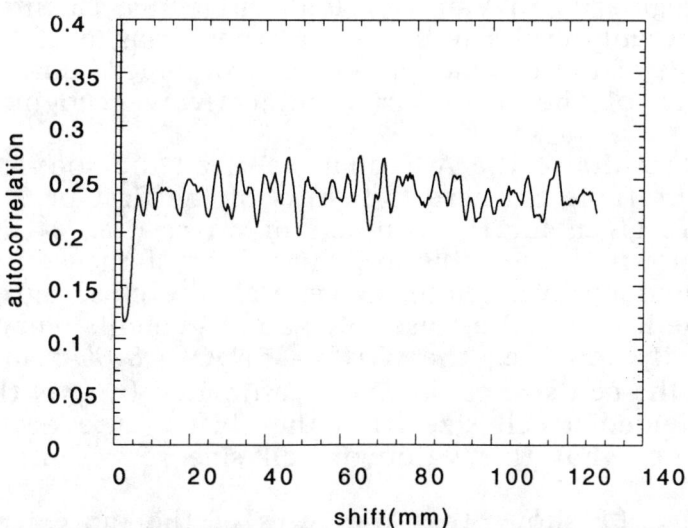

Fig. 8b The average ACF of the triple-point trajectories in the θ− direction for the soot-foil in Fig. 1:
sh 20; $(C_2\ H_2 + 2.5O_2\ ) + 0\%Ar.$

subjectively filters out information when the experimenter tries to obtain only a single cell size from a given foil.

5. Results and Discussion

The results of the histogram analysis of all the soot-foils of Table 1 are shown in Table 2. In all cases, the mean cell size given by the histogram was found to be very close to the value of cell size estimated by eye within experimental error. For the special case of the $(C_2H_2 + 5N_2O) + 80\%Ar$ mixture, the cellular structure was too irregular for the cell size to be estimated by eye. The histogram, however, managed to yield mean cell size of approximately 7 mm.

Figure 9 shows that the variance of the cell size decreases as the percentage of argon dilution in the acetylene oxygen mixture increases, confirming the observations that the cellular pattern becomes more regular as the argon dilution is increased. Fig. 10 shows that for Van Tiggelen's soot-foils, the variance was also found to increase as the regularity of the cellular pattern became poorer. In fact, it was found to increase by a factor of almost 40 spanning the range of good regularity to very poor regularity in Van Tiggelen's subjective classification. The present results indicate that the variance of the line spacings of the cellular pattern can provide a quantitative measure of the irregularity subjectively recognized by eye.

The results of the ACF analysis of all the soot-foils are given in Table 3 where the shifts of the first peak in the ACF are given for the two sets of waves (i.e., θ+ and θ-). Also shown in the table are the values of the cell size as estimated by eye. Note, in general, the close agreement obtained. For the case of Van Tiggelen's very poor regularity foil [i.e., the $(C_2H_2 + 5N_2O) + 80\%Ar$ mixture] where the cell size could not be estimated by eye, the ACF still yielded a cell size from the shift of the first peak. However, what this value of cell size represents is not clear.

Figure 11 shows that the ACFs of the more irregular cellular patterns (e.g., Van Tiggelen's poor regularity foil) contain one or more large peaks located at shifts larger than the cell size estimated by eye. For this case, the first peak is located at a shift of approximately 6 mm, but significantly larger peaks occur at shifts of approximately

Table 2 Histogram analysis results

Foil name	Mixture	λ By eye, mm	Mean λ θ+, mm	Mean λ θ-, mm	Variance θ+, mm	Variance θ-, mm
sh 25	(C$_2$H$_2$ + 2.5 O$_2$) + 75%Ar	6	7.06	6.16	4.88	4.25
sh 10	(C$_2$H$_2$ + 2.5 O$_2$) + 50%Ar	7	7.91	6.96	11.99	6.29
sh 20	(C$_2$H$_2$ + 2.5 O$_2$) + 0%Ar	7	7.45	7.40	13.48	11.61
Van Tiggelen good	(H$_2$ + 0.5 O$_2$) + 50%He	13	14.34	13.82	1.23	1.24
Van Tiggelen poor	(C$_2$H$_2$ + 1.25 O$_2$) + 0%Ar	7	6.88	6.71	8.38	9.48
Van Tiggelen very poor	(C$_2$H$_2$ + 5N$_2$O) + 80%Ar	---	7.05	7.36	38.39	51.05

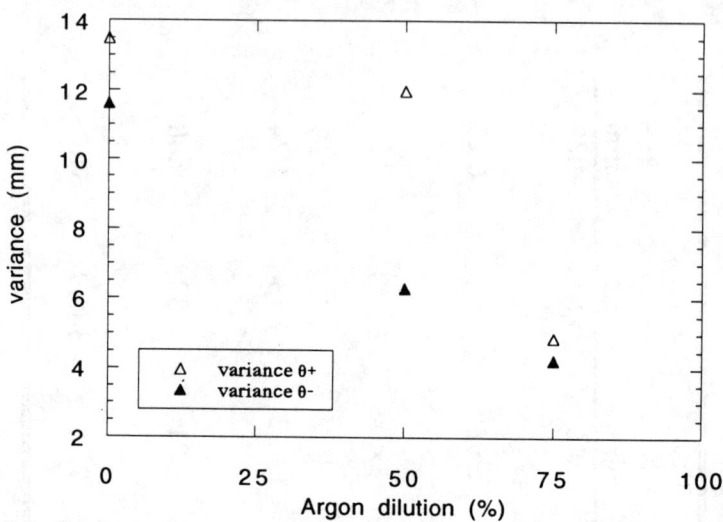

Fig. 9 Plot of the variance of the cell size vs argon dilution.

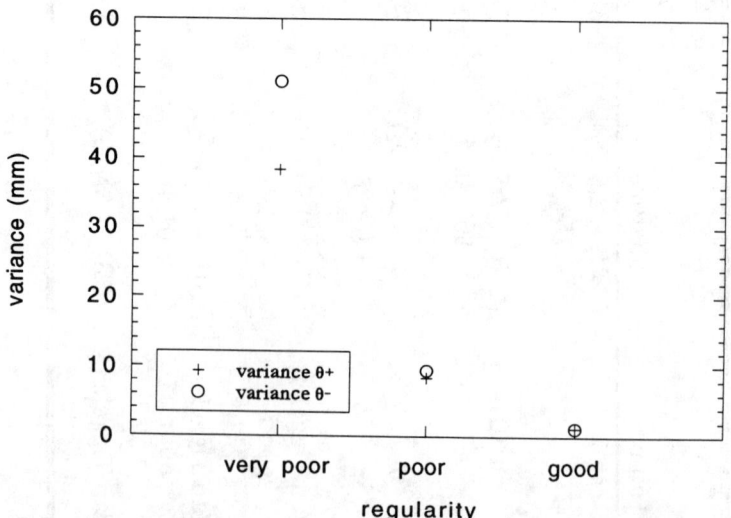

Fig. 10 Plot of the variance of the cell size vs regularity for Van Tiggelen's foils.

Table 3 ACF analysis results

Foil name	Mixture	λ By eye, mm	Shift of 1st peak θ+, mm	Shift of 1st peak θ-, mm
sh 25	(C_2H_2 + 2.5 O_2) + 75%A_R	6	7.12	6.64
sh 10	(C_2H_2 + 2.5 O_2) + 50%A_R	7	7.49	7.87
sh 20	(C_2H_2 + 2.5 O_2) + 0%A_R	7	6.61	6.16
Van Tiggelen good	(H_2 + 0.5 O_2) + 50%He	13	14.6	14.07
Van Tiggelen poor	(C_2H_2 + 1.25 O_2) + 0%A_R	7	6.68	5.3
Van Tiggelen very poor	(C_2H_2 + 5N_2O) + 80%A_R	--	3	2.58

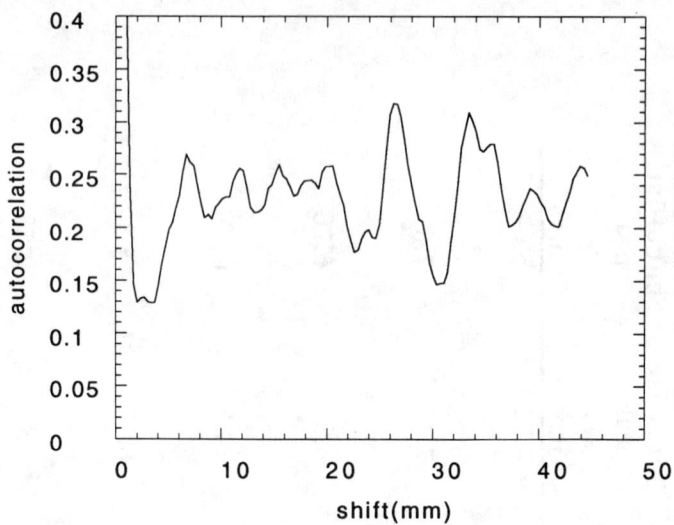

Fig. 11a The average ACF of the triple point trajectories in the θ+ direction for Van Tiggelen's poor regularity foil.

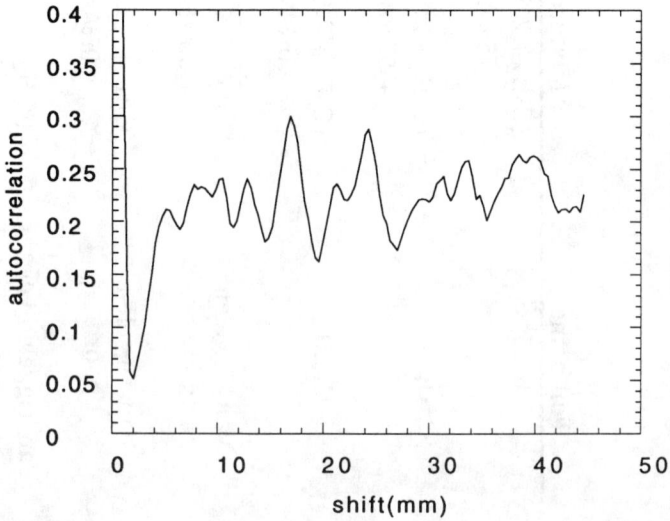

Fig. 11b The average ACF of the triple point trajectories in the θ− direction for Van Tiggelen's poor regularity foil.

17 and 24 mm as well. This suggest that there are dominant modes in the transverse waves of the detonation having characteristic spacings differing significantly from cell size estimated by eye.

6. Conclusion

Despite the somewhat tedious process of hand drawing the cellular pattern, the digital analysis techniques described in this paper provide useful results. Calculating the variance of the cell size provides an experimental method of quantitatively measuring the level of irregularity in the cellular pattern of soot-foils. The mean cell size obtained from the histogram as well as the shift of the first peak of the ACF provide a less subjective measure of the characteristic cell size.

The present methods can also be applied to end-on soot-foils, which give a better instantaneous representation of the detonation structure. The digital techniques could, in fact, be improved by using the photographic technique of recording the end-on structure developed by Presle et al.[15] This method, which uses light reflected off an aluminized sheet of mylar to record the cellular pattern of the detonation structure, would provide a more noise-free image eliminating the need for hand tracing the cellular pattern.

Although the human eye (and brain) are still needed for the initial noise elimination of the cellular pattern, this does not impart any subjectivity in the interpretation of the foil itself. It may be concluded that digital signal processing techniques can improve on the traditional methods of analyzing soot-foils.

Acknowledgments

The authors would like to thank Constant Maton and Sarah Brun for their dedicated technical contributions and McGill Shock Wave Physics Group.

References

[1]Edwards, D. H., Thomas, G. O., and Nettleton, M. A., "The Diffraction of a Planar Detonation Wave at an Abrupt Area Change," *Journal of. Fluid Mechanics*, Vol.95, 1979, pp.79-96.

[2] Mitrofanov, V. V., and Soloukhin, R. I., "The Diffraction of Multi-Front Detonation Waves," *Soviet Physics-Doklady*, Vol. 9, 1965, p. 1055.

[3] Knystautas, R., Lee, J. H. S., and Moen, I., "Determination of Critical Tube Diameter for C2H2-air and C2H4-air Mixtures,"*Christian Michelsens Institute Report*, Bergen, Norway, 1981.

[4] Lee, J. H. S., "Initiation of Gaseous Detonation," *Annual Review of Physics and Chemistry*, Vol. 28, 1977, pp.75-104.

[5] Lee, J. H. S., "Dynamic Parameters of Gaseous Detonations," *Annual Review of Fluid Mechanics*, Vol. 16, 1984, pp. 311-336.

[6] Moen, I. O., Sulmistras, A., Thomas, G. O., Bjerketvedt D., and Thibault, P. A., "The Influence of Cellular Regularity on the Behavior of Gaseous Detonations", *Progress in Astronautics and Aeronautics*, Vol. 105, AIAA, NY, 1985, p.220.

[7] Dupré, G., Joannon, J., Knystautas R., and Lee, J. H. S., "Unstable Detonations in the Near-Limit Regime in Tubes," *The 23rd International Symposium on Combustion*, The Combustion Institute, Pittsburg, PA, 1990, p.1813.

[8] Libouton, J. -C., and Van Tiggelen, P. J., "Cinétique, Structure, et Entretien des Ondes de Détonation," *Proceedings of the First Specialist Meeting (International) on Combustion*, Bordeaux, France, 1981, p. 437.

[9] Shepherd, J. E., Moen, I. O., Murray, S. B., and Thibault, P. A., "Analysis of the Cellular Structure of Detonations," *21st Symposium (International) on Combustion*/The Combustion Institute, 1986, pp.1649-1658.

[10] Shepherd, J. E., and Tieszen, S. R., "Detonation Structure and Image Processing," Sandia National Laboratories Report. SAND86-0033, Albuquerque, NM, 1986.

[11] Tufillaro, N. B., Ramshankar, R., and Gollub, J. P., "Order-Disorder Transition in Capillary Ripples," *Physics Review Letters*, Vol 62, N°. 4, 1989, pp.422-425.

[12] Van Tiggelen, P. J., private communication,1983.

[13] Schwartz, M., and Shaw, L., Signal Processing: Discrete Spectral Analysis, Detection and Estimation, McGraw -Hill, 1975.

[14] Oppenheim A. V., and Schafer R. W., Discrete-Time Signal Processing, Prentice Hall, NJ, 1989.

[15] Presles, H. N., Desbordes, D., and Bauer, P., "An Optical Method for the Study of the Detonation Front Structure in Gaseous Explosive Mixtures," *Combustion and flame*, Vol. 70, N°. 2, 1987, p. 204.

Cylindrical Detonations in Methane-Oxygen-Nitrogen Mixtures

Miloud Aminallah* and Jacques Brossard†
Université d'Orléans, Bourges, France
and
A. Vasiliev‡
Siberian Academy of Sciences, Novosibirsk, Russia

Abstract

The characteristics of the cylindrical detonation considered as an intermediate situation compared with planar and spherical cases are investigated. The gaseous mixtures $CH_4 + xO_2 + zN_2$ ($1 \leq x \leq 3$ and $0 \leq z \leq 1$) and $x\, CH_4 + (1-x)\, C_3H_8 + (5-3x)\, O_2 + zN_2$ ($0 \leq x \leq 1$ and $0 \leq z \leq 2$) are confined in a flat square chamber (50 x 50 cm²) with different thicknesses (e = 10-20 and 29 mm). The initial conditions are room temperature T_0 and variable pressure $0.3 \leq p_0$ (bar) ≤ 1.7. Detonation of the mixture is initiated by an exploding wire located either at the center of the square chamber or at the apex of the 90^{deg} sector. The energy deposition is variable. The transition from deflagration to detonation and the propagation are investigated with ionization probes and piezoelectric pressure transducers. The cellular structure and the DDT are observed by the soot film technique. To discuss and compare the detonability limits, we have evaluated the energy efficiency of the linear electrical ignitor as a function of the energy release and the thickness of the experimental chamber. The detonability limits are expressed as

---

Copyright © 1992 by the American Institute of Aeronautics and Astronautics, Inc. All rights reserved.
*Research Assistant, Laboratoire de Recherche Universitaire.
†Professor, Laboratoire de Recherche Universitaire.
‡Research Scientist, Institute of Hydrodynamics.

function of the fuel concentration, the nitrogen dilution, and the initial pressure. The transition from deflagration to detonation is observed and the predetonation radius is clearly correlated with the cell size. Independent of the initial conditions, the predetonation radius is always lower than a critical maximum value (~ 75 mm). All of the experimental results obtained with cylindrical diverging detonation waves are well correlated with different previous results obtained with plane and spherical symmetries. In particular, the critical energy of initiation of cylindrical detonation is closely related to the critical energy of spherical detonation.

Introduction

In recent years it was shown[1-6] that it is possible to correlate many detonation parameters particularly with chemical induction times. The transition of detonation from planar to spherical geometry, the cell size, the induction length, and the critical energy for direct initiation are well correlated.

From the practical point of view, for hazard analysis, for instance, the prediction of the initiation energy which characterizes the detonability of a given gaseous mixture appears as absolutely essential.

Numerous models[3,4] now are capable of estimating quantitatively the critical initiation of unconfined detonation. They are based on experimental observations and thermochemical properties of the mixture. Experimentally the critical energy is probably the parameter easiest to investigate in spite of the diversity of the energy deposition devices. It is clear[5,6] that the direct initiation and the self-initiation with transition from deflagration to detonation need a minimum energy deposited in the gaseous medium. The initiation occurs only if the energy deposition which generates the shock wave is at least of a certain strength and duration.[7,8]

Numerous experimental data are obtained from spherical and/or hemispherical symmetries[6,9,10]. They concern a wide variety of gaseous mixtures with oxygen and air. The mixtures based on methane particularly appear as very interesting cases for two major reasons. The first one is the presence of this fuel in the industrial activity ; and the second one is the important chemical stability of the molecule. This second reason necessitates a very large energy deposition for generating the direct initiation of unconfined detonation. The limits of detonability are narrow and experiments are difficult.

The purpose of the present work is to contribute to the knowledge of the detonation ability and the cell structure of mixtures contained in cylindrical vessels. Although the cylindrical symmetry is considered as the intermediate case between planar and spherical symmetries, the experimental data are not numerous[8,11-13] because of the apparent difficulty to avoid the wall effect. Our investigations concern gaseous methane/oxygen mixtures diluted with propane and/or nitrogen. Taking into account that it is easier to create the direct initiation by means of the cylindrical symmetry, a large range of experimental data is collected and correlated with different earlier results obtained with plane or spherical propagation. The possibility of predicting the initiation energy for spherical propagation is confirmed; and the critical cylindrical radius of detonation is clearly related to the critical spherical radius and to the cell size.

Experimental

Mixtures

The experiments are performed in

$$CH_4 + xO_2 + zN_2$$

mixtures with various values of x and z, particularly

$$1 \leq x \leq 3 \text{ and } z = 0$$

and

$$xCH_4 + (1 - x) C_3H_8 + (5 - 3x) O_2 + zN_2$$

mixtures with

$$0 \leq x \leq 1 \text{ and } 0 \leq z \leq 2$$

more sensitive by means of propane dilution. The initial conditions are room temperature $T_0 \sim 290$ K and pressure p_0 in the range of 0.3-1.7 bar.

Detonation Chamber

The mixtures are confined in a very rigid flat square chamber (50 x 50 cm²) with different thicknesses e = 10-20 and 29 mm. Few experiments are conducted in the 4-mm-thick chamber.

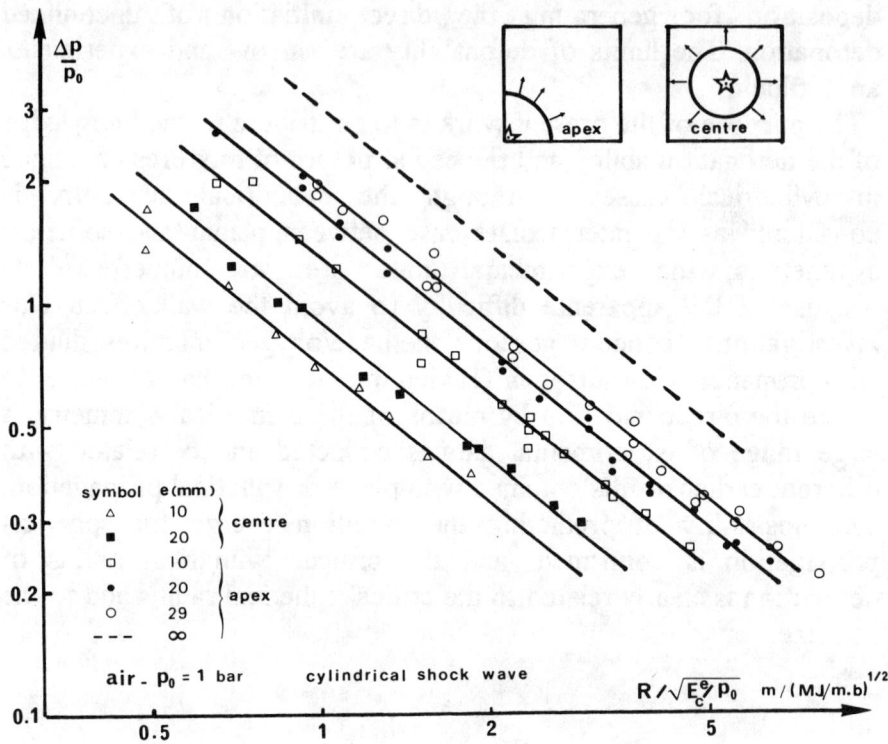

Fig. 1 Relative overpressures of the cylindrical shock wave vs dimensionless radial distance and bomb thicknesses.

The detonation is initiated by means of a copper exploding wire (0.18 mm in diameter, the length of the wire equals the thickness) located either at the center of the square chamber or at the apex of the 90^{deg} angle sector. The nominal electrical energy is accumulated in C = 8.33 µF capacitor with the variable potential in the range of 4-8.5 kV (300 J the maximum nominal electrical energy).

The initiation and the propagation of the detonation wave are investigated simultaneously by means of ionization probes, piezoelectric pressure gauges, and the soot-film technique.

Experimental Technique

The first procedure is to qualify the electrical initiation device. The nominal electrical stored energy $E = 0.5 \, CV^2$ is deposited either at the center of the experimental chamber or at the apex. The time duration of the deposition is approximately 3.5 µs. For the given thickness e of the bomb, the same nominal energy deposition creates the cylindrical shock wave which propagates either inside the 2π or the $\pi/2$ solid angle, respectively. The relative shock strength $\Delta p/p_0$ is directly related to the dimensionless similitude radius $R(E_c/p_0)^{1/2}$ which characterizes the cylindrical propagation. E_c (MJ/m) is the nominal electrical energy per unit length and is written Ecc and Eca according to the exploding wire location. Figure 1 shows the experimental results of the cylindrical shock strength for the thickness range of 10-29 mm and for the two cases of propagation in air at room pressure. The shock strength is clearly dependent on the thickness e of the chamber. Apparently the efficiency of the electrical device is inversely related to the thickness. When the energy release Eca is located at the apex it is easy to plot the shock strength vs the 1/e values and to extrapolate the curves toward the infinite thickness. In this way one arbitrary limiting curve is defined (Fig.1) which corresponds to the energy of reference E_{ca}^{∞}. Then the energetic efficiency coefficient

$$\eta_{ca}^e = E_{ca}^{\infty} / E_{ca}$$

is calculated for each thickness e (Table 1) independently of the shock strength which ranges from 0.2 to 2.0 bar.

A more detailed analysis of the different curves (Fig.1) based upon the constant shock strength at one fixed radius shows that : 1) the energetic efficiency coefficient increases with the thickness increase and 2) the experimental results obtained with the energy deposition located at the center compared with those obtained from the apex show that the relative efficiency coefficient E_{ca}/E_{cc} is approximately 0.33 independent of the length of the wire. This value does not agree with the predicted coefficient deduced from the geometrical analysis ; that is to say, the solid angle ratio $(\pi/2)/2\pi = 0.25$.

To connect the critical initiation energies in both the cylindrical and spherical detonation cases, it appears necessary to estimate the energy efficiency of the same initiator generating a hemispherical shock wave. Equipped with a U-shaped wire (20 mm long) the relative overpressure of the shock wave is investigated and compared with previous results[14] obtained from chemical gaseous detonation.

Therefore the chemical energy equivalency can be deduced. Figure 2 shows the experimental results and from the comparison of the two curves the efficiency coefficient $E_s/E = 0.16$ is evaluated. This

Table 1 Energetic efficiency coefficient of the initiation device

e, mm	10	20	29
$E_{ca}^{\infty}/E_{ca}^{e}$	0.22	0.36	0.46
$E_{ca}^{\infty}/E_{cc}^{e}$	0.34	0.32	

value is in good agreement with the value (0.14) proposed by Desbordes[6] for a similar device, (E_s the equivalent point energy).

In fact, it is clear that the energy efficiency depends on the nature and the initial pressure of the gaseous mixture in which the amount of energy is deposited. But, in first approximation and for the general discussion of our experimental results, it is supposed that the efficiency coefficients are not truly different in air or methane-oxygen mixture and that the relative situation of the set of curves is preserved.

Results

Now it is well known that the initiation of the detonation wave gaseous mixtures occurs according two different modes: 1) the direct initiation, which necessitates a large energy deposition and 2) the self-initiation characterized by a predetonation radius R_c through which an acceleration process of the front flame is developed. This radius is independent of the amount of energy. In numerous experiments we have observed a predetonation length and our results are compared with two sets of values. First are the values collected from previous works and related to the plane, cylindrical, and spherical symmetries by means of the characteristic radius of explosion (see Lee[4,8]) defined by $R_0 = (E/p_0)^{1/(j+1)}$ where $j = 0, 1,$ and 2, respectively, and E the amount of energy deposition. R_0 is the same for all geometries. Second are the values deduced from the measured cell size and the critical initiation energy

$$E_c = K_j \rho_0 I D^2 \left(\frac{j+3}{2}\right)^{j+1} b^{j+1}$$

based upon the Zeldovitch criterion and where K_j and I are numerical constants dependent on the geometry and the blast wave integral, D the CJ detonation velocity, and b the cell width.

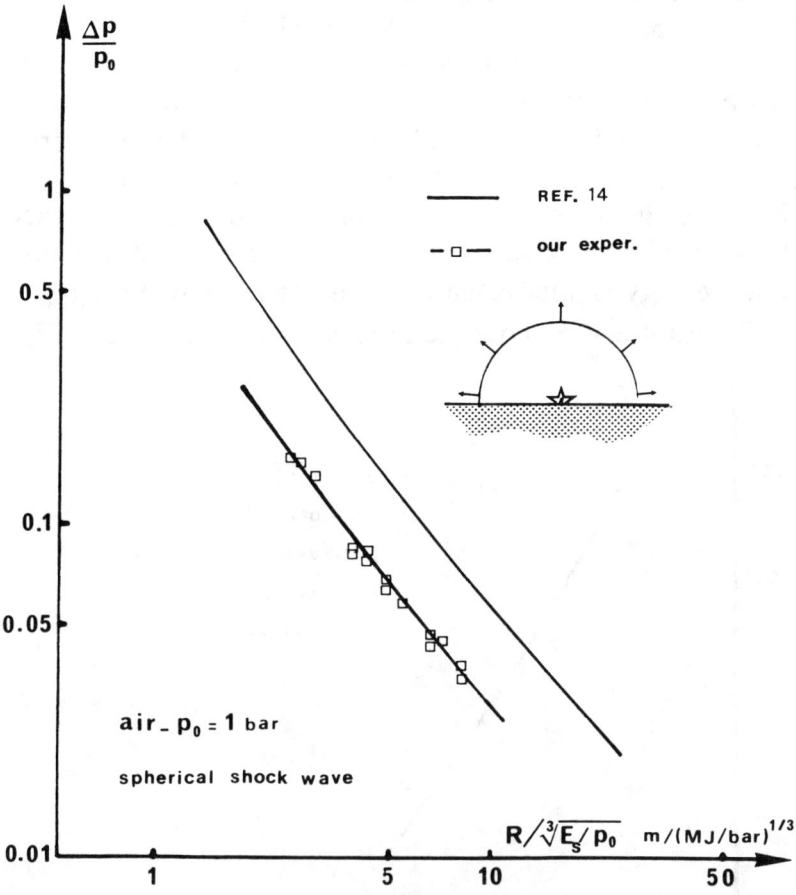

Fig. 2 Relative overpressure of the hemispherical shock wave vs dimensionless radial distance from the exploding wire.

Energy of initiation

In the range of our experimental conditions the initiation of the detonation wave always was obtained by transition from deflagration to detonation. It was observed that independent of the fuel percentage, the inert dilution, and the initial pressure the maximum value of the critical radius R_c was approximately 75 mm. The transition zone was detected by means of the average velocity of propagation deduced from the transit time duration between ionization probes or pressure gauges ; and the critical radius R_c was evaluated by measuring the average traces on the soot-film records.

Following the initiation the detonation wave propagation is uniform and the velocity agrees well with the CJ calculated one.

Figure 3 shows a typical frontier curve limiting the domain of self-initiation of a stoichiometric methane-oxygen mixture at initial pressure $p_0 = 1.4$ bar. Apparently, the nominal initiation energy E^e_{cc} (J/cm) needed for fully cylindrical ignition is decreasing with the thickness e increase. In fact, applying the above efficiency coefficients of the initiation device we observe that the nominal initiation energy is approximately constant $E^\infty_{cc} \sim 30$ J/cm ($p_0 = 1.4$ bar). Figures 4 and 5 show the nominal initiation energies E^e_{cc} and

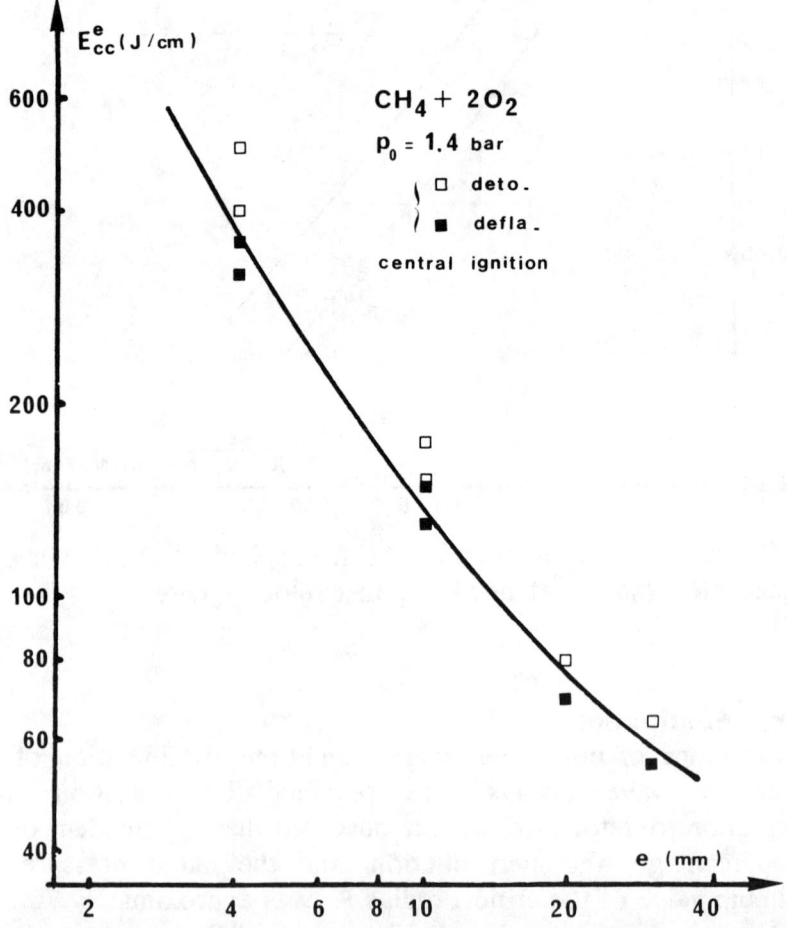

Fig. 3 Minimum nominal initiation energy for cylindrical detonation in $CH_4 + 2O_2$ mixture vs the thickness of the chamber; central ignition.

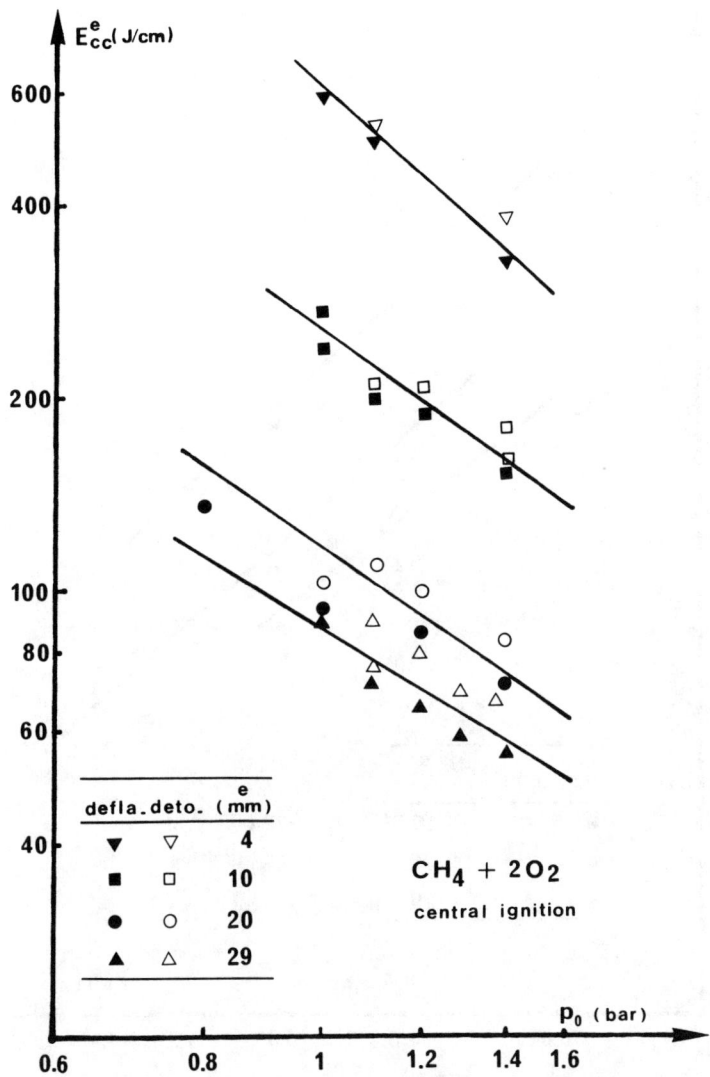

Fig. 4 Minimum nominal initiation energy for cylindrical detonation in CH_4+2O_2 mixture vs nitial pressure; central ignition.

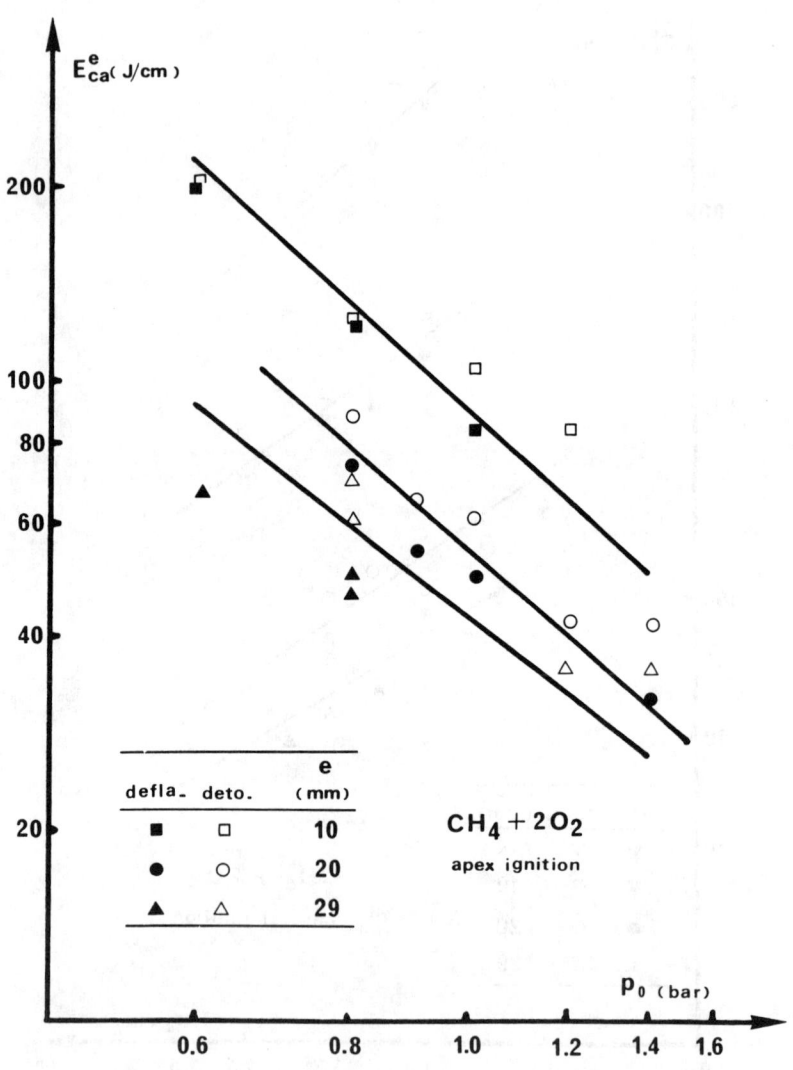

Fig. 5 Minimum nominal initiation energy for cylindrical detonation in CH_4+2O_2 mixture vs the initial pressure ; apex ignition.

Table 2 Nominal and corrected initiation energies for cylindrical detonation in CH_4+O_2 mixture (full symmetry)

p_0, bar	0.6	1.0	1.4
E_C^∞, J/cm	100-130	45-60	27-33
E, J/cm	16-21	7-10	4-5

E_{ca}^e initial pressure po and the thickness e as a parameter. The energy is clearly decreasing as po increases. For a fixed value po, the application of the efficiency coefficients leads to the same order of magnitude of the energy independently of the thickness and the ignitor location effects (Table 2). The approximate function is (full cylindrical symmetry)

$$E_c^\infty \text{ (J/cm)} \approx 51 \, p_o^{-1.6} \text{ (}p_o \text{ bar)}$$

Figures 6 and 7 show the minimum nominal initiation energy as a function of the fuel equivalence ratio of CH_4-O_2 mixtures and the initial pressure p_0 (e = 20 mm). The pressure effect on the detonability is very sensitive and this sensitivity is dependent on the fuel concentration. Few experiments are conducted using a different ignition device including a lower value of the storage capacitor C = 3 µF. The results (see Fig.7) show that the lower the capacitor is, the lower the amount of energy to ignite the detonation wave is and the higher the energy efficiency coefficient is.

It seems important to note the difficulty in defining the frontier curves which define the minimum nominal initiation energy. The most important reasons are 1) the large scattering of the experimental results; 2) the limited maximum value of the nominal stored electrical energy which limits the range of initial pressures near the lower boundary values; and 3) the mechanical resistance of the experimental chamber which limits the initial pressure near the upper-boundary values.

Several authors[2-4] have demonstrated the significant effects arising from dilution by fuel and/or inert component. Particularly the minimum initiation energy of mixtures based on gaseous methane is very affected by the presence of propane and/or nitrogen. The propane substitution decreases the characteristic induction length of the mixture and the nitrogen dilution increases the induction length. Figure 8 plots the minimum nominal initiation energy as a function of initial pressure and composition in stoichiometric methane-propane-oxygen-nitrogen mixtures. This initiation energy is directly correlated with the induction length.[4] The main results of our experimental investigation are as follows.

1) Taking into account the range of experimental initial conditions, the initiation of detonation in $CH_4 + O_2$ diluted by nitrogen was not observed even if the initial pressure $p_0 = 1.6$ bar.

2) The initiation energy decreases by factor of two with 5% propane substitution ($0.95CH_4 + 0.05\ C_3H_8 + 2.15O_2$).

3) The initiation energy increases by a factor of four with 13% nitrogen dilution ($0.95CH_4 + 0.05C_3H_8 + 2.15O_2 + 0.5N_2$).

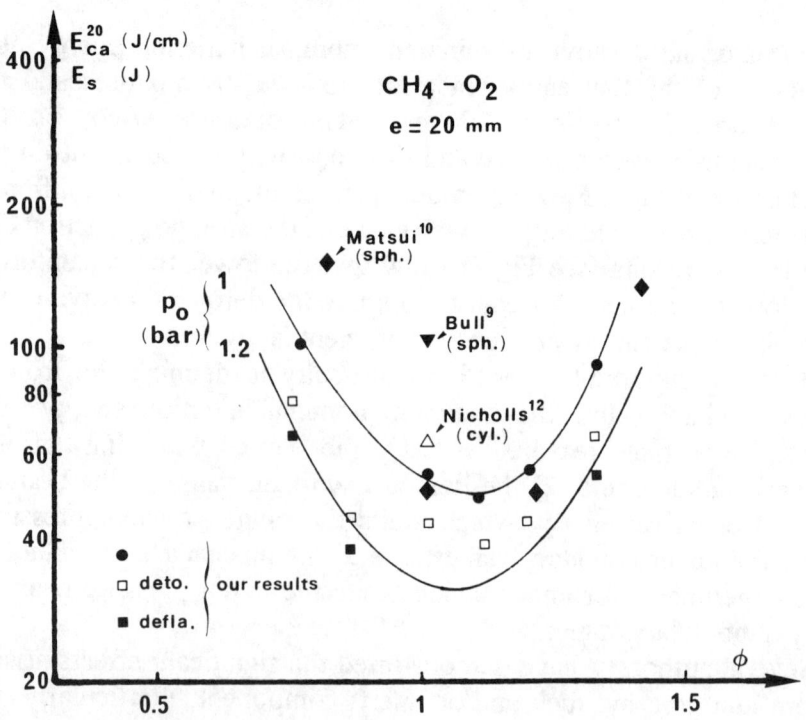

Fig. 6 Minimum nominal initiation energy as function of fuel equivalence ratio of CH_4-O_2 mixtures.

Cellular Structure

The cellular structure is investigated by means of the classical soot-film technique. The soot film is deposited on a brass foil maintained in contact with one of the parallel plane walls of the chamber. The quality of the records is not excellent because the successive moving reflected pressure waves in the closed volume affect the traces and sometimes the metal sheet. However, numerous results are obtained and correlated with the initial characteristics of the detonable mixture. Simultaneously the critical radius R_c of the predetonation zone is clearly observed and measured. The cellular structure is generally stable and the

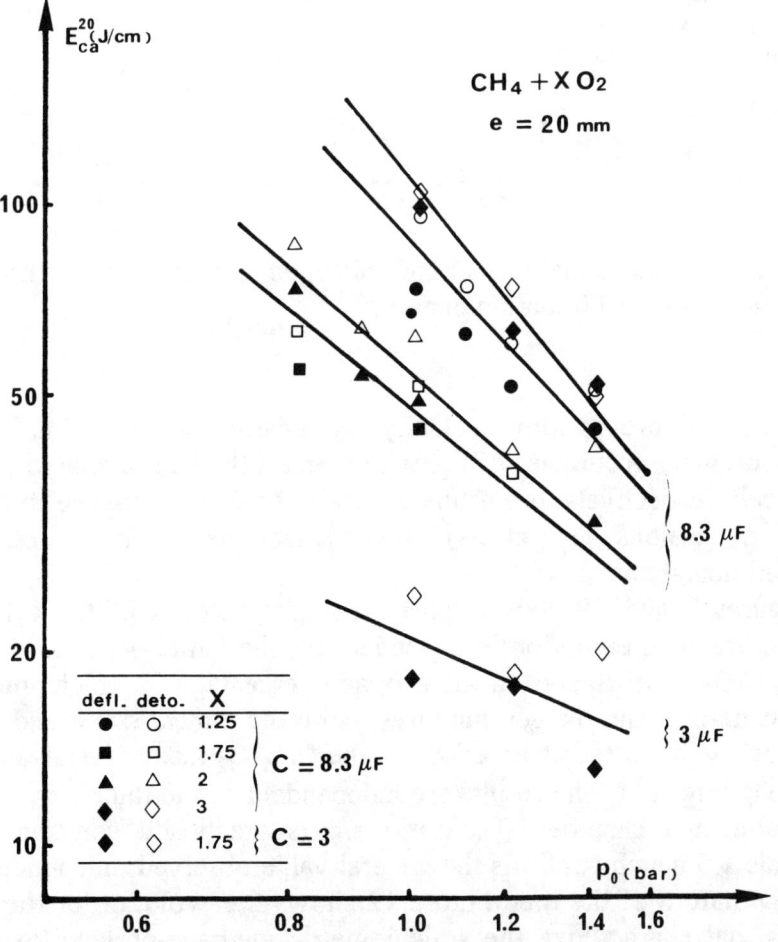

Fig. 7 Minimum nominal initiation energy as function of initial pressure in CH_4-O_2 mixtures; effect of the electrical characteristics of the ignition device.

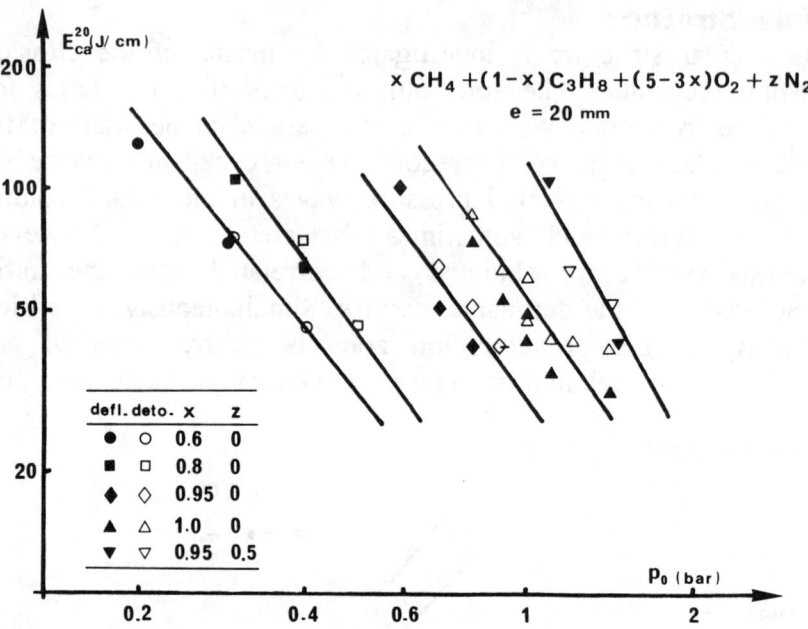

Fig. 8 Propane substitution and nitrogen dilution effects on the minimum nominal initiation energy.

detonation propagation velocity is nearly the CJ one. The measurement accuracies of R_c, with a and b the length and width of the cell, respectively, are estimated to be 15%. We observe that the cell dimensions are clearly lower than the thickness of the experimental chamber.

Figures 9 and 10 show typical soot-film records of the cellular structure. The evolution is dependent on the initial pressure of the CH_4+2O_2 mixture and on the propane percentage in stoichiometric methane-propane-oxygen mixtures. Both the dimensions a and b of the cells are plotted as functions of the CH_4-O_2 mixture equivalence ratio ϕ (Fig. 11). The results are independent of the thickness of the experimental chamber. The ratio a/b is practically constant and equals 1.6 which confirms the general value observed independently of the nature of the fuel. Figure 12 shows the evolution of the cell sizes that characterize the stoichiometric methane-propane-oxygen mixtures at different initial pressures. The results agree very well with previous results.[6,15-17] We have deduced the dependence of the

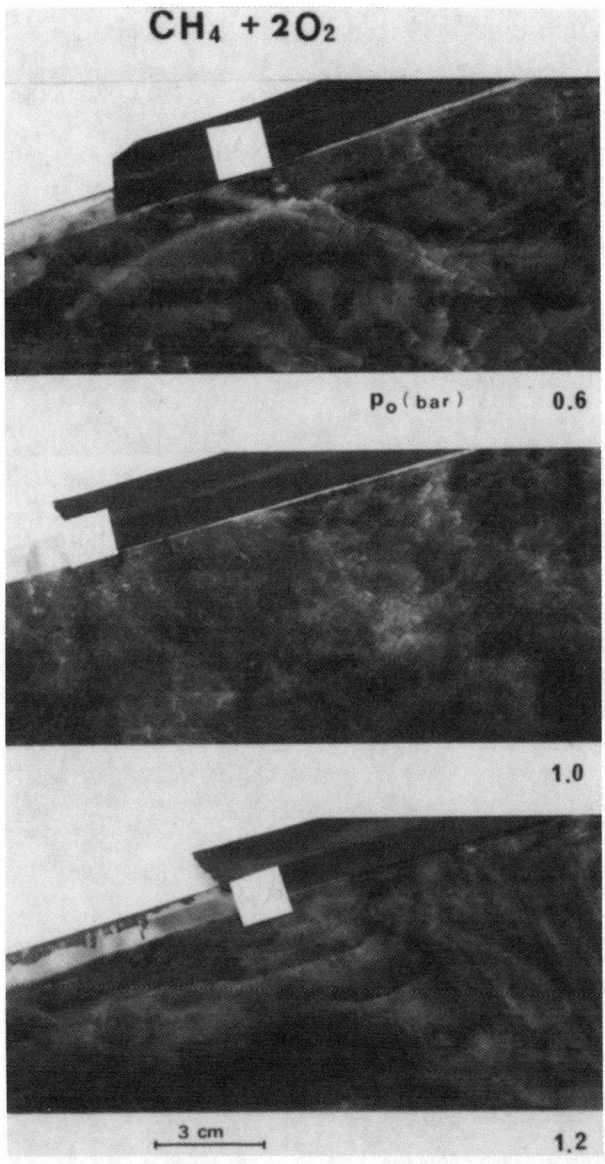

Fig. 9 Cellular structure of detonation in CH_4+2O_2 mixture; initial pressure effect e = 10 mm.

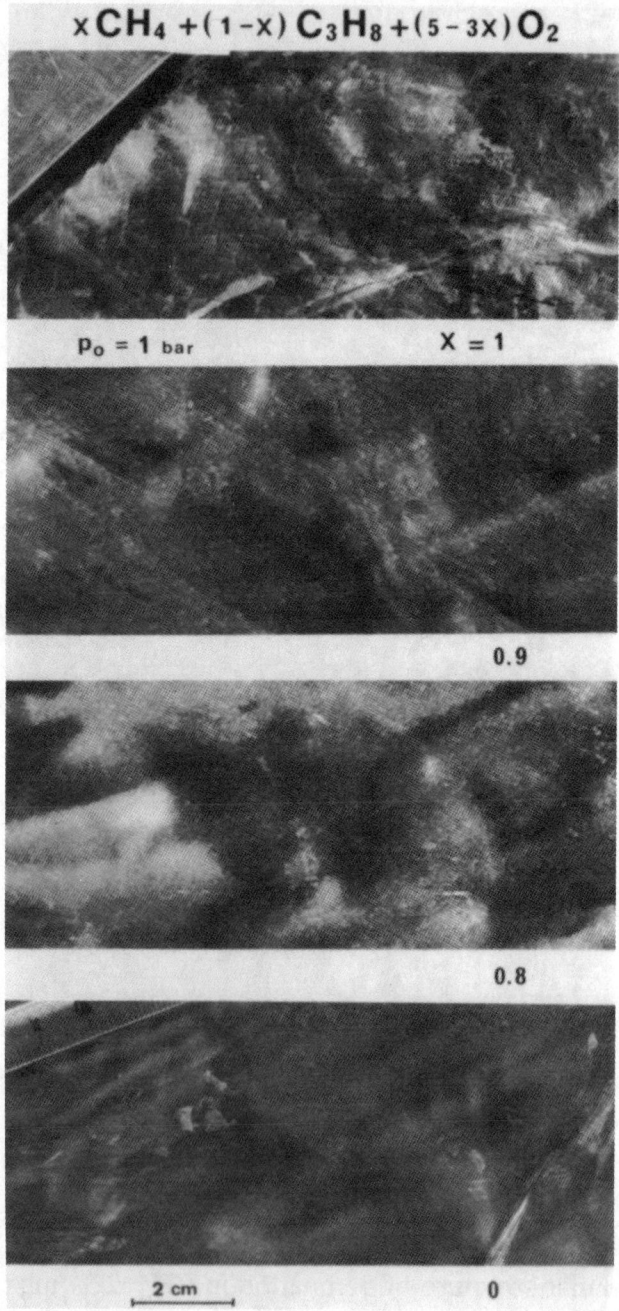

Fig.10 Cellular structure of detonation in stoichiometric methane-propane-oxygen mixtures: propane substitution effect, $p_0 = 1$ bar, $e = 20$ mm.

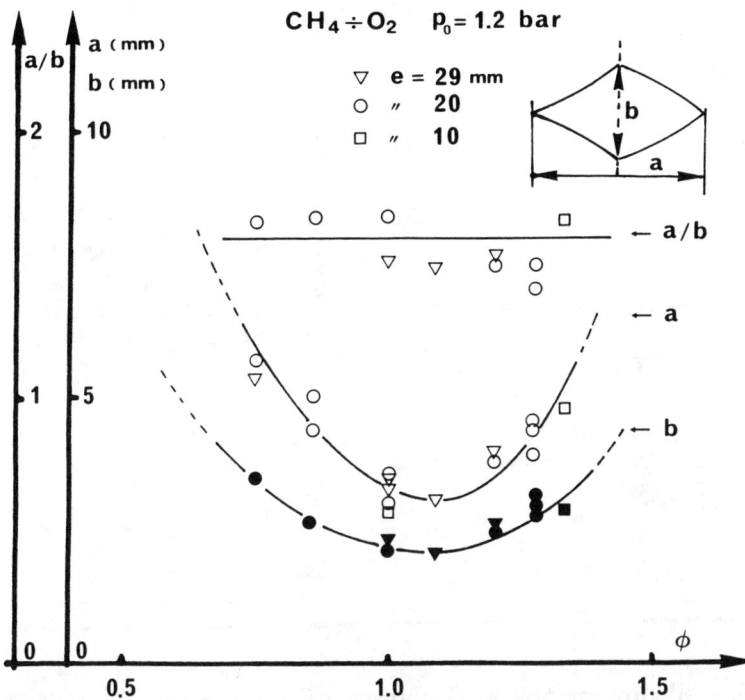

Fig.11 Cell dimensions as functions of the fuel equivalence ratio in CH_4-O_2 mixtures.

cell width b on the initial pressure p_0, written as $b = k\, p_0^{-n}$. The comparison (see Table 3) with results collected by Urtiew[1] and Vasiliev[18] agrees very well. The exponent n decreases with the sensitivity of the mixture increase.

General Discussion

Applying the criterion based upon the characteristic radius R_0 (~ 10 mm) defined by the balance of the chemical energy delivered by the reaction and the amount of energy of initiation, it was correct to consider the detonation wave as a self-initiated one and the propagation in the R range 10-30 cm as a self-sustained one characterized by the CJ values parameters. In the ranges of our experimental conditions we have never observed the direct initiation of the detonation wave. More or less large the critical radius R_c of

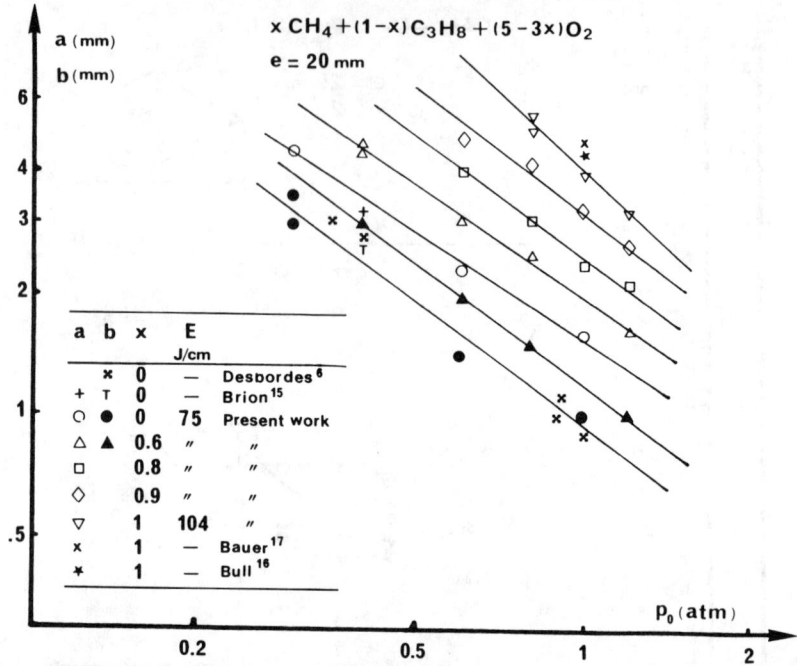

Fig. 12 Cell dimensions as functions of initial pressure of methane-propane-oxygen mixtures.

initiation is always observed and, consequently the effective value of the initiation energy is lower than the critical one.

The evolution of the initiation energy vs the fuel equivalence ϕ (see Fig. 6) is characterized by the typical U curve. For CH_4-O_2 mixtures the minimum energy corresponds to the slightly rich mixture $\phi = 1.1$. The slope of the branches of the U curve is smooth for lean mixtures and sharp for rich ones. We note that the cylindrical detonation wave was obtained in $CH_4 + 3.5O_2$ ($\phi = 0.66$) but observed in $CH_4 + O_2$ ($\phi = 1.5$) even if the initial pressure is $p_0 = 1.6$ bar.

Experimental data which concern methane-oxygen mixture detonability in cylindrical symmetry are very rare in the literature. However Nicholls et al.[12] have conducted experiments in a cylindrical sector (angle of 20^{deg} and 52.1 mm thick). They investigated methane-oxygen-nitrogen stoichiometric mixtures initiated by means of condensed explosive. The extrapolation of the initiation energy curve to no nitrogen indicates that only 14 J/cm

Table 3 Cell width b dependence on pressure for CH_4+O_2 mixture

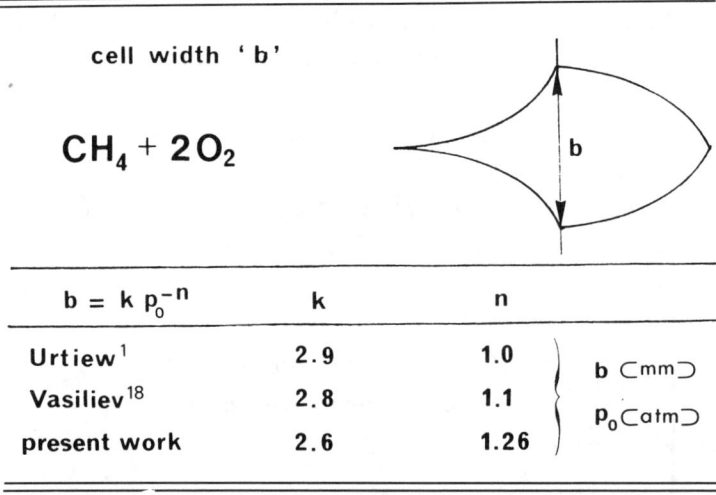

would be required to initiate $CH_4 + 2O_2$ at $p_0 = 1$ bar initial pressure. Then applying the ratio that relates the geometry of both the experimental devices, the equivalent energy would be $(90/20^{deg}) \times 14 = 63$ J/cm, in good agreement with our results, 54 J/cm (see Fig. 6).

Then it was interesting to correlate the results obtained in the case of cylindrical wave with those of spherical case. To facilitate the comparison, in the same graph (Fig. 6) we have plotted the critical initiation energy E_s (J) of spherical detonation. The values are those of Bull[9] and Matsui et al.[10] The last ones are close to the calculated values proposed by Westbrook et al.[2] We note that the experimental value $E_s = 1200$ J proposed by Desbordes[19] is 10 time higher than that of Bull. Probably this value is not a critical one for the stoichiometric methane-oxygen-mixture.

A comparison of the various but not numerous data which concern the minimum initiation energy of fully cylindrical detonation will be made in the E_{cc}-ϕ plane. The first approach to the comparison considers the constant value of the characteristic radius defined by either the cylindrical energy E_c (J/cm) or the spherical energy E_s (J):

$$(E_c/p_0)^{1/2} = (E_s/p_0)^{1/3}$$

Few experimental and calculated[10] E_s values are found in the literature and the equivalent E_c energies are deduced. The second approach to the comparison considers the relation between the critical energy E_c and the cell width. This relation is

$$E_c = 15.725\, \rho_0\, D^2\, b^2$$

where ρ_0 and D are the initial density and the CJ velocity, respectively.

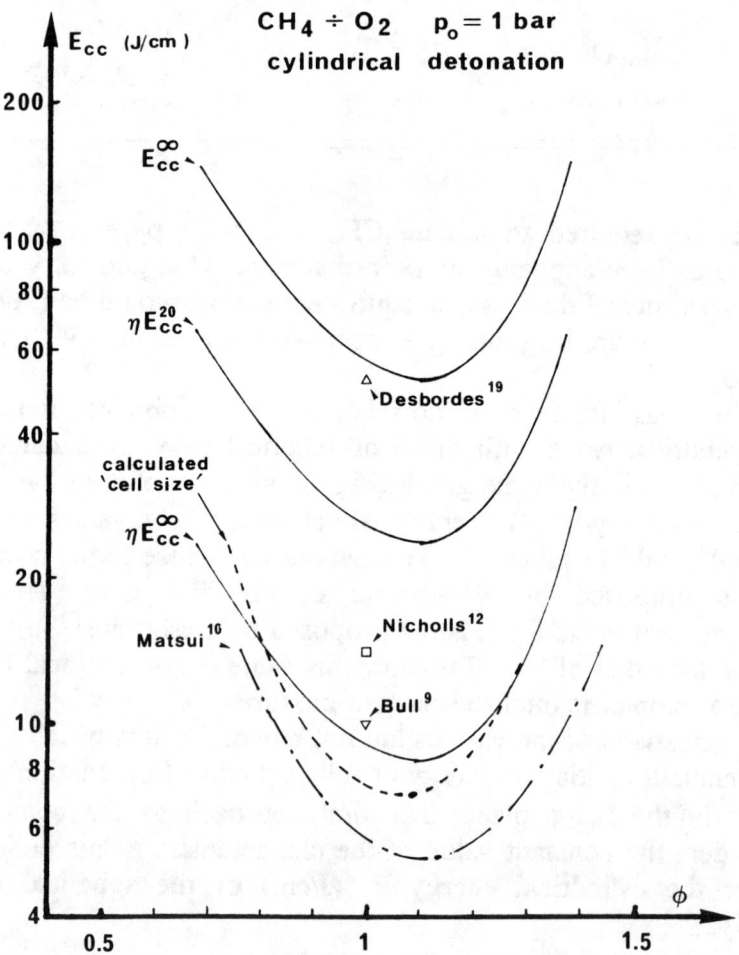

Fig. 13 Initiation energy of fully cylindrical detonation vs fuel equivalence ratio for CH_4-O_2 mixtures, at $p_0 = 1$ bar: our results, solid line ; computed, dotted line.

Figures 13 and 14 show the comparisons with our experimental results which are not critical values. They are represented by three different curves : the first one corresponds to the E_{cc}^{∞} energy independent of the thickness e of the bomb and based upon an arbitrary extrapolation to the infinite thickness. The second one corresponds to the E_{cc}^{20} energy on which is applied the energy efficiency coefficient $\eta = 0.16$ defined above. The last one corresponds to the E_{cc}^{∞} energy combined with the efficiency coefficient $\eta = 0.16$.

Taking into account the minimum value proposed by Nicholls et al.[12], the conclusion would be that the initiation energy for fully

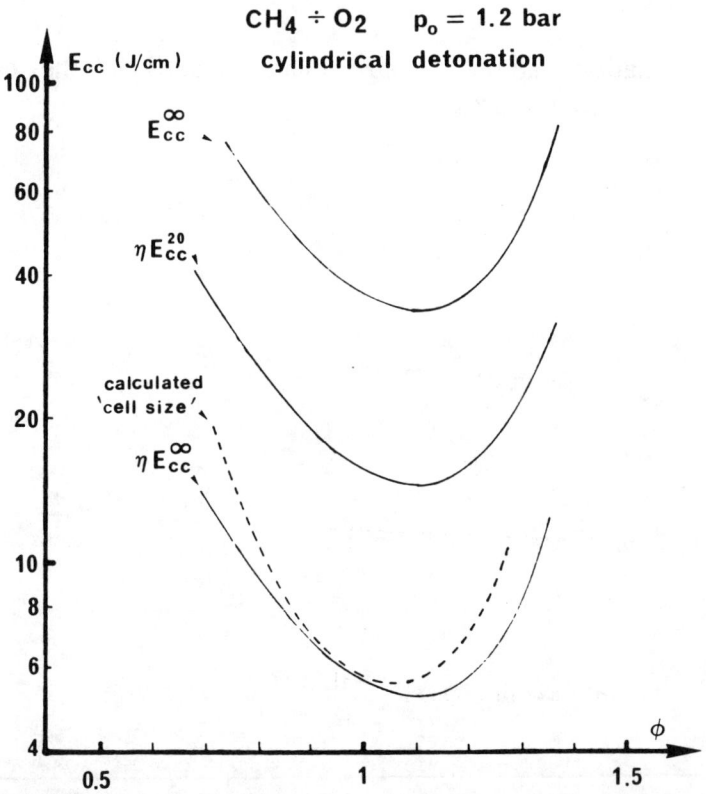

Fig.14 Initiation energy of fully cylindrical detonation vs fuel equivalence ratio for CH_4-O_2 mixtures at $p_0 = 1.2$ bar: our results, solid line ; computed, dotted line.

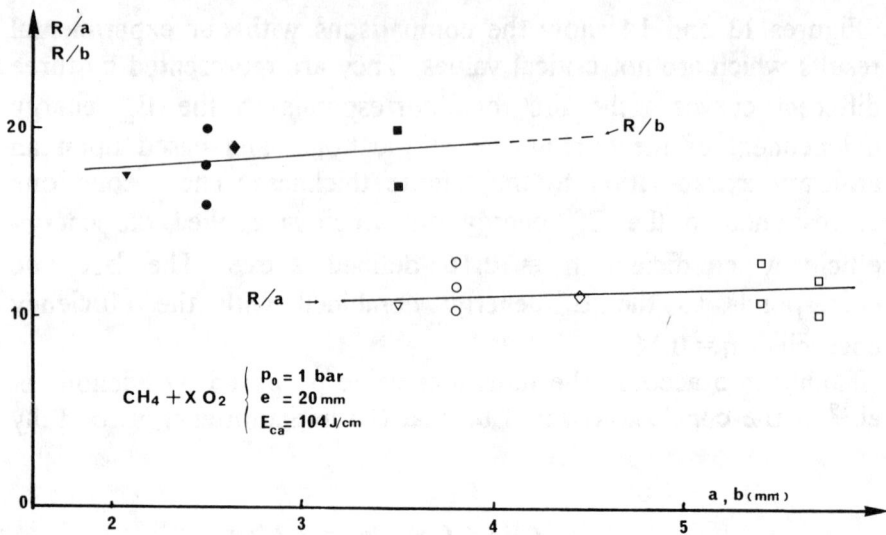

Fig.15 Initiation radius-cell size characteristics in the case of methane-oxygen mixtures.

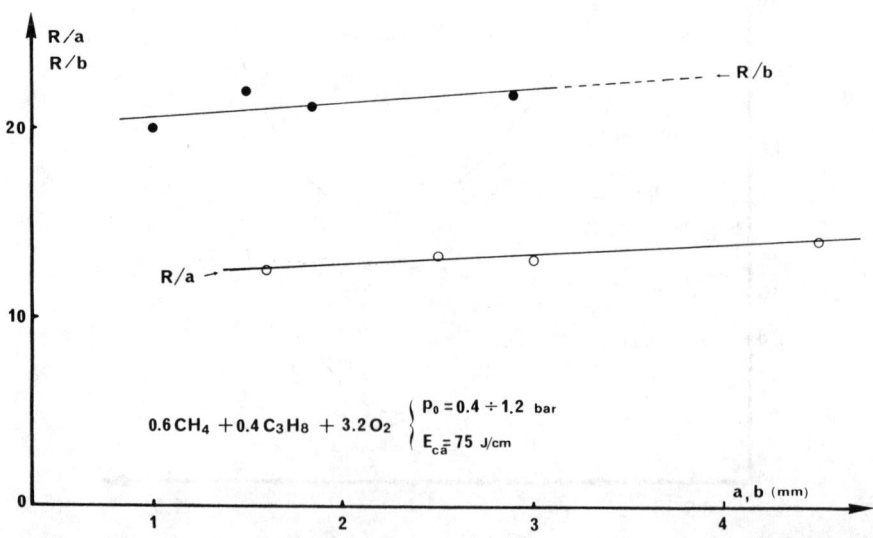

Fig.16 Initiation radius-cell size characteristics in the case of methane-propane-oxygen mixtures.

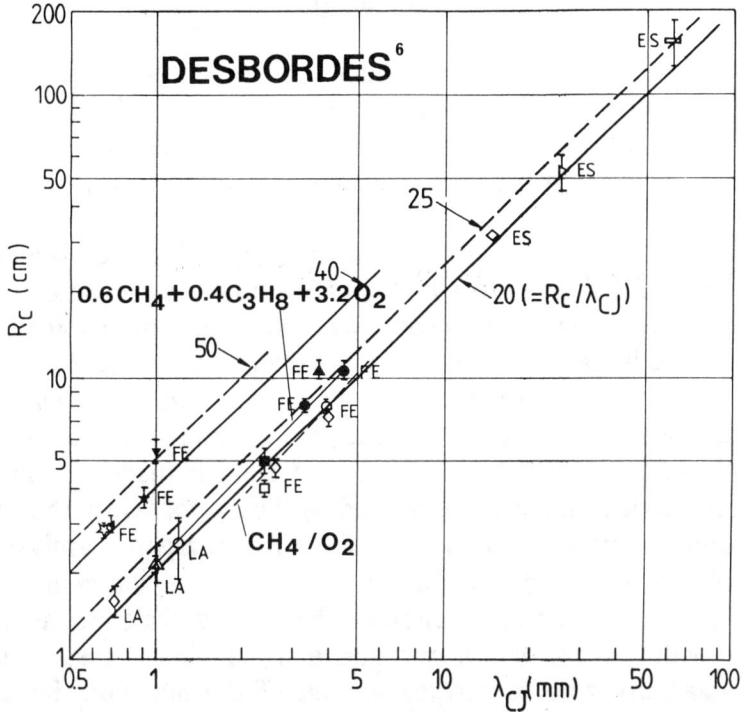

Fig. 17 Critical initiation radius vs cell width.

cylindrical detonation of stoichiometric methane-oxygen mixture at 1-atm initial pressure is 5-14 J/cm (see Table 2).

If we carry on the observations on the cellular structure of the detonation wave we note that the ratio a/b of the cell sizes is approximately constant (~ 1.6) independent of the mixture and the initial pressure. But Figures 15 and 16 show that the ratios R/a and R/b, where R is considered as the critical initiation radius R_c, are dependent on the cell size. The slopes are the same for the methane-oxygen and methane-propane-oxygen mixtures but the values are different. Our R/b (or R_c/λ_c) experimental results are shown in Fig. 17 which plots numerous previous data.[6] In particular for methane-oxygen mixtures, the ratio R/b is always lower than 20. This maximum value corresponds to the maximum observed radius R - 70-80 mm.

Conclusion

Despite the difficulty in conducting experiments in severe mechanical conditions and the large scattering of the results inherent in the physical phenomena, the investigation has led to numerous data which characterize the methane-oxygen nitrogen mixtures especially.

It is easier to create a detonation wave in cylindrical symmetry than in a spherical one, but a preliminary study of the characteristics of the electrical initiation device was necessary to eliminate the effects of the initiator location or the thickness of the flat bomb. We have put in evidence that the nominal minimum initiation energy is practically independent of both these parameters but it is difficult to know the effective absolute value. An attempt is made to compare the hemispherical shock wave created by the ignitor with the shock wave created by a gaseous explosion. An energetic equivalency coefficient is then defined. Consequently, the minimum initiation energy is investigated as function of the fuel equivalence ratio, the dilution with propane and/or nitrogen, and the initial pressure. From the cellular structure, a new evaluation of the minimum initiation energy appears to be correct. The comparison with previous investigations (cylindrical and spherical cases) leads to 5-15 J/cm, the necessary amount of energy to ignite the detonation of stoichiometric methane-oxygen mixture. Numerous new data has been assembled on the ability of methane to detonate and confirms the relation of the critical radius with the cell size.

References

[1] Urtiew, P. A., and Tarver, C. M., "Effects of Cellular Structure on the Behavior of Gaseous Detonation Waves under Transient Conditions," *Dynamics of Explosions and Reactive Systems*, Progress in Astronautics and Aeronautics, Vol. 75, AIAA, New York, 1980, pp. 370-384.

[2] Westbrook, C. K., and Urtiew, P. A., "Chemical Kinetic Prediction of Critical Parameters in Gaseous Detonations," *Nineteenth Symposium (International) on Combustion*, The Combustion Institute, Pittsburgh, PA, 1982, pp. 615-623.

[3] Westbrook, C. K., Pitz, W. J., and Urtiew, P. A., "Chemical Kinetics of Propane Oxidation in Gaseous Detonations," *Dynamics*

of Explosions and Reactive Systems, Progress in Astronautics and Aeronautics, Vol. 94, AIAA, New York, 1984, pp. 151-174.

[4] Benedick, W. B., Guirao, C. M., Knystautas, R., and Lee, J. H., "Critical Charge for the Direct Initiation of Detonation in Gaseous Fuel-Air-Mixtures," *Dynamics of Explosions and Reactive Systems*, Progress in Astronautics and Aeronautics, Vol. 106, AIAA, New York, 1986, pp. 181-202.

[5] Desbordes, D., "Correlation Between Shock Flame Predetonation Zone Size and Cell Spacing in Critically Initiated Spherical Detonations," *Dynamics of explosions and Reactive Systems*, Progress in Astronautics and Aeronautics, Vol. 106, AIAA, New York, 1986, pp. 166-180.

[6] Desbordes, D., "Aspects stationnaires et transitoires de la détonation dans les gaz : relation avec la structure cellulaire du front," Thèse Docteur ès Sciences Physiques, Université de Poitiers, July, 1990.

[7] Kailasanath, K., and Oran, E. S., "Power-Energy Relations for the Direct Initiation of Gaseous Detonations," *Dynamics of Explosions and Reactive Systems*, Progress in Astronautics and Aeronautics, Vol. 94, AIAA, New York, 1984, pp. 38-54.

[8] LEE, J. H., "Initiation of Gaseous Detonation," Annual *Review of Physical Chemistry*, Vol. 28, 1977, pp. 75-104.

[9] Bull, D. C., Elsworth, J. E., Hooper, G., and Quinn, C. P., "A Study of Spherical Detonation in Mixtures of Methane and Oxygen Diluted by Nitrogen," *Journal of Physics*, D.Appl.Physics, Vol. 9, 1976, pp. 1991-2000.

[10] Matsui, H., and Lee, J. H., "On the Measure of the Relative Detonation Hazards of Gaseous Fuel Oxygen and Air Mixtures," *Seventeenth Symposium (International) on Combustion*, The Combustion Institute, Pittsburgh, PA, 1979, pp. 1269-1280.

[11] Brossard, J.,Manson, N., and Niollet, M., "Propagation and Vibratory Phenomena of Cylindrical and Expanding Detonation Waves in Gases," *Eleventh Symposium (International) on Combustion*, The Combustion Institute, Pittsburgh, PA, 1967, pp. 623-633.

[12] Nicholls, J. A., Sichel, M., Oza, R. D., Gabrijel, Z., and Vandermolen, R., "Detonability of Unconfined Natural Gas-Air Clouds," *Seventeenth Symposium (International) on Combustion*, The Combustion Institute, Pittsburgh, PA, 1979, pp. 1223-1234.

[13] Vasiliev, A. A., "Initiation Criteria of Gaseous Detonations," *Fizika Goreynia i Vzryva*, Novosibirsk, 1983, Vol. 19, No. 1, pp. 121-131.

[14] Brossard, J., Leyer, J. C., Desbordes, D., Saint-Cloud, J. P., Hendrickx, S., Garnier. J. L., Lannoy, A., and Perrot, J.,"Air Blast Unconfined Gaseous Detonations," *Dynamics of Explosions and Reactive Systems*, Progress in Astronautics and Aeronautics, Vol. 94, AIAA, New York, 1984, pp. 556-566.

[15] Brion, B., "Etude de la structure de la detonation marginale," Mémoire de Diplôme d'Etudes Approfondies, Université d'Orléans, March, 1978.

[16] Bull, D. C., Elsworth, J. E., Shuff, P. J., and Metcalfe, E., "Detonation Cell Structure in FuelAir Mixture," *Combustion and Flame*, Vol. 45, No. 1, 1982, pp. 7-22.

[17] Bauer, P., "Contribution à l'étude de la détonation des mélanges explosifs gazeux à pression initiale élevée, " Thèse Doctorat ès Sciences Physiques, Université de Poitiers, June, 1985.

[18] Vasiliev, A. A., Private communication, Novosibirsk, 1990.

Desbordes, D., "Célérités de propagation des détonations sphériques divergentes dans les mélanges gazeux," Thèse 3è cycle, Université de Poitiers, June, 1973.

Chapter III. Initiation of Detonation Waves

Structure of Reaction Waves Behind Oblique Shocks

C. Li,* K. Kailasanath,* and E. S. Oran*
Naval Research Laboratory, Washington, DC 20375

Abstract

In this study, the structure of reaction waves behind wedge-induced, oblique shocks in hydrogen-oxygen-nitrogen mixtures were investigated by time-dependent numerical simulations. Both standing (stable) and moving (unstable) oblique detonation waves were observed. In both cases, there is an induction region immediately behind the shock where combustion radicals are produced. After the radical accumulation reaches a certain level, the associated energy release and temperature increase generate a set of deflagration waves through which the pressure rises smoothly. This set of deflagration waves converge into and steepen the original oblique shock to form an oblique detonation. When the temperature increase is high enough to choke the flow, the reactive shock becomes unstable and moves upstream. The basic structure of this unstable detonation is similar to that of the stable detonation.

Introduction

The concept of using oblique detonations for propulsion-engine design is an attractive alternative approach for achieving efficient energy conversion.[1,2] However, to maximize performance, the oblique detonation has to be generated and stabilized at appropriate locations in the system. Therefore, successful development of such systems depends on knowledge

---

This paper is declared a work of the U.S. Government and is not subject to copyright protection in the United States.
* Laboratory for Computational Physics and Fluid Dynamics

of the structure of these waves and understanding the related physical and chemical processes.

When a fuel-oxidizer mixture passes through a shock, the high temperatures and pressures behind the shock may be sufficient to induce chemical reactions and subsequent energy release. The structures that arise due to this energy release may be either detonations or deflagrations, depending on the magnitude of the flow velocity, the rates of radical production and accumulation, and the heat released from the chemical reactions. There have been studies[3-6] showing that standing detonation waves can be established on or near a wedge or blunt body under the appropriate flow conditions. However, the structure of oblique detonation waves needs to be investigated and the conditions under which stable oblique standing detonation waves exist need to be determined.

In this study, the coupled continuity equations for density, momentum, energy, and individual species densities are solved using the Flux-Corrected Transport algorithm[7] in conjunction with a two-step model for chemical reactions. In this two-step reaction model, the first step represents the radical formation and accumulation and the second step represents energy release.[8,9] The transport calculation and the reaction model are coupled together by the time-step splitting technique.

Structure of Oblique Detonations

We present here a numerical simulation of a stoichiometric hydrogen-air mixture (hydrogen:oxygen:nitrogen/2:1:3.76) flowing at Mach 8 around a wedge of 23 deg. The inlet temperature and pressure were 300 K and 101.3 kPa, respectively. A schematic of the 10.0×3.75 cm^2 computational domain is shown in Fig. 1. The inflow conditions were applied on the upper and left boundaries. The outflow conditions were imposed to the right and the first part of the lower boundary. The second part of the lower boundary,

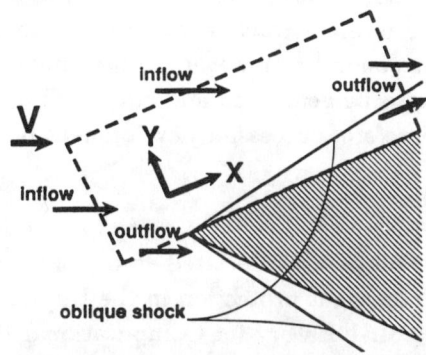

Fig. 1 Schematic of the computational domain attached to the wedge surface.

starting at 1.0 cm, is attached to the wedge surface which is represented by a solid wall. The computational domain is extended in the y direction to insure the proper implementation of the boundary conditions.

Figure 2 shows contours of density, temperature, water mass fraction, and induction parameter from the simulation ($\Delta x = 0.25$ mm). Note that the induction parameter represents a normalized, effective radical concentration.[8-10] These contours show significant energy release and water formation behind the induction region near the wall. Figure 3 shows profiles of temperature, pressure, Mach number, and water mass fraction at 1.0 mm above the wall along the x direction. The induction zone behind the shock can be clearly observed, where the properties remain nearly constant. After the water formation starts at the end of the induction zone, temperature increases monotonically and the Mach number decreases accordingly. Pressure first rises then falls to a fairly constant value. The initial pressure increase is due to the rapid energy release in the early stage of the water-formation reaction.

Figure 4, an enlargement of the induction zone and a portion of the region behind it, shows a set of deflagration waves corresponding to the gradual pressure rise shown in Fig. 3. These deflagration waves propagate upward at the local Mach angle and gradually converge into and steepen

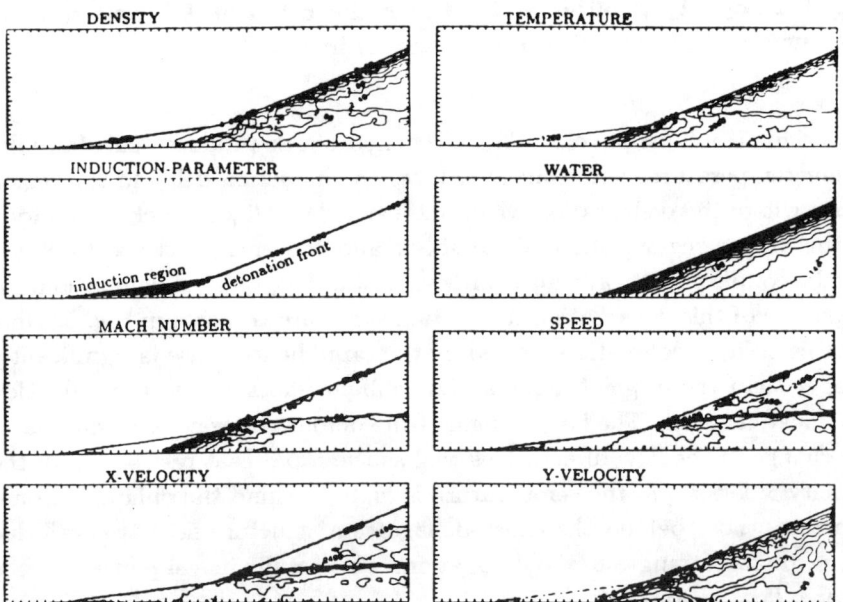

Fig. 2 A detonation structure stabilized on a 23-deg wedge in a Mach 8, stoichiometric hydrogen-air mixture.

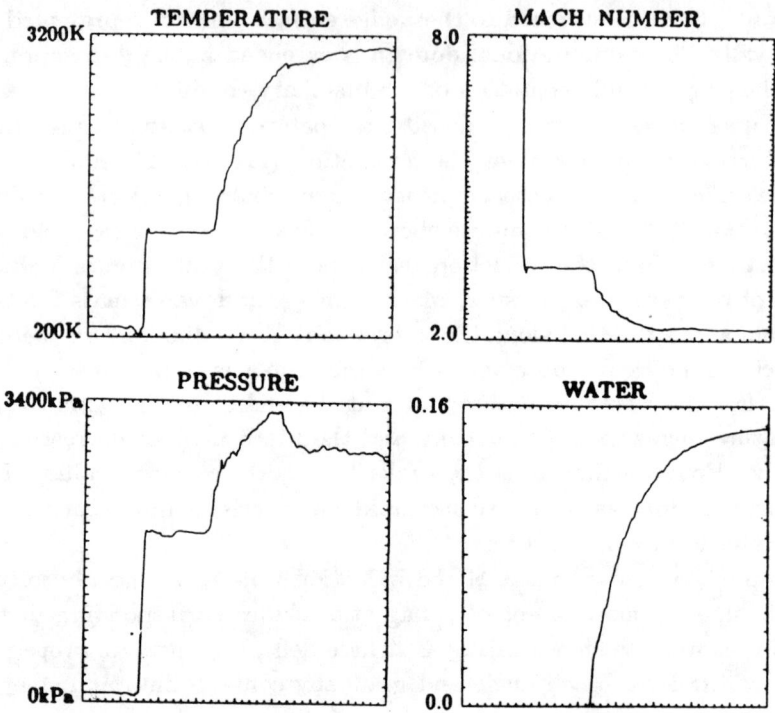

Fig. 3 Property profiles along the x direction at 1.0 mm above the surface from the simulation shown in Fig. 2.

the original oblique shock. This steepened shock produces substantially higher temperatures and sharply reduces the induction delay in the region above where the deflagration waves intersect the oblique shock. Therefore, in this upper region, the oblique shock and the energy-release front are closely coupled and form an overdriven, standing oblique-detonation. A schematic of this deflagration-detonation structure is shown in Fig. 5. Note the this oblique detonation formed by the rapid heat release is significantly steeper than the original, nonreactive, oblique shock before the induction zone near the wall. The larger angle of the detonation wave accommodates the change in the flow direction as well as the rapid heat release. Also, the velocity is lower and the temperature is higher behind the oblique detonation than those behind the shock-deflagration structure near the wall due to the greater strength of the detonation. This results in a slip line between the two flow regions.

The structure of the induction zone is crucial to the generation and stabilization of any reaction waves behind the shock. However, the time scale

REACTION WAVES BEHIND OBLIQUE SHOCKS 235

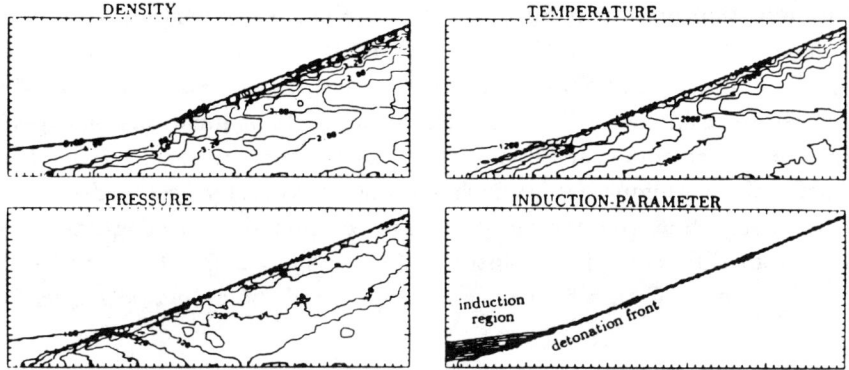

Fig. 4 An enlarged domain showing the detailed wave structures behind the induction zone from the simulation shown in Fig. 2.

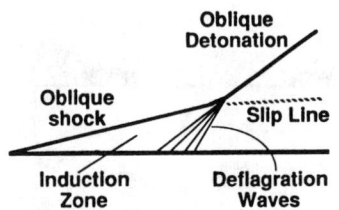

Fig. 5 Schematic of the deflagration-detonation structure behind the induction zone.

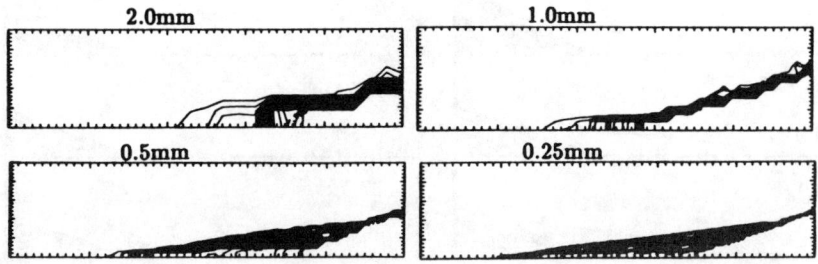

Fig. 6 Computed induction zone from simulations with different resolutions.

of the induction process is fairly short under usual combustion conditions and depends nonlinerly on the composition of the mixtures and the temperature and pressure behind the shock. This time-scale decreases rapidly as temperature increases.[10] Unless the scale of the simulations or experimental measurements is small enough to resolve this distance, the detailed structure of the oblique detonations will not be correctly observed.

To assure that the computations are conducted with adequate resolutions, four different grid resolutions, $\Delta x = \Delta y = 2.0, 1.0, 0.5$ and 0.25 mm, were tested. Figure 6 shows enlargements of the induction region for

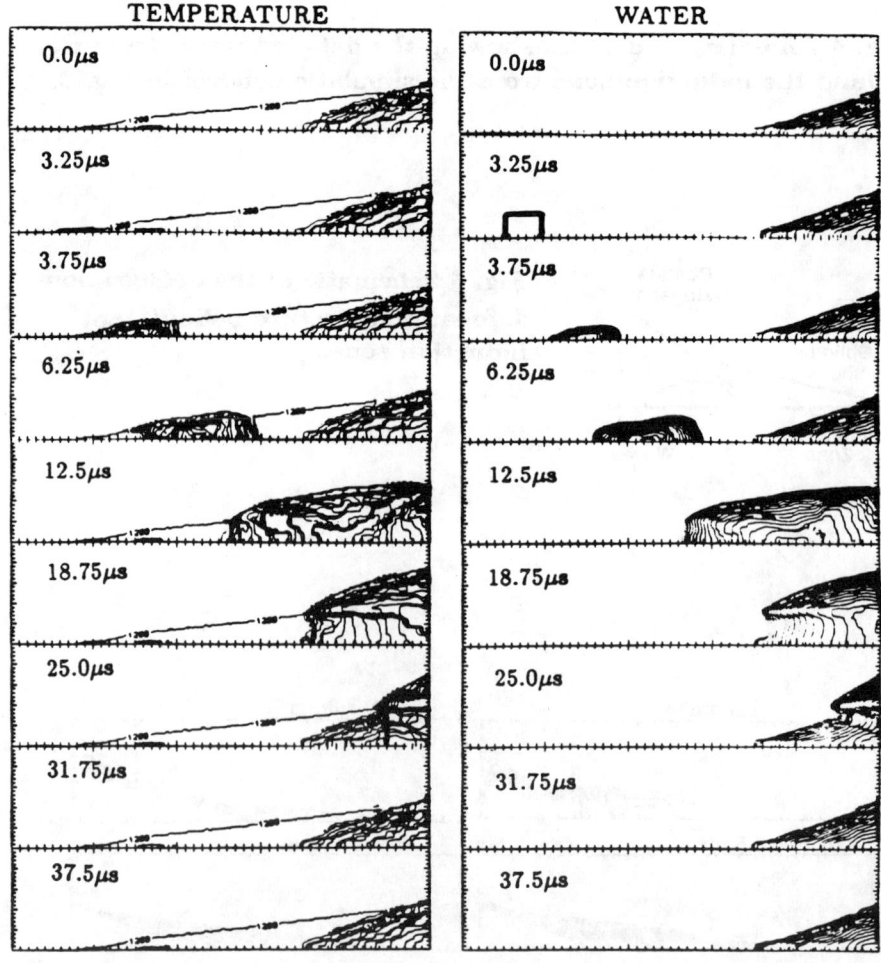

Fig. 7 Development of a distrubance on an oblique detonation stabilized on a 23-deg wedge in a Mach 8, stoichiometric hydrogen-air mixture.

all four cases. The two coarser resolutions (2.0 and 1.0 mm) show virtually instantaneous combustion within one or two cells. However, the two finer resolutions (0.5 and 0.25 mm) properly resolve the induction zone. It is evident here that an inadequate resolution not only produces quantitatively inaccurate results but also generates qualitatively incorrect physical features.

Stability of the Oblique Detonation

In the previous section, we demonstrated that an oblique detonation can be stabilized on a wedge. How stable are such oblique detonations and will disturbances significantly change the detonation structure? To answer this question, a computation was performed in which a burning pocket is artificially introduced as a disturbance into the flowfield originally computed for the mixture of $H_2:O_2:N_2/2:1:3.76$, the same conditions used in the previous section. Now we are interested in the development of the perturbation and its effects on the detonation structure. Figure 7 shows a separate detonation structure forms in and near the disturbed area if the burning pocket is large enough to provide sustainable energy release. This new detonation structure is convected downstream and increases in size. It then merges with the original detonation structure, creating a large and stronger detonation structure. However, the changes in the detonation structure caused by the disturbance is eventually convected out of the computational domain and the original flowfield is recovered. In this case, the detonation structure is quite stable.

If the nitrogen dilution is reduced, the oblique detonation becomes steeper and stronger due to the related increase in the specific heat release from the chemical reactions. However, for mixtures up to $H_2:O_2:N_2/2:1:1$, the detonation structure remains stable and similar to that in the early cases. If the nitrogen dilution is completely eliminated (the stoichiometric hydrogen-oxygen mixture, $H_2:O_2/2:1$) for the flow and geometric conditions used in the previous cases, the detonation structure is no longer stable. A simulation conducted for this case shows that the entire shock-detonation structure moves upstream (Fig. 8). It is interesting to note that overall features such as the induction zone and deflagration waves remain similar to those in the previous cases, however, increases in the shock strength and shock angle are more dramatic. Again, the properties of mixture are quite different due to the different strength of the shock near the wall and the oblique detonation above the induction zone. In this case, the velocity shear is quite strong and some vortical structures can be observed.

The detonation structure shown in Fig. 8 is destabilized due to the choking condition behind the reactive shock. Properties of the mixture

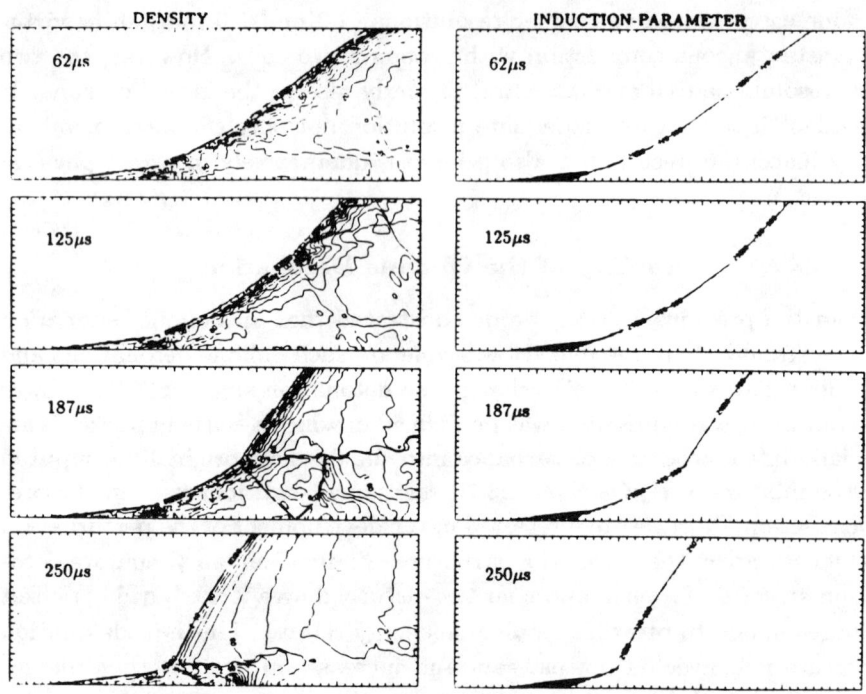

Fig. 8 An unstable detonation generated by a 23-deg wedge in a Mach 8, stoichiometric hydrogen-oxygen mixture.

behind this reactive shock depend on the amount of energy released and on other flow properties in front of the shock. For every flow condition, there exists a maximum amount of heat that can be absorbed without choking the flow. When the energy released exceeds this value, the heated flow will no longer be able to pass through the area behind the shock and, therefore, the shock-detonation structure is driven upstream to create a larger area to accommodate the heated flow. If a sufficiently large area of outflow cannot be generated, the entire structure will be driven of out the computational domain in a simulation or of out the test apparatus in an experiment. In addition, like a regular, nonreactive shock, the detonation structure can be destabilized by increase in the wedge angle or change in the inflow conditions.

Summary and Conclusions

Our numerical simulations demonstrate that a steady oblique detonation can be stabilized behind an oblique shock in supersonic flow of hydrogen-oxygen mixture diluted with nitrogen. However, the detonation structure of this kind will be stable only if the flow behind the shock-detonation structure are not choked. Otherwise, the whole structure will be driven upstream. It has been demonstrated that both stable and unstable detonations exhibits a similar structure. Such structures have also been observed in other forms of detonations such as the overdriven detonation in a partially sealed channel and detonation cells propagating in an open channel.[9] However, the basic detonation features may be affected by the boundary layers and further studies are currently been conducted.

Acknowledgment

This work is supported by the Air Force Office of Scientific Research and the Naval Research Laboratory. Authors gratefully acknowledge discussions with Dr. J. P. Boris.

References

[1] Hertzberg, A., Bruckner, A. P., Bogdanoff, D. W., and Knowlen, C., "The RAM Accelerator and Its Applications: A New Approach for Reacting Ultrahigh Velocities," *16th Symposium on Shock Tubes and Shocks Waves*, Aachen, West Germany 1987.

[2] Hertzberg, A., Bruckner, A. P., Bogdanoff, D. W., and Knowlen, C., "Thermodynamics of the RAM Accelerator," *17th Symposium on Shock Tubes and Shocks Waves*, Bethlehem, Pennsylvania, 1989.

[3] Glenn, D. E. and Pratt, D. T., "Numerical Modeling of Standing Oblique Detonation Waves," AIAA Paper 88-0440, Jan. 1988.

[4] Cambier, J. L., Adelman, H. and Menees, G. P, "Numerical Simulations of an Oblique Detonation Wave Engine," AIAA Paper 88-0063, Jan. 1988.

[5] Fujiwara, T. , Matsuo, A., and Nomoto, H., "A Two-Dimensional standing Detonation Supported by a blunt body or a Wedge," AIAA Paper 88-0089, Jan. 1988.

[6] Wang, Y., Fujiwara, T., Aoki, T., and Arakawa, H., "Three- Dimensional standing Oblique Detonation Wave in a Hypersonic Flow," AIAA Paper 88-0478, Jan. 1988

[7] Boris, J. P. and Book, D. L., "Solution of the Continuity Equations by the Method of Flux-Corrected Transport," *Methods in Computational Physics*, Vol. 16, 1976, pp. 85–91.

[8] Oran, E.S., Boris, J.P., Young T., Flanigan, M., Burk, T., and Picone, M., "Numerical Simulations of Detonations in Hydrogen-air and Methane-air Mixtures," *the 18th International Symposium on Combustion*, 1981

[9]Kailasanath, K., Oran, E.S., Boris, J.P., and Young, T.R., "Determination of Detonation Cell Size and the Role of Transverse Waves in Two-Dimensional Detonation," *Combustion and Flame* Vol. 61, 1985, pp. 199-209.

[10]Burks, T. L. and Oran, E. S., "A Computational Study of the Chemical Kinetics of Hydrogen Combustion," Naval Research Laboratory Memo. Rept. 4446, May, 1982.

Ignition in a Complex Mach Structure

E. S. Oran* and J. P. Boris†
Naval Research Laboratory, Washington, DC 20375
D. A. Jones‡
Materials Research Laboratory, Victoria, Australia
and
M. Sichel§
University of Michigan, Ann Arbor, Michigan 48109

Abstract

Highly resolved, two-dimensional, time-dependent numerical simulations are used to to isolate and study ignition that occurs when a curved shock reflects from a wall in a highly reactive mixture. In particular, we investigate the ignition that occurs in a complex Mach reflections. Previous nonreactive simulations showed the details of the structure such as the turbulence and heating in the slip lines behind the Mach stem. Previous less-resolved reactive numerical simulations showed that ignition occurs in the vacinity of Mach structures. In the highly-resolved simulations presented here, ignition occurs in the hotter, vortical structure behind the Mach stem even though the pressures there are somewhat lower. This is because the chemical induction times have an exponential temperature dependence but a much weaker pressure dependence. In one case after the initial ignition, a volumentric explosion occured in the turbulent region behind the Mach stem that over drives the leading shock and quickly produces a propagating detonation in the mixture.

This paper is a work of the U.S. Government and is not subject to copyright protection in the United States.
  * Senior Scientist for Reactive Flow Physics
  † Chief Scientist, Laboratory for Computational Physics and Fluid Dynamics
  ‡ Senior Research Scientist, Explosive Ordnance Division
  § Professor, Department of Aerospace Engineering

Introduction

Ignition or reignition of detonations often appears in the region where a curved shock, sometimes followed by a decoupled reaction wave, reflects from a boundary or another shock. Even if the conditions of regular reflection are not severe enough to produce instant ignition, the conditions of a curved shock leading to Mach reflection and complex shock structure can cause additional local temperature rise and ignition. In this paper, we isolate and study such ignition phenomena in complex shock structures by a series of two-dimensional numerical simulations.

Over the last several years, we have performed an extensive series of numerical simulations of detonation ignition, transmission, decay, and reignition. A common feature seen in a number of these simulations is reignition in the region of a Mach stem formed when a curved shock reflects from a bounding surface. As a specific example, consider Fig. 1, a series

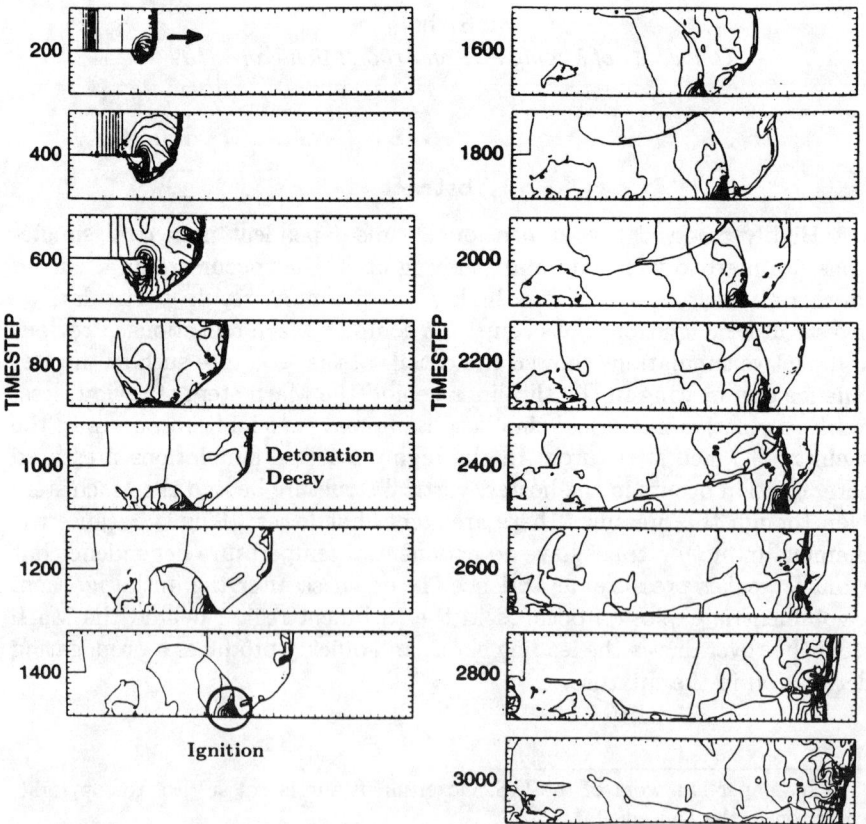

Fig. 1 Pressure contours from a simulation of the lateral transmission of a detonation from a stoichiometric mixture of hydrogen and oxygen into lean mixture of hydrogen and oxygen.[1]

of pressure contours showing a detonation that expands laterally into a layer of less energetic material.[1] The detonation, initially traveling at the Chapman-Jouget velocity, decays into a decoupled shock and propagating reaction zone in both the upper and the lower materials. In particular, we focus on the reflection process from the lower wall shown in steps 800–1800. The decaying shock expands into the lower material and is curved when it hits the bottom wall. After it reflects, soon after step 600, the reflected shock propagates upward. Eventually, through its interaction with the incident shock ahead of it, the reflected shock transitions into a Mach stem structure by step 1400. Some time after this, in the region of the Mach stem, the detonation is reignited.

A similar ignition process is shown in Fig. 2, which shows contours of the primary physical variables for a late time in a computation of ignition from a shock impinging on a canister of propane that has been leaking for some time into the ambient air.[2] This is an even more complex problem both physically and chemically than the one shown in Fig. 1, and this simulation shows several separate regions of ignition by several different mechanisms. However, the ignition source on the far right definitely occurs after the Mach stem passes through the region where the stoichiometry is appropriate for ignition.

Neither of the computations shown above were extremely well resolved, and neither focused on reignition in the region of the Mach stem. However, within the limits of the resolution tests done on both of these problems, ignition near the Mach stem appeared to occur at the same place and time. These and similar tests indicate that the computations must be fundamentally correct, even though the details of the process are not clearly resolved.

Teodorczyk et al.[3] reported experiments that studied the propagation of quasidetonations (detonations that propagate in very rough tubes at speeds substantially below the CJ velocity). In a series of experiments showing the propagation of a detonation over an obstacle and the diffraction of a detonation from a corner, they noted the importance of shock reflections from the wall in the overall reinitiation process. When they delayed the reflections by changing the spacing of obstacles or walls, the transition to the quasidetonation state was delayed. These experiments also showed that if the Mach stem formed on shock reflection is strong enough, the detonation can reignite. The exact morphology and dynamics of the mechanism by which this reignition occurs are not clear and can vary depending on the energetic material and the shock strength. The detonation can reignite, for example, by autoignition as a consequence of adiabatic shock heating in the region of the Mach stem, or it can occur in the region of vortex mixing in the shear layer near the wall behind the Mach stem. It is clear from many computations that the wave structure behind the Mach stem is complex and that there are regions where shocks intersect and the temperature reaches

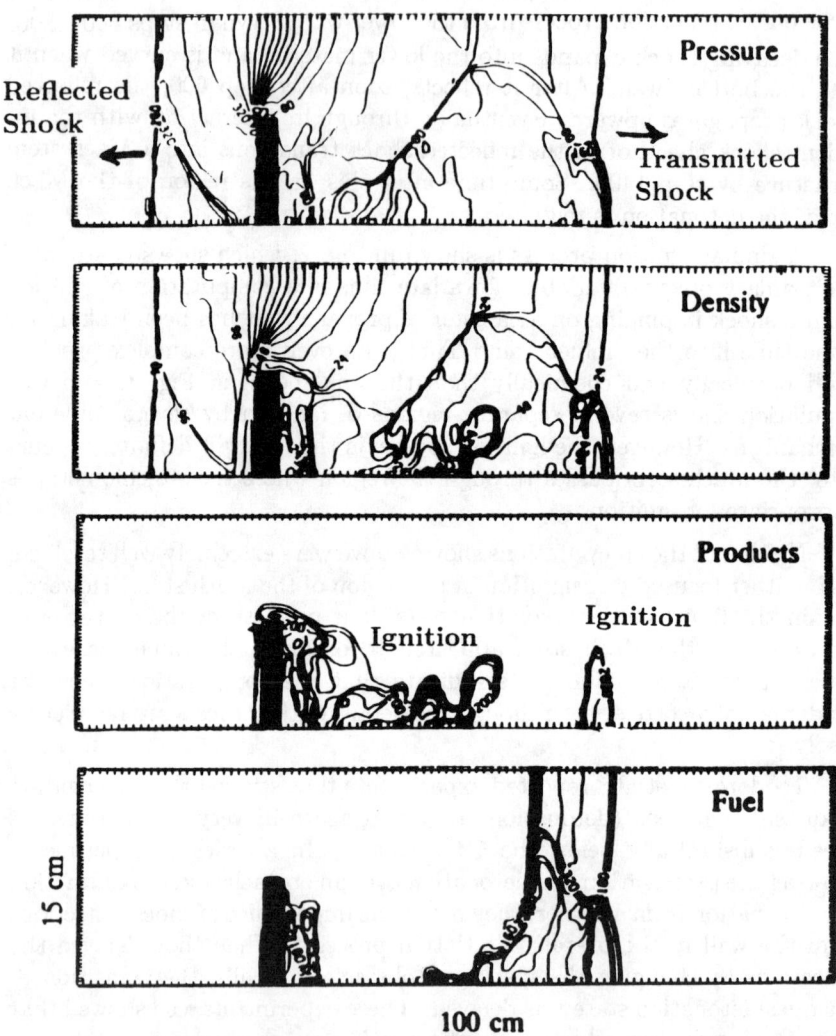

Fig. 2 Contours of pressure, total density, and product and fuel densities from a simulation of the reflection and transmission of a detonation in a propane gas; the maximum propane concentration is around the obstacle.[2]

ignition temperature. There are also shear layers entraining and mixing material that has chemically reacted in varying amounts.

In this paper, we isolate and study this ignition phenomenon with enough resolution to see exactly how and where the ignition occurs in the region of the complex Mach structure. There have been extensive experimental, theoretical, and numerical studies of Mach reflection.[4-6] However, beyond some very preliminary computations,[7] there has been relatively little theoretical or computational work focusing on the detonation ignition process in complex Mach structures. This problem is tractable now that new computational capabilities are available.

Model and Method of Solution

Physical Model

The numerical simulations are based on solutions of the compressible, time-dependent, conservation equations for total mass density ρ, momentum $\rho\mathbf{v}$, and energy E. In a multispecies fluid in chemical reactions, we also need individual species number densities $\{n_i\}$. The effects of molecular diffusion, thermal conduction, and radiative diffusion have been omitted. The first two effects are generally insignificant on the time scales of interest for detonations; the last is not significant for the hydrogen-oxygen systems of interest here. An ideal-gas equation of state is assumed for the gas-phase calculations, although the heat capacities and enthalpies of the individual species are allowed to be different and temperature dependent.

In the calculations described here, the full set of elementary chemical reactions describing the elementary reactions are not included in the model. Instead, we use the induction parameter model which reproduces the essential features of the chemical reaction and energy release process.[1] In the earliest form of this model, three quantities are tabulated as a function of temperature, pressure, and stoichiometry: the chemical induction time, the time during which energy release actually takes place, and the amount of energy released. These quantities may be obtained by integrating the full set of elementary chemical reactions, or they may be gathered from experimental data. In this model, a quantity called the *induction parameter* is defined and convected with the fluid in a Lagrangian manner. This parameter records the temperature history of a fluid element and, when the element has been heated long enough, energy release is initiated. Such a model works because it reproduces the temperature dependence of the detailed chemical reactions. It is valid as long as the computational timestep is smaller than any of the important fluid-dynamic fluctuations and for fast flows in which the convective time scales are significantly faster than those for physical diffusion. The model was described previously[8] and subsequently further developed[9,10] for studies of the cellular structure of detonations. A similar approach was used for hydrogen combustion[11,12] and propane combustion,[2] and to model ignition delay in vaporized sprays.[13]

The convective transport terms in the equations are solved using the nonlinear, fully compressible flux-corrected transport (FCT) algorithm.[14,15] The FCT algorithm is an explicit, conservative, finite-volume method designed to insure that all conserved quantities remain monotonic and positive. It is particularly effective in maintaining steep gradients and generally accurate solutions in both supersonic and subsonic flow calculations. There is a documented procedure for using the one-dimensional version of this algorithm, with direction and timestep splitting, to produce two-dimensional or three-dimensional calculations. Fully two-dimensional and three-dimensional versions are also available.[11,15] Specific applications to detonations are described in several references.[8–10,16,17]

Those parts of the reactive-flow equations which describe the chemical reactions are solved both separately and by a combination of implicit and explicit methods. These results are combined with the FCT solutions for convective transport by timestep-splitting methods.[15]

Computations on a Massively Parallel Computer

These computations of multidimensional, highly compressible, time-dependent reacting flows were carried out on the massively parallel computer, the Connection Machine (CM).[18] A CM consists of thousands of individual scalar processors, connected by hypercube communications, and all synchronized to a single instruction stream in a single-instruction multiple-data (SIMD) architecture. At NRL, we are currently using a CM with 16,384 (16 K) processors. Communication to and control of these processors is through a front-end computer that may be either a VAX, a Symbolics, or a Sun. Each individual processor can be reconfigured by software to obtain powers of two increase in the number of virtual processors. The actual number of virtual processors is limited by the storage the problem solution requires. Floating-point arithmetic is carried out by Weitek chips, each of which does pipelined processing of the floating-point operations for 64 of the scalar processors. Online color graphics are available to display the evolution of a calculation and to make video tapes. The programs are written in one of several languages specially adapted to parallel processing, *Lisp, C*, or a special version of Fortran 8X.

We converted the recent one-dimensional and multidimensional versions of the FCT algorithms into C* and PARIS. The kernel of the algorithms consists of about 38 lines of code well-suited for parallel instructions. We implemented the induction parameter model described above using timestep-splitting techniques to combine the results with the results of convective transport.

A fundamental concern with the CM is how to avoid paying a severe penalty (factors of two, four, or more in computational time) for computations which do not have periodic boundary conditions. Periodic boundary conditions are built into the CM hardware and cost no time or program-

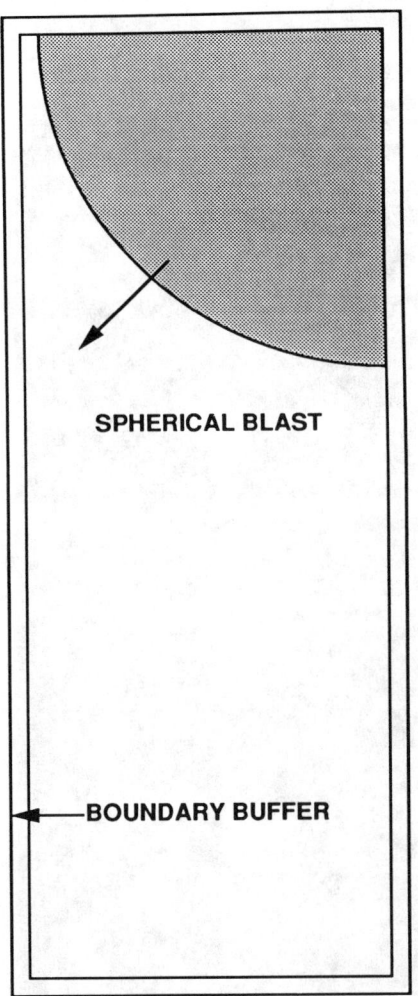

Fig. 3 Initial conditions for the simulations described in this paper.

ming effort to implement for problems in any dimension, as long as the number of computational cells in each dimension is a power of two. We have tried or considered several approaches to programming more realistic boundary conditions.[18] For the computations performed for the Mach stem studies, we implemented inflow, outflow, and reflecting wall boundary conditions through a method we call *uniform boundary-condition algorithms*. The basis of this approach is to perform the same computations for all cells at all times, whether or not they are boundary cells, and to choose the "correct" value for each cell. The particular algorithms used here are based on ideas presented by Li et al.[19] Such an approach is particularly useful in two-dimensional computations and for supersonic flow problems with relatively straightforward internal or external boundary conditions.

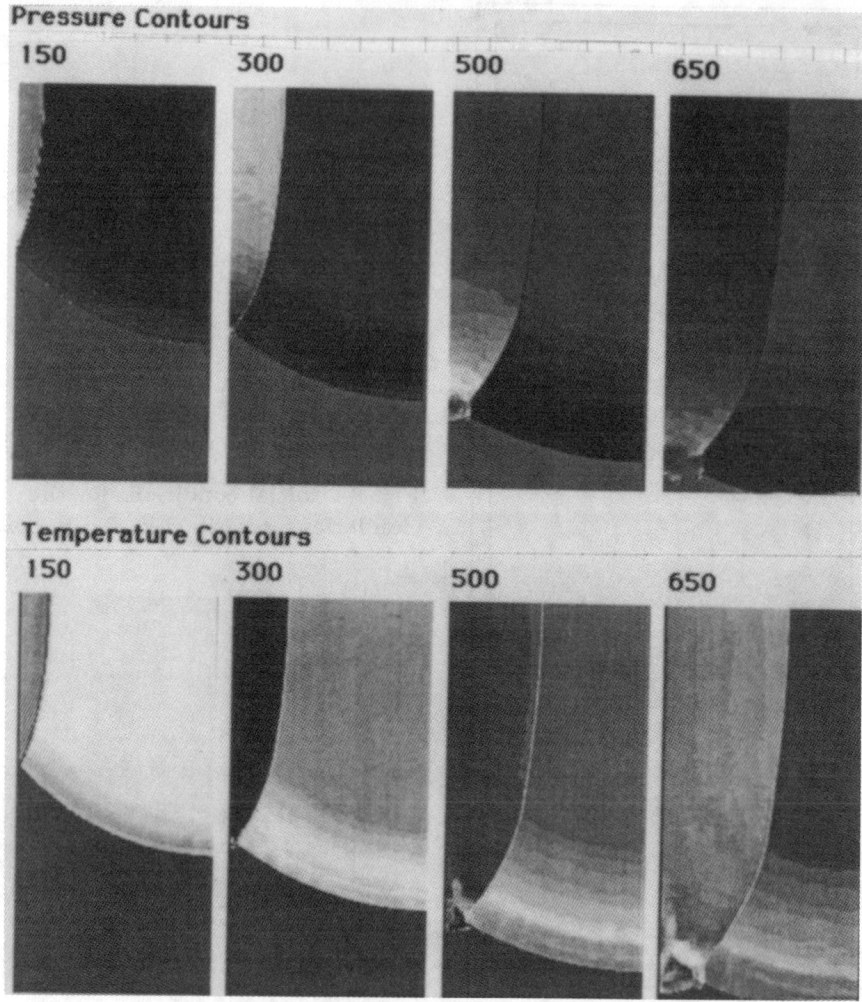

Fig. 4 Pressure and temperature contours for a simulation in which reaction has not yet occurred.

Discussion of Simulations

In the computations and experiments where reignition appears to occur in the region of a Mach stem, the conditions some time prior to ignition consist of a curved shock followed at some distance by a decoupled reaction zone. This situation arises as a confined detonation expands past an obstacle or boundary into a less confined space. To simulate this situation in a controlled environment, we chose, as an initial condition, a cylindrical shock with no reaction behind it impinging on a flat reflecting wall. The jump conditions across the shock are initialized such that the expanding,

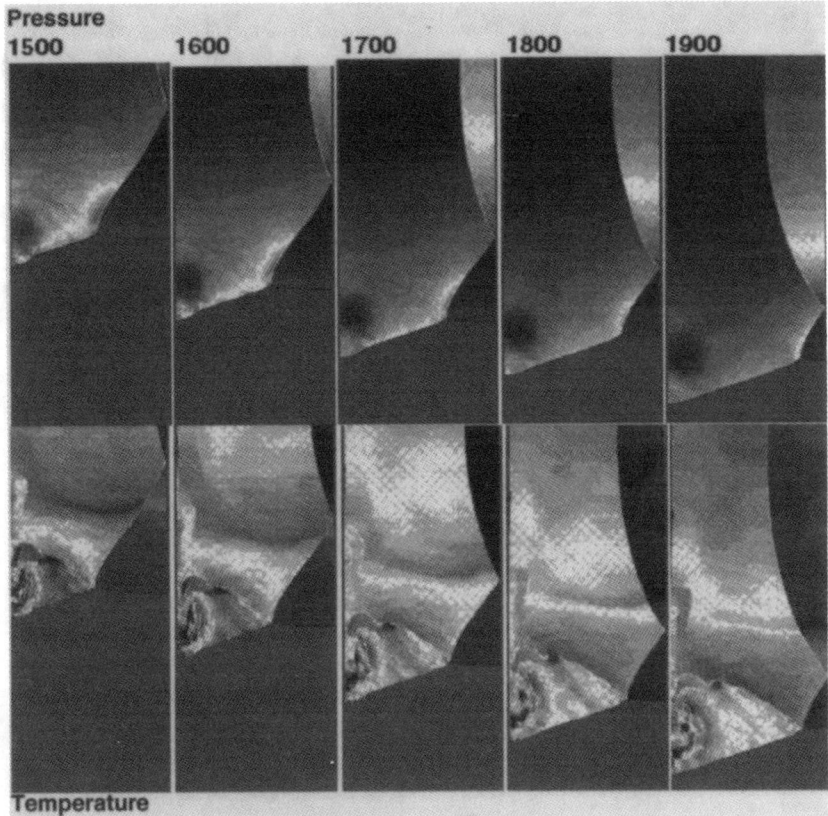

Fig. 5   Pressure and temperature contours for a simulation in which reaction has not yet occurred.

constant-density, constant-pressure domain inside the shock radius is not hot enough to ignite the mixture during the time of the experiment. Since the region behind the shock is expanding like a blast wave, the shocked gas never gets hotter than the initial conditions until the cylindrical blast wave reflects from the flat wall. Figure 3 is a schematic of the initial conditions of the simulations.

When the leading shock reflects from the wall at the point of first contact directly below the center of the blast wave, the pressure and temperature rise due to normal shock reflection. As the cylindrical shock expands, however, the point of reflection moves and meets the incident shock at oblique angles. At a critical angle, which is reached before the incident shock has decayed appreciably, a Mach stem forms at the end of the region where only regular reflection has occurred previously. Figures 4 and 5 show temperature and pressure contours from two such simulations. Initially we see regular reflection, which then develops into a double Mach or a complex Mach reflection.

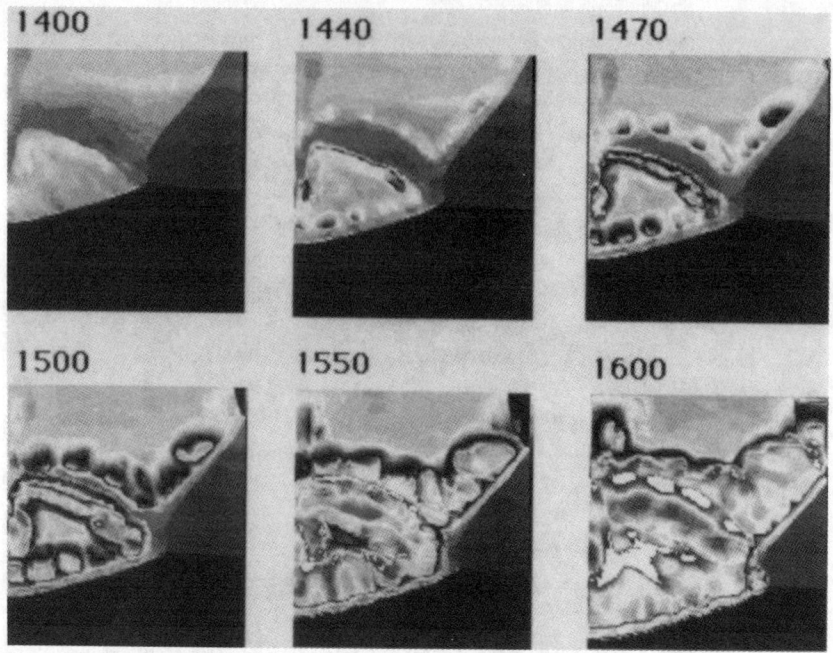

Fig. 6 Ignition in the complex Mach structure system.

We performed simulations using kinetic parameters characteristic of stoichiometic mixtures of $H_2:O_2$ (undiluted) at standard temperature and pressure (298 K and 1 atm). To isolate the Mach-structure ignition process, we allowed the Mach structure to propagate from an initially unreactive mixture into this highly energetic hydrogen mixture. As we have seen in the figures above, the highest temperatures occur from heating along the slip line and at the wall where the vortex rolls up. When ignition is allowed to occur, as shown in Fig. 6, it therefore occurs along this slip line and around the vortex, effectively forming an imploding detonation. At the point of implosion, the temperature and pressure become very high, forming a hydrodynamic explosion, which then moves outward and ignites the entire mixture. Tonello's experiments demonstrate this explosion effect.[20]

Acknowledgments

This work was sponsored by the Naval Research Laboratory through the Office of Naval Research. The authors would like to acknowledge the advice and suggestions given by Allen Kuhl and C. Richard DeVore, the input of Carolyn Kaplan, Chiping Li, and K. Kailasanath, and the valuable help of Diana Yap.

References

[1] Oran, E. S., Jones, D. A., and Sichel, M., "Numerical Simulations of Detonation Transmission," *Proceedings of the Royal Society A*, Vol. 436, 1992. pp. 267-297.

[2] Kaplan, C. R., and Oran, E. S., "Spontaneous Ignition and Detonation of Propane-Air Mixtures," *Combustion Science and Technology*, Vol. 80, 1991, pp. 185-205.

[3] Teodorczyk, A., Lee, J. H. S., and Knystautas, R., "Propagation Mechanism of Quasi-Detonations," *Twenty-Second Symposium (International) on Combustion*, The Combustion Institute, Pittsburgh, PA., 1989, pp. 1723-1731.

[4] Bashenova, T. V., Gvozdeva, L. G., and Nettleman, M. A., "Unsteady Interactions of Shock Waves," *Progress in Aerospace Science*, Vol. 21, 1984, pp. 249-331.

[5] Lee, J.-H., and Glass, I. I., "Pseudo-Stationary Oblique-Shock-Wave Reflections in Frozen and Equilibrium Air," *Progress in Aerospace Science*, Vol. 21, 1984, pp. 33-80.

[6] Kuhl, A. L., Ferguson, R. E., Chien, K.-Y., Glowacki, W., Collins, P., Glaz, H., and Colella, P., "Simulation of a Turbulent Wall Jet in a DMR Flow," to appear in *Progress in Aeronautics and Astronautics*, 1992.

[7] Book, D., Boris, J., Kuhl, A., Picone, M., Oran, E., and Zalesak, S., "Simulation of Complex Shock Reflections from Wedges in Inert and Reactive Mixtures," *Proceedings of the 7th International Conference on Numerical Methods in Fluid Dynamics*, Springer-Verlag, New York, 1981, pp. 84-90.

[8] Oran, E. S., Boris, J. P., Young, T. R., and Picone, J. M., "Numerical Simulations of Detonations in Hydrogen-Air and Methane-Air Mixtures," *Proceedings of the 18th Symposium (International) on Combustion*, The Combustion Institute, Pittsburgh, PA, 1981, pp. 1641-1649.

[9] Kailasanath, K., Oran, E. S., and Boris, J. P., "Determination of Detonation Cell Size and the Role of Transverse Waves in Two-Dimensional Detonations," *Combustion and Flame*, Vol. 61, 1985, pp. 199-209.

[10] Guirguis, R., Oran, E. S., and Kailasanath, K., "Numerical Simulations of the Cellular Structure of Detonations in Liquid Nitromethane — Regularity of the Cell Structure," *Combustion and Flame*, Vol. 65, 1986, pp. 339-366.

[11] DeVore, C. R., "Flux-Corrected Transport Algorithms for Two-Dimensional Compressible Magnetohydrodynamics," NRL Memorandum Rep. 6544, Naval Research Laboratory, Washington, DC, 1989.

[12] Taki, S., and Fujiwara, T., "Numerical Simulation of Triple Shock Behavior of Gaseous Detonation," *Eighteenth Symposium (International) on Combustion*, The Combustion Institute, Pittsburgh, PA, 1981, pp. 1671-1681.

[13] Gubin, S. A., and Sichel, M., "Calculation of the Detonation Velocity of a Mixture of Liquid Fuel Droplets and a Gaseous Oxidizer," *Combustion Science and Technology*, Vol. 17, 1977, pp. 109-117.

[14] Boris, J. P., and Book, D.L., "Solution of the Continuity Equation by the Method of Flux-Corrected Transport," *Methods in Computational Physics*, Vol. 16, 1976, pp. 85-129.

[15] Oran, E. S., and Boris, J. P., *Numerical Simulation of Reactive Flow*, Elsevier, New York, 1987, pp. 264-313.

[16] Jones, D. A., Sichel, M., Guirguis, R., and Oran, E. S., "Numerical Simulation of Layered Detonations," Vol. 133, *Progress in Astronautics and Aeronautics*, AIAA, NY, 1991, pp. 202-220.

[17] Oran, E. S., Kailasanath, K., and Guirguis, R. H., "Numerical Simulations of the Development and Structure of Detonations," *Progress in Astronautics and Aeronautics*, Vol. 114, 1988, pp. 155–169.

[18] Oran, E. S., Boris, J. P., Brown, E. F., and Whaley, R. O., "Exploring Fluid Dynamics on a Connection Machine," *Supercomputing Review*, May 1990, pp. 52–60.

[19] Li, C., Oran, E. S., and Boris, J. P., "A Uniform Algorithm for Boundary and Interior Regions and Its Application to Compressible Flow Simulations," *Parallel Computational Fluid Dynamics*, 1992, MIT Press, Cambridge, MA, pp. 87–96.

[20] Tonello, N., and Sichel, M., private communication.

Photographic Study of the Direct Initiation of Detonation by a Turbulent Jet

M. Inada,* J. H. Lee,† and R. Knystautas†
McGill University, Montreal, Quebec, Canada

Abstract

The phenomenon of direct initiation of detonation by a turbulent jet of combustion products mixing in an explosive gas medium has been studied by high-speed Schlieren photography. High-pressure combustion gases were generated by the reflection of a detonation wave from a thin diaphragm in the upstream chamber. The rupture of the diaphragm then released a jet of product gases through an orifice to initiate the detonation in the downstream chamber. High-speed Schlieren photographs were taken with a Barr and Stroud framing camera. For comparison, Schlieren photographs were also taken of the phenomenon of transmission of detonation through an orifice which represents the critical tube diameter phenomenon. The results indicate that the flow structure in the jet initiation process differs drastically from that associated with a decaying detonation in the critical tube diameter case. In the critical tube situation, a well-defined transverse wave pattern already exists in the decaying detonation as it emerges from the orifice. Reinitiation occurs in the "enlarged cells" of the decaying front where the transverse waves of the substructure are amplified. In jet initiation, no transverse waves are initially present

---

Copyright © 1992 by the American Institute of Aeronautics and Astronautics, Inc. All rights reserved.
 *Project Engineer, Department of Mechanical Engineering, Nuclear Power Engineering Test Center, Tokyo, Japan.
 †Professor, Department of Mechanical Engineering.

and they have to be generated from the amplification of the high-frequency acoustic waves that always accompany high-speed reacting flows. Initiation therefore requires a much more sensitive mixture (for the same orifice opening) than the critical tube diameter case. The present results suggest that all initiation phenomena are essentially identical. The difference lies in the nature of the initial flowfield where the generation and amplification of transverse waves has to occur.

Introduction

Perhaps the most probable mechanism for the formation of a detonation wave in an accidental unconfined vapor cloud explosion is due to jet initiation. That a jet of combustion products can lead to direct initiation of a spherical detonation has been demonstrated by Knystautas et al[1] in 1978. Since then, the jet initiation problem has been studied by Schildnecht et al.[2] Ungut and Shuff,[3] Moen et al.,[4] Mackay et al.,[5] and Carnasciali et al.[6] Except for the work of Carnasciali et al.,[6] where the sudden rupture of a diaphragm was used to start the jet, the problem investigated by the others was essentially that of the transition to detonation from an accelerating flame as it emerged from a confined pipe into a sudden expansion. It may appear that the differences are minor. However, in the flame jet initiation (or more appropriately, the transition to detonation), the gasdynamic flow structure in the flame jet is a characteristic of its acceleration history in the pipe prior to its exit into the unconfined cloud. For example, in most all of these studies, obstructions were placed in the pipe (and even right at the exit plane of the jet[4,5]) to achieve a high flame speed to promote transition. Not only the turbulence parameters are strongly dependent on the geometry of these obstructions, they also induce a set of strong transverse shocks in the flame jet. These strong transverse shocks in the jet may in fact play the dominant role in the initiation process rather than the usual proposed mechanism of initiation by "turbulent mixing". Indeed, numerical simulation of the flame jet initiation problem by Thibault[7] with a central blockage at the exit plane of the pipe has revealed that a high temperature and pressure region exists at the axis due to the focusing of a cylindrical converging shock. This "hot spot" becomes the site of the subsequent initiation phenomenon. These studies of transition in flame jets are somewhat similar to the previous investigations by Lee et al.[8] and Knystautas et al.[9] of accelerating flames from a rough pipe exiting into a smooth pipe. In all of these transition problems, the flowfield (turbulence and transverse wave patterns created by the obstacles) is characterized by the pipe and obstacle geometries. Therefore it is difficult to isolate the mechanisms of the initiation process. Thus far, no quantitative correlation of the critical conditions required for jet initiation has been achieved.

In the study of Carnasciali, a diaphragm was used to ensure that neither unburned gases nor products were vented through the opening prior to complete combustion in the first chamber. The diaphragm was timed to rupture at the peak constant volume explosion pressure. Although the flowfield of the transient jet is influenced by the diaphragm opening characteristics, the use of different thicknesses and diaphragm materials may permit these effects to be assessed. The present study is essentially a follow up on the work of Carnasciali who did not make photographic observations of the initiation phenomena. Thus much of details of the initiation process remain obscure. The use of high-speed Schlieren photography in the present study permits the details of the initiation process to be revealed.

Experimental Details

The basic apparatus in this study consists essentially of two combustion chambers isolated from one another initially by a thin diaphragm at the orifice between the two chambers. The diaphragm permits different explosive mixtures to be used in each of the two chambers if required. Rather than use constant volume combustion in the first chamber to create the high-pressure combustion products, the mixture in the first chamber was detonated. The detonation wave reflects from the diaphragm and the reflected shock propagates into the Taylor expansion wave in the products, bringing the gas to rest. By the time the diaphragm has ruptured, most of the transient gas dynamic processes have decayed and the properties of the product gases that exit the orifice are quite similar to that from a constant volume explosion. Using the detonation mode, a much thinner diaphragm can be employed. The rapid detonation process also permits a more precise synchronization of the various events in the photographic process than the slower constant volume explosion mode.

The experimental apparatus used is shown in Fig. 1. It consists of a detonation tube 2 m long and 5 cm diameter (i.e., the first chamber) connected to a larger rectangular flame chamber 0.6 m long and 0.3 x 0.3 m cross section. Two opposite sides of this rectangular flame chamber are equipped with glass plate windows for Schlieren photography. The detonation tube and the flame chamber are connected through an orifice opening. Orifice plates of different diameters (73, 47, and 30 mm) were used in the experiments. Various diaphragm materials and thicknesses were used and in most of the experiments a very thin sheet of aluminum foil, lightly scribed by a sharp needle along 4 diameters 45 deg. apart was found to be most consistent. Most of the photographic studies were carried out with acetylene-oxygen mixture because of its sensitivity, thus permitting the initial pressure of the experiment to be kept low. Other less sensitive mixtures (C_2H_2 - O_2 with high argon dilution, H_2 - O_2, CH_4 - O_2, C_3H_8 - O_2) were also used. For the less sensitive mixtures which necessitate the use of higher initial pressures, the glass windows were replaced by steel plates and only "go-no go" type of experiments were carried out to establish the critical conditions for initiation. No photographic observations were made for these less sensitive mixtures.

For the high-speed Schlieren photography, a Barr and Stroud rotating mirror framing camera was used. The double-pass Schlieren system employed a 25-cm-diam mirror and a linear xenon flash tube (Xenon Corp. FPA8100C) powered by a high-voltage capacitor bank was used as the light source. Pressure transducers (PCB 113A24, 5 mV/psi) mounted on the detonation tube (chamber A) and the flame chamber, permitted the detonation velocities to be determined and thus the go or no go condition to be deduced from the time-of-arrival at the various gauge locations. A schematic diagram of the apparatus and

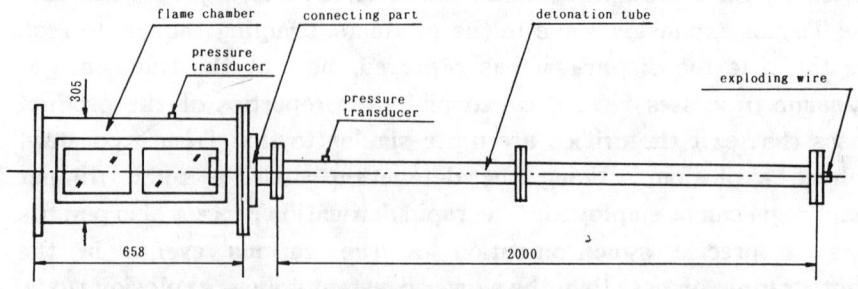

Fig. 1 Schematic diagram of experimental apparatus.

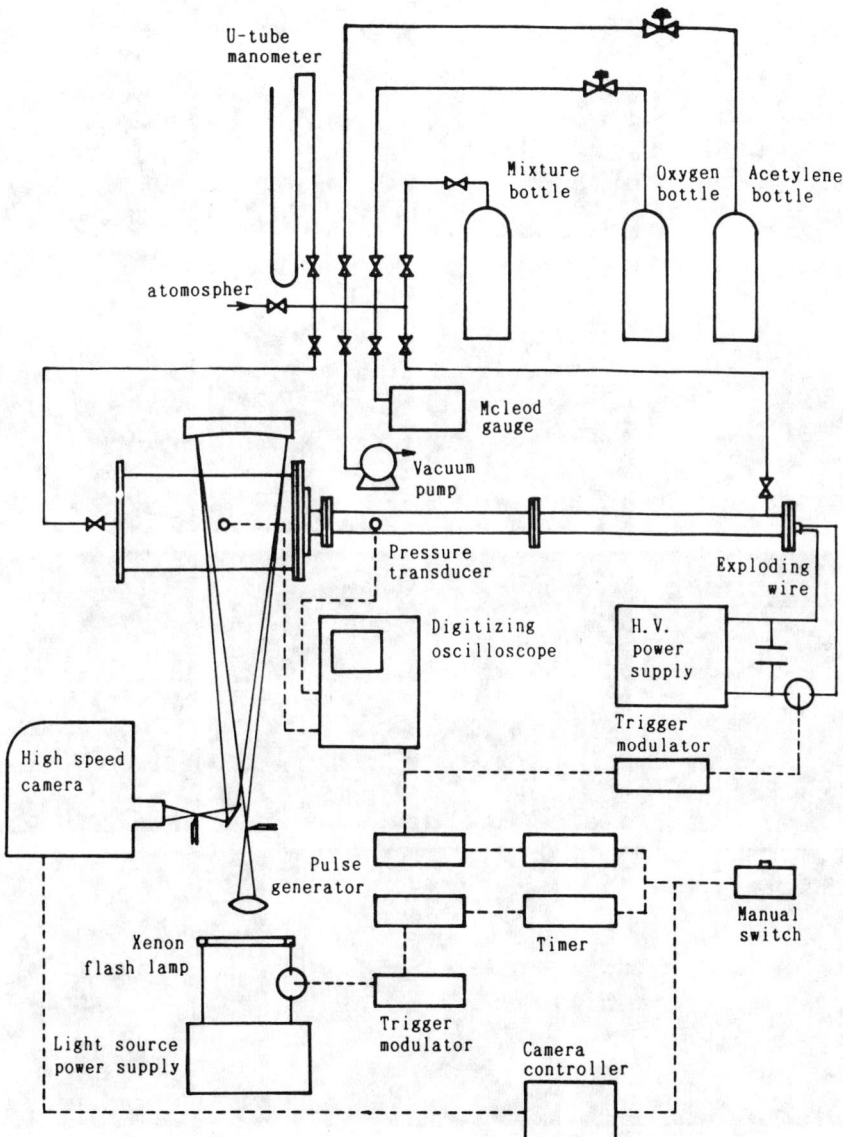

Fig. 2 Schematic diagram of experiment and diagnostics.

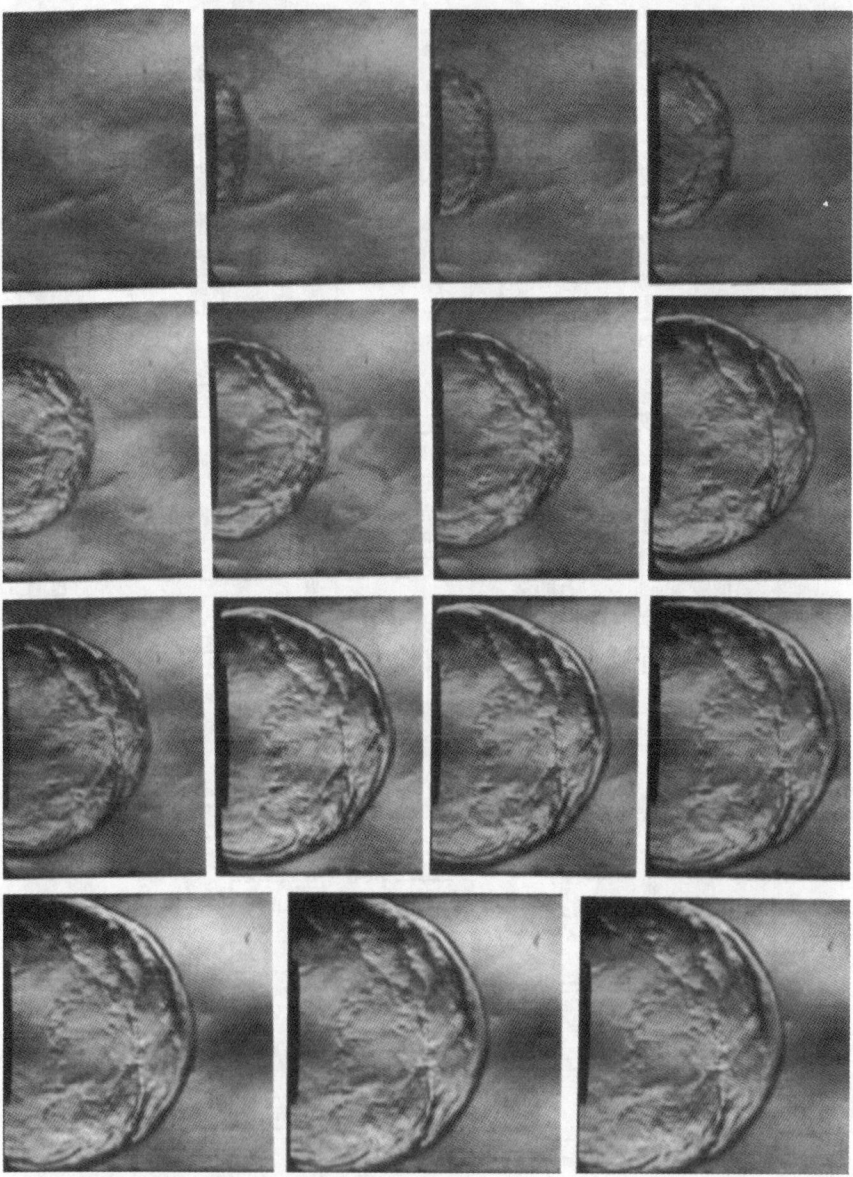

Fig. 3 Diffraction of a planar detonation wave emerging from a confined tube: failure to transmit into the unconfined.

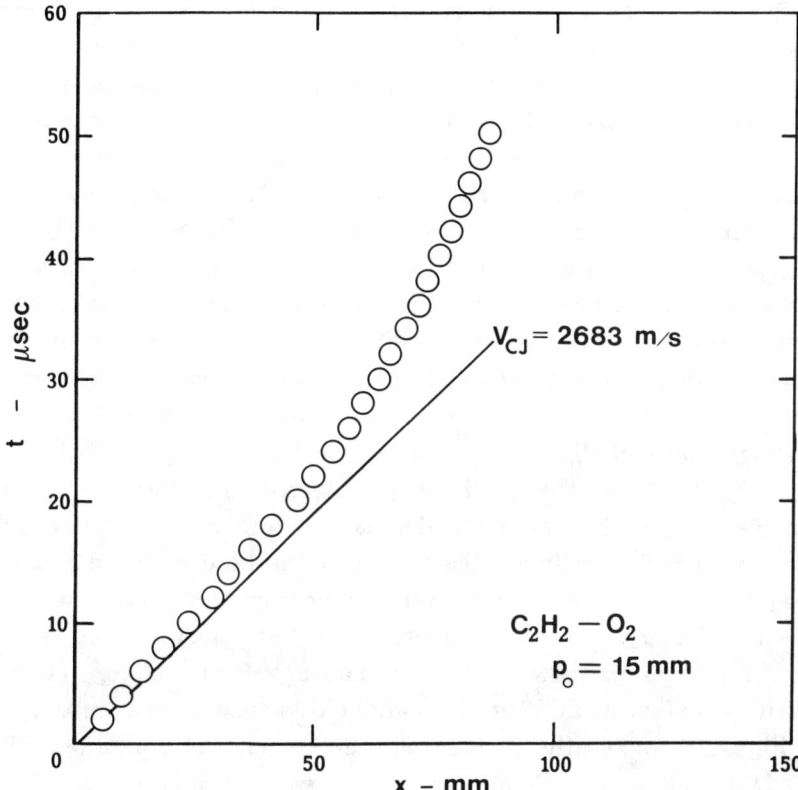

Fig. 4 Trajectory of a failed detonation wave emerging from a confined tube.

associated diagnostics is shown in Fig. 2. In certain experiments, smoke foils were used downstream of the orifice in the flame chamber to record the evolution of the detonation structure during the initiation process.

Results and Discussions

Since the mechanisms of failure and reinitiation of a detonation wave emerging from a circular pipe into unconfined space (i.e., the critical tube diameter case) are well known, it is of interest to use this as a basis for comparison with the jet initiation phenomenon. In the critical tube diameter experiments, no diaphragm is used to separate the two chambers. Figure 3 shows the high-speed photographic

Schlieren sequence of a detonation emerging from the orifice under subcritical conditions. The initial cellular planar wave becomes spherical as the lateral expansion waves propagate towards the axis. Eventually, the spherical shock separates from the reaction zone. It should be noted that the cellular structure is present throughout the attenuation process and cells grow coarser as the wave decays. The spherical shock front and its trailing reaction zone become progressively smoother. The decaying detonation front trajectory (along the axis) is plotted in Fig. 4. It decays from a value of 2253 m/s (the theoretical C-J value is 2683 m/s) to approximately one-half this initial velocity subsequent to the expansion waves reaching the axis. The ratio of d/λ = 6.9 for this case is in accord with the measured value of $d/\lambda \cong 13$ for the critical condition.

Figure 5 shows the Schlieren photographs for the supercritical case (i.e., $d/\lambda \geq 13$) where the detonation does not fail. It can be observed that fine cellular structure near the head of the detonation along the axis is always preserved, indicating that new transverse waves are being generated to counteract the lateral expansion wave of the diffraction process. The detonation velocity along the axis remains constant at 2569 m throughout this supercritical case.

Figure 6 shows the phenomenon at the critical condition. The wave attenuates as the lateral expansion waves converge towards the axis. The cellular structure of the front gets coarser as the wave attenuates as in the subcritical case. However, detonation reinitiation occurs in the attenuated front and a "detonation bubble" from a local initiation site can be seen in late frames of the sequence. The detonation bubble eventually grows to engulf the entire attenuated front recovering the initial finer scale cellular structure of a normal detonation in this mixture. The value of d/λ in this case is around 11, in accord with the usual empirical correlation of $d/\lambda \cong 13$ for the critical tube diameter. It is important to note that in the critical tube diameter problem, a set of transverse waves is always associated with the attenuating detonation front emerging from the orifice. Hence, the reinitiation process occurs in a flow structure where strong transverse shocks are already present. We may interpret the reinitiation process in the critical tube diameter problem as a recoupling of the transverse waves with the shock front to regain its former cellular structure. It is more of a recovery process from a perturbed state to its initial state. This is to be contrasted to a jet initiation problem (or a transition process) where a set of transverse waves is not originally present.

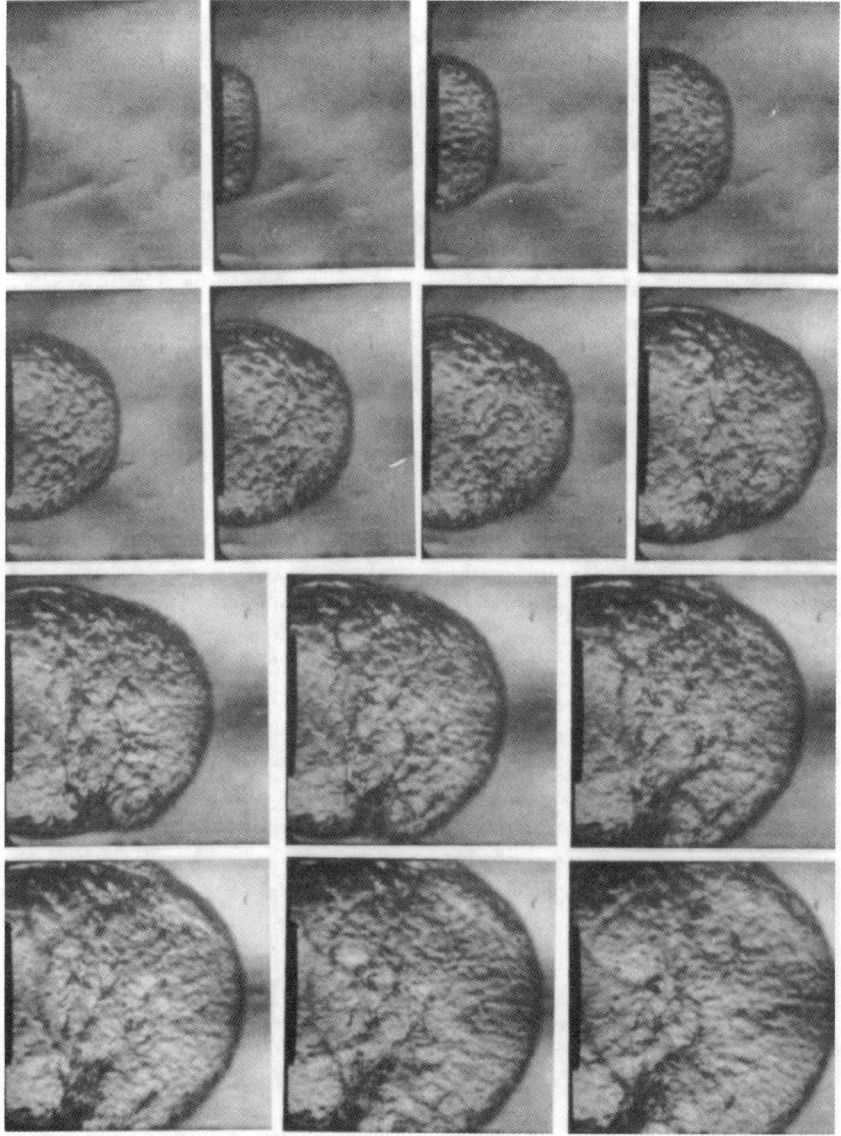

Fig. 5 Diffraction of a planar detonation wave emerging from a confined tube: successful transmission into the unconfined.

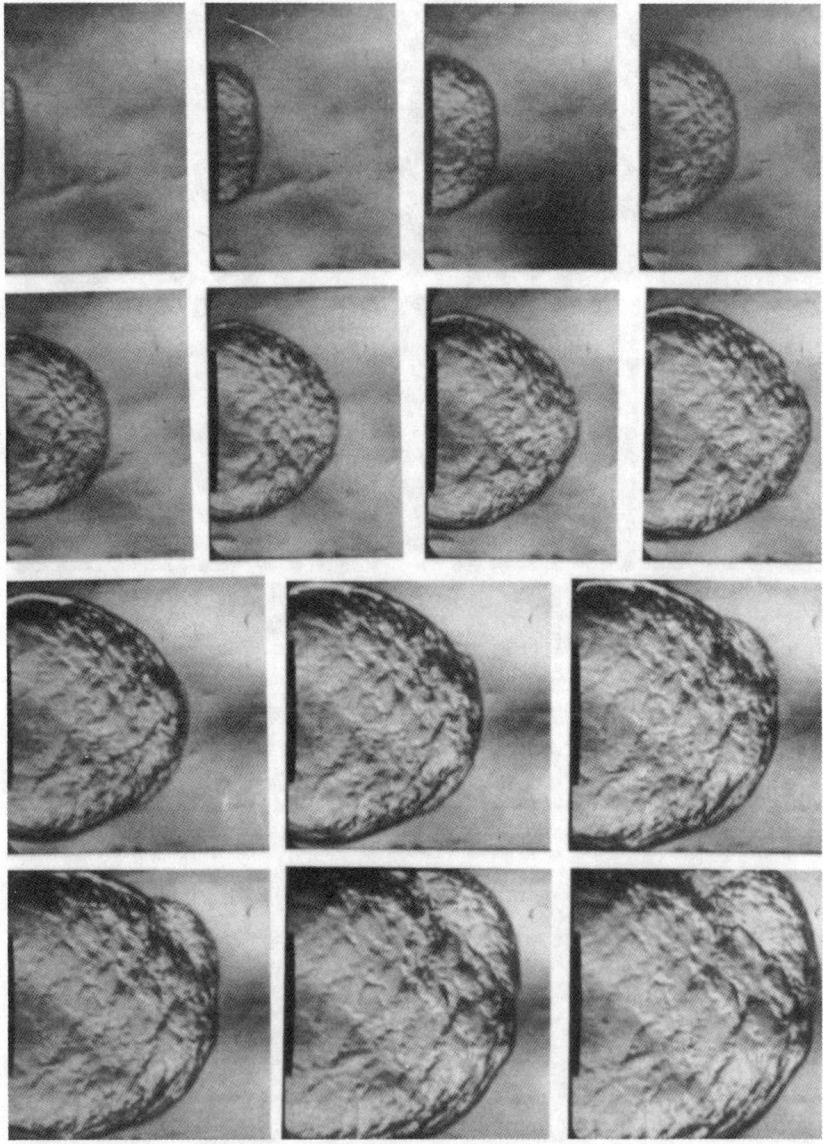

Fig. 6 Diffraction of a planar detonation wave emerging from a confined tube: reinitiation of detonation from a failing wave.

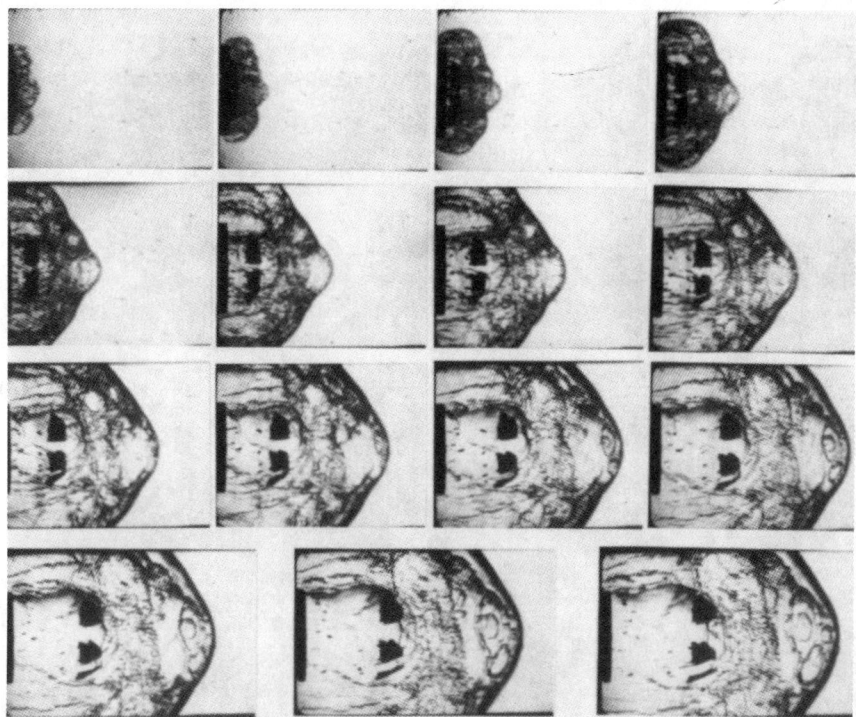

Fig. 7 Turbulent hot jet emerging into an explosive gas mixture: failure to initiate detonation.

Thus jet initiation requires both the formation of transverse waves and their resonant coupling to form a cellular detonation.

Figure 7 shows a sequence of Schlieren photographs for the subcritical case of jet initiation. Comparing with Figs. 3, 5, and 6, we can see the distinct differences in the flow structure of the product gases that emerge from the orifice. A well-defined leading shock front with a set of strong transverse shock waves is now absent. It is of interest to note that even the use of an extremely thin piece of plastic foil as a diaphragm will completely change the flow structure of the emerging flow. The reflection of the detonation from the diaphragm destroys the normal cellular transverse wave structure of the incident detonation front. Thus downstream of the orifice, the upstream flow characteristic is not retained. We feel that this is the reason why the critical conditions required for jet initiation involves values of d/λ of the order of five times that of the critical tube diameter problem.

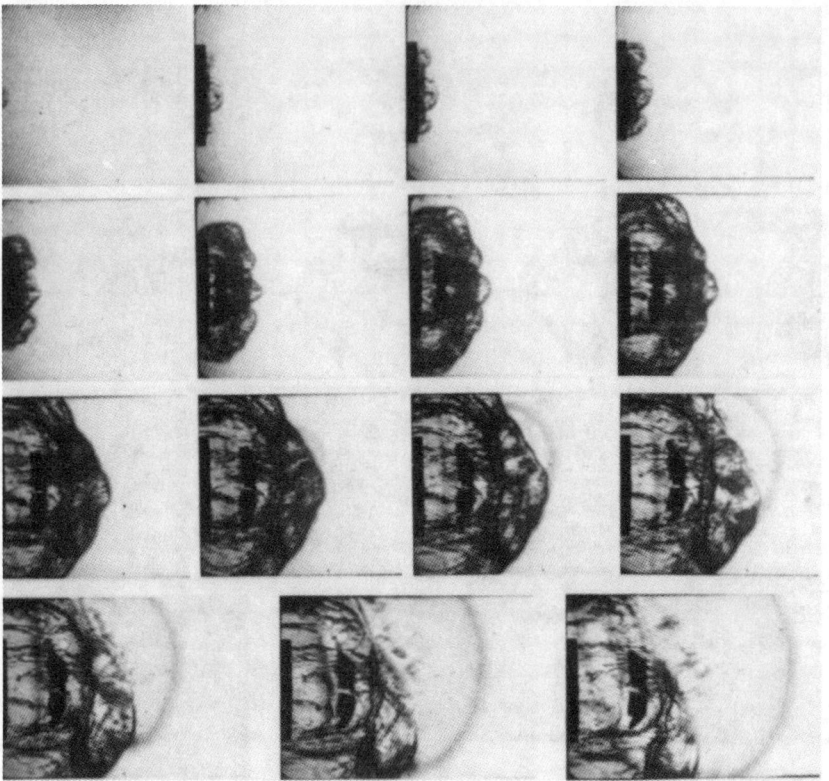

Fig. 8 Turbulent hot jet emerging into an explosive gas mixture: initiation of detonation.

Without the established transverse wave structure of the upstream detonation, it is then necessary to develop such a transverse wave pattern for the formation of a detonation. The transverse wave pattern is formed from the high-frequency acoustic waves that always accompany a high-speed flow, particularly when chemical reactions are present. It is then the resonant coupling of these high-frequency waves with the chemical processes at some localized region in the turbulent jet (i.e., hot spot) that eventually amplifies to form the detonation. In the critical tube diameter problem, the localized region is usually within an enlarged detonation cell of the attenuated front. It is essentially the growth of the substructure within the cell that reinitiates the detonation. In jet initiation, it appears more difficult to achieve localized regions where conditions are favorable for the amplification and coupling of the background high-frequency acoustic waves to become transverse shocks which eventually evolve to a detonation. In Fig. 7, it appears that a detonation bubble is formed near the axis of the jet. However, in subsequent frames, it is clear that the shock and reaction zone are also decoupled in this region to form a deflagration. The pieces of diaphragm material in the jet flow structure are clearly evident in Fig. 7.

Figure 8 shows the sequence where jet initiation is successful. The initial flowfield is similar to the subcritical case. However, in later frames of Fig. 8, a detonation bubble emerges from within the jet which propagates at a much higher velocity away from the jet at late times. If the conditions are above the critical values (i.e., mixtures that are most sensitive), more than one initiation site is usually observed and they also occur earlier in time when the jet emerges. These jet initiation photographs are similar to those we obtained earlier.[1] However, in the previous study only single-shot Schlieren pictures were taken and we were not able to follow the time history of the subsequent development.

Figure 9 shows the trajectories of the jet and the subsequent detonations formed. Both subcritical and supercritical cases are plotted. It can be observed that in all of the various cases in the present study, the initial jet velocity is of the order of 1100 m/s, quite close to the sound speed of the combustion products indicating that the jet is a "choked" jet. For the subcritical case, the jet front velocity slowly decays. When initiation occurs, the detonation propagates at 2807 m/s close to the corresponding C-J velocity of the mixture. Unlike the previous investigations of Ungut and Shuff,[3] Moen et al.,[4] and Mackay et al.,[5] where flame jets of different velocities and flow structures that are widely different because of the different obstacle configurations used to accelerate the flame to the different velocities in the tube, the initial conditions, (i.e., jet velocity and flow structure) of the present study are the same. Thus it may be possible to seek a correlation between the jet diameter and the mixture sensitivity (as characterized by the cell size λ). Figure 10 shows the results for the case of $C_2H_2 + O_2$ using various orifice diameters. The $d/\lambda = 13$ for the critical case is plotted for comparison. The results of Carnasciali are also shown in Fig. 10. It can be observed that in general the d/λ value for jet initiation is about five times that for the critical tube diameter case. The data are not sufficient to conclude on a general universal criterion of the d/λ value for jet initiation. Even in the case

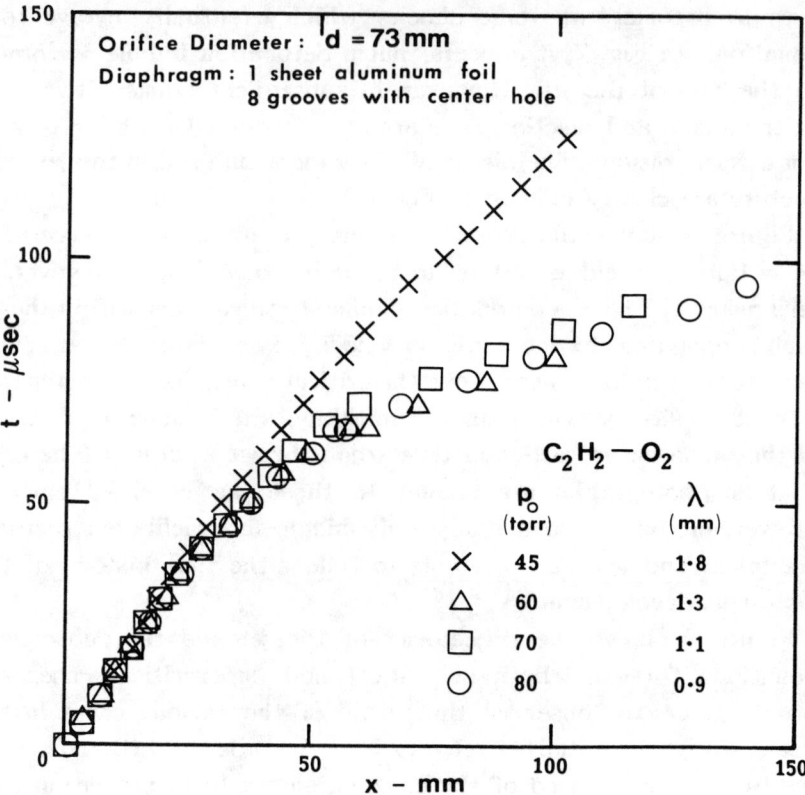

Fig. 9 Trajectories of leading edge of hot turbulent jet for successful and unsuccessful initiation of detonation.

of the critical tube diameter problem, it is found that mixtures with a regular cell structure (e.g., highly argon diluted C_2H_2 - O_2 or H_2 - O_2 mixtures) require a much higher value of d/λ for successful transmission.[10] In fact, the value of d/λ for the critical tube diameter problem for highly diluted mixtures with argon is about the same as for the jet initiation case. To test the influence of cell regularity on the jet initiation problem, mixtures diluted with argon were also used. Using C_2H_2 + 5/2 O_2 with 75% argon dilution, it was found that the d/λ ratio is 100 as compared to the case with no argon dilution where $d/\lambda \cong 40$. For the case of C_3H_8 + 5O_2 where cells are somewhat more irregular because of a higher activation energy, it is found that $d/\lambda \cong 30$ which is slightly less than the value of $d/\lambda \cong 40$ for the case of C_2H_2 + O_2 where the cells are more regular. More definitive experiments must be carried out before concrete conclusions can be made regarding the influence of cell regularity on the critical conditions of jet initiation. However, the few tests carried out support the observations obtained in the critical tube diameter problem, that is, mixtures with a more regular cell pattern require a higher value of d/λ for initiation. Cell regularity is correlated with the stability of the mixture.[11]

Conclusions

It appears that jet initiation (like the critical tube diameter problem or the transition problem) essentially involves the establishment of a transverse wave pattern that is coupled with the chemical processes. In a general sense, all initiation or transition phenomena require the formation of such a coupled transverse wave pattern because the structure of a self-sustained detonation is essentially that of a transverse wave pattern in resonant coupling with the chemical processes. The major differences in the various modes of initiation arise from the initial flow structure from which the coupled transverse waves are to be developed. The present photographic study clearly demonstrates the similarities and differences between two modes of initiation, that is, in the critical tube diameter and jet initiation situations. In the critical tube diameter case, the flow structure is essentially that of a decaying cellular detonation where the cell pattern gets progressively coarser due to the attenuation by the lateral expansion wave. Reinitiation occurs locally in an enlarged detonation cell when the substructure amplifies to form a coupled transverse wave pattern. Thus unstable mixtures with highly irregular cells with substructure are more prone to reinitiation than mixtures with a highly regular cell pattern where the substructure is absent. The present study of jet initiation demonstrates a similar initiation behavior and indicates that reinitiation occurs locally, but the flow structure of the jet differs drastically from that of the critical tube diameter case. With the transverse wave pattern of the detonation destroyed by reflection at the diaphragm in the jet initiation case, it is necessary to regenerate a new transverse wave pattern from the ever present high-frequency acoustic radiation in a high-speed compressible reacting flow. The precise local conditions where the amplification and coupling of transverse waves in the jet flow occur

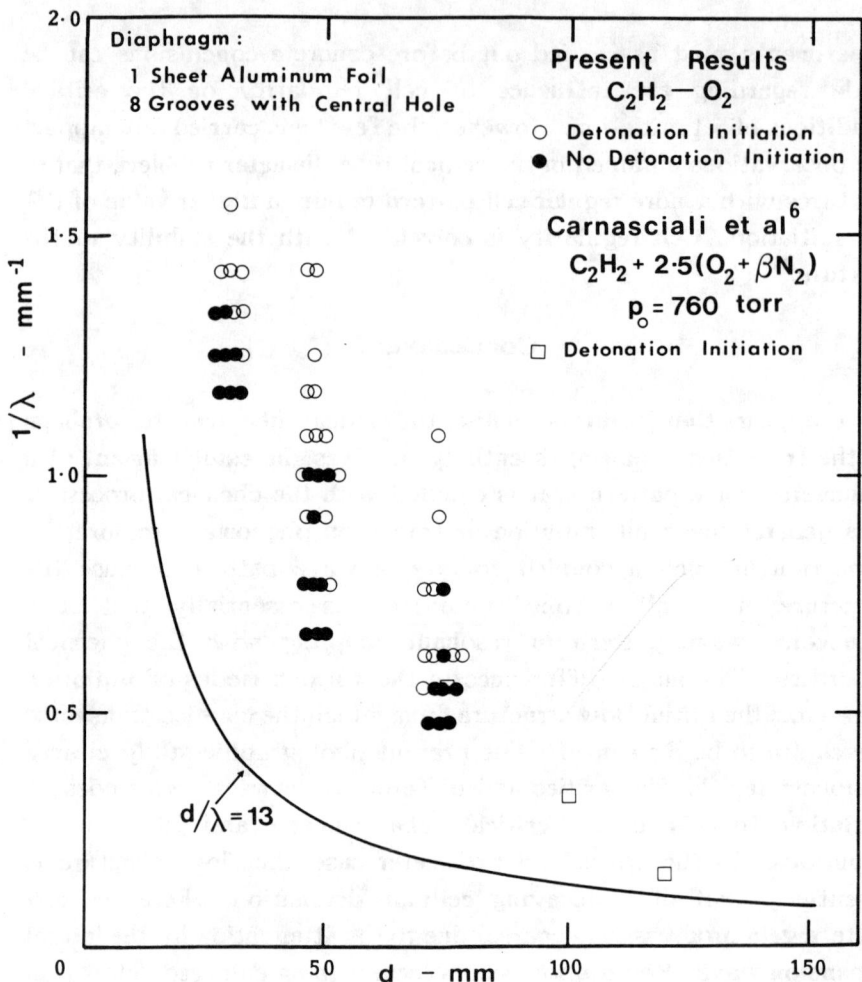

Fig. 10 Correlation of hot jet initiation results as a function of jet size and mixture sensitivity (λ).

are not clear. It is suspected to occur in large eddies that autoignite when unburned gases are entrained and mixed with the products. It would be extremely difficult to determine the exact local conditions for initiation due to turbulent mixing.

It appears that the jet initiation problem is strongly dependent on the flow structure of the jet. Initial and boundary conditions are likely to play an important role and it would be difficult to seek a general criterion for direct initiation by turbulent jet. Even in the relatively well-defined case of the critical tube diameter problem, it is not possible to obtain a general criterion that is universally valid (e.g., the $d/\lambda = 13$ correlation). Cell regularity and orifice geometry play an important role and these factors have not been clearly resolved.

Acknowledgments

This work was carried under the sponsorship of the Natural Sciences and Engineering Research Council of Canada under Grants A-3347 and A-7091 and of the Department of National Defense of Canada (DRES) under Contract O1SG.W7702-6-2521-A. The authors would like to thank Mr. Aris Makris for carrying out some of the experiments with argon diluted mixtures.

References

[1] Knystautas, R., Lee, J. H. S., Moen, I. O, and Wagner, H. G., "Direct Initiation of Spherical Detonation by a Hot Turbulent Gas Jet," *Proceedings of the 17th Symposium (International) on Combustion*, The Combustion Institute, Pittsburgh, PA, 1979, pp. 1235-1244.

[2] Schildknecht, M., Geiger, W., and Stock, M., "Flame Propagation and Pressure Buildup in a Free Gas-Air Mixture Due to Jet Ignition," *Progress in Astronautics and Aeronautics*, Vol. 94, 1984, pp. 474-490.

[3] Ungut, A., and Shuff, P. J., "Deflagration to Detonation Transition from a Venting Pipe," *Combustion Science and Technology*, Vol. 63, 1989, pp. 75-87."

[4] Moen, I. O., Bjerketvedt, D., Jenssen, A., Hjertager, B. H., and Bakke, J. R., "Transition to Detonation in a Flame Jet," *Combustion and Flame*, Vol. 75, 1989, pp. 297-308.

[5] Mackay, D. S., Murray, S. B., and Moen, I. O., "Flame Jet Ignition of Large Fuel-Air Clouds," *Defense Research Establishment Suffield Canada/DRES Memo*. 1274, 1990.

[6] Carnasciali, F., Lee, J. H. S., Knystautas, R., and Fineschi, F., "Turbulent Jet Initiation of Detonation," *Combustion and Flame*, Vol. 84, 1991, pp. 170-180.

[7] Thibault, P. A., private communication.

[8] Lee, J. H. S., Knystautas, R., and Freiman, A., "High Speed Turbulent Deflagrations and Transition to Detonation in H_2-Air Mixtures," *Combustion and Flame*, Vol. 56, 1984, pp. 227-239.

[9] Knystautas, R., Lee, J. H. S., Peraldi, O., and Chan, C. K., "Transmission of a Flame from a Rough to a Smooth-Walled Tube," *Progress in Astronautics and Aeronautics*, Vol. 106, 1986, pp. 37-52.

[10] Moen, I. O., Sulmistras, A., Thomas, G. O., Bjerketvedt, D., and Thibault, P. A., "Influence of Cellular Regularity on the Behavior of Gaseous Detonations," *Progress in Astronautics and Aeronautics*, Vol. 106, 1986, pp. 220-243.

[11] Shepherd, J. E., Moen, I. O., Murray, S. B., and Thibault, P. A., "Analyses of the Cellular Structure of Detonations," *Proceedings of the 21st Symposium (International) on Combustion*, The Combustion Institute, Pittsburgh, PA, 1988, pp. 1644-1658.

Transition from Fast Deflagration to Detonation Under the Influence of Wall Obstacles

R. S. Chue,* J. H. Lee,† T. Scarinci,* A. Papyrin,‡ and R. Knystautas†
McGill University, Montreal, Quebec, Canada

Abstract

In the present study we investigate the transition of a fast deflagration (planar shock followed by reaction front propagating at about half the C-J detonation velocity) induced by perturbing it with transverse pressure waves generated by a series of small obstacles placed periodically along the channel wall. To obtain the fast deflagration an initial cellular detonation wave is caused to fail by damping the transverse waves associated with the cellular structure by acoustic absorbing walls. The experiments were carried out with three different mixtures: $2C_2H_2 + 5O_2 + 75\%Ar$, $C_3H_8 + 5O_2$ and $CH_4 + 2O_2$; and four different obstacle configurations were used to control the perturbation transverse waves frequency. The initial pressure varies between 25 Torr and 160 Torr. The transition process is studied using Schlieren framing and streak photography. The results indicate that for a given mixture with a given obstacle geometry, there is a minimum pressure below which transition is not observed. Above this pressure, transition is observed and the distance it takes for transition to occur is dependent on the initial pressure. The results also indicate that there is an optimal obstacle spacing (or, equivalently, transverse wave perturbation frequency) of the order of the channel height at which transition is most efficient, i.e., transition can be induced for less sensitive mixtures. The results demonstrate the essential role of transverse pressure perturbation on the formation of the detonation. The present results also show that transition can be induced much easier for mixtures having irregular cells ($C_3H_8 + 5O_2$ and $CH_4 + 2O_2$) than for mixtures having regular cells ($2C_2H_2 + 5O_2 + 75\%Ar$).

Copyright © 1992 by the American Institute of Aeronautics and Astronautics, Inc. All rights reserved.

* Graduate Student, Department of Mechanical Engineering.

† Professor, Department of Mechanical Engineering.

‡ Professor, permanent address: Institute of Pure and Applied Mechanics, USSR Academy of Science, Novosibirsk.

1. Introduction

The transition from deflagration to detonation is an important problem but remains unresolved todate. It is not possible to predict a priori for a given system, i.e., for a prescribed mixture in a prescribed tube geometry, if a deflagration can accelerate to detonation or not. Neither is it possible to predict the time (or distance) it would take for transition to occur for systems where transition is known to be possible. Experimental measurements in one system (e.g., fixed-tube diameter) cannot be correlated with another, nor can transition data for one mixture be linked to other mixtures.

Shepherd and Lee[1] have classified the transition phenomenon into two regimes, the initial flame acceleration and the onset of detonation. The initial flame acceleration regime is strongly influenced by initial and boundary conditions. It is unlikely that a general quantitative description of this regime may be achieved. However, the phenomenon of the onset of detonation appears to have certain universal characteristics. In general, the deflagration speed prior to the onset of detonation is of the order of 1000 m/s, about half the Chapman-Jouguet detonation value. In smooth tubes the onset of detonation is also seen to originate from localized regions in the turbulent flame brush (so-called hot spots). Localized explosions from these hot spots become spherical detonation "bubbles" which grow to catch up with the leading shock front of the deflagration. Urtiew and Oppenheim[2-5] and Meyer and Oppenheim[6] have elucidated on the phenomenon of the genesis of detonations in their excellent photographic studies. However, the mechanisms for the formation of the hot spot and the subsequent amplification of the shock wave from the hot spot to form the spherical bubbles (i.e., SWACER[7]) are not understood quantitatively as yet. One of the major difficulties in making further progress is due to the random nature of the formation of the explosion centers. Hence, it is very difficult to experimentally obtain repeatable and controllable initial conditions at which the onset of detonation can be studied.

The use of wall obstacles (or so-called Shchelkin's spiral) to induce transition from deflagration to detonation had been observed many years ago by Laffite[8] and Shchelkin.[9] The mechanism by which transition is facilitated was credited to the generation of turbulence by the obstacles, hence promoting flame acceleration. However, more recent experiments by Teodorczyk[10] have demonstrated that it is due to the effect of the pressure waves generated by the obstacles rather than to turbulence. By placing acoustic absorbing materials underneath wall obstacles, Teodorczyk[10] observed that the damping of the transverse pressure waves inhibited transition. Hence turbulence alone, without transverse pressure waves, is demonstrated to be insufficient for detonation to form.

In the present paper we use a better controlled initial condition of a fast deflagration and re-examine the effects of transverse pressure waves on the transition to detonation more systematically. Using a recent result obtained by Dupre et al.,[11] an initially self-sustained C-J detonation wave is quenched by damping out its transverse waves using acoustic absorbing walls to obtain a fast deflagration consisting of a planar shock followed by a flame. This fast deflagration shares some similar features with the ones that precede the onset of detonation as they both propagate at about half the C-J detonation speed.[12] This fast deflagration will serve as our initial condition where transition is induced by generating transverse pressure waves to perturb it. To change the frequency of the transverse perturbation, we varied the obstacle spacing. However, no attempt has been made to change the amplitude of the pressure perturbation by changing the obstacle height. Three different mixtures were used: stoichiometric oxy-acetylene with 75% argon dilution, stoichiometric propane-oxygen, and stoichiometric methane-oxygen.

2. Experimental Details

The experiments are performed in a rectangular detonation tube of approximately 140 cm in length with a cross section of 3 x 1.6 cm. Two large glass windows extending over the entire length of the detonation tube are mounted on the side walls to facilitate flow visualization. The tube consists of three sections: the ignition section where a C-J detonation is formed, followed by a damping section to remove the transverse waves to obtain a fast deflagration, and the obstacle-filled test section to perturb the deflagration to induce transition to detonation. The beginning of the damping section is located approximately 40 cm from the ignition source. The length of the damping section in the present experiments is 12 cm. The obstacle section has a length of 30 cm. The initiation of the detonation is achieved using a powerful spark discharge. The acoustic absorbing walls of the damping section are constructed with layers of wire screens (nine layers of 1- x 1-mm mesh) to simulate a porous medium. The transverse pressure perturbations in the transition section are generated by periodically spaced obstacles on the walls. The obstacles are made of small sections of round solid rods of 3 mm diameter, and the spacings of the obstacles are varied to change the perturbation frequency. A sketch of the damping and obstacle sections is shown in Fig. 1.

The phenomenon of transition to detonation is observed using laser stroboscopic Schlieren photography as well as streak Schlieren photography. The laser pulses for the framing photographs are generated at 18.6 μsec between frames with a ruby laser. The streak photographs are taken with a Cordin 330 camera using a xenon flash tube as the light source. For streak photography the image is taken along the centerline of the channel. The field of view for photographing the transition process covers about 20 cm of the obstacle-filled test section as well as the end of the damping section so that the initial structure of the deflagration prior to transition can be confirmed. As a further diagnostic, photodiodes are used to check the velocity of the deflagration before being perturbed by the obstacles as well as at the end of the obstacle section (beyond the field of view for photography) to monitor the detonation velocity.

3. Results and Discussions

The attenuation of a detonation wave in stoichiometric acetylene-oxygen mixture with 75% argon dilution initially at 100 Torr is shown in Fig. 2 by the sequence of laser stroboscopic Schlieren photographs. It is observed that the detonation fails progressively as it traverses the damping section, as the transverse waves are attenuated by the acoustic absorbing walls. The reaction

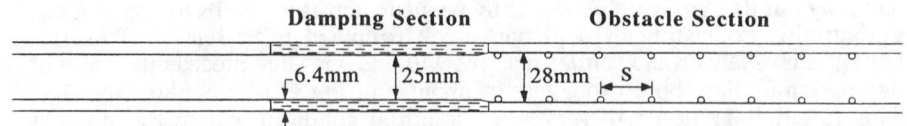

Fig. 1   Sketch of the damping and test sections: channel width is 1.6 cm, and S is obstacle spacing.

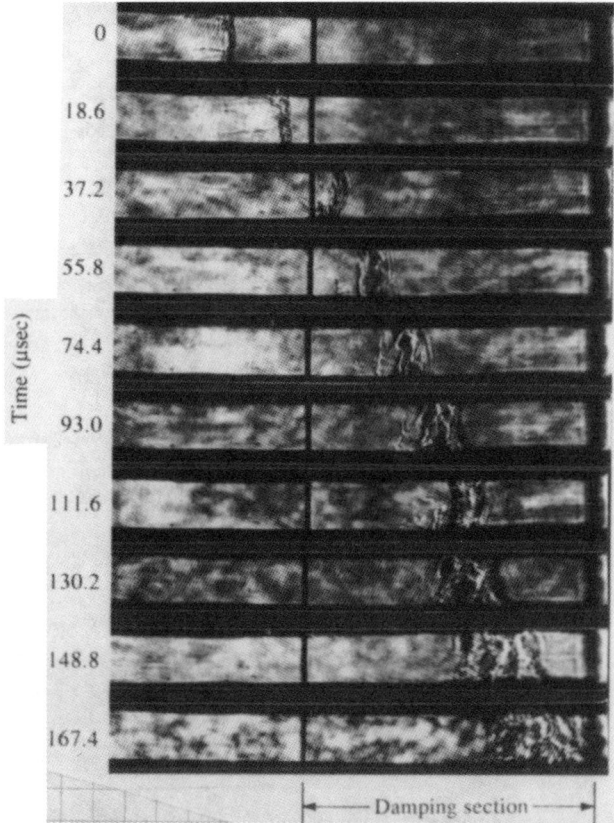

Fig. 2 Laser stroboscopic Schlieren photographs of the attenuation of detonation wave in the damping section, $2C_2H_2 + 5O_2 + 75\%\,Ar$ at 100 Torr: black vertical marker at the left denotes the beginning of the damping section, 18.6 μs between frames, the first frame is arbitrarily labeled as time 0.

zone gradually decouples from the leading shock and forms a much thicker structure. Eventually a fast deflagration (shock front-reaction zone complex) is formed and this served as the initial condition for the transition to occur when it is induced by transverse pressure perturbations.

Figure 3 shows the framing photographs as the deflagration enters the section with obstacles (obstacle spacing equal to the channel height, 28 mm). It can be seen that as the leading shock of the deflagration interacts with the wall obstacles, reflected transverse pressure waves are generated and propagate away from the obstacles as a pair of circular fronts (e.g., third and fifth frames). The forward propagating fronts of these reflected waves intersect with the leading shock while the rearward portion propagates into the reaction zone. These events are repeated as the leading shock interacts with a new pair of wall obstacles. At the same time, the reaction zone becomes more turbulent as it enters the obstacle section and interacts with the transverse pressure waves. The transverse pressure perturbations can increase the rate of burning by the production of vorticity

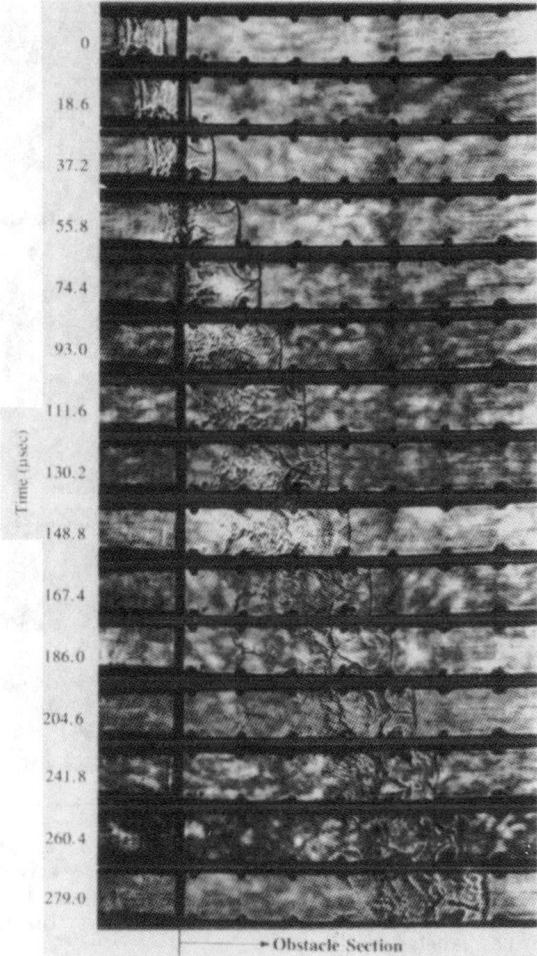

Fig. 3 Laser stroboscopic Schlieren photographs of the propagation of the fast deflagration in the obstacle section, obstacle spacing equals channel height (28 mm), $2C_2H_2 + 5O_2 + 75\%Ar$ at 100 Torr, transition not observed: black vertical marker at the left denotes the beginning of the obstacle section, 18.6 μs between frames, the first frame is arbitrarily labeled as time 0.

through two mechanisms. The three-shock Mach interaction of the transverse pressure waves with the leading shock produces shear layers in the unburned mixture ahead of the turbulent flame brush. Also the pressure gradient from the transverse pressure perturbations interacts with the density gradient in the flame zone and generates vorticity through the baroclinic mechanism ($\nabla p \times \nabla 1/\rho$). Note that these mechanisms are to be distinguished from the shear flow turbulence generated by the wall roughness and obstacles, as already pointed out by Teodorczyk.[10] The increase in burning rate of the reaction zone then generates pressure waves of its own. If the self-generation of the pressure waves

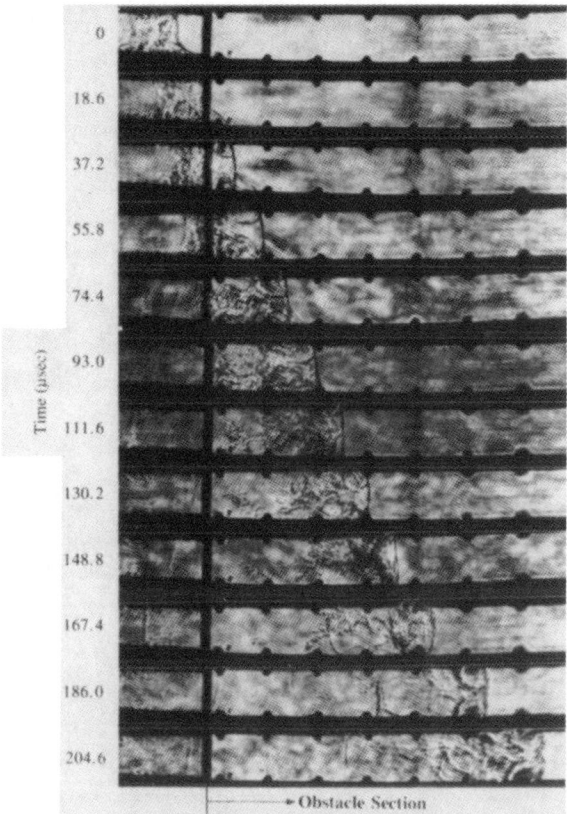

Fig. 4   Laser stroboscopic Schlieren photographs of the propagation of the fast deflagration in the obstacle section, obstacle spacing equals channel height (28 mm), $2C_2H_2 + 5O_2 + 75\%Ar$ at 140 Torr, transition observed: black vertical marker at the left denotes the beginning of the obstacle section, 18.6 µs between frames, the first frame is arbitrarily labeled as time 0.

is coherent with the induced perturbation, then coupling is facilitated and transition to detonation occurs. The conditions for coherence are a function of the sensitivity of the mixture (i.e., initial pressure) as well as the frequency of the induced perturbation. For the initial pressure tested in Fig. 3 coherence is not achieved and the deflagration remains unaccelerated with the leading shock and reaction zone propagating at constant velocities. Since the reaction zone propagates at a slightly lower velocity than the leading shock, the relatively uniform region separating the shock and the reaction zone grows continuously and the reaction zone is unable to couple with the shock to cause transition.

When the initial pressure is increased to above 130 Torr, the deflagration is observed to transit to detonation. Figure 4 displays the time sequence of framing photographs of the transition process for the mixture at an initial pressure of 140 Torr. The initial interaction of the leading shock with the obstacles is very similar to that illustrated in Fig. 3, however, the interaction of

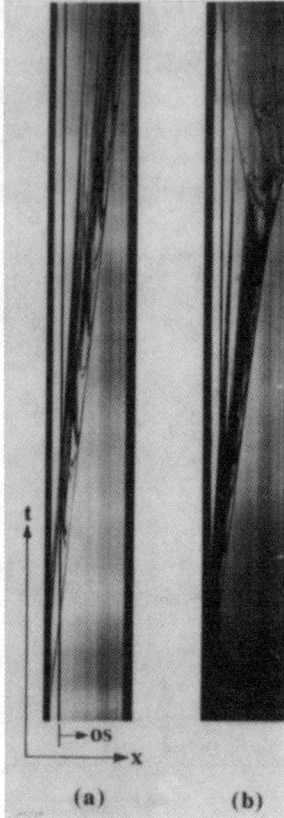

Fig. 5 Streak photographs illustrating the effect of initial pressure on transition, $2C_2H_2 + 5O_2 + 75\%Ar$ at a) 100 Torr, and b) 135 Torr. The thin black vertical line on the left of each photograph denotes the beginning of the obstacle section (os).

the transverse pressure waves with the reaction zone results in the acceleration of it as indicated by the separation between the leading shock and the reaction zone becoming progressively smaller. The stronger interaction of the pressure waves with the fast deflagration also results in a more turbulent structure behind the leading shock. At the 10th frame transition is observed to occur. In the last two frames, a detonation is generated which propagates at about the C-J velocity of the mixture.

The effect of initial pressure on transition can also be demonstrated in the streak photographs shown in Fig. 5. The orientations of the time and distance axes are indicated in the figure. The thin vertical dark line on the left of each streak photograph marks the beginning of the obstacle section. In the photographs, the trajectories of the leading shock and flame front can be clearly identified and they propagate at quite constant velocities (slopes of the trajectories) as they exit from the damping section. For an initial pressure of 100 Torr (Fig. 5a), as the deflagration enters the obstacle region, the reflections

TRANSITION FROM DEFLAGRATION TO DETONATION

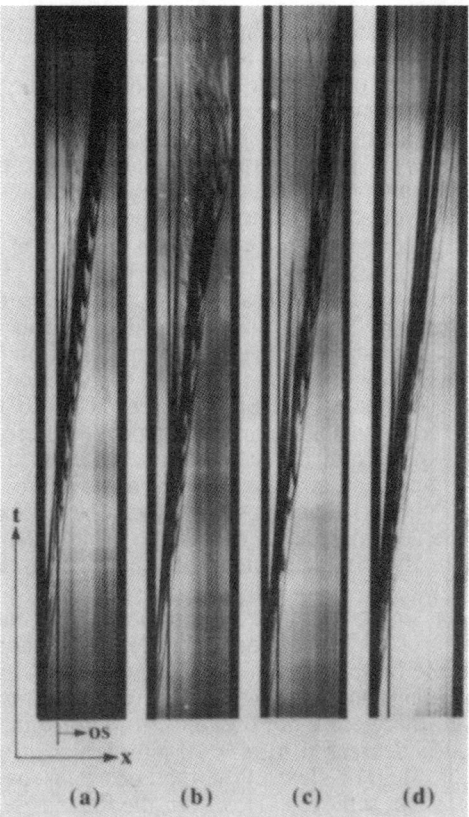

Fig. 6 Streak photographs illustrating the effect of obstacle spacing on transition, mixture is $2C_2H_2 + 5O_2 + 75\%Ar$: a) s = 10 mm at 122 Torr, b) s = 20 mm at 125 Torr, c) s = 28 mm at 127 Torr, d) no obstacles at 160 Torr. The thin black vertical line on the left of each photograph denotes the beginning of the obstacle section (os).

of the leading shock at the obstacles generate a series of transverse pressure waves propagating away from the locations of reflection (at the obstacles). The forward and backward propagating waves form a thick band of V-shaped trajectories near each obstacle. The average velocity of the structure remains fairly constant throughout the distance traveled and transition to detonation is not observed. As the initial pressure is increased to 135 Torr (Fig. 5b) the initial deflagration appears very similar to the previous case for 100 Torr. However, as time progresses, the reaction zone remains close to the leading shock and the trajectory of the leading shock displays an abrupt change in slope as the flame zone exhibits an abrupt acceleration and overtakes the leading shock.

A series of tests were performed systematically for each of the three mixtures for different pressures. The initial pressure tested for stoichiometric acetylene-oxygen with 75% argon ranges from 100 to 160 Torr, for stoichiometric propane-oxygen the initial pressure tested ranges from 28 to 46 Torr, and for stoichiometric methane-oxygen the pressure tested ranges from 60 to 160 Torr.

The results clearly indicate that high initial pressure (i.e., high mixture sensitivity) favors transition, as would be expected.

To examine the effect of obstacle spacing or frequency of the transverse pressure perturbation on transition, the experiments are performed for obstacle spacing of 10, 20, and 28 mm, corresponding to spacing over channel height ratio (s/D) of 0.36, 0.71, and 1. To further compare the results with no pressure perturbation, the experiments are repeated when the obstacles are removed. Figure 6 shows a series of streak photographs for the C_2H_2 - O_2 - Ar mixture for different obstacle spacings. Figures 6a-6c are obtained at approximately the same initial pressure near 125 Torr. By comparing Figs. 6a and 6b, it can be seen that the frequency of the generated reflected waves for s = 10 mm is doubled over that for s = 20 mm due to the decrease in obstacle spacing. Although there are more obstacles to perturb the deflagration for s = 10 mm, the mixture remains as a steady deflagration with a propagation velocity of about 794 m/s. When the obstacle spacing is increased to 20 mm, transition is observed (Fig. 6b). The figure shows that, amongst the different spacings tested, the obstacle spacing of 20 mm is the most favorable for transition to occur. When no obstacles are present, no transition was observed for the mixture at the pressures tested. Figure 6d illustrates the shock-flame structure of the fast deflagration with the leading shock and flame front having quite constant velocities (slopes) in the streak photograph obtained at the maximum test pressure.

Figure 7 displays a similar set of streak photographs for the propane-oxygen mixture with different obstacle sections used. These photographs are taken at approximately the same initial pressure of 35 Torr. Again, by comparing Figs. 7a and 7b, which have obstacle spacings of 10 mm and 20 mm, respectively, it appears that although there are more obstacles to perturb the deflagration, the rapidity of transition is decreased for s = 10 mm as transition appears to take place at the end of the field of view. When the obstacles are removed, Fig. 7c shows that transition is not achieved. However, the flame is seen to accelerate to about the same velocity as the shock as indicated by the almost parallel trajectories of the two fronts. The shock-flame complex is seen to be similar to the quasisteady regime which occurs at critical conditions in many other initiation experiments such as direct initiation,[13,14] incident shock initiation,[15] and critical tube diameter experiments.[16] Given sufficient distance, it is quite possible that transition would occur later on. When the initial pressure is increased to 44 Torr, transition is achieved within the test section, although the transition distance is drastically increased when there are no obstacles.

In comparing the transition processes for $2C_2H_2 + 5O_2 + 75\%Ar$ and $C_3H_8 + 5O_2$ in Figs. 6 and 7, it appears that the transition process for the argon-diluted oxy-acetylene mixture (regular cell structure) is accompanied with a rather abrupt change in slope of the leading shock trajectory. For propane-oxygen (irregular cell structure), instead of a sharply defined onset, the transition is accomplished over several "steps," as the shock trajectory experiences a series of more gradual accelerations.

The minimum initial pressures above which transition is observed for the three mixtures tested and for different obstacle spacings are tabulated in Table 1. The minimum pressure required for transition to occur in the available tube length for the configurations tested is lowest (i.e., lowest sensitivity) for obstacle spacing of 20 mm, indicating that an optimal spacing for transition for the present experiment lies near s/D = 0.71 (i.e., of order 1). In all of the conditions tested, the generation of transverse pressure waves with obstacles always leads to transition at a lower initial pressure than the smooth wall case.

TRANSITION FROM DEFLAGRATION TO DETONATION

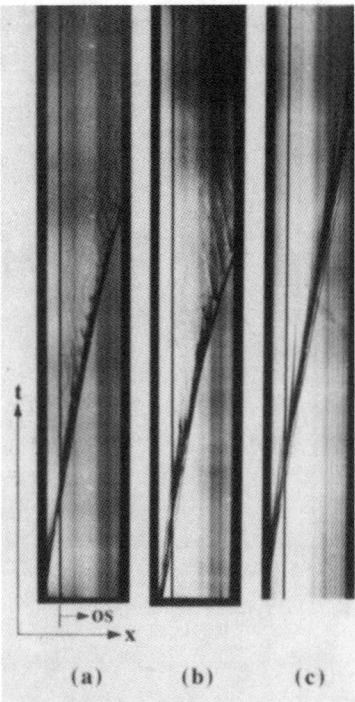

Fig. 7 Streak photographs illustrating the effect of obstacle spacing on transition, mixture is $C_3H_8 + 5O_2$ near the initial pressure of 35 Torr: a) s = 10 mm at 38 Torr, b) s = 20 mm at 35 Torr, c) no obstacles at 35 Torr. The thin black vertical line on the left of each photograph denotes the beginning of the obstacle section (os).

Table 1 Minimum initial pressures above which transition is observed

	Minimum pressure for transition, Torr			
	Obstacle spacing, mm			
Mixture	10	20	28	No obstacle
$2C_2H_2 + 5O_2 + 75\%Ar$	125	110	130	---
$C_3H_8 + 5O_2$	40	34	34	44
$CH_4 + 2O_2$	100	60	65	125

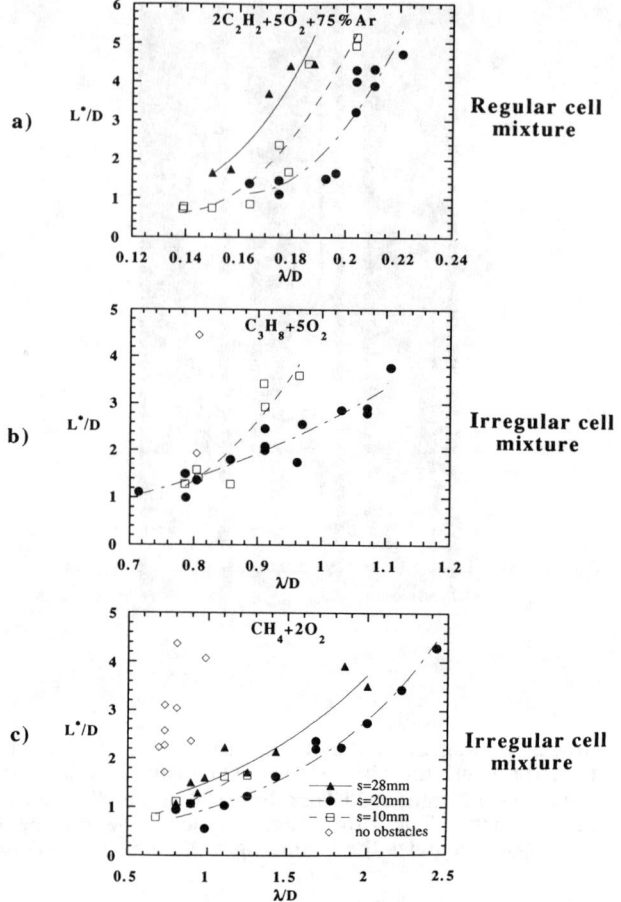

Fig. 8 Transition distance L^* vs mixture sensitivity (cell size λ), nondimensionalized with channel height D: a) $2C_2H_2 + 5O_2 + 75\%Ar$, regular detonation cell structure, b) $C_3H_8 + 5O_2$, irregular detonation cell structure, c) $CH_4 + 2O_2$, irregular detonation cell structure.

This clearly demonstrates the influence of transverse pressure perturbation in the formation of detonation.

The above results are summarized in Fig. 8 which plots the transition distance obtained from the streak photographs for the various cases. In the figure, the transition distance L^* is plotted against the mixture sensitivity as characterized by the cell size λ, both being normalized with respect to the channel height D. The figure shows that the transition distance is strongly dependent on the mixture sensitivity. As the cell size increases (or equivalently, decreasing the initial pressure), the transition distance increases. The results show that above a certain sensitivity (e.g., about $\lambda/D < 0.16$, for $2C_2H_2 + 5O_2 + 75\%Ar$ with s = 10 mm) transition occurs consistently after about one channel height ($L^*/D \sim 1$). This distance seems to be the minimal distance for onset to take place. However, there appears to be no correlation between the transition distance with λ/D as it is clearly dependent on other factors, such as mixture regularity. As seen on the figure, the transition distance is also influenced by the obstacle spacing, which corresponds to the frequency of the transverse wave perturbation. The transition distance for the obstacle spacing of 20 mm ($s/D = 0.71$) is consistently lower than for the other obstacle spacings tested and when no obstacles are used. This again indicates that there is an optimal transverse perturbation frequency for transition to occur. Another important feature noted is that for the same transition distance, the cell size for the propane-oxygen and methane-oxygen mixtures (Figs. 8b and 8c) is generally an order of magnitude higher than that for the argon-diluted acetylene-oxygen mixture (Fig. 8a). Propane-oxygen and methane-oxygen are known to have much more irregular cell structure than the argon-diluted mixture. It indicates that transition is greatly facilitated for irregular cell mixtures.

4. Concluding Remarks

In the present study, the results show that transverse pressure waves play an important role in the transition from deflagration to detonation. This suggests that the transition phenomenon is one of resonance coupling between the gasdynamic processes and the chemical reactions that drive the pressure oscillation. The results also indicate that there is an optimal obstacle spacing, or alternatively a transverse wave frequency, that facilitates transition the most. Preliminary results indicate that this optimal obstacle spacing is of the order of the tube dimension (wall spacing). However, the details of the interaction of the transverse pressure waves with the deflagration has not been completely elucidated in the present study. More systematic experiments are currently being performed to examine this more closely.

The results also conclusively show that sensitivity with respect to transition is strongly affected by the mixture's detonation cell regularity. For the highly argon diluted mixture, a value of λ/D of the order of 0.1 is needed for transition to occur, whereas for the irregular mixtures of propane-oxygen and methane-oxygen, transition can occur with a λ/D of order 1, i.e., a near-limit wave. In fact, transition is observed even with no obstacles in the tube for the irregular mixtures and it thus seems that irregular systems have more facility to undergo transition after failure. This is in fact supported by the existence of galloping waves which are more readily observed in irregular systems than in regular ones. For the highly argon-diluted mixtures, the galloping mode (i.e., failure and self-reinitiation with no external "help") is very difficult, if not impossible, to observe as demonstrated in the near limit detonation study of Dupre.[17] Since cell regularity is related to the ease in which pressure perturbations can be amplified, this result therefore reinforces the notion that the formation of a detonation is a consequence of generating a self-organizing structure from the gasdynamic and chemical processes.

Acknowledgments

This work was supported by NSERC Grants A-7091 and A-3347. T. Scarinci is supported by an NSERC graduate fellowship. We would like to gratefully acknowledge the important contribution of A. Teodorczyk who carried out the preliminary photographic observations of transition in rough tubes.

References

[1] Shephard, J. D., and Lee, J. H. S., "On the Transition from Deflagration to Detonation," *Major Research Topics in Combustion*, ICASE/NASA LaRC Series, edited by M. Y. Hussaini et al., Springer-Verlag, New York, 1992, pp. 439-487.

[2] Urtiew, P. A. and Oppenheim, A. K., "Onset of Detonation," *Combustion and Flame*, Vol. 9, No. 4, Dec. 1965, pp. 405-407.

[3] Urtiew, P. A., and Oppenheim, A. K., "Experimental Observations of the Transition to Detonation in an Explosive Gas," *Proceedings of the Royal Society of London, Ser. A*, Vol. 295, No. 1440, Nov.1966, pp. 13-28.

[4] Urtiew, P. A., and Oppenheim, A. K., "Detonative Ignition Induced by Shock Merging," *11th Symposium (International) on Combustion*, The Combustion Institute, Pittsburgh, PA, 1967, pp. 665-670.

[5] Urtiew, P. A., and Oppenheim, A. K., "Transverse Flame-Shock Interactions in an Explosive Gas," *Proceedings of the Royal Society of London, Ser. A*, Vol. 304, No. 1478, Apr. 1968, pp. 379-385.

[6] Meyer, J. W., Urtiew, P. A., and Oppenheim, A. K., "On the Inadequacy of Gasdynamic Processes for Triggering the Transition to Detonation," *Combustion and Flame*, Vol. 14, No. 1, Feb. 1970, pp. 13-20.

[7] Lee, J. H., Knystautas, R., and Yoshikawa, N., "Photochemical Initiation of Gaseous Detonations," *Acta Astronautica*, Vol. 5, No. 11, Nov. 1978, pp. 971-982 .

[8] Laffite, P., *Comptes Rendus Hebdomadaires des Seances de l'Academie des Sciences*, Vol. 186, 1928, p 95.

[9] Shchelkin, K.I., *Soviet Physics Journal of Experimental and Technical Physics*, Vol. 10, 1940, p. 823.

[10] Teodorczyk, A., private communication, 1989.

[11] Dupre, G., Peraldi, O., Lee, J. H., and Knystautas, R., "Propagation of Detonation Waves in an Acoustic Absorbing Walled Tube," *Dynamics of Explosions*, Progress in Astronautics and Aeronautics, Vol. 114, edited by A. L. Kuhl et al., AIAA, New York, 1988, pp. 248-263 .

[12] Chue, R. S., Clarke, J. F., and Lee, J. H., "Chapman-Jouguet Deflagrations," submitted for publication in the *Proceedings of the Royal Society of London*, September 1992.

[13] Bach, G. G., Knystautas, R., and Lee, J. H., "Initiation Criteria for Diverging Gaseous Detonations," *13th Symposium (International) on Combustion*, The Combustion Institute, Pittsburgh, PA, 1954, pp. 1097-1110.

[14] Edwards, D. H., Hooper, G., Morgan, J. M., and Thomas, G. O., "The Quasi-Steady Regime in Critically Initiated Detonation Waves," *Journal of Physics D: Applied Physics*, Vol. 11, No. 15, Oct. 1978, pp. 2103-2217.

[15] Sutton, P., PhD. Thesis, Dept. of Physics, University College of Wales, Aberystwyth, U.K., 1985.

[16] Edwards, D. H., Thomas, G. O., and Nettleton, M. A., "The Diffraction of a Planar Detonation Wave at an Abrupt Area Change," *Journal of Fluid Mechanics*, Vol. 95, Part 1, Nov. 1979, pp. 79-96.

[17] Dupre, G. Joannon, J., Knystautas, R., and Lee, J., "Unstable Detonations in the Near-Limit Regime in Tubes," *23rd Symposium (International) on Combustion*, Orleans, The Combustion Institute, Pittsburgh, PA, 1990, pp. 1813-1820.

Simulations for Detonation Initiation Behind Reflected Shock Waves

Yasunari Takano*
Tottori University, Tottori, Japan

Abstract

Numerical simulations are carried out for detonation initiation behind reflected shock waves in a shock tube. The two-dimensional thin-layer Navier-Stokes equations with chemical effects are numerically solved by use of a combined method consisting of the Flux-Corrected Transport scheme, the Crank-Nicolson scheme, and a chemical calculation step. Effects of chemical reactions occurring in a shock-heated hydrogen, oxygen, and argon mixture are estimated by using a simplified reaction model: two progress parameters are introduced to take account of induction reactions as well as exothermic reactions. Simulations are carried out referring to several experiments: generation of multidimensional and unstable reaction shock waves; strong and mild ignitions; and reacting shock waves in hydrogen and oxygen diluted in argon mixture.

Introduction

A reflected shock technique has been often utilized to investigate initiation of detonation, a series of processes such that a reaction front, generated behind a reflected shock wave, develops to a reaction shock wave which overtakes the reflected shock wave. Remarks from some experimental investigations, relating to the present numerical investigations, can be found in the following. Gilbert and Strehlow[1] explained the detonation initiation process as a one-dimensional phenomenon, comparing their one-dimensional analyses with experimentally-determined distance-time diagrams of the detonation initiation behind reflected shock waves in a hydrogen, oxygen, and

---

Copyright ©1993 by the American Institute of Aeronautics and Astronautics, Inc. All rights reserved.
* Professor, Department of Mechanical Engineering

argon mixture. However, Strehlow[2] indicated that a three-dimensional structure appears at the reaction shock wave before it overtakes the reflected shock wave for some conditions. Meyer and Oppenheim[3] distinguished strong ignition as one-dimensional phenomena from mild ignition as multidimensional phenomena, visualizing reflected shock flowfields in stoichiometric hydrogen and oxygen mixture. In the present study, numerical simulations are carried out to reveal multidimensional and unstable features of the detonation initiation. Also simulations are conducted for experiments by Takano and Akamatsu[4] who compared pressure measurements behind reflected shock waves in hydrogen and oxygen diluted in argon mixture with their analysis.[5]

Simplified Reaction Model

To estimate chemical effects rigorously in simulations, it is preferable to employ a detailed reaction model that includes essential elementary reactions. However, in the present simulations, a simplified reaction model is used to save computation cost. Taki and Fujiwara[6] succeeded in simulating unstable two-dimensional detonation by considering a simplified model for induction and exothermic reactions. In the present model, two progress parameters α and β are introduced to take into account induction reactions as well as exothermic reactions:

$$\dot{\alpha} = -\rho K_1 exp(-\frac{E_1}{RT}) \tag{1}$$

$$\dot{\beta} = -\rho K_2 exp(-\frac{E_2}{RT})\beta(\beta - \frac{1-\beta}{K_{eq}}) \tag{2}$$

$$\frac{1}{K_{eq}} = K_3 exp(-\frac{E_3}{RT}) \tag{3}$$

Here, $\dot{\alpha}$ and $\dot{\beta}$ denote rates of these progress parameters. Set at one in the initial region, α decreases when the gas is heated by a reflected shock wave. When α attains zero, β decreases from one to its equilibrium value defined by the equilibrium constant, K_{eq}. Rate coefficients and activation energies in Eqs. (1-3) are written for stoichiometric hydrogen and oxygen diluted in argon mixture as follows:

$$K_1 = \frac{\sigma_{O_2}}{4.0 \times 10^{-11}} \quad (\frac{m^3}{kgs}) \tag{4}$$

$$K_2 = \frac{\sigma_{O_2}}{5.6 \times 10^{-9}} \quad (\frac{m^3}{kgs}) \tag{5}$$

$$K_3 = 400. \quad (\frac{m^3}{kgs}) \tag{6}$$

$$E_1/R = 9200. \quad (K) \tag{7}$$

$$E_2/R = 3000. \quad (K) \tag{8}$$

$$E_3/R = 20000. \quad (K) \tag{9}$$

$$Q = 0.48 \times 10^9 \sigma_{O_2} \quad (J/kg) \tag{10}$$

Here, σ_{O_2} denotes moles of oxygen molecules per unit mass. Q represents combustion heat per unit mass. These coefficients are determined so that induction times, exothermic characteristic times, and amounts of released combustion heat can be fitted to those obtained from a detailed reaction model. Figure 1 shows a typical result of chemical kinetics simulations employing a detailed reaction model used in previous investigations of Takano and Akamatsu[4,5] and Takano.[7] The induction times t_i and the exothermic characteristic times t_c are decided from profiles of released chemical energy as shown in Fig. 1. Also, equilibrium values of β are estimated from ultimate values of released chemical energy. Figure 2 shows comparisons for several characteristic times between the simplified model and the detailed model. The rate equations (1) and (2) based on the simplified model are used for simulations of reflected shock waves in an $H_2/O_2/Ar$ mixture in which temperature conditions are at above 1100 K. However, for simulations of strong and mild ignitions, in which temperature conditions are at about 1000 K, the following formulas for the induction time given by Meyer and Oppenheim[3] are used instead of Eq. (1):

$$\dot{\alpha} = -\rho K_1 exp[-\frac{E_1}{RT} - Cp^2 exp(\frac{D}{T})] \tag{11}$$

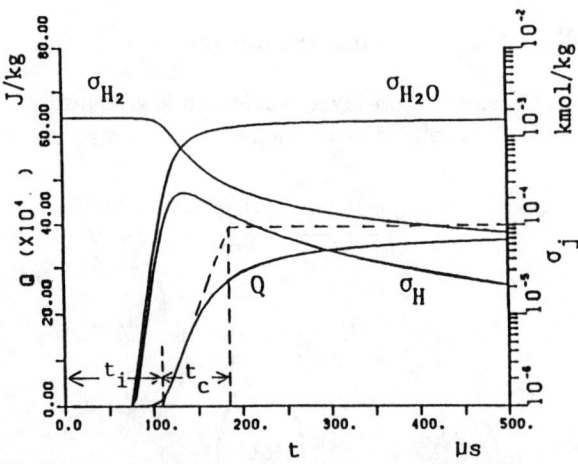

Fig.1 Released chemical energy Q and variations of moles per unit mass for chemical species ($\sigma_{H_2}, \sigma_H, \sigma_{H_2O}$) as functions of time elapsed after heating to a condition at the temperature of 1100 K and the density of 1 kg/m³; initial concentrations are $\sigma_{Ar} = 0.0243$, $\sigma_{H_2} = 0.00162$, $\sigma_{O_2} = 0.0081$ kmol/kg.

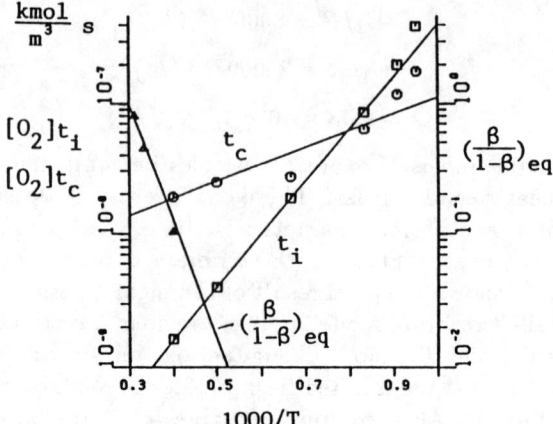

Fig.2 Comparisons for induction times and exothermic characteristic times between the simplified model (solid lines) and the detailed model (symbols).

$$K_1 = \frac{\sigma_{O_2}}{2.25 \times 10^{-11}} \quad \left(\frac{m^3}{kgs}\right) \tag{12}$$

$$E_1/R = 9130. \quad (K) \tag{13}$$

$$C = 3 \times 10^{-9} \quad (atm^{-2}) \tag{14}$$

$$D = 19000. \quad (K) \tag{15}$$

Basic Equations

The two-dimensional thin-layer Navier-Stokes equations with reaction terms given by the simplified reaction model are employed as basic equations:

$$\frac{\partial Q}{\partial t} + \frac{\partial F}{\partial x} + \frac{\partial G}{\partial y} = \frac{\partial R}{\partial x} + \frac{\partial S}{\partial y} + H \tag{16}$$

$$p = \frac{1}{\gamma - 1}[E - \frac{\rho}{2}(u^2 + v^2)] = \rho \mathcal{R} T \tag{17}$$

$$Q = \begin{pmatrix} \rho \\ \rho u \\ \rho v \\ E \\ \rho \alpha \\ \rho \beta \end{pmatrix} \tag{18}$$

$$F = \begin{pmatrix} \rho u \\ \rho u^2 + p \\ \rho u v \\ u(E+p) \\ \rho u \alpha \\ \rho u \beta \end{pmatrix} \qquad (19)$$

$$G = \begin{pmatrix} \rho v \\ \rho u v \\ \rho v^2 + p \\ v(E+p) \\ \rho v \alpha \\ \rho v \beta \end{pmatrix} \qquad (20)$$

$$R = \begin{pmatrix} 0 \\ \frac{4}{3}\mu u_x \\ \mu v_x \\ R_4 \\ \rho D_\alpha \alpha_x \\ \rho D_\beta \beta_x \end{pmatrix} \qquad (21)$$

$$S = \begin{pmatrix} 0 \\ \mu u_y \\ \frac{4}{3}\mu v_y \\ S_4 \\ \rho D_\alpha \alpha_y \\ \rho D_\beta \beta_y \end{pmatrix} \qquad (22)$$

$$H = \begin{pmatrix} 0 \\ 0 \\ 0 \\ \rho Q \dot{\beta} \\ \rho \dot{\alpha} \\ \rho \dot{\beta} \end{pmatrix} \qquad (23)$$

$$R_4 = u\frac{4}{3}\mu u_x + v\mu v_x + \lambda T_x \qquad (24)$$

$$S_4 = u\mu u_y + v\frac{4}{3}\mu v_y + \lambda T_y \qquad (25)$$

Here, x and y are the distances from the end wall and from the side wall of a shock tube, respectively; ρ the density; u and v the velocity components in the x and in the y coordinates, respectively; E the total energy density; μ and λ the coefficients of viscosity and of heat conductivity, respectively; D_α and D_β the diffusion coefficients; T the temperature; $\Re$ the gas constant; and Q the combustion heat. The thin-layer approximations with respect to the side wall ($y = 0$) and to the end wall ($x = 0$) are made for R and S, respectively.

Finite Difference Method

Employing the splitting technique, the basic equations are numerically solved by a combined method consisting of the Flux-Corrected Transport (Lax-Wendroff-FCT) scheme, the Crank-Nicolson scheme, and a chemical calculation step. When Q at $t = t_n$ is expressed as Q^n, the present algorithm can be written as follows:

$$Q^{n+2} = L_{dx} L_x L_{dy} L_y L_r L_r L_y L_{dy} L_x L_{dx} Q^n \tag{26}$$

Here, L_x, L_y, L_{dx}, L_{dy}, and L_r are finite difference operators which, respectively, map from $Q(t)$ to $Q(t + \Delta t)$ satisfying splitted equations as follows:

$$\frac{\partial Q}{\partial t} + \frac{\partial F}{\partial x} = 0 \tag{27}$$

$$\frac{\partial Q}{\partial t} + \frac{\partial G}{\partial y} = 0 \tag{28}$$

$$\frac{\partial Q}{\partial t} = \frac{\partial R}{\partial x} \tag{29}$$

$$\frac{\partial Q}{\partial t} = \frac{\partial S}{\partial y} \tag{30}$$

$$\frac{\partial Q}{\partial t} = H \tag{31}$$

The Lax-Wendroff-FCT scheme is employed for inviscid gasdynamic calculations for L_x and L_y because it can be applied to reacting gas straightforwardly. The Crank-Nicolson scheme is used to calculate the transport equations in the L_{dx} and L_{dy} steps. An analytical solution for Eq. (2), which is obtained assuming constant temperature during the chemical step, is used to evaluate L_r.

Computations are conducted in two-dimensional reflected shock region with a half-width (radius) of 2 cm and a length of 15 cm. Flowfields are assumed to be axially symmetric. Computations are carried out over a 321×61 point grid system consisting of uniform grids of $\Delta x = \Delta y = 0.5$ mm located in the middle, and clustering grids located in narrow zones with a width of 5 mm adjacent to the walls.

Initial and Boundary Conditions

Boundary conditions at the walls used in the simulations are nonslip for the velocity: $u = v = 0$, room temperature: $T = T_w$, and perfect catalytic to quench ignitions: $\alpha = \beta = 1$.

Computations are started from initial situations just before reflection of incident shock waves on the end wall. Behind the incident shock waves, uniform flows are generated and also unsteady boundary layers develop over the side wall. These boundary layers, investigated by Mirels,[8] are steady in coordinates fixed with the incident shock waves at the speed of u_S. In the present simulations, to estimate initial data in part of the boundary layers, a polynomial of Pohlhausen is used to approximate normalized profiles of the velocity $\tilde{u}$ and the temperature $\tilde{\theta}$ as functions of a similarity variable η based on Howarth-Dorodnitsyn transformation[9]:

$$\eta = \sqrt{\frac{\rho_2(u_S - u_2)}{2\mu_2(x - x_S)}} \int_0^y \frac{\rho}{\rho_2} dy \qquad (32)$$

$$\tilde{u}(\eta) = \begin{cases} \tilde{u}_w + (1 - \tilde{u}_w)\Phi(\eta/\eta_\delta), & 0 \leq \eta \leq \eta_\delta \\ 1, & \eta > \eta_\delta \end{cases} \qquad (33)$$

$$\tilde{\theta}(\eta) = \begin{cases} \tilde{\theta}_w + (1 - \tilde{\theta}_w)\Phi(\eta/\eta_\delta), & 0 \leq \eta \leq \eta_\delta \\ 1, & \eta > \eta_\delta \end{cases} \qquad (34)$$

$$\Phi(\eta) = \eta(2 - 2\eta^2 + \eta^3) \qquad (35)$$

Here, $\tilde{u}_w = u_S/(u_S - u_2)$, $\tilde{\theta}_w = T_w/T_2$. The subscript 2 denotes the properties in the freestream behind the incident shock wave. Values of η_δ which correspond to the boundary-layer edge are determined so that the approximate profiles may fit to exact solutions of the Mirels boundary layer as shown in Fig. 3. Here, Y is a normalized coordinate of y which relates

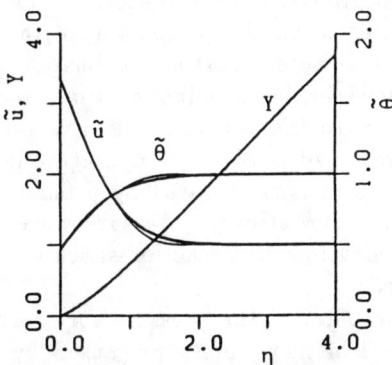

Fig.3 Comparisons of Mirels boundary layer between exact solutions (thick lines) and approximate profiles (thin lines); ratio of specific heat $\gamma = 1.4$, Prandtl number $Pr = 0.75$, and molecular weight $M_w = 12$ are used for an incident shock wave at Mach number $M_S = 2.5$ in stoichiometric H2/O2; $\eta_\delta = 1.8$ is determined.

Table 1 Conditions for simulations

Case	p_1, kPa			T_1	u_S	T_F
	Ar	H_2	O_2	K	Km/s	K
1	2.8	0.80	0.40		0.800	1146
2	0	1.33	0.67		1.340	1042
3	0	3.56	1.78	296	1.300	993
4	12	0.80	0.40		0.735	1202
5	12	1.33	0.67		0.735	1147

to η as follows:

$$Y = y\sqrt{\frac{\rho_2(u_S - u_2)}{2\mu_2(x - x_S)}} \int_0^\eta \tilde{\theta}(\eta) d\eta, \qquad x > x_S \qquad (36)$$

Results and Discussions

Table 1 shows conditions for simulations where p_1 and T_1 are the pressure and the temperature in initial state, u_S the velocity of incident shock waves, and T_F the temperature in frozen state behind reflected shock waves.

To compare chemical effects of the simplified model on gasdynamics with those of the detailed model, simulations are carried out for case 4 over a uniform grid system of $\Delta x = 0.5$ mm, numerically solving one-dimensional Euler equations with reaction terms given by the simplified model and by the detailed model, respectively. Case 4 is a condition in which the detailed simulation is shown to predict pressure profiles almost identical to measured pressure histories by Refs. 4 and 7.

Figure 4 shows profiles of the pressure for the detailed model. Here, t is the time elapsed after the shock reflection. The pressure behind the reflected shock wave is shown almost constant within 120 μs elapsed after the reflection. This is because almost no combustion heat is released during induction period. However, in an induction time, an ignition occurs and a reaction front is generated from the end wall. Pressure waves, generated by heat-release, propagate and raise the temperature in front of the reaction front. The reaction front is accelerated by promoted ignition due to the temperature rise and then stronger pressure waves are generated in its front. As a result, it develops to a reaction shock wave which overtakes the reflected shock wave.

Figure 5 shows profiles of the pressure for a simplified model. It is observed that profiles of rising pressure caused by accelerating reaction front are not as sharp in Fig. 5 as those in Fig. 4. This fact means that the release of chemical energy within the reaction front does not occur as rapidly in the simplified model as that in the detailed model. However, the simplified model reproduces equivalent detonation structure to that in the detailed model after the reaction shock wave overtakes the reflected shock wave.

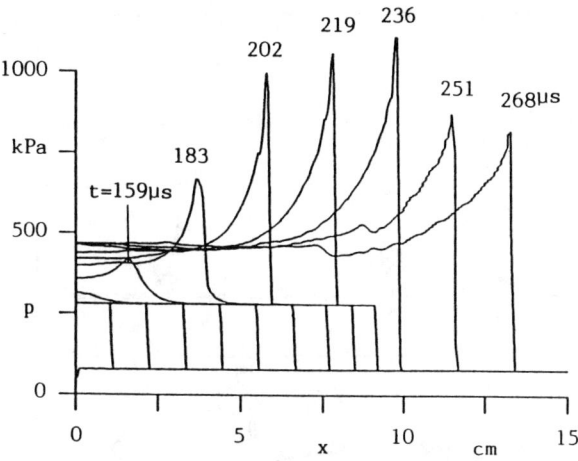

Fig.4 Pressure profiles in the reflected shock region obtained from a one-dimensional simulation employing the detailed model (case 4).

Figure 6 shows density contour diagrams in the reflected shock region for case 1. It is noticed that the reflected shock wave is curved near the side wall due to interactions with a side-wall boundary-layer and the separation of the boundary layer takes place at the foot of the reflected shock wave. The temperature in a gaseous region near the side wall processed by the bifurcated shock waves is a little lower than that in the middle behind the normal reflected shock wave. As a result, a reaction shock wave is curved near the side wall due to delayed ignition. The reaction shock wave is observed to reflect on the side wall regularly as shown at 157.7 μs. Then, the transition of the regular to the Mach reflection takes place in the reaction

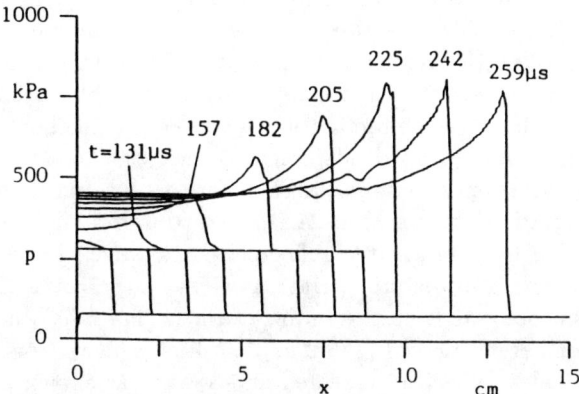

Fig.5 Pressure profiles in the reflected shock region obtained from a one-dimensional simulation employing the simplified model (case 4).

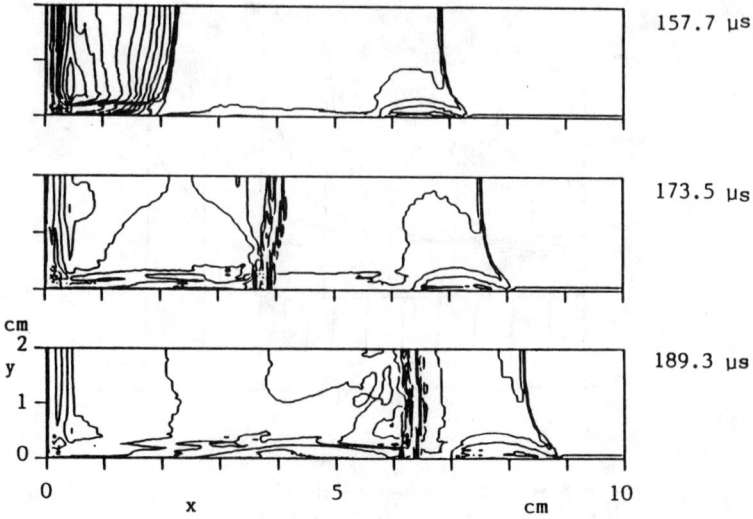

Fig.6 Density contour diagrams in the reflected shock region (case 1).

shock wave as shown at 173.5 μs. Afterward a plane unstable reaction shock wave follows after the reflected shock wave. Thus, this simulation reveals a transition of a reaction front to a multidimensional and unstable reaction shock wave behind a reflected shock wave.

Figure 7 shows density contour diagrams for case 2 referring to an experiment of strong ignition visualized by Meyer and Oppenheim.[3] It is observed that a plane reaction front is formed several millimeters apart from the end wall and the side wall in the reflected shock region at 71.7 μs. This is probably due to effects of a boundary layer on the end wall and a lower temperature region processed by bifurcated reflected shock wave. A burnt gas region spreads out as a reaction front propagates in its front as shown at 74.2 μs and 76.3 μs. The reaction front develops to a reaction shock wave (80.4 μs, 84.2 μs) which overtakes the reflected shock wave (90.6 μs). The transmitted reaction shock wave propagates as an unstable detonation wave (96.9 μs, 103.4 μs). These computed flowfields are in excellent qualitative agreement with visualized flowfields (Fig. 1 in Meyer and Oppenheim[3]).

Figure 8 shows density contour diagrams for case 3 referring to an experiment of mild ignition. It is noticed that separation bubbles are generated over the side wall by a bifurcated reflected shock wave. In this case, the ignition starts from two distinct points as shown at 173.1 μs. Separated reaction fronts propagate from the distinct kernels (174.3 μs) and separated burnt gas regions are coalesced (175.2 μs, 177.4 μs). A reaction shock wave is observed to follow after the reflected shock wave as shown at 189.0 μs. This simulation predicts a feature characterizing the mild ignition that the ignition starts from distinct kernels. Comparing the computed flowfields

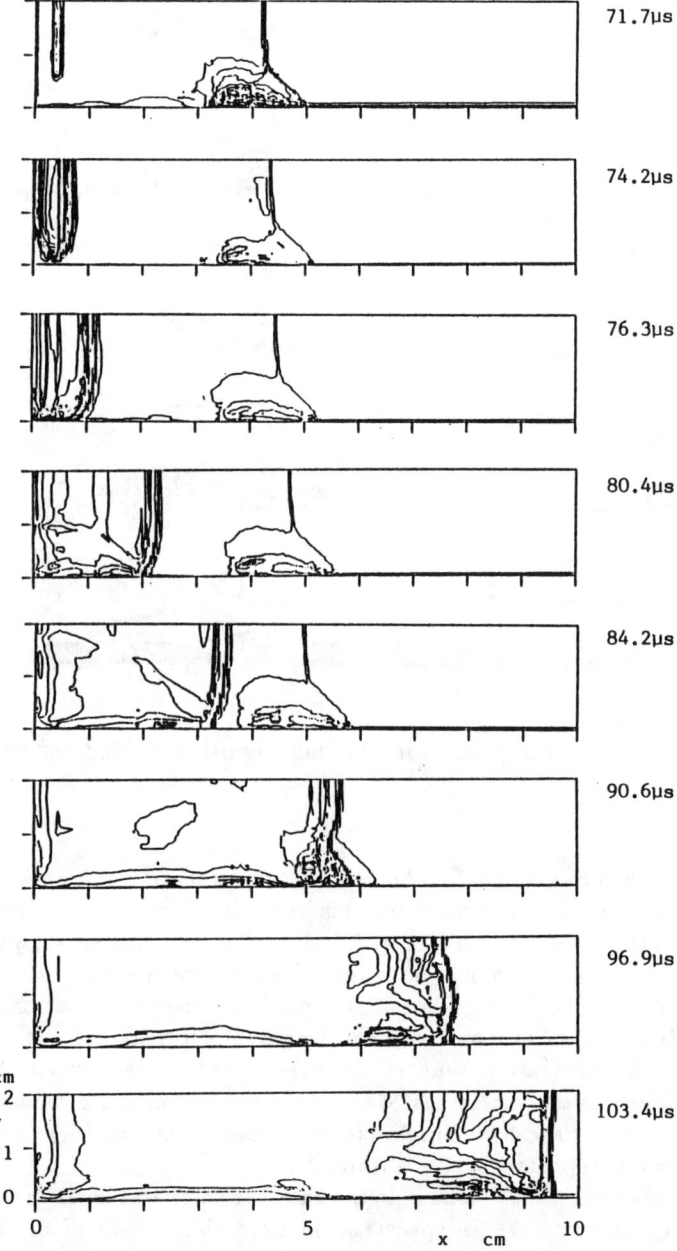

Fig.7 Density contour diagrams for strong ignition behind reflected shock wave (case 2).

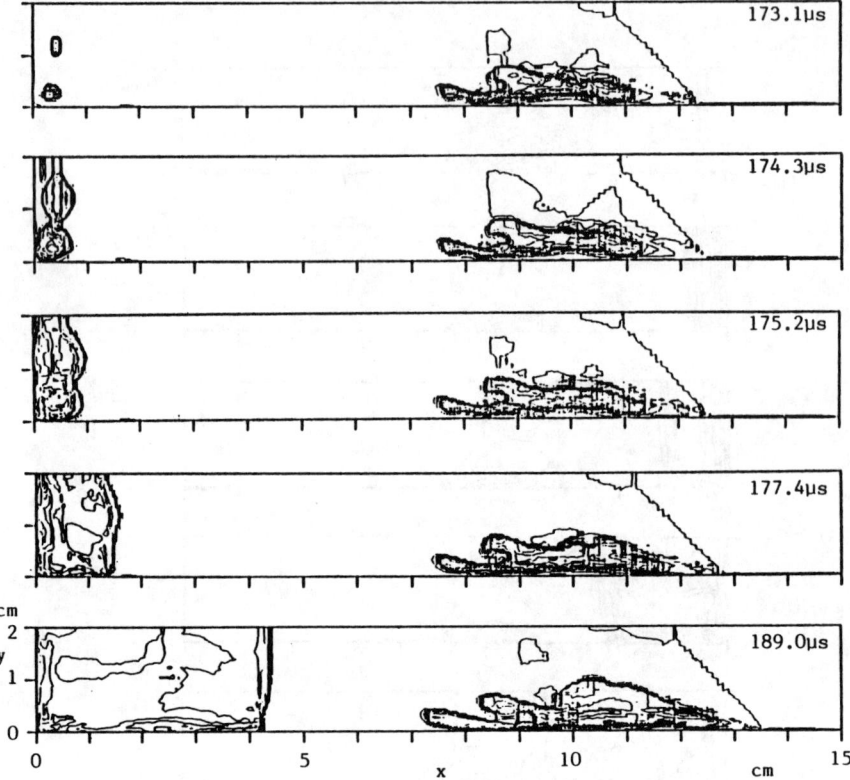

Fig.8 Density contour diagrams for mild ignition behind reflected shock wave (case 3).

with visualized ones (Fig. 3 in Meyer and Oppenheim[3]), some discrepancies are found between them: more wrinkled reaction fronts are observed in the experiment; the height of bifurcated limbs of the reflected shock wave in the simulation is much higher than that in the experiment. Three-dimensional computations with finer grids might reproduce wrinkled reaction fronts in the mild ignition. Reasons why the height of bifurcated limbs are much higher in the simulation than in the experiment are not clear. It should be mentioned that simulations of bifurcated reflected shock waves in ideal gas by use of a procedure similar to the present one can yield reasonable predictions as reported by the author.[10]

Results of the simulations indicate that the temperature in the reflected shock regions fluctuates in space and in time. Variations of the temperature are observed to be at most 5 K. Considering that sensitivities of the induction time to temperature fluctuations in the reflected shock region are 0.6 μs/K and 5.4 μs/K for case 2 and case 3, respectively, it can be considered that spotty ignitions appearing in the simulation for case 3 are

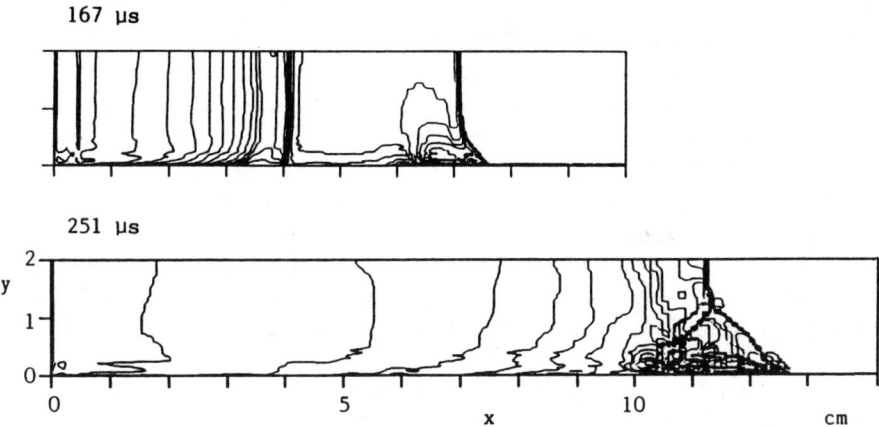

Fig.9 Density contour diagrams in the reflected shock region (case 4).

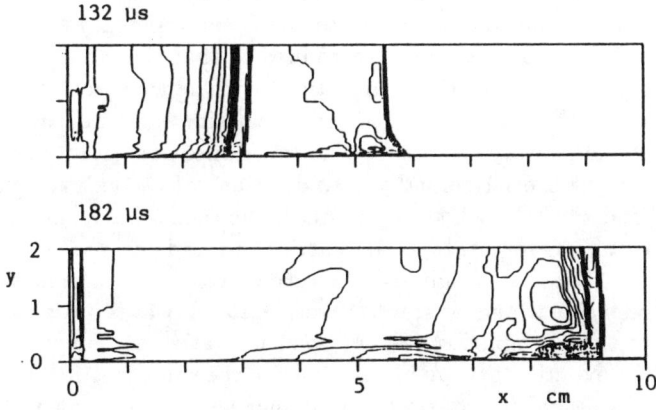

Fig.10 Density contour diagrams in the reflected shock region (case 5).

due to temperature fluctuations. In the framework of the present simulations, the temperature fluctuations are probably due to weak pressure waves caused by the reflection of rear oblique compression waves in the bifurcated reflected shock wave on the separation bubble over the side wall.

Figure 9 shows density contour diagrams for case 4 referring to an experiment given by Takano and Akamatsu.[4] After a reaction shock wave overtakes a reflected shock wave, a bifurcation structure on the side wall is noticed to develop remarkably in a transmitted detonation wave. It is difficult to understand that such a phenomenon occurs when the reflected shock wave is reinforced by heat release due to exothermic reactions. Consulting computational results, it is found that the combustion heat is released immediately behind a normal reflected shock wave, but it is not released behind an oblique shock wave of the bifurcated reflected shock wave where induction reactions are still dominant and the exothermic reactions are not yet switched on. Accordingly, the pressure in separation bubbles under the bifurcated limbs remains so low that the separated flows cannot penetrate behind the normal reflected shock wave. Hence the separation bubbles grow because the separated flows continue to accumulate in them.

Figure 10 shows density contour diagrams for case 5. In this case, bifurcations do not develop in transmitted detonation waves. This is because the concentration of combustible gas is higher and the ignition time is shorter so that the heat-release can occur immediately, even near the side wall.

Conclusions

Simulations are performed for detonation initiation behind reflected shock waves in a hydrogen, oxygen, and argon mixture in a shock tube. Chemical effects are estimated by applying a simplified reaction model: two progress parameters are introduced to take into account induction reactions as well as exothermic reactions. Rate coefficients and activation energies used in the simplified model are determined so that the induction time, the exothermic characteristic time, and amount of released combustion heat can be fitted to those obtained from chemical kinetics simulations containing essential elementary reactions. The two-dimensional thin-layer Navier-Stokes equations are numerically solved by use of a combined method of the Lax-Wendroff-FCT scheme, the Crank-Nicolson scheme, and a chemical calculation step. Computations are carried out and compared with several experiments. Results of simulations reveal transitions of reaction fronts to multidimensional and unstable reaction shock waves. Also, computed flowfields for strong ignition are in good qualitative agreement with those visualized in experiments. A simulation for the mild ignition predicts the feature that the ignition starts from distinct kernels. It is also shown that chemical effects enhance interactions of reflected shock wave with side-wall boundary layer depending on whether the ignition occurs behind bifurcated reflected shock wave or not.

Acknowledgment

This work was partly supported by a Grant in Aid for Scientific Research No. 01550148 of the Ministry of Education, Science and Culture, Japan.

References

[1]Gilbert, R. B., and Strehlow, R. A., "Theory of Detonation Initiation Behind Reflected Shock Waves," *AIAA Journal*, Vol. 4, 1966, pp. 1777-1783.

[2]Strehlow, R. A., "Detonation Initiation," *AIAA Journal*, Vol. 2, 1964, pp. 783-784.

[3]Meyer, J. W., and Oppenheim, A. K., "On the Shock-Induced Ignition of Explosive Gases," *Thirteen Symposium (International) on Combustion*, 1971, pp. 1153-1164.

[4]Takano, Y., and Akamatsu, T., "Chemical Effects on Reflected-Shock Region," *Progress in Astronautics and Aeronautics*, Vol. 105, 1986, pp. 347-364.

[5]Takano, Y., and Akamatsu, T., "Analysis of Chemical Effects on Reflected-Shock Flowfields in Combustible Gas," *Journal of Fluid Mechanics*, Vol. 160, 1985, pp. 29-45.

[6]Taki, S., and Fujiwara, T., "Numerical Analysis of Two-Dimensional Nonsteady Detonations," *AIAA Journal*, Vol. 16, 1978, pp. 73-77.

[7]Takano, Y., "An Application of the Random Choice Method to Reactive Gas with Many Chemical Species," *Journal of Computational Physics*, Vol. 67, 1986, pp. 173-187.

[8]Mirels, H., "Boundary-Layer behind Shock or Expansion Wave Moving into Stationary Fluid," NACA TN 3712, 1956.

[9]Takano, Y., "Simulations for Effects of Side-Wall Boundary-Layer on Reflected-Shock Flowfields in Shock Tubes," *Proceedings of the Sixteenth International Symposium on Shock Tubes and Waves*, 1988, pp. 645-651.

[10]Takano, Y., "Simulations of Reflected-Shock Waves in Shock Tubes Taking Account of Side-Wall Boundary-Layer Effects," *Transactions of the Japan Society of Mechanical Engineers*, Vol. 531B, 1990, pp. 3205-3209 (in Japanese).

Limiting Tube Diameter of Gaseous Detonation

S. M. Frolov* and B. E. Gelfand†
Russian Academy of Sciences, Moscow, Russia

Abstract

The currently used criterion of detonation limit in tubes is based on experimental observations indicating transition of multifront detonation structure to spinning wave configuration near the limit. According to the Zel'dovich theory the physical mechanism of limit consists in increasing momentum and energy losses as far as channel diameter decreases. In this paper an attempt is made to varify the empiric criterion in the frame of a modified one-dimensional theory of detonation limits. The principal feature of the present model is that properly enlarged values of drag and heat transfer coefficients are used in momentum and energy conservation equations as compared to the values in a steady stabilized flow. Experimental values of the CJ detonation velocity and correlations for ignition delay are also incorporated into the model. For the whole variety of fuel-oxydizer mixtures the calculated limiting tube diameter correlates well with the value given by the empirical criterion. Also stated is a good quantitative agreement between calculated and measured dependencies of the limiting tube diameter on mixture composition and initial

---

Copyright © 1992 by the American Institute of Aeronautics and Astronautics, Inc. All rights reserved.
*Senior Scientist, Institute of Chemical Physics.
†Chief Scientist, Institute of Chemical Physics.

pressure. Results of the comparison indicate that the empirical criterion based on a cell size of multifront detonation is being confirmed by one-dimensional model of limits.

Introduction

According to Zel'dovich[1] detonation limits in tubes exist due to interaction of the flow in the reaction zone with tube walls. Decrease in tube diameter leads to increase in losses caused by mixture stagnation and cooling at tube walls. In a tube of diameter less than a certain limiting value losses dominate and the self-sustaining detonation wave propagation becomes impossible.

At present the empiric criterion of limit is used based on experimentally observed transition from multifront detonation structure to the structure containing only one transverse wave[2,3]. Since the single-head spinning detonation seems to occur when $\lambda \approx \pi d$ (Ref.3), the determination of the limiting tube diameter reduces to measuring or calculating the cell size λ (d is the tube diameter). The values of λ have been measured by Knystautas et al.,[2] Vasiliev et al.,[3] and Bull et al.[4] for a whole variety of fuel-air and fuel-oxygen mixtures. The comprehensive review of detonation cell models and results of numerical calculations of multifront detonation structure in wide channels presented by Vasiliev et al.[3] shows that a good agreement between predicted and measured results is achieved for some combustible mixtures while for other mixtures the agreement remains poor. Experimental observations of Refs.2-4 show that in general for tubes of circular cross section $\lambda = (1 : 4)d$. Recent publication by Murrey and Moen[5] shows that the lack of success in identifying strict universal correlations between the channel width and the detonation cell size like $\lambda \approx \pi d$ is probably due to dependence of the cell size on the detonation velocity deficit (or tube diameter) and the contribution of viscous boundary layers at tube walls.

Thus it stands to reason that the present theoretical investigations of detonation limits are out of touch with the physical mechanism of the limits. It seems useful to consider the problem of detonation limits in tubes with due regard for momentum and energy losses in the reaction zone of the detonation wave and compare results with the currently used empirical criterion. Such a problem was formulated for the first time by Zel'dovich[1] in the frame of one-dimensional approach. The results of Zel'dovich[1] agree qualitatively with experimental observations. One may anticipate that the one-dimensional approach allows the basic governing parameters of the problem to be distinguished. The question then consists in correctly setting the governing parameters or their functional dependencies. When formulating the one-dimensional problem of detonation limits the two key questions arise concerning the chemical kinetics modeling and the determination of the drag coefficient in a highly nonuniform flow behind the detonation front.

There are two ways for modeling chemical energy release in the detonation wave. One is based on ignition delay data obtained in shock tubes, the other concentrates on detailed mechanisms of fuel oxydation. When calculating momentum losses the problem arises of taking into account boundary layer incipience behind the lead shock front. We shall use here the first approach to the chemical energy release modeling and take into account our recent results on shock wave attenuation in tubes[6] which allow important conclusions about the question of drag coefficient in the induction zone of a detonation wave to be made.

Formulation

The kinetic parameters of fuel-air and fuel-oxygen mixtures have been choosen by comparison of the well-known expression for an adiabatic induction period

$$\tau_i = \frac{c_v RT^2}{kQE} \beta_0^{-(n-1)} \exp\frac{E}{RT}$$

with an empirical correlation for ignition delay of

$$\tau = A[F]^a[Ox]^b[In]^c \exp\frac{E_*}{RT} \qquad (1)$$

Here, c_v is the specific heat at constant volume, R the gas constant, T the temperature, k the pre-exponential factor in the Arrhenius type reaction, Q the heat release, E the activation energy, β_0 the initial concentration of a consumed component, n the reaction order; $[F]$, $[Ox]$, and $[In]$ are the molar concentrations of fuel, oxydizer, and inert, respectively; and E_* is the experimental value of the activation energy. Such a comparison results in the following relations:

$$n = 1 - (a + b + c)$$

$$E = E_*$$

$$k = \frac{c_v RT_f^2}{A\rho_f^{1-n}QE} \left[\left(\frac{Y_F}{\mu_F}\right)_0^a \left(\frac{Y_{Ox}}{\mu_{Ox}}\right)_0^b \left(\frac{Y_{In}}{\mu_{In}}\right)_0^c\right]^{-1}$$

where Y and μ are the mass fraction and the molecular mass, respectively, subscript 0 denotes the undisturbed state upstream of the leading shock, and T_f, and ρ_f are the values of temperature and density at the shock front, defined by the detonation velocity D. The ratio of the specific heats γ is assumed to be constant and equal to the average value between the corresponding quantities for the initial mixture and combustion products; the composition of the latter being calculated through the use of a stoichiometric formula neglecting dissociation. For simplicity γ is assumed to equal 8/7 in hot CO_2 and

H_2O. The heat release Q is then determined from the experimental value of the ideal detonation velocity

$$Q = D_0^2/2(\gamma^2 - 1)$$

Thus, the kinetics of the chemical processes can be expressed by

$$\frac{d\beta}{dx} = -\frac{k\beta^n}{u} \exp\left(-\frac{E}{RT}\right) \qquad (2)$$

where x is the distance to the lead shock front of the steady detonation wave, and u is the velocity.

Our recent study of shock wave attenuation in rough tubes[6] revealed the following important feature. It was found that the drag coefficient C_f in nonstationary flow conditions behind a shock wave correlated satisfactorily with the hydraulic drag coefficient ς measured in steady stabilized flow conditions. We assume now that the experimental correlation $C_f \approx \varsigma$ for the drag coefficient in a rough tube holds for a tube with smooth walls, i.e. in accordence with Blausius law

$$C_f = 0.3164/Re^{0.25} \qquad (3)$$

where $Re = w_f d/\nu$, with $w_f = D - u_f$ the gas velocity at the wave front and ν the kinematic viscosity determined from the following expression derived from kinetic theory of gases[7]:

$$\nu = 26.7 \cdot 10^{-6} (\mu T_f)^{0.5}/r_M \rho_f$$

Here r_M is the radius of a mixture molecule in a model of hard spheres (9). In a tube with smooth walls the Reynolds analogy holds and hence the local dimensionless heat-transfer coefficient $C_H = \alpha/c_p \rho(D - u) = C_f/2$ (Ref.8). The detonation velocity is then calculated by an iterative process involving integration of Eq.(1) together with the following conservation equations written in a frame of references attached to the detonation wave front

$$\frac{d}{dx} \rho u = 0$$

$$\frac{d}{dx} [p + \rho u^2] = \frac{4\sigma}{d}$$

$$\frac{d}{dx}\left[\rho u \left(h + \frac{u^2}{2}\right)\right] + Q \frac{d\beta}{dx} = -\frac{4q}{d} + \frac{4\sigma D}{d}$$

The drag force is given by the expression

$$\sigma = C_f \rho |D - u|(D - u)/2$$

The heat flux to the wall is calculated from the formula

$$q = \alpha(T_{st} - T_0)$$

where T_{st} is the gas stagnation temperature. The solution of the problem must meet the following boundary conditions:

$$x = 0: \quad \rho = \rho_f, \quad p = p_f, \quad \beta = 1$$
$$x \to \infty: \quad \rho = \rho_0, \quad p = p_0$$

where we assume that reaction occurs without a change in the mole number. The procedure of integration is described by Zel'dovich et al.[8]

Note that Eq.(2) represents a higher value of the local drag coefficient compared with the commonly used value $C_f = c/4$. It is worth emphasizing that the latter expression is adequate only for a chemically inert flow at distances > 10-15 tube diameters, i.e., after closure of boundary layers at the centerline. In the growing boundary layer the drag coefficient is larger than in the stabilized flow; hence, the use of $c/4$ rather than c in the expression of C_f for the flow behind a detonation front clearly leads to an underestimate of momentum losses.

One other important circumstance should be mentioned. It seems reasonable to assume that momentum and heat losses become apparent mostly in the induction zone, where one may neglect chemical processes. After reaction runaway the losses have almost no effect on the rate of chemical energy release, except for the 'tail' of the reaction zone where the chemical reaction rate is small and losses become dominant. Thus, the expression (3) which is valid for the unreacted flow may be taken with some reserve as the basis of the present model.

Results and Discussion

Table 1 presents predictions for some stoichiometric hydrocarbon-air and hydrocarbon-oxygen mixtures. The calculated value of the limiting tube diameter d_l, is compared with the quantity λ/π, which has been recommended for determination of the limiting diameter on the basis of experimental observations[3]. The values of A, a, b, c, E_* taken from Refs.9-14 result in satisfactory agreement between d_l and λ/π, especially for such readily detonable mixtures as C_2H_2, H_2, and C_2H_4 with either oxygen or air. Also shown in the table are predictions of the reduction in velocity at the limit of propagation $D_l D_0^{-1}$, the ratio of the reaction zone length l_l, at the limit to the predicted limiting diameter l_l/d_l, and the amount of mixture remaining unburnt at the CJ plane β_* at the detonation limit. The quantity $(D_0 - D_l)D_0^{-1}$ does not exceed 10-15% in accord with Zel'dovich[1]. The quantity β_l attains 5-6% at the CJ plane. Attention should be drawn to the fact that the reaction zone length at the limit is close to the limiting tube diameter, i.e., $l_l \approx d_l$ and is much larger than the induction zone length calculated in the absence of losses ($l_l \gg l_i$). In fact, the results indicate that the length of the reaction zone at the limit is of the order of cell size λ.

Figure 1 and Table 2 show dependencies of the limiting tube diameter on equivalence ratio Φ for C_2H_2- and C_2H_4-air mixtures at the initial pressure $p = 0.1$

Table 1 Limiting tube diameter of gaseous detonations in stoichiometric fuel-air and fuel-oxygen mixtures

N	Mixture	$A, 10^{11}$ (*)	a	b	c	E kkal/mole	D_i/D_0	l_i/d_i	β_i, %	d_i, mm	λ, mm
1	$C_2H_2+2.5O_2$	1.549	-0.67	-0.33	0	17,3	0.86	0.9	5	0.14	0,3
2	C_2H_2+air	1.549	-0.67	-0.33	0	17,3	0.90	1.5	5	2.10	5,7
3	$C_2H_4+3O_2$	0.355	0	-1.	0	27,5	0.92	0.8	4	0.19	0,56
4	C_2H_4+air	0.355	0	-1.	0	27,5	0.94	1.2	2,5	8.5	25
5	$H_2+0.5O_2$	2.254	0	-1.	0	15,2	0.91	1.0	3	1.5	1,57
6	H_2+air	2.254	0	-1.	0	15,2	0.92	0.9	5	4	15
7	C_2H_6+air	2.350	0.46	-1.26	0	34,2	0.96	0.9	0.8	52	60
8	C_3H_3+air	28.0	0.29	-1.19	0.39	36,8	0.98	0.7	0.7	80	63
9	CH_4+air	12590	0.48	-1.94	0	46,2	0.98	1.0	0.7	500	330

Note: The error of l_i, D_i and β_i calculations does not exceed 5% of the values shown in the table. The kinetic parameters have been chosen from the consideration of the best fitting between the conditions of ignition delay measurements and conditions in the induction zone of a detonation wave.

*) $\sec(\text{mole/l})^{-(a+b+c)}$

Table 2 Dependence of limiting parameters of gaseous detonations on initial equivalence ratio at p = 0.1 MPa

mixture	Φ	D_0 m/s	μ kg/kmole	γ	Q, MJ/kg	n	E kkal/mole	D_l/D_0	l_l mm	λ mm	β_l %
C_2H_2	0,50	600	28,73	1,358	1,517	2	17,3	0,925	20	29	4
	0,63	700	28,70	1,349	1,762	2	17,3	0,905	10	12,6	6
	0,76	770	28,67	1,341	1,962	2	17,3	0,895	6	8,5	6
	1,0	864	28,62	1,328	2,276	2	17,3	0,906	3,4	5,7	5
	1,32	1940	28,56	1,320	2,538	2	17,3	0,870	2,4	4	6,5
	2,1	2020	28,42	1,303	2,923	2	17,3	0,890	1,3	4,9	6
C_2H_4	0,52	1500	28,81	1,353	1,354	2	27,5	0,967	218	250	1,1
	0,63	1610	28,80	1,348	1,588	2	27,5	0,960	59	100	1,6
	0,75	1700	28,80	1,340	1,816	2	27,5	0,950	26	57	2,1
	1,0	1822	28,78	1,327	2,183	2	27,5	0,940	10	25	2,5
	2,34	1775	28,72	1,293	2,345	2	27,5	0,960	23	140	1,5

Note: Presented are the experimental values of D_0 and λ.

MPa. A good correspondence of predicted d_l values with the available values λ/π (Ref.3) is worth noting for lean mixtures. Note, that empirical correlations for τ (Eq.(1)) used in the present calculations were derived for lean mixtures which may serve as the explanation for the discrepancy between the results for $\Phi > 1$.

Figure 2 shows a comparison of calculated values of d_l (curve 1) and l_l (curve 2) with measured values of cell width λ (points 3) and cell length L (points 4) in the detonation wave propagating through $C_2H_2 + 7.6O_2 + Z$ Ar mixture at $p = 0.013$ MPa. The quantity $\theta = 2/(Z + 8.6)$ represents the volumetric content of inert (Ar) in the mixture. Experimental values of D and L are taken from Vandermeiren and Van Tiggelen[15]. We have taken into account that $\lambda \approx 0.62L$ (see Refs.2 and 3). It follows from Fig. 2 that $d_l \approx \lambda$ and $l_l \approx L$ within the wide range of θ. Note that the calculated amount of unburnt mixture at the CJ plane achieves 8%.

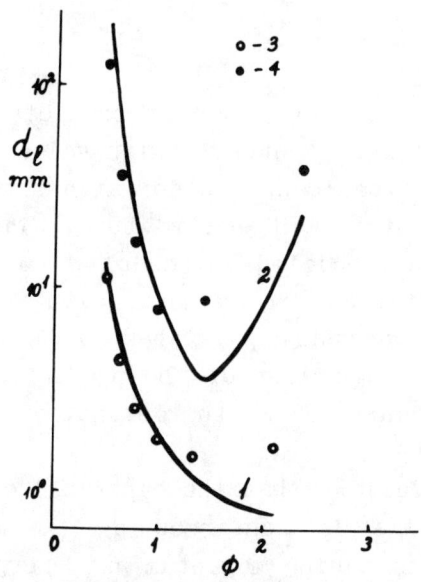

Fig. 1 Predicted (1, 2) and measured (3, 4) dependencies of the limiting tube diameter on the equivalence ratio. 1, 3 correspond to C_2H_2-air mixture, and 2, 4 correspond to C_2H_4-air mixture.

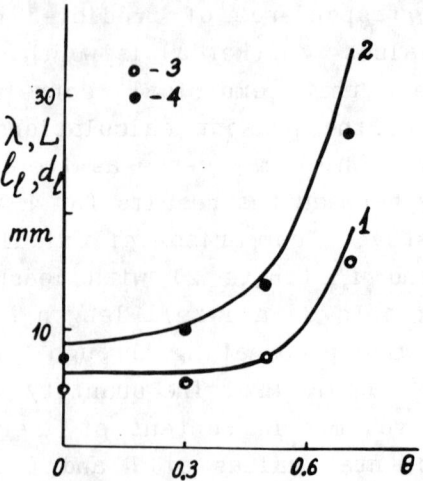

Fig. 2 Dependencies of the limiting values of dynamic detonation parameters of $C_2H_2 + 7.6\ O_2 + Z$ Ar mixture on the volumetric content of inert (Ar): 1, d_l (prediction); 2, l_l (prediction); 3, λ (experiment); and 4, L (experiment).

Figure 3 is a comparison of predicted values of the limiting diameter d_l (curve 1) with values of λ/π (points 3) taken from experimental observations of Bull et al.[4] for a stoichiometric C_2H_2-air mixture. Close agreement is evident between predicted values of d_l with measurements of λ/π. From Fig. 3, for example, with a tube of $d = 10$ mm the critical pressure p is about 0.025 MPa. Also shown in Fig. 3 is a comparison of the predicted length of the reaction zone (curve 2) at the detonation limit with cell size λ (points 4).

Thus, it follows that the empirical criterion widely used in practice is confirmed by the one-dimensional theory of limits taking momentum and energy losses into account. The predicted results indicate that for a whole variety of fuel-oxydizer mixtures the limiting tube diameter and the reaction zone length at the detonation limit are close enough to the transversal cell size of

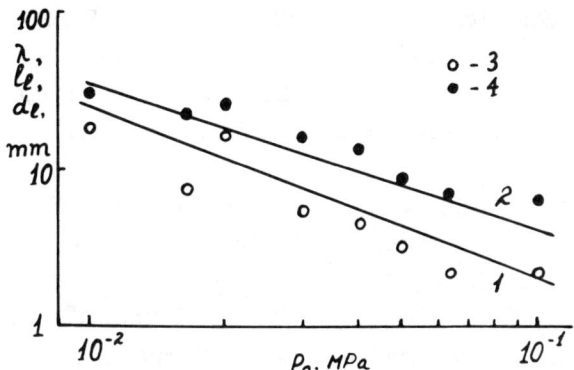

Fig. 3 Dependences of the limiting values of dynamic detonation parameters of stoichiometric mixture C_2H_2-air on initial pressure: 1, d_l (prediction); 2, l_l (prediction); 3, λ/π (experiment); 4, λ (experiment).

a multifront detonation wave. The approximate approach to the description of chemical processes adopted here is not valid in all cases because experimental investigations of the ignition delay are carried out, as a rule, with the use of highly diluted mixtures at relatively low pressure. In view of this there exists the essential requirement to study specific features of the ignition process at pressures corresponding to the values typical for detonations. For most mixtures of practical interest, such information may be obtained mainly from experimental studies since the development of the reliable detailed kinetic schemes seems to be problematic in the observable future.

References

[1] Zel'dovich, Y.B., "On the Theory of Detonation Propagation in Gaseous Systems.", Sov. J. Exp. Theor. Phys., Vol.10, No.5, 1940, pp.542-568.

[2] Knystautas, R., Guirao, C., Lee, J. H., and Sulmistras, A., "Measurement of Cell Size of Hydrocarbon-Air Mixtures and Prediction of Critical Tube Diameter, Critical Initiation Energy, and Detonability

Limits", *Dynamics of Shock Waves, Explosions and Detonations*, edited by J. R. Bowen, N. Manson, A.K. Oppenheim, and R.I. Soloukhin, Progress in Astronautics and Aeronautics, Vol. 94, AIAA, New York, 1984, pp.23-37.

[3] Vasiliev, A. A., Mitrofanov, V. V., and Topchiyan, M. E. "Detonation Propagation in Gases", *Fizika Goreniya i Vzriva*, Vol.23, No.5, 1987, p.109-125.

[4] Bull, D. C., Elsworth, J. E., Shuff, P. J., and Metcalfe, E., "Detonation Cell Structure in Fuel-Air Mixtures", *Combustion and Flame*, Vol.45, 1982, pp.7-22.

[5] Murray, S. B. and Moen, I. O. "The Influence of Confinement on the Structure and Behavior of Gaseous Detonation Waves", *Shock Tubes and Waves, Proceedings of the 16th Symposium (International) on Shock Tubes and Waves*, Aachen, Germany, 1988, pp.751-757.

[6] Frolov, S. M., Gelfand, B. E., Medvedev, S. P., and Tsyganov, S. A. "Quenching of Shock Waves by Barriers and Screens", *Proceedings of the 17th Symposium (International) on Shock Waves and Shock Tubes*, Lehigh Univ., Bethlehem, PA., American Institute of Physics Conference Proceedings, Vol.208, 1989, pp.314-320.

[7] Reid, R. C., Prousnitz, J. M., and Sherwood, T. K. "The Properties of Gases and Liquids", Third ed., McGraw-Hill, 1977, Chapter 9.

[8] Zel'dovich, Y. B., Gelfand, B. E., Kazdan, Y. M., and Frolov, S. M. "Propagation of Detonation in a Rough Tube with Deceleration and Heat Transfer Taken into Account", *Fizika Goreniya i Vzriva*, Vol.43, No.3, pp.103-112.

[9] White, D. R. "Density Induction Times in Very Lean Mixtures of D_2, H_2, C_2H_2, with O_2", *Proceedings of the 11th Symposium (International) on Combustion*, The Combustion Institute Pittsburgh, PA., 1967, pp.147-151.

[10] Hidaka, Y., Kataoka, T., and Suga, M. *Bulletin of the Chemical Society of Japan*, Vol.47, No.9, pp.2166-2174.

[11] Schott, G. L. and Kinsey, J. L. "Kinetic Studies of Hydroxyl Radicals in Shock Waves II, Induction Times in Hydrogen-Oxygen Reactions", *Journal of Chemical Physics*, Vol.29, 1958, pp.1177-1181.

[12] Burcat, A., Crossley, R. W., and Scheller, K. "Shock Tube Investigation of Ignition in Ethane-Oxygen-Argon Mixtures", *Combustion and Flame*, Vol.18, 1972, p.115-128.

[13] Zamanskii, V. M. and Borisov, A. A. "The Mechanism and Promotion of Ignition of Advanced Fuels", *Sov. Ser. Progress in Science and Technology, All-Union Institute of Scientific and Technical Information (VINITI)*, Moscow, Vol.18, 1989.

[14] Cheng, R. K. and Oppenheim, A. K. "Autoignition in Methane-Hydrogen Mixtures", *Combustion and Flame*, Vol.58, No.2, 1984, pp.125-134.

[15] Vandermeiren, M. and Van Tiggelen, P. J. "Cellular Structure in Detonation of Acetylene-Oxygen Mixture", *Dynamics of Shock Waves, Explosions and Detonations*, edited by J.R. Bowen, N.Manson, A.K.Oppenheim, and R.I.Soloukhin, Progress in Astronautics and Aeronautics, Vol.94, 1986, pp.104-117.

Effect of Flame Inhibitors on Detonation Characteristics of Fuel-Air Mixtures

A. A. Borisov,* V. V. Kosenkov,† A. E. Mailkov,‡ V. N. Mikhalkin,§ and S. V. Khomik¶

Russian Academy of Sciences, Moscow, Russia

Abstract

Propagation of flames and detonation waves in fuel-air mixtures with inhibitor additives is studied. Propane and hydrogen are chosen as fuels and tetrafluorodibromoethane as an inhibitor. Experiments are performed in a tube. The detonation and flammability limits are determined as functions of the inhibitor concentration in the mixture. For propane-air mixtures the additive decreases the detonation velocity beyond the value characteristic of detonation limits of noninhibited mixtures. The flammability limits narrow faster than the detonability limits as the inhibitor concentration grows, so that detonation propagates in mixtures that can not support flame. In inhibited hydrogen-air mixtures detonation limits are narrower than the flammability limits. A comparison of the calculated and measured detonation velocities reveals that the inhibitor additives enhance to some extent the heat release behind the detonation wave. For hydrogen-air mixtures the measured velocities exceed the calculated ones at

---

Copyright © 1992 by the American Institute of Aeronautics and Astronautics, Inc. All rights reserved.
*Head of Laboratory, N. Semenov Institute of Chemical Physics.
†Graduate Student, Moscow Physical Engineering Institute.
‡Junior Researcher, N. Semenov Institute of Chemical Physics.
§Engineer, Moscow Physical Engineering Institute.
¶Senior Researcher, N. Semenov Institute of Chemical Physics.

high inhibitor concentrations. The detonation front cells grow with the inhibitor concentration in both mixtures. The thermodynamic and kinetic aspects of the processes governing the detonation wave propagation in inhibited mixtures are discussed.

Introduction

It is well known that halides of paraffins are good flame extinguishers reducing appreciably the regions of gas-mixture flammability (see, for example, Refs. 1 - 4. These investigations demonstrated that halogenated compounds inhibit oxidation reactions in flames both chemically and thermally, with the chemical factor prevailing in many cases, and that the amount of the additive required to arrest the flame is quite high (commensurate with that of the fuel concentration).

Supression of detonation by homogeneous additives is studied less extensively, virtually no information is available about the effect of additives on detonability of fuel-air mixtures.

There are papers indicating that the effect of these additives on detonation in gaseous mixtures differs from that on flames.[5] The additives may both facilitate detonation propagation slightly decreasing the characteristic reaction times behind the detonation front, and suppress it acting as a mixture diluent. This was illustrated by taking as an example CF_3Br and $C_2F_4Br_2$ as inhibitors and hydrogen-oxygen and acetylene -oxygene mixtures diluted with Ar and ethylene-air mixtures as detonating media.

Such additive behavior in the two modes of reactive wave propagation can be anticipated because the limiting reaction stages in both types of reactive waves are quite different. In flames with chain reactions and active species of various mobility the low-temperature zone plays an important role in further development of the reaction and, hence, inhibitors converting H atoms and other active radicals into less active ones may drastically reduce the rate of fuel conversion and change the mixture flammability. High-temperature self-ignition, which is the main pathway of fuel conversion in detonation waves, is also affected by additives of various chemicals. But in this case the brominated compounds can no longer be considered as stable species leading to chain termination, since they decompose at a rather high rate, recovering partly the chain carriers. Moreover, halogenated compounds usually decompose faster than hydrocarbon fuels thus enhancing chain initiation. Therefore it is not surprising that these compounds may behave as inhibitors (both thermal, due to their high heat capacity, and chemical, at low enough temperatures, due to scavenging hydrogen atoms) and promoters (also both thermal, due to possible increase of the overall heat release in the CJ plane, and chemical, due to enhanced chain initiation).

In previous investigations no attempts were made to compare the limits of flame and detonation propagation in inhibited fuel-air mixtures. This comparison is of practical importance since halides are often used as extinguishers of gaseous flames, which naturally raises the question of whether these inhibited mixtures can detonate or not. The present paper is intended to answer this question, i.e., to show how the flammability and detonability limits are affected by additives.

Experimental

Experiments were carried out in two types of the set ups. The flammability limits were determined in a vertical steel tube 60 mm i.d. and 220 cm long following a procedure close to the standard one. The tube was evacuated before introducing the combustible mixture. The mixture was ignited at the bottom end of the tube by a pyrotechnical igniter MB-2 augmented by a 0.5-g black powder charge. The flame speed was measured with four ionization gauges evenly spaced along the tube. The signal from the gauges was recorded on a mirror-galvanometer oscillograph. The central electrode (a steel wire 0.6 mm in diameter) of the ionization gauge spanned the tube cross section. If the flame quenched before it reached the upper end of the tube we assumed that the mixture of a given composition was incapable of supporting combustion. This made it possible to measure the flammability limits by a simple and reliable technique.

The detonation limits were measured in a horizontal tube 70 mm i.d. and 25 m long. The detonation velocity was measured along last 12 m of the tube with four piezoelectric pressure gauges. Detonation waves were considered to be stable when no definite trend of their decay was observed and the spread of the velocities as measured at various bases did not exceed ± 1%. Recorded also were soot prints on a 300-mm-long foil inserted in the end section of the detonation tube.

Detonation was initiated by blowing up a stoichiometric propane-oxygen mixture filled a tube section comprising of no less than 6% of the total tube length. Mixtures were prepared in a vessel 80 liter in volume by partial pressures of the components. The inhibitor (tetrafluorodibromoethane, TFDBE) was measured with a hypodermic syringe and injected into the vessel. After that the gases and vapor were thorughly mixed with a fan. To minimize the effect of errors in mixture composition on the results of measurements the procedure of mixture preparation was the same in flame and detonation experiments. The mixtures studied are: stoichiometric hydrogen-air, stoichiometric propane-air, lean propane-air (3 vol.% propane in the mixture). Some experiments were also performed with heptane-air mixtures.

Results and Discussion

The flame speeds measured in the vertical tube along a 1.25-m base in the middle of the tube are shown in Fig. 1. Since the flame shape was not studied, the aforementioned speeds give only an idea about their order of magnitude under one-dimensional conditions. But what is important in Fig. 1 are the ends of the graphs which indicate the limiting (or quenching) inhibitor concentrations. As is seen, the relative (with respect to fuel) limiting inhibitor concentrations in the case of hydrocarbon-air mixtures are much larger than for hydrogen-air mixtures, although the ratio of the absolute inhibitor contents in these mixtures is reverse. The mixtures tested ceased to burn at the following inhibitor concentrations: > 13%, 3.6%, and 3% for the stoichiometric hydrogen-air, 4% propane-air, and 3% propane-air mixtures, respectively. This is consistent with the rate of the flame-speed decay with increasing inhibitor content, e.g., 3% additive of TFDBE to a stoichiometric propane-air mixture decreases

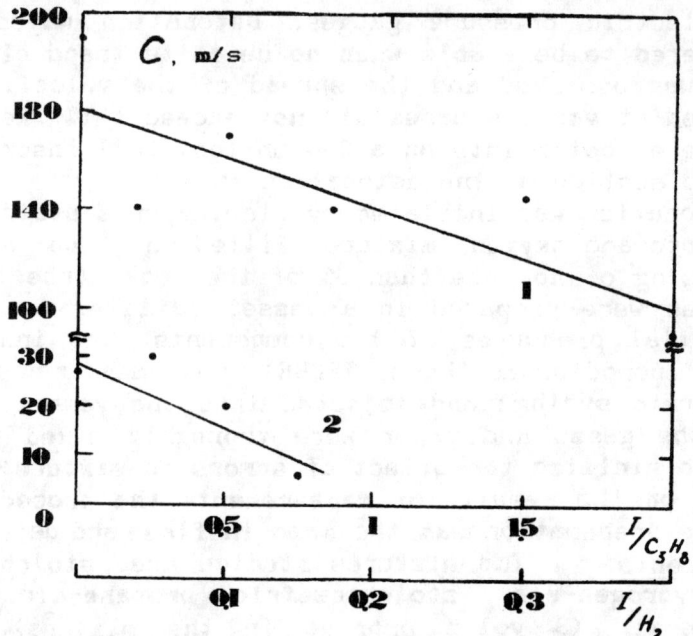

Fig. 1 Flame speed in fuel air mixtures in a vertical tube as a function of the inhibitor concentration: 1) 29.5% hydrogen + air; 2) 4% propane + air.

the flame speed four-fold, whereas in the hydrogen-air mixture the flame speed drops by less than 30%.

Experiments with heptane-air mixtures yielded a typical peninsula-like dependence of the flammability limits on the inhibitor concentration shown in Fig. 2. The minimum inhibitor concentration suppressing the flame in this mixture is 1.5%, which is twice as low as that for the propane mixture. There are indications in the literature that some mixed inhibitors may suppress flames more effectively than individual compounds. PCl_3 and $(C_2H_5)_2NH$ additives to TFDBE, which are believed to increase its effect, showed no synergetic effect, i.e. these additives do not enhance the effect of TFDBE.

Figure 3 displays detonation velocities in the same mixtures as functions of the inhibitor content. The vertical bars in Fig. 3 indicate the minimum flame suppressing inhibitor concentrations. The straight lines present the data of equilibrium thermodynamic calculations of the detonation velocities. Both calculations and measurements show that detonation velocities reduce

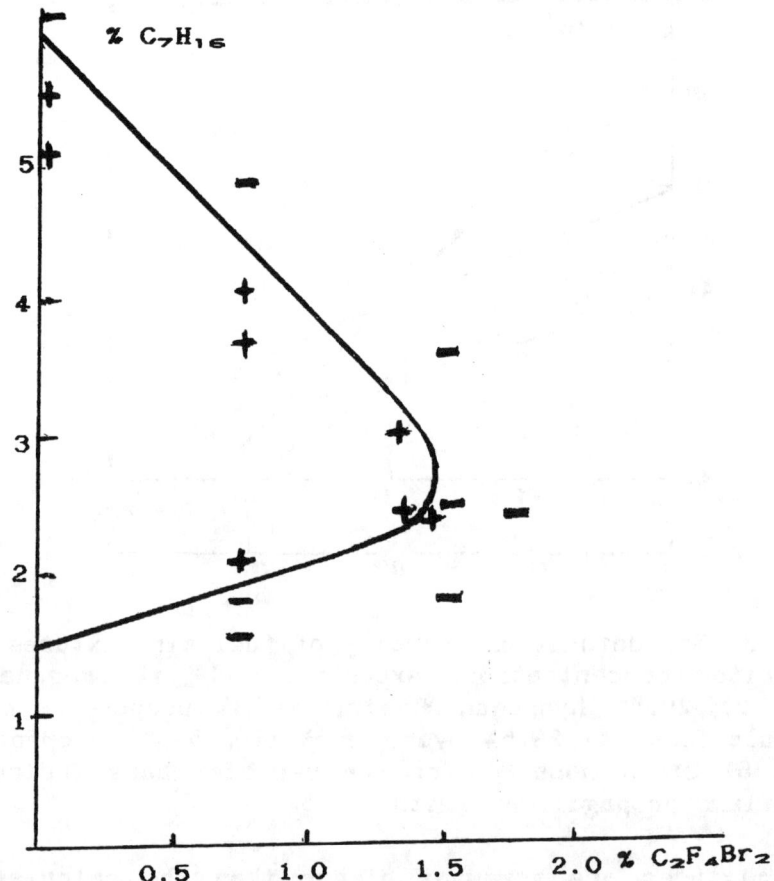

Fig. 2 Flammability limits of heptane-air mixtures at various concentrations of $C_2F_4Br_2$. + go, - no go.

monotonically as the inhibitor concentration grows. As is seen, the measured detonation velocities in hydrocarbon-air mixtures are slightly lower than the calculated ones (by about 50-70 m/s). This is not surprising since the detonation cell size in propane-air mixtures is quite large (comparable with the tube diameter). The energy and momentum losses in such essentially multidimensional waves may account for the observed velocity deficits. In the hydrogen-air mixtures, where the cell size is much less, the velocity deficit amounts only to 20 m/s. It should be noted that at large inhibitor concentrations the experimental velocities in hydrogen-

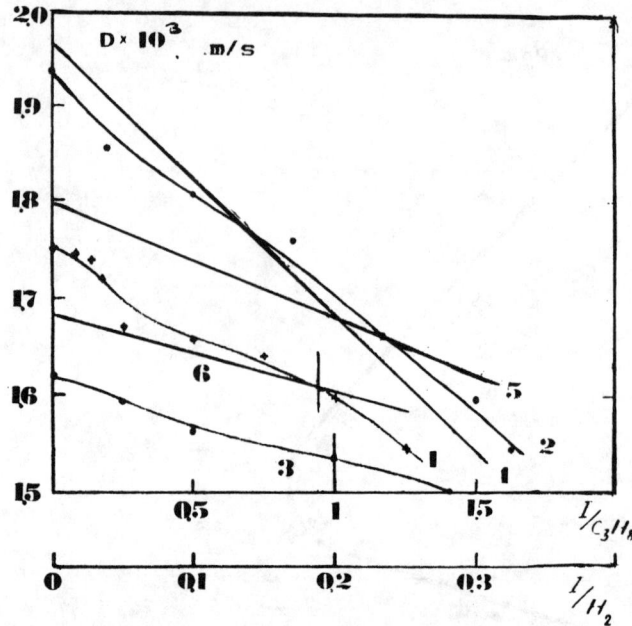

Fig. 3 The detonation velocity of fuel-air mixtures vs inhibitor concentration; experiment: 1) 4% propane + air, 2) 29.5% hydrogen + air, 3) 3% propane + air; calculations: 4) 29.5% hydrogen + air, 5) 4% propane + air; 6) 3% propane + air; the vertical bars indicate the flame propagation limits.

air mixtures are somewhat higher than the calculated ones. This is just opposite to what would be expected from the measured detonation front cells.

The absolute values of the inhibitor concentration at which the mixtures cease to detonate are 12% for the hydrogen-air mixture and about 5% for the propane-air mixtures. Hence in hydrogen-air mixtures quenching of flame by TFDBE also means suppression of detonation. The situation is different in hydrocarbon-air mixtures since in a certain range of inhibitor concentrations they may detonate but do not burn. This is the concentration range to the right of the vertical bars in Fig. 3. We did not attempt to find the limiting inhibitor concentration for heptane-air mixtures, but detonation in these mixtures was initiated and propagated at a velocity of 1700 m/s at absolute TFDBE concentrations of 1.9%, which is higher than the minimum flame extinguishing concentration.

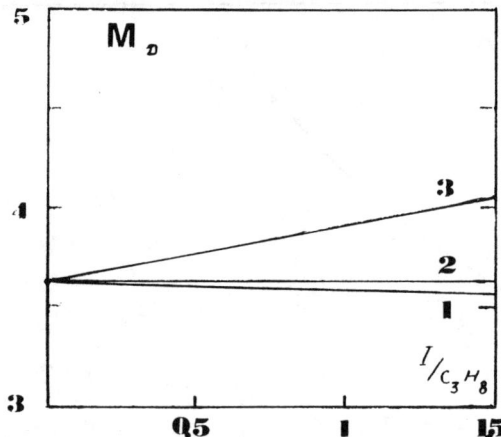

Fig. 4 Calculated detonation wave Mach number for a stoichiometric propane air mixture with additives: 1) Ar, 2) CO_2, 3) $C_2F_4Br_2$

One more important feature of detonations in the inhibited mixtures is an additional energy release at the expence of more complete oxidation of the reaction products by both oxygen and halogens. This is demonstrated by straight lines in Fig. 4 representing Mach numbers of detonation waves in propane-air mixtures with various additives: Ar, CO_2, and TFDBE. The first two additives reduce the detonation wave Mach number, as should be anticipated from the thermodynamic detonation theory, whereas TFDBE increases M_D. It may be the additional energy release which is responsible for this anomaly.

The susceptibility of fuel-air mixtures to initiation impulse may be characterized by the detonation cell size. The longitudinal cell size for the propane-air and hydrogen-air mixtures is plotted as a function of the relative inhibitor content in Fig. 5. It follows from Fig. 5 that TFDBE additives increase monotonically the cell size and, consequently, reduce the mixture detonability. The minimum energy of unconfined-detonation initiation should grow, as the inhibitor concentration increases, roughly proportionally to the cell size cubed; which means that the minimum energy of direct detonation initiation in both mixtures at the

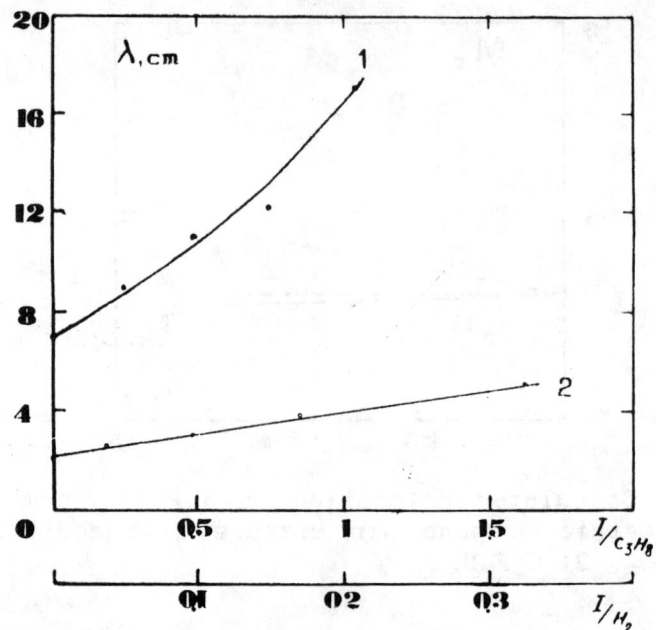

Fig. 5 The longitudinal detonation cell size in fuel-air mixtures vs inhibitor concentration: 1) 4% propane + air, 2) 29.5% hydrogen + air.

inhibition limit grows by about an order of magnitude as compared to that of the uninhibited mixtures.

This estimate was verified by direct experiments on initiation of plane detonation in a tube. Two types of the initiator were used: a sheet solid high explosive and detonating gaseous fuel-oxygen mixtures. Experiments were conducted in a vertical tube 2 m long and 142 mm in diameter and in a horizontal tube 25 m long and 70 mm in diameter. The minimum energy of detonation initiation in the propane-air-inhibitor mixtures grew by about a factor of 2.5 as the inhibitor concentration rose from zero to the limiting one. For the heptane-air mixtures the initiation energy growth is seen from the data presented in Table 1. The data are obtained for a mixture containing 2.5% C_7H_{16}, the TFDBE concentration is taken with respect to the fuel. These changes in the initiation energy of plane detonation conform to a respective increase in the initiation energy of unconfined detonation by about an order of magnitude.

Table 1 Detonation initiation parameters

% of TFDBE	$E_{init.}$, MJ/m²	D, m/s	Go/no go
50	2.5	1720	go
75	2.5	–	no go
75	5.0	1700	go
75	3.5	1675	go
0	1.3	–	no go
0	2.5	1770	go

Thus the aforementioned provides evidence of two opposite trends in the TFDBE effect on detonation of fuel-air mixtures. On the one hand, the additives increase slightly the overall heat release in the wave and, hence, facilitate detonation propagation and; on the other, they decrease the temperature behind the lead shock wave (due to their high specific heat) and, hence, impair detonability of the mixture. That it is predominantly specific heat effect that increases the reaction time behind detonation waves is proved by shock tube experiments, in which the respective inhibited hydrocarbon-air mixtures are ignited behind reflected shock waves. These experiments demonstrated that the additives either did not affect the ignition delays or slightly decreased them.

The fact that the detonation velocity of the inhibited hydrogen-air mixture at the limit is higher than that in the uninhibited mixture (1420 m/s) at the lean limit may be explained by a change in the reaction mechanism observed in shock tube studies. Indeed, the ignition delays of hydrogen-oxygen mixtures at high pressures grow steeply at about 1200 K due to the changeover from the branched-chain ignition to nearly thermal with short chains. Replacement of highly reactive H, O, and OH radicals by halogenated radicals, that are much less reactive, shifts the point where the slope of the Arrhenius plot of the ignition delays for hydrogen-air mixtures steepens toward higher temperatures. This makes the detonation waves less stable at higher velocities.

The only explanation of a peculiar fact of abnormally high detonation velocities (as compared to the equilibrium calculated ones) observed in inhibited hydrogen-air mixtures at high inhibitor concentrations should be sought for in some kind of disequilibrium in the detonation products. It is unlikely that this is due to the finite rate of the vibrational equilibration in the products (as has been suggested to explain a similar behavior of hydrogen-chlorine detonation waves) since the relaxation times in the presence of such triatomic molecules as water should be very small. Hence, one can suspect that the minimum of the thermodynamic potential functional corresponding to the equilibrium state of the products is not the deepest one in this system, so that the system may pass through a deeper intermediate minimum when moving from the unburnt mixture to the end products.

Thus the studies performed in the present work revealed the following:

1) The addition of halogenated compounds, which effectively suppress fuel-air flames for both thermal and chemical reasons, to hydrocarbon-air mixtures changes detonability of these mixtures to a lesser extent than their flammability, the ratio of the detonation limits to flammability limits in the inhibited mixtures reverses, i.e., in a certain range of inhibitor concentrations the mixture detonates but does not burn.

2) In hydrogen-air mixtures this ratio is traditional, i.e., the range of inhibitor concentrations where these mixtures burn is wider than that where they detonate.

3) Addition of inhibitors to both mixtures increases the minimum energy of direct detonation initiation and the detonation cell size, most likely due to the high specific heat of the additive and temperature drop behind the lead shock front.

4) Inhibitor additives slightly augment the heat release behind the detonation front which shows up as an increase in the detonation Mach number with increasing inhibitor content in the mixture.

5) Measured detonation velocities in hydrogen-air-inhibitor mixtures are higher than the calculated ones at large inhibitor concentrations, a fact which requires further investigation of the possible reaction routes in these detonation waves.

References

[1] Lask, G., and Wagner, H. Gg., "Influence of Additives on the Velocity of Laminar Flames", *VIIIth Symposium on Combustion*, Williams and Wilkins, Baltimore, MD, 1962, pp. 432-438.

[2] Rosser, M. A., Wise. H., and Miller, J., "Mechanism of Combustion Inhibition by Compounds Containing Halogen", *VIIth Symposium on Combustion*, Butterworth, London. England, 1959, pp. 175-182.

[3] Vandermeiren, M., Safieh, H. Y. and Van Tiggelen P.J., "Action d'inhibiteurs halogénés sur la propagation des flammes", *1st Specialist Meeting of the Combustion Institute*, Bordeaux, France, 1981, pp. 178-183.

[4] Baratov, A. N., Karaulov, F. A., and Makeev, V. I., "Investigations into Inhibition of $H_2-O_2-N_2$ Flames by Halogenated Hydrocarbons", *Fizika Goreniya i Vzryva*, vol. 6, No. 1, 1970, pp. 18-26.

[5] Vandermeiren, M., and Van Tiggelen, P.J., "Structure of Gas Phase Detonations in Acetylene Mixtures-Role of an Inhibitor", *Proceedings of the Internat. Symposium on Intense Dynamic Loading and its Effects*, Beijing, China, 1986, pp. 219-224.

Propagation of Gaseous Detonations Through Regions of Low Reactivity

T. Engebretsen*
Norwegian Defense Construction Service, Oslo, Norway
D. Bjerketvedt†
Christian Michelsen Institute, Bergen, Norway
and
O. K. Sønju‡
Norwegian Institute of Technology, Trondheim, Norway

Abstract

Two types of detonation experiments have been performed, as follows: 1) Detonation propagation limits for acetylene–air and ethylene–air mixtures in a square tube with internal dimensions of 125 x 125 mm have been studied. In the experimental apparatus, two regions of homogeneous gas mixtures were separated by a fast-opening slide valve. Chapman–Jouguet detonations were established in a 5-m-long section with different acetylene–air concentrations, and propagated into acetylene–air or ethylene–air mixtures with lower reactivity in a section of approximately 3 m length. Detonation propagation limits were found to be independent of Chapman–Jouguet parameters in the donor section. For acetylene–air detonations, the propagation limit was found to be 4.0%. This corresponds to a cell width approximately equal to the width/height of the channel. For ethylene the

---

Copyright © 1992 by the American Institute of Aeronautics and Astronautics, Inc. All rights reserved.
* Research Scientist.
† Senior Scientist, Department of Science and Technology.
‡ Professor, Thermal Energy Division.

propagation limit was 5.0%, a cell width of about 0.6 times the width/height of the channel. 2) Reinitiation of detonation across gaps of 100 or 150 mm length containing air and acetylene-air and ethylene-air mixtures, with concentrations below the detonability limits as defined by the experiments above, has also been studied. The same tube was used, with two slide valves to separate the gap from the detonable gases on both sides during gas filling. Reinitiation was found to be enhanced by increasing values of the Chapman-Jouguet parameters in the donor gas, decreasing the length of the gap, and increasing the reactivity in the acceptor gas. Reinitiation was also found to be enhanced by increased reactivity in the gap. For example, reinitiation in stoichiometric acetylene-air across a 100-mm gap occurred approximately 1 m earlier (and almost instantaneously) with 3.0% acetylene-air in the gap than with air in the gap. This effect was observed even with gas mixtures in the gap far below the lower flammability limit for standard atmospheric conditions (1 atm and 25 °C).

Introduction

In accidental situations, gas clouds will consist of all kinds of compositions between pure air and pure fuel gas. Detonation propagation in such media is not well understood at the present time. Previous studies related to this problem are briefly discussed below.

Marginal detonation waves have been studied by Moen et al.[1] They studied detonation propagation in circular tubes of varying diameter. In marginal ethylene-air detonations, that is, with a transverse wavelength in the detonation front on the order of the tube diameter, spinning and galloping regimes were observed. Vasil'ev[2] employed a hydraulic analog to find criteria for the relation between cell width and tube geometry for detonation propagation limits. Cell width is defined as the propagation distance of two transverse waves in the detonation front before collisions with the neighboring transverse waves or the confining walls. For circular and square tubes this limit was calculated to be $L=D=S/\pi$, where L is the width of a square channel, D the diameter of a circular tube, and S the cell width. Experimentally, a limiting detonation with a cell width equal to the width of a rectangular channel was realized for a channel with $L/\delta=6$, where L is the width and δ the height of the channel. This was found to agree well with the hydraulic analog.

Detonation propagation across concentration gradients have been studied by Strehlow et al.[3] It was found that transverse waves in detonations adjusted more slowly to the new Chapman-Jouguet-conditions when a detonation propagated from a strong to a weak mixture than from a weak to a strong mixture.

Reinitiation of detonation has been studied by Bull et al.,[4] who studied "sympathetic detonations" over air gaps in stoichiometric ethylene-air and propane-air mixtures. Necessary lengths of air gaps to prevent propagation of spherical detonations were found to be 150 mm for ethylene-air and 125 mm for propane-air. Bjerketvedt[5] and Bjerketvedt et al.[6] studied reinitiation of acetylene-air and ethylene-air detonations across an inert region (air) of varying length. The detonation tube was divided into donor, inert, and acceptor sections. Reinitiation of detonation in the acceptor section was characterized by a time delay and a corresponding distance into the acceptor section before transition to detonation. Reinitiation was enhanced by: 1) increasing shock strength, i.e., higher CJ pressure and CJ velocity for the detonation in the donor section; 2) decreasing length of the inert region; and 3) increasing reactivity in the acceptor section. Shock propagation prior to reinitiation was also found to be influenced by boundary-layer effects. Numerical calculations, including a factor taking heat transfer and friction into account, agreed well with experimental data for detonation propagation into air.

Thomas et al.[7] studied reinitiation of detonation across an inert region in acetylene-oxygen mixtures. Two different cross sections were used, a rectangular tube with internal dimensions of 22 x 10 mm and a circular tube with 50-mm internal diameter. The transition mechanisms were assumed to be the same for the two tubes; that is, flame acceleration and where the shock propagating into the acceptor gas only preheated the unburned gas. This was because the shock velocity generally was below 60% of the CJ value, which had been found earlier to be the limit for direct shock initiation of detonations in smooth tubes. Transition to detonation occurred much faster in the tube with largest cross-sectional area, due to less influence of boundary-layer effects. Increasing diffusion across the interfaces also enhanced transition to detonation.

In the present tests detonation propagation into gases with low reactivity and reinitiation of detonation across a gap of inert gas or low-reactivity gas were studied. This work is also presented in a doctoral thesis by Engebretsen, which was completed in 1991.[8]

Experimental Setup

Detonation Tube

The detonation tube was designed to create regions with homogeneous gas mixtures, separated by concentration gradients as abrupt as possible. It consisted of a steel tube with square cross section and internal dimensions of 125 x 125 mm. The total length was approximately 9 m including a window section for high-speed photography. The wall thickness was 6 mm, and the inner surface was untreated. The tube could be divided into two or three sections by means of fast-opening slide valves, and the sections could be filled with gas mixtures individually. This setup was mainly the same as used by Bjerketvedt[5] and Bjerketvedt et al.[6]

Figure 1 shows the detonation tube with the slide valves and the window section. The three sections of the square tube are denoted 1, 2, and 3, corresponding to the detonation propagation direction. The length of section 2 could be varied by mounting short sections between the flanges. In the present tests section 2 was 100 or 150 mm long. The regions were filled with acetylene–air or ethylene–air mixtures. Detonation was initiated by a spark plug in equimolar oxy–acetylene in the "Booster section". Figure 1 also shows locations of pressure transducers and a microwave Doppler radar. The two slide valves consisted of 1 mm thick stainless steel plates, each connected to a pneumatic cylinder. The slide valves had to operate fast to minimize mixing at the interfaces caused by diffusion, buoyancy and turbulence. The steel blades moved from closed to fully open position in 0.1 s. The time from the blades started to move to the ignition system was triggered was 0.25 s. In earlier tests[5,6] the mixing zone between the regions caused by removal of the steel blades was found to be less than 60 mm long. Using a section 2 that is 100 mm long, there should be at least 40 mm of undisturbed gas between the mixing zones. The air and fuel gases were premixed from commercial cylinders and purged through each section for approximately 10 times the volume of each section before testing. The air used was of grade 2 to minimize the influence of moisture and other impurities, the other gases used were commercial grade. The gas concentration was monitored continuously with an infrared gas alalyzer of type Wilkins Miran 1A. The accuracy of the analyzer was +/- 0.05 vol. %. All sections were vented to atmospheric pressure prior to ignition.

Diagnostics

The diagnostic system was designed to monitor pressure–time and velocity–time histories for detonations and shocks. To monitor pressure, piezo–electric pressure transducers of type Kistler 603B were used. The transducers with adapters were mounted flush with the inner walls of the

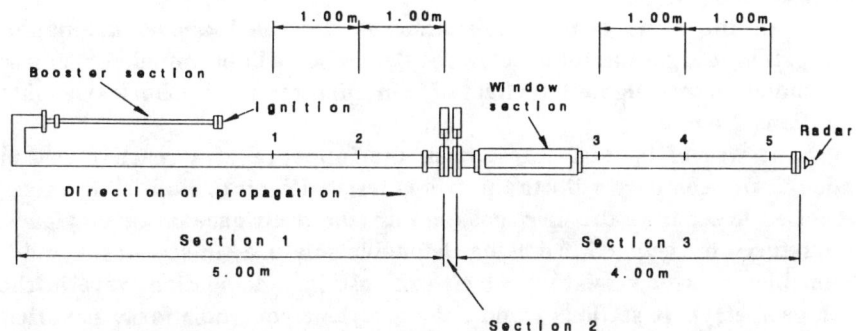

Fig. 1 Experimental setup: 1–5 denote pressure transducers.

tube with an accuracy of +/- 0.1 mm. For continuous monitoring of the velocity of the detonations, the flames and the hot combustion products, an X-band microwave Doppler transceiver of type M/A-COM MA 86656 Series D was used. The operating frequency was 10.525 GHz. The microwave signal from the transceiver (radar) reflects on ionized gases as in detonations, flames and hot combustion products. The scaling factor for the radar used was 70.2 Hz/(m/s). The radar pointed straight into the tube from the end through a 40 mm thick polyethylene "window". It was connected to an amplifier with approximately 200 times amplification and a 20-700 kHz band pass filter.

Measurements from the pressure transducers and the microwave doppler radar were recorded on a digital high-speed data logger (transient recorder) with a sampling rate of up to 10^7 samples/s. The data logger was triggered by time of arrival of the detonation at transducer no. 1. It was connected to an IBM compatible personal computer (PC) for control and presentation of results.

For some selected tests schlieren photography was also performed. A setup with two mirrors, 0.50 m in diameter, and a high-speed camera of type Hycam II was used.

Experimental Program and Results

Experimental Program

The first series of tests were performed with only one slide valve used. In these tests detonation propagation into acetylene-air and ethylene-air mixtures was studied to determine the lean limit for detonation propagation. Only acetylene was used as donor gas in section 1. The acetylene-air concentration in section 1 was varied in the following steps: 5.00, 6.50, 7.75 (approximately stoichiometric) and 9.00% to observe an eventual change of the lean propagation limit as the CJ-parameters in section 1 changed.

In the same test configuration as described above, detonation propagation into gas mixtures below the detonability limit (found above) was also studied to investigate the effect of lean mixtures on the shock strenghts in sections 2 and 3.

It was of interest to establish the influence of a reactive gas in section 2 and compare with the previous tests with air[5,6]. Both slide valves were used to separate the inert gas (air) or lean acetylene-air or ethylene-air mixtures in section 2 from the detonable mixtures in sections 1 and 3 (detonable: a gas-air mixture which can sustain a detonation wave in the given geometry). In sections 1 and 3 the acetylene concentration was varied in the same steps as in the tests with detonation propagation into lean mixtures. That is 5.00, 6.50, 7.75 and 9.00%.

Table 1 Detonation propagation limit in acetylene/air mixtures

Section 1 (% C$_2$H$_2$-air)	Sections 2 and 3 (% C$_2$H$_2$-air)	Detonation propagation	$d_c/13$, mm	Detonation vel., m/s	CJ-velocity, m/s
5.00	3.50	No	–	–	–
5.00	4.00	Yes	135	1571.5	1594.0
6.60	3.10	No	–	–	–
6.55	3.60	No	–	–	–
6.50	4.00	Yes	135	1568.5	1594.0
6.55	4.20	Yes	108	1598.1	1615.5
7.70	3.80	No	–	–	–
7.70	3.90	No	–	–	–
7.75	3.90	Yes	–	–	–
7.70	4.00	No	135	–	–
7.75	4.10	Yes	119	1608.8	1604.7
7.75	4.30	Yes	92	1623.4	1626.2
7.75	4.50	Yes	71	1650.2	1647.7
7.75	5.00	Yes	40	1697.2	1701.4
9.00	3.20	No	–	–	–
9.00	3.50	No	–	–	–
9.00	4.05	Yes	131	1591.7	1599.4

Table 2 Detonation propagation limit in ethylene/air mixtures

Section 1 (% C_2H_2-air)	Sections 2 and 3 (% C_2H_4-air)	Detonation propagation	$d_c/13$, mm	Detonation vel., m/s	CJ-velocity, m/s
7.75	4.05	No	–	–	–
7.75	4.50	No	112	–	–
7.75	4.80	No	83	–	–
7.75	5.00	Yes	69	1710.3	1724.9
9.00	4.80	No	83	–	–
9.00	5.05	No	68	–	–
9.00	5.35	Yes	54	1736.0	1749.9

Results for Detonation Propagation Limits

Tables 1 and 2 give an overview of the results in a Yes/No representation whether a detonation propagated or not. The lean limit for detonation propagation was found to be very close to 4.00% for acetylene-air and 5.00% for ethylene-air. For the acetylene tests the propagation limit did not change with varying concentrations in section 1, i.e., varying CJ-parameters in the donor detonation. This behavior was also verified by the few ethylene tests run. Tables 1 and 2 also show average velocities between transducers 4 and 5 for detonations with low reactivity mixtures in sections 2 and 3. These transducers were used because the detonation was slightly overdriven just after crossing the interface between the two gas mixtures. Comparisons with CJ-velocities, also shown, give negligible velocity deficits. The largest deviation between measured detonation velocity and CJ-velocity was 1.6 %. Included in Tables 1 and 2 are also previousely measured values for $d_C/13^9$, where d_C is "critical tube diameter," which is a length scale used which may characterize the detonation sensitivity of a gas.[9] This will be discussed later.

The pressure and radar records for a typical test with no detonation propagation is shown in Fig. 2. In Fig. 2a the pressure records show the decay of the shock front when a detonation propagates from a 9.00% C_2H_2 mixture into a 3.20% C_2H_2 mixture. The radar record in Fig. 2b shows the detonation in section 1 at the left, then the shock and the flame decouple, and the radar then measures the decay of the contact surface, i.e., combustion products from the detonation in the 9.00% C_2H_2 mixture. Typical records for a successful detonation propagation are shown in Fig. 3. In Fig. 3a the pressure records show a detonation propagating into a 4.05% C_2H_2 mixture from a 9.00% C_2H_2 mixture. The radar record in Fig. 3b shows detonation velocities, both velocities are close to the CJ velocity on the average. The apparent velocity drop after the original detonation in section 1 has propagated into the leaner mixture is because the reflection of microwaves at the contact surface (combustion products from the original detonation) dominates in this region. The average velocity calculated from the transducer measurements in this region shows no evidence of such a velocity drop. The apparent velocity drops later in this plot are due to a periodic behavior of the reflected microwaves (the signal is modulated).

Results for Detonation Propagation into Gas Mixtures Below the Detonability Limit

Figure 4a shows average shock velocities measured between the interface between section 1 and 2 and transducer no. 3 for acetylene. Two different donor detonation strengths are shown. A systematic increase in shock velocity with increasing reactivity of the gas in section 2 and 3 can be observed, even with 1.00% C_2H_2-air in these sections. Similar comparisons for the peak pressures in the shock front arriving at transducer 3, shows the same behavior in Fig. 4b.

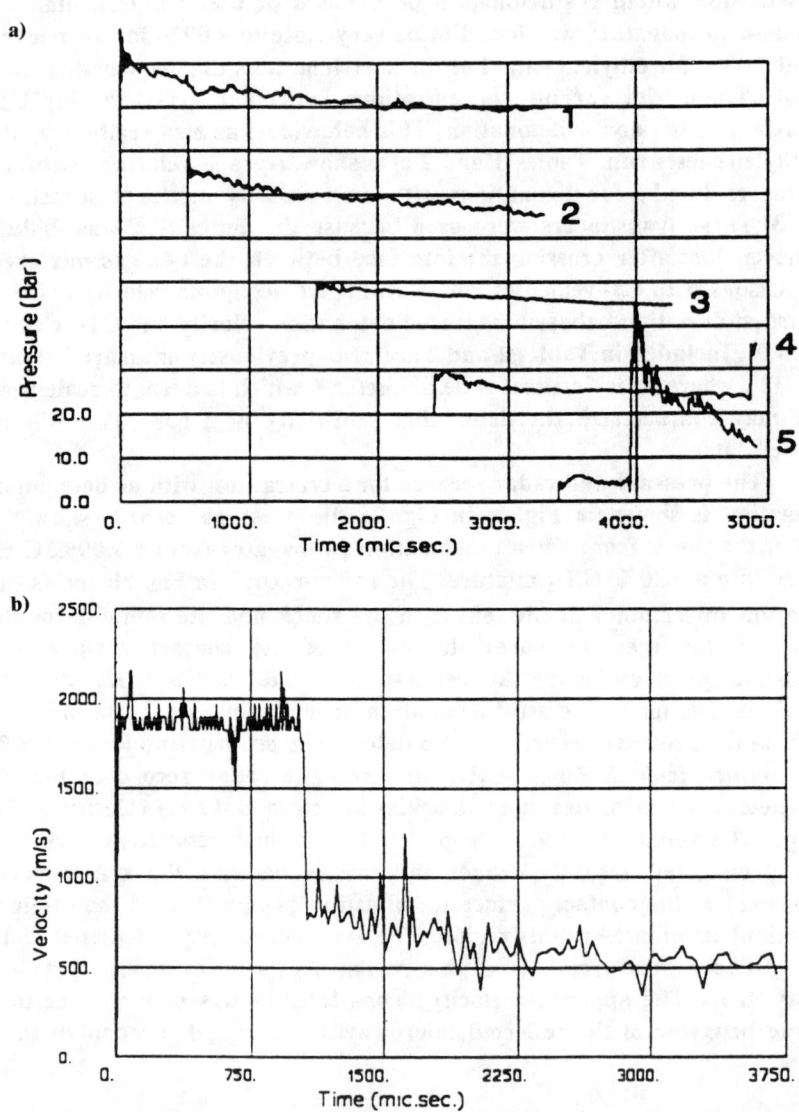

Fig. 2 a) Pressure records and b) velocity from radar record showing a detonation propagating from 9.00% C_2H_2–air into 3.20% C_2H_2–air. Transducers 1 and 2 show detonation in section 1, transducers 3–5 show decay of shock in section 3; the second pressure peak for transducers 4 and 5 is reflected shock from a local explosion at the tube end; radar record shows failure of detonation.

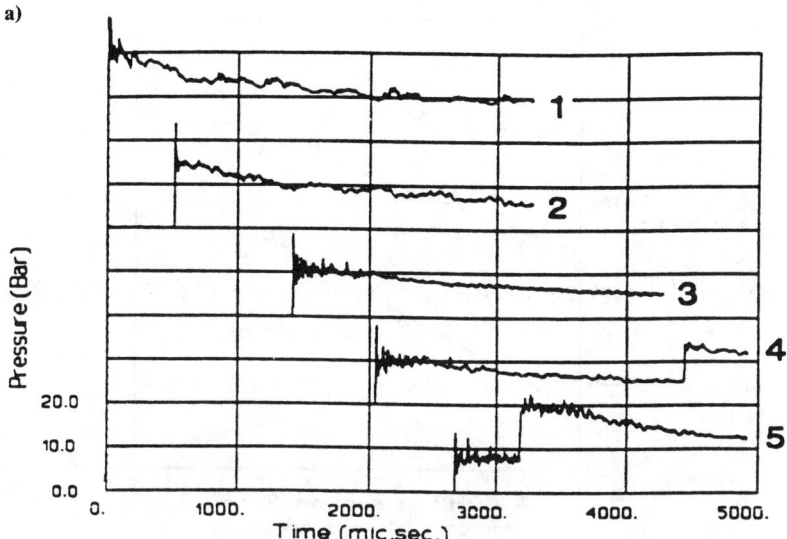

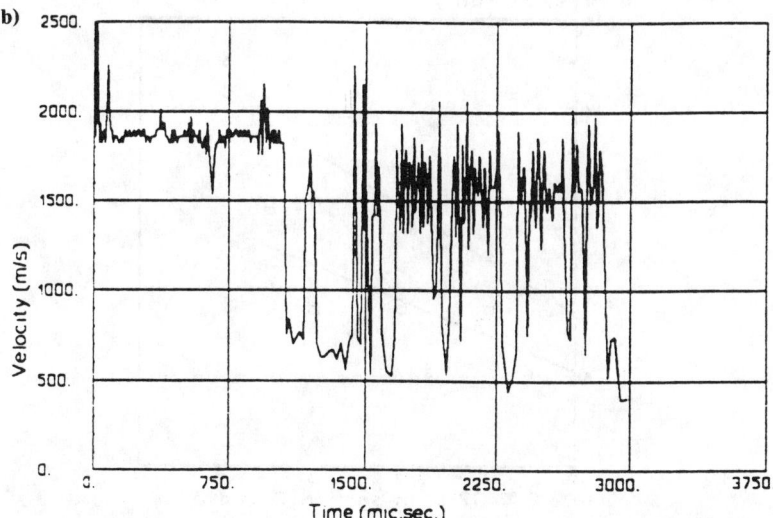

Fig. 3 a) Pressure records and b) velocities from radar record showing a successful detonation propagation in 4.05% C_2H_2–air. Donor detonation in 9.00% C_2H_2–air; the second pressure peak for transducers 4 and 5 in a) is the reflected shock from the tube end.

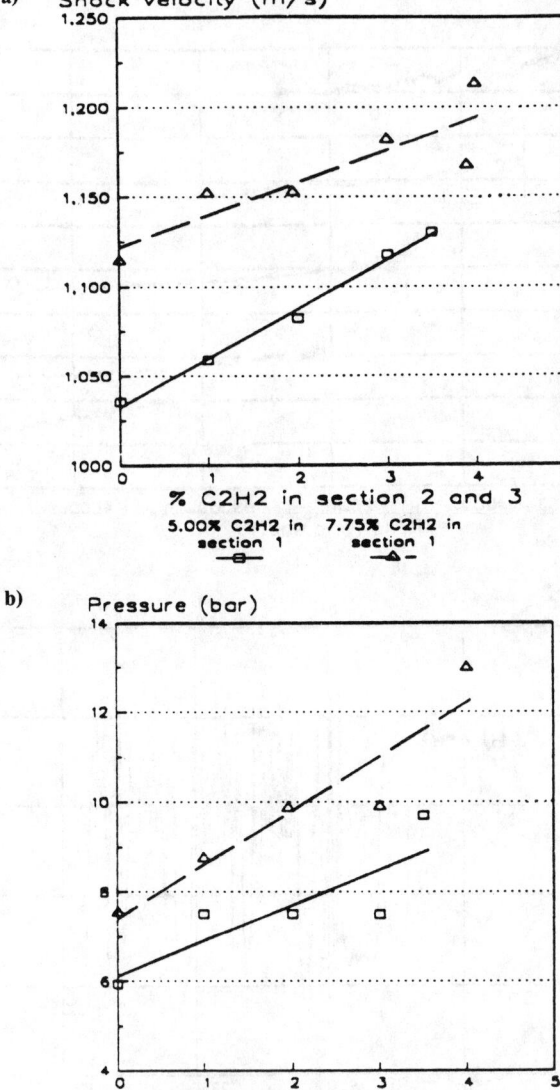

Fig. 4 Detonation propagation into lean acetylene–air mixtures: a) variation of average shock velocities with varying reactivity in the lean mixtures and b) the corresponding variation in peak pressure shown for two different donor detonation strengths.

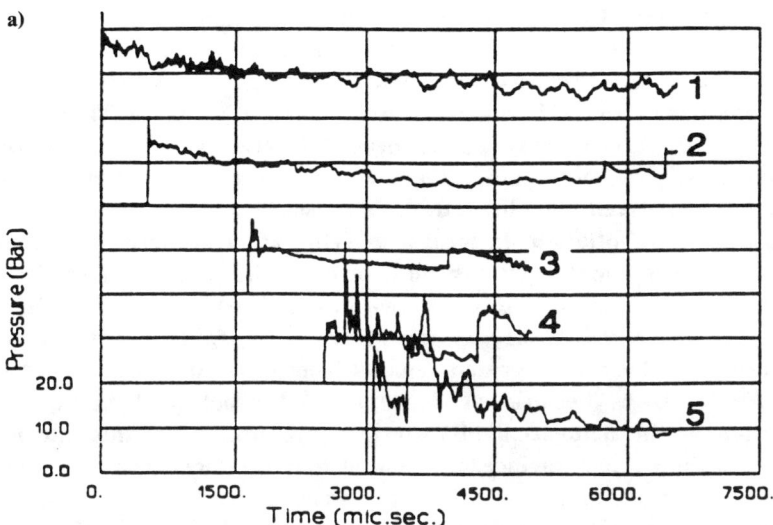

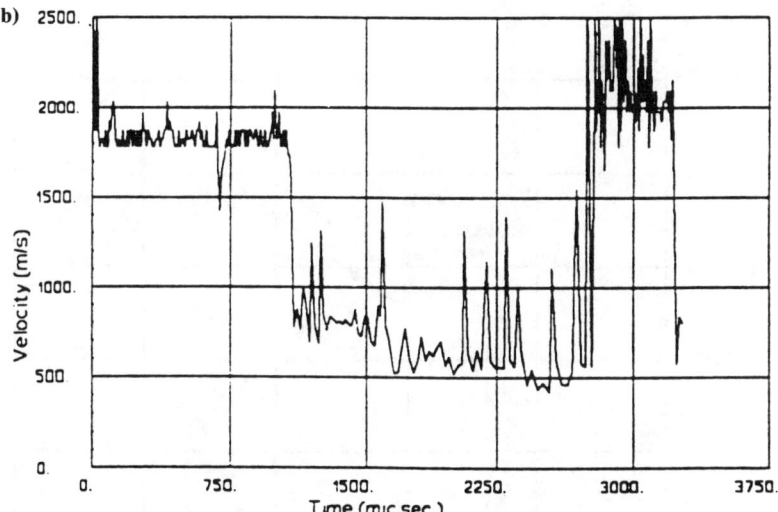

Fig. 5 a) Pressure records and b) velocities from the radar record showing reinitiation at 2784 μs in 7.75% C_2H_2–air, across a 150–mm gap of 0.95% C_2H_2–air; the second pressure peak for transducers 3 and 5, and the last peak for transducer 4 in a) is the reflected shock from the tube end.

Results for Reinitiation of Detonation Across Regions of Low Reactivity

Figure 5 shows pressure (5a) and radar (5b) records for a typical reinitiation test. The experimental conditions are: 7.75% C_2H_2-air in sections 1 and 3 and 0.95% C_2H_2-air in section 2 (0.15 m length). The radar record in Fig. 5 also shows an overdriven detonation after reinitiation.

The criterion chosen for when reinitiation has occurred is that a detonation has at least CJ velocity or greater and stays stable. That is, rapid fluctuations in the velocity up to near CJ level, or shock and flame complexes with velocities not far below CJ, are not encountered. Time-distance plots derived from the radar and pressure records were used to find the position of reinitiation. In most tests reinitiation occurred as a distinct change in velocity making it easy to determine the position of reinitiation occurrence. An example of this is shown in Fig. 6, which shows reinitiation in 7.75% C_2H_2 across a 0.10 m gap of 1.95% C_2H_4. In Fig. 6 the curve starting from 0,0 is the integrated velocity from the radar record. At 2.00 m the velocity drops because of decoupling of shock and flame in the detonation as the detonation hits the lean mixture. After this, the radar reflects on the contact surface, i.e., combustion products from the original detonation. Time of arrival of the leading shock at the pressure transducers is shown by dots. The shock travels at a higher velocity than the contact surface. Reinitiation occurs somewhere behind the leading shock and ahead

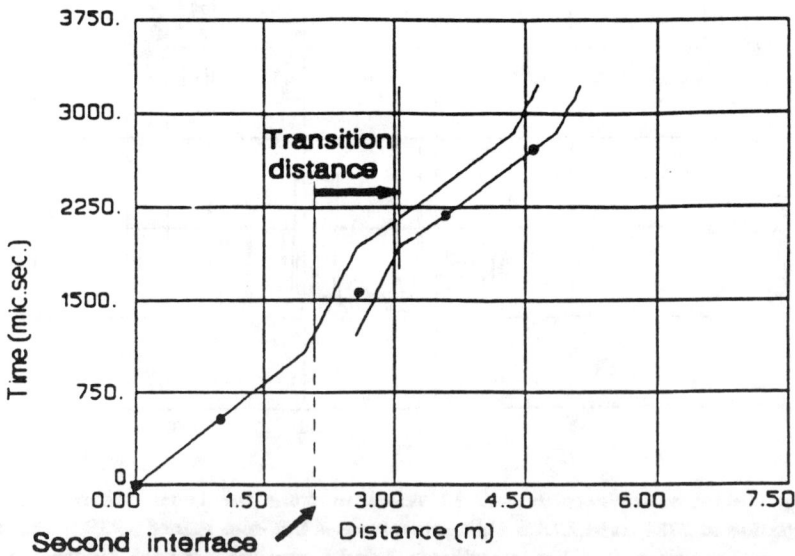

Fig. 6 Time–distance plot created by integrating the velocities measured by the Doppler radar: time of arrival at the pressure transducers are shown as dots; the two last transducers are used to correct the time–distance plot after reinitiation has occurred.

of the contact surface. Therefore, the original curve from the radar has to be corrected by time of arrival of the detonation after reinitiation at the two last pressure transducers in the tube, to determine the position where reinitiation occurred.

Figures 7–9 show comparisons between the different series of tests. Transition distance is measured from the interface between sections 2 and 3 to the position where a new detonation is established. Figure 7 shows the influence of varying CJ parameters in section 1, as the acetylene concentration in section 2 varies. In the reinitiation tests with 7.75% and 9.00% C_2H_2 in section 1, the variation of the transition distance showed a regular behavior. This was not the case with the two lower detonation strengths in section 1. For example, with 5.00% C_2H_2 in section 1 reinitiation occurred only in one test. In that case, reinitiation occurred almost immediately for 3.50% C_2H_2 in section 2. In two more tests with 6.50% C_2H_2 in section 1, reinitiation occurred close to the tube end, that is, with transition distances of about 4 m (after the end of the radar record). Figure 8 shows the effect of varying reactivity in section 3, as the acetylene concentration in section 2 varies. Again, the behavior is quite regular for the two highest concentrations and erratic for the weakest concentrations. Figure 9 shows the influence of using two different gases, acetylene and ethylene, in section 2. Heat of reaction is used as a basis for comparison. The best-fitted straight lines to the data for the two gases are almost overlapping,

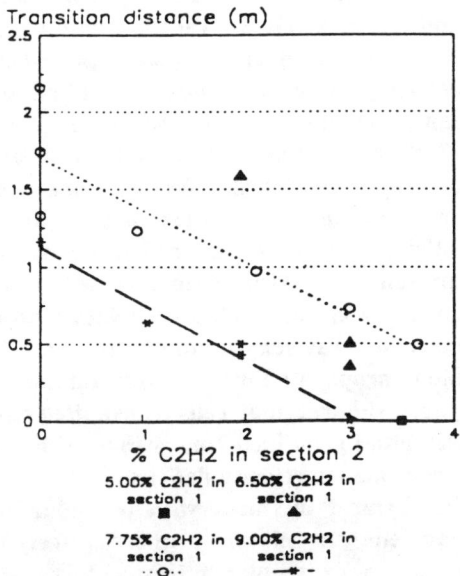

Fig. 7 Transition distance comparisons: experimental conditions: varying acetylene–air concentrations in sections 1 and 2; length of section 2 is 100 mm, 7.75% C_2H_2–air in section 3.

which indicates that the same energy release in section 2 gives about the same influence on the transition distance.

Discussion

Detonation Propagation Limits in a Square Tube

The tests gave propagation limits close to 4.00% for acetylene-air and 5.00% for ethylene-air. Mitrofanov and Soloukhin[10] found that a certain number of cells in the detonation front was required to make a detonation continue to propagate from a channel and into an unconfined cloud, depending on the geometry. If the critical diameter or channel width for detonation propagation is denoted d_C and the cell width S, Mitrofanov and Soloukhin found that $d_C \approx 10\text{-}13S$ in acetylene-oxygen mixtures. Moen et al.[9] found that $d_C = 13S$ was a good correlation as well for detonation in fuel-air mixtures in tubes. It has been found that this correlation breaks down for fuel-air mixtures with very regular cells, where d_C on the order of 20 has been reported. The critical tube diameter d_C is a property which can be reliably measured with a high degree of repeatability. This means that $d_C/13$ probably is a better measure for an average cell width than using cell widths measured from smoked foils. In fuel-air mixtures the cells are quite irregular, and it is somewhat up to the scientist to define a representative cell. In addition, cells often contain substructures which complicates the definition of a cell. In Tables 1 and 2 $d_C/13$ are included using d_C measurements by Moen et al.[9] Using the $d_C/13$ correlation, the propagation limit for acetylene corresponds to $d_C/13$ of about 130 mm, that is, of about the same size as the tube width. For ethylene the limit corresponds to $d_C/13$ of about 70 mm. These cell widths also agree with cell widths measured by Knystautas et al.[11] for the same concentrations. The present propagation limits observed were not in agreement with the calculations by Vasil'ev[2]; a much smaller cell was required for detonation propagation in the present tests. Also, comparing with tests by Moen et al.,[1] detonation did propagate in circular tubes even with a tube diameter on the order of one transverse wavelength (half the cell width). Spinning detonations were able to propagate far below this limit. Moen et al.[12] showed that detonations in gases with irregular cells transmitted easier through open tube ends and orifices into unconfined clouds than with more regular cells. Irregular cells also gave smaller velocity deficits, that is, less influence by the confining tube walls. Typical mixtures with irregular cell structure are acetylene-air and ethylene-air mixtures, whereas acetylene-air mixtures diluted with argon have very regular cell structures. Also, referring to Shepherd et al.,[13] who analyzed the spectrum of cell widths for different gases, the difference between limiting $d_C/13$ for acetylene and ethylene might be caused by a difference in cellular regularity.

Reinitiation of Detonation Across Regions of Low Reactivity

Comparing previous tests with air[5,6] and present tests with air, one observes quite good agreement, even though there is some scatter in the results. In general, the same behavior was observed, namely, transition distance decreased with increasing CJ parameters in section 1, with decreasing length of section 2, and with increasing reactivity in section 3. In addition, reinitiation was enhanced by the presence of a reactive gas in section 2, and transition distance decreased with increasing reactivity in this section. The influence of a gas mixture with low reactivity was also verified by the tests with detonation propagation into gas mixtures below the detonability limit (Fig. 4) (detonability limit: the lower limit for the gas concentration, where it is possible for a detonation wave to propagate in the given geometry). Even if a detonation fails when it propagates into section 2, the low-reactivity mixture can still burn. There are no data available for lower flammability limit (LFL) at such high temperatures. But LFL measurements by Hustad and Sønju [14] indicate that LFL can be scaled linearly from LFL at standard atmosphere to 0% at adiabatic flame temperature for LFL at standard atmosphere (LFL varies very little with pressure). It should be added that Hustad and Sønju measured LFL for gases other than the two used in the present tests. But several different gases showed this linear behavior. Considering the shock propagating into section 2, the transmitted shock strength was never below 10–15 bars overpressure.

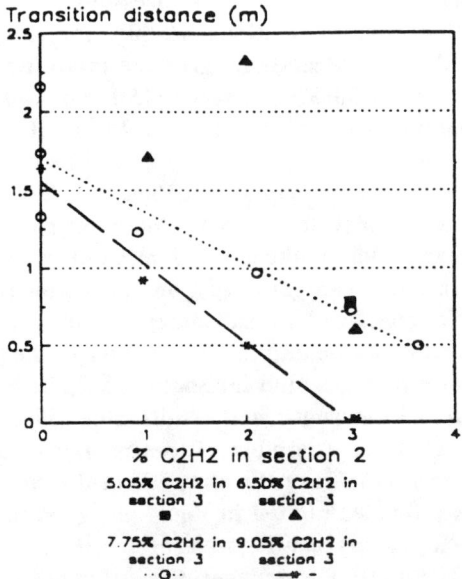

Fig. 8 Transition distance comparisons: experimental conditions: 7.75% C_2H_2-air in section 1, varying reactivities (acetylene concentrations) in sections 2 and 3.

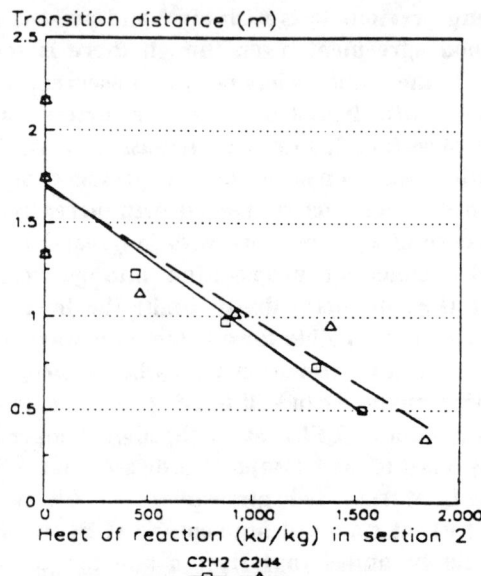

Fig. 9 Transition distance comparisons: reinitiation in 7.75% C_2H_2–air across a 100-mm gap of acetylene–air and ethylene–air mixtures below the detonability limit.

Using 15 bars as an estimate, gasdynamic tables give about 3.5 times temperature rise for a normal shock. Starting with standard temperature, 298.16 K, the temperature after shock compression is about 1044 K. The LFL for standard atmospheric conditions is 2.50% for acetylene and 2.70% for ethylene.[15] Equilibrium calculations give the following adiabatic flame temperatures: 1264 K for 2.50% C_2H_2 and 1383 K for 2.70% C_2H_4. Scaling of the lower flammability limit gives LFL = 0.4% for C_2H_2 and 0.7% for C_2H_4 at 1044 K.

The results indicate (see Fig. 4 and Figs. 7–9) that even the leanest mixtures used in the present tests would burn or partly burn. This is supported by the linear scaling of the lower flammability limits above. When a shock propagates into the lean gas in section 2, the flow behind the shock, relative to the shock front, will be subsonic; i.e., the shock will then be influenced by the rarefaction behind it. Combustion in the lean gas mixture will give a weaker rarefaction behind the shock front, the transmitted shock will not lose as much of its strength, and reinitiation will occur earlier than with inert gas in section 2. Therefore, it seems that the influence of a reactive gas in section 2 is to vary the strength of the shock traveling into section 3. The shock velocities plotted in Fig. 4 are good indications of this. The increase in shock velocity can be seen, even with the leanest mixtures in sections 2 and 3. Figure 9 compares the influence of acetylene and ethylene on the transition distance. Transition distances vs heat of reaction (equilibrium calculations) are plotted. The nearly overlapping, best-fitted lines indicate that the energy release behind the leading shock front is of importance. This energy release gives a certain pressure increase superimposed on the rarefaction behind the leading shock. This pressure increase is also dependent on the reaction rate of the gas. For the two gases used the reaction rates were likely sufficiently high, indicated by the successful scaling with heat of reaction.

Figure 7 shows an interesting behavior in the tests with the largest cell widths in section 1. In addition to the four tests with transition shown in Fig. 7, three tests with 6.50% C_2H_2-air and one with 5.00% C_2H_2-air, some more tests with similar conditions gave no reinitiation for 5.00% C_2H_2-air and reinitiation near the tube end for 6.50% C_2H_2-air. In these tests reinitiation ceased to occur, occurred late, or occurred very fast. Three of the reinitiations shown in Fig. 7, with 5.00% C_2H_2 and 6.50% C_2H_2, occurred faster than in tests with 7.75% C_2H_2-air in section 1. Tests with smaller cell widths, i.e., more reactive mixtures in section 1, gave a much more gradual variation in transition distance as the concentration in section 2 varied. This may be related to the experiments by Strehlow et al.[3] They studied the behavior of the transverse waves when a detonation propagated across a concentration gradient. In the present tests the transverse waves appeared to be of great importance. Reinitiation was likely dependent on how the transverse waves entered the detonable mixtures after section 2. In the tests with smaller cell widths the detonations probably behaved more like one-dimensional waves, whereas for larger cell widths reinitiation was likely dominated by the transverse waves.

The tests with the lowest reactivities in section 3, shown in Fig. 8, and some tests with reinitiation across 0.15-m gaps in stoichiometric acetylene-air which are not shown, gave a considerable amount of scatter in transition distance. These transition distances were large. Large transition distances gave more scatter, probably due to a greater sensitivity to perturbations at the walls as the shock and flame complex prior to reinitiation was pushed far downstream in the tube.

Early in the present tests it was observed that pressure transducers that were not accurately flush with the tube walls could trigger transition to detonation in section 3 instantaneously, showing that reinitiation is very sensitive to wall effects.

Recently an experimental study by Chue et al.[16] on fast deflagrations has been performed. Flames were accelerated in channels with rough walls or repeated obstacles. After propagating sufficiently far, the velocity of both the leading shock and the flame behind it approached a constant value. This value was approximately 0.5 times the CJ detonation velocity for the same mixtures. The flame was propagating slightly slower than the leading shock. When this velocity level was reached, the deflagration seemed to become relatively independent of the confinement, analogous to the behavior of multiheaded detonations. However, further downstream transition to detonation was often observed. Chue et al.[16] also tested with smooth tubes, where the transition to detonation was found to occur almost instantaneously.

In the present study evidence of a flame behind the leading shock prior to reinitiation, similar to the deflagrations observed by Chue et al.,[16] were made by means of fast Fourier transform (FFT) of the radar signals[8] prior to reinitiation. Figure 10 shows an example of a radar record (Fig. 10a) and its FFT (Fig. 10b), prior to reinitiation. Three echoes are seen, corresponding to the radar reflecting off three surfaces propagating with different velocities. The numbers along the X-axis in Fig. 10b are the number of periods in the time interval chosen, which in turn give the dominating frequencies in the radar record. Comparisons with pressure records and high-speed photographs, have verified that these three echoes are reflection at the contact surface, a flame following the leading shock, and the leading shock front itself. Normally the shock front should not reflect the microwaves because the shocked fuel-air mixture does not have different dielectric properties compared to the unshocked fuel-air mixture. The only explanation so far is that the radar reflected at dust particles from the tube walls dispersed into the gas by the flow behind the shock front.

Figure 11 shows velocities from FFTs for a test where reinitiation occurred downstream of transducer 5, and after the end of the record. After failure of the original detonation, the flame behind the leading shock seems to approach a constant value, whereas the shock is slightly accelerated. The

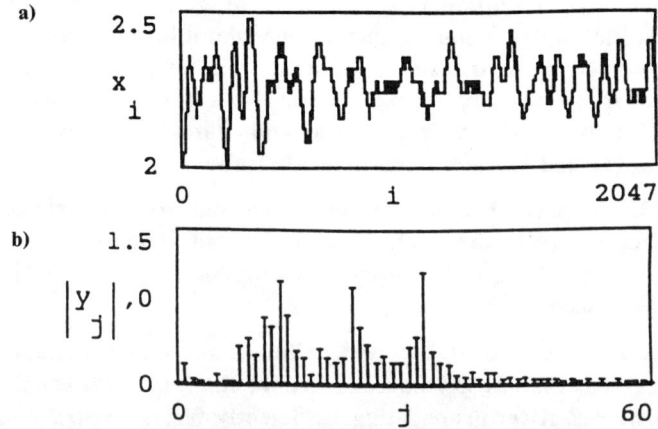

Fig. 10 a) radar record prior to reinitiation (with poor resolution due to the plotting routine included in the FFT program) and b) fast Fourier transform shows three dominating frequencies: i, sample number; x_i, value of sample in volts; j, number of periods in the chosen time interval; and $|y_j|$, absolute, real value of the Fourier amplitudes.

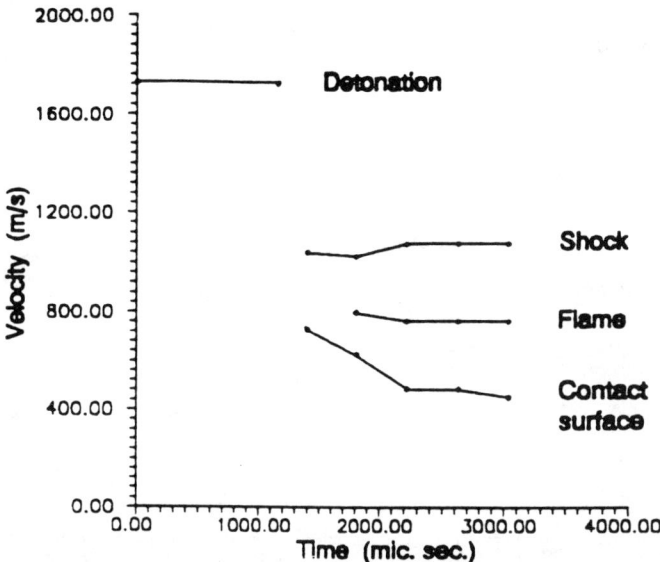

Fig. 11 Velocities from fast Fourier transforms showing development of shock and flame complex prior to reinitiation. Experimental conditions: 5.05% C_2H_2–air in section 1, air in section 2 (100 mm length), and 7.75% C_2H_2–air in section 3.

velocities of the shock and flame complex are approximately 0.55–0.6 times the corresponding CJ–detonation velocity for the shock, and 0.4 times CJ–detonation velocity for the flame. This indicates that the phenomenon is similar to the observations by Chue et al.[16] A difference in the present tests compared to the tests by Chue et al. was that expansion of the combustion products from the original detonation in the present tests pushed the shock and flame complex forward.

It was of interest to investigate the velocities of the shock and flame complex immediately prior to reinitiation for the cases where reinitiation did not occur instantaneously. The results are shown in Fig. 12. The shock velocities appear to be almost the same (on the average) prior to reinitiation, regardless of the donor detonation strengths, the lengths of section 2, or the reactivity in the gas in this region. This suggests that reinitiation under critical conditions requires a certain shock strength to occur. Chue et al. found the transition process hard to explain, because it is a transition from one extreme to another. In the present tests transition to detonation occurred as a localized explosion, associated with peak pressures ranging up to 60 bars in stoichiometric acetylene–air. Schlieren photographs showed reinitiation occurring between the leading shock and the flame, as a very luminous and localized region.[8]

The discussion of the reinitiation tests can be summarized as follows: Reaction in the region of low reactivity varied the shock strength propagating into the detonable mixture further downstream. Prior to

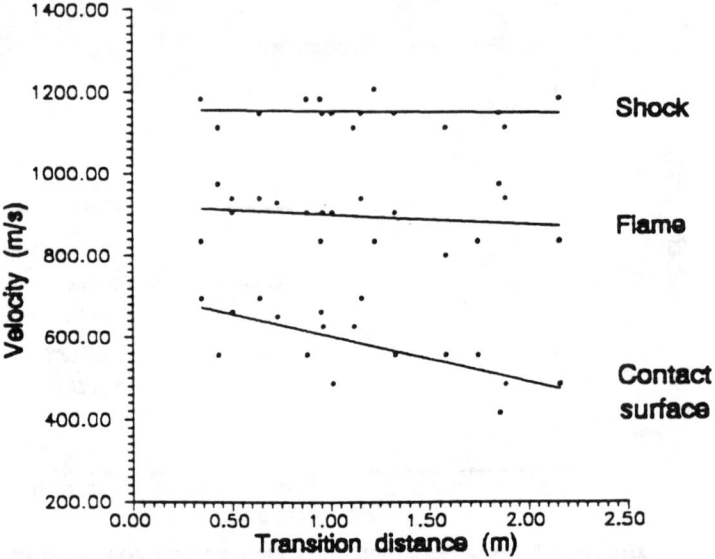

Fig. 12 Velocities from fast Fourier transforms showing velocities of shock and flame complexes immediately before reinitiation in 7.75% C_2H_2–air.

reinitiation a shock and flame complex propagated with velocities similar to velocities reported by Chue et al.,[16] however, with a larger difference between the shock and flame velocities.

Conclusions

Detonation propagation limits in a square tube did not change with varying CJ parameters in section 1. For acetylene, which is the most reactive gas used, the limit observed corresponded to a $d_C/13$ value about equal to the tube width; that is, an average cell width of about this size. Earlier tests with propagation limits in round tubes,[1] gave detonation for C_2H_2 with internal diameters smaller than one single transverse wave (half the cell width); that is, propagation limits in square and circular tubes are very different. The main reason for this is that a detonation is able to spin in a circular tube. Neither spinning nor galloping regimes were observed in the present tests in a square tube.

An important observation in the tests with detonation propagation across regions of low reactivity was that the presence of a reactive gas even at low reactivities enhanced reinitiation downstream of it. This was the case even with concentrations far below the lower flammability limit for standard atmospheric conditions. Figure 12 also shows that shock and flame velocities are fairly constant prior to reinitiation. The results show that great care has to be taken if detonable gas mixtures are to be separated by some kind of an inerting system. A fuel/air mixture, even far below the lower flammability limit for standard atmosphere, can not be treated as an inert gas.

Acknowledgments

This work was sponsored by The Norwegian Defence Construction Service (NDCS). The authors are grateful to I. O. Moen, G. O. Thomas and J. Bakken for their contribution in discussing the results, and to A. Jenssen for his support which made this work possible.

References

[1] Moen, I. O., Donato, M., Knystautas, R., and Lee, J. H., "The Influence of Confinement on the Propagation of Detonations near the Detonability Limits," Eighteenth Symposium (International) on Combustion, The Combustion Institute, Pittsburg, PA, 1981, pp. 1615–1622.

[2] Vasil'ev, A. A., "Near-limiting Regimes of Gaseous Detonation," translated from Fizika Goreniya i Vzryva, Vol. 23, No. 3, 1987, pp. 121–126.

[3] Strehlow, R. A., Adamezyk, A. A., and Stiles, R. J., "Transient Studies of Detonation Waves," Astronautica Acta, Vol. 19, 1972, pp. 509–527.

[4] Bull, D. C., Elsworth, J. E., McLeod, M. A., and Hughes, D., "Initiation of Unconfined Gas Detonations in Hydrocarbon–Air Mixtures by a Sympathetic Mechanism," Progress in Astronautics and Aeronautics, Vol. 75, American Institute of Aeronautics and Astronautics, Inc., New York, NY, 1980, pp. 61–72.

[5] Bjerketvedt, D., "Re-initiation of Detonation Across an Inert Region," Doktor Ingeniør Thesis, Norwegian Institute of Technology, Department of Mechanical Engineering, May 1985.

[6] Bjerketvedt, D., Sønju, O. K., and Moen, I. O., "The Influence of Experimental Condition on the Reinitiation of Detonation Across an Inert Region," Progress in Astronautics and Aeronautics, Vol. 106, American Institute of Aeronautics and Astronautics, Inc., New York, NY, 1985, pp. 109–130.

[7] Thomas, G. O., Sutton, P., and Edwards, D. H., "The Behavior of Detonation Waves at Concentration Gradients," Combustion and Flame, Vol. 84, Nos. 3 and 4, 1991, pp. 312–322.

[8] Engebretsen, T., "Reinitiation of Detonation Across Regions of Low Reactivity," Doktor Ingeniør Thesis, Norwegian Institute of Technology, Department of Mechanical Engineering, December 1991.

[9] Moen, I. O., Funk, J. W., Ward, S. A., Rude, G. M., and Thibault, P. A., "Detonation Length Scales for Fuel–Air Explosives," Progress in Astronautics and Aeronautics, Vol. 94, American Institute of Aeronautics and Astronautics, Inc., New York, NY, 1983, pp. 55–79.

[10] Mitrofanov, V. V., and Soloukhin, R. I., "The Diffraction of Multifront Detonation Waves," Soviet Physics–Doklady, Vol. 9, No. 12, 1965, pp. 1055–1058.

[11] Knystautas, R., Guirao, C., Lee, J. H., and Sulmistras, A., "Measurements of Cell Size in Hydrocarbon–Air Mixtures and Predictions of Critical Tube Diameter, Critical Initiation Energy, and Detonability Limits," Progress in Astronautics and Aeronautics, Vol. 94, American Institute of Aeronautics and Astronautics, Inc., New York, NY, 1984, pp. 23–37.

[12]Moen, I. O., Sulmistras, A., Thomas, G. O., Bjerketvedt, D., and Thibault, P. A., "The Influence of Cellular Regularity on the Behavior of Gaseous Detonations," Progress in Astronautics and Aeronautics, Vol. 106, American Institute of Aeronautics and Astronautics, Inc., New York, NY, 1986, pp. 220–243.

[13]Shepherd, J. E., Moen, I. O., Murray, S. B., and Thibault, P. A., "Analyses of the Cellular Structure of Detonations," Twenty-first Symposium (International) on Combustion, The Combustion Institute, Pittsburg, PA, 1986, pp. 1649–1658.

[14]Hustad, J. E., and Sønju, O. K., "Experimental Studies of Lower Flammability Limits of Gases and Mixtures of Gases at Elevated Temperatures," Combustion and Flame, Vol. 71, No. 3, 1988, pp. 283–294.

[15]Strehlow, R. A., Combustion Fundamentals, McGraw-Hill, New York, 1985, p. 373.

[16]Lee, J. H., private communication (Chue, R. S., Clarke, J. F., and Lee, J. H., "Chapman-Jouguet Deflagrations," draft paper), McGill University, Montreal, Canada, 1990.

Failure of the Classical Dynamic Parameters Relationships in Highly Regular Cellular Detonation Systems

D. Desbordes,* C. Guerraud,† L. Hamada,‡ and H. N. Presles§
Ecole National Supérieure de Mécanique et d'Aérotechnique, Poitiers, France

Abstract

Critical initiation conditions for the onset of spherically expanding detonation propagating in $C_2H_2 + 2.5\, O_2 + ZI$ mixtures, highly diluted ($Z \geq 10.5$) by a monoatomic inert gas I = He, Ar, or Kr, were determined by observing the build up to detonation produced by two direct modes of initiation. The spherical detonation was initiated in the first mode of initiation by diffracting a planar detonation from a 52-mm-i.d. tube into a larger volume, and in the second mode by exploding a wire with an energy source E. Large discrepancies between our measured values of the critical initiation parameters and those predicted by the classical empirical relationships for much mixtures ($d_c = 13\lambda_{CJ}$ and $E_c \sim \rho_0 D_{CJ}^2 \lambda_{CJ}^3$), were obtained, similar to those previously reported for mixtures heavily diluted with Ar. For large values of Z (corresponding to 75 and 81% dilutions by volume), and independent of the nature of I and the mode of initiation, the mean radius of curvature R_c of the shock front for a critically initiated wave was found to be 40-45λ_{CJ}, i.e., the shock front radius contained about twice the number of cells as generally observed ($R_c \cong 20\lambda_{CJ}$). Consequently, the critical initiation parameter for our mixture can be obtained

---

Copyright © 1993 by the American Institute of Aeronautics and Astronautics, Inc. All rights reserved.
*Professor, Laboratoire d'Energétique et de Détonique.
†Research Engineer, Laboratoire d'Energétique et de Détonique.
‡Research Scientist, Laboratoire d'Energétique et de Détonique.
§Senior Research Scientist, Laboratoire d'Energétique et de Détonique.

from the classical relationship by 1) changing the critical diameter rule to $d_c \cong 24\text{-}28\lambda_{CJ}$, and 2) increasing the critical energy by an order of magnitude ($\sim 2^3$). For mixtures with different values of Z, λ_{CJ} (Z) was found to remain proportional to the chemical induction length L_i. Consequently, λ_{CJ} (Z) varies as D_{CJ} (Z), a_{CJ} (Z) or a_0 (Z) (where a is the velocity of sound) and the self-sustained detonation wave becomes more stable and propagates at lower velocities as the weight of I is increased. The mixture with 81% of Kr, supports a very stable detonation propagating at a velocity ~ 1100 m/s which can only be obtained in other mixtures containing heavier monatomic gases such as Xe and Ra. The role of the three-dimensional structure (size, regularity) on the existence of the detonation wave in gaseous mixtures is also discussed.

1. Introduction

It is now widely recognized that direct initiation of a spherical expanding detonation wave in gases needs : 1) a minimum size of the tube diameter, $d = d_c$, when a plane detonation (CJ or overdriven) propagates in this round tube and diffracts into a large volume,[1-3] or 2) a minimum energy deposited in the medium, $E = E_c$, when a powerful point source of explosion is used.[1-4] In these two modes of initiation, at criticality, the minimum thermodynamical conditions are achieved behind the curved shock wave produced, when its radius of curvature attains a critical value $R = R_c$.[3-5] The first mode corresponds to a nonideal source of explosion because critical surface of the tube exit (or diameter d_c) depends on the strength of the diffracting planar detonation wave.[6] The second mode, when the pressure generated near the source is very high, can be considered as an ideal source of explosion.[7] Powerful exploding wires belong to that kind of sources. In both cases, locally at the beginning of the solicitation, the detonation regime exists in the gaseous mixture as 1) the planar detonation wave at the tube exit or 2) the spherical overdriven detonation wave created instantaneously in the neighbors of the explosion point source.[8]

Such detonations are completely destroyed, in critical conditions, 1) by the lateral expansion moving from the edge of the tube to the axis or 2) by the central expansion in spherical geometry, respectively. The detonation quickly turns into a decaying spherical shock wave followed by a gradually decoupling flame front. When the radius of curvature R has grown to R_c, onset of detonation suddenly occurs and the detonation from that radius maintains its propagation, and tends closely to the CJ conditions. This critical radius of the detonation front is in a general way linked with the cell width λ_{CJ} of the mixture by a constant factor of 20, (see Ref. 5). This corresponds to the source of critical conditions required, that is, 1) $d_c = 13 \lambda_{CJ}$ for the critical diameter of transmission of a CJ plane detonation and 2) $E_c \sim \rho_0 D_{CJ}^2 \lambda_{CJ}^3$ for the critical energy of the point source of explosion.

A few years ago, Moen et al.[9] showed clearly that the classical law, $d_c = 13\lambda_{CJ}$, no longer holds with large dilution by argon (75% in volume) of an explosive gaseous mixture ($C_2H_2 + 2.5O_2$). This behavior was confirmed later.[6,10]

The distinctive feature of the progressive dilution by a monoatomic inert gas of a reactive mixture lies in the increase and the maintenance of the temperature behind the ZND shock wave, except for very large dilutions (cf. Table 1). So, inversely, the reduced activation energy Ea/RT of the mixture decreases. Moreover, a large dilution with monoatomic inert gas contributes to enhance the "regularity" of the three-dimensional structure and, in fact, to favor the coupling of the cells with the transverse modes of vibration of the tube.[11] If the tube is of rectangular or square section, for instance, cells appears surprisingly very regular[12] (octahedrons, tetrahedrons), because of the modifications of shapes and sizes by boundary conditions. During its self-sustained propagation in tubes, the detonation wave is more dependent on confinement and during its direct initiation in spherical expanding geometry, more dependent on the radius of curvature of the front.

Besides Ar, observations are extented in this paper to a few monoatomic inert diluents such as He and Kr in the $C_2H_2 + 2.5O_2$ mixtures. The same dilution (Z = conste in $C_2H_2 + 2.5O_2 + Z\,I$ mixture) by such an inert gas gives the same Mach number of the CJ wave and, obviously, the same thermodynamic conditions behind the ZND and the CJ planes except the density and the sound velocity, regardless of the inert I (see Table 1). Critical conditions of direct initiation of detonation are systematically investigated for large dilutions with these three diluents. Cellular structures of self-sustained detonations at room temperature T_o and at different initial pressures p_o are recorded and measured, and critical conditions of direct initiation (critical diameter of diffraction d_c and minimum initiation energy E_c) are explored.

2. Experimental Devices

Two experimental devices are used for the study of critical initiations of detonation :

1) The first one is devoted to the measurement of the critical diameter of transmission of a plane detonation, consisting in a tube of about 4 m long and 52 mm i.d., connected with a cylindrical vessel of larger sizes, essentially described in Ref. 6.

2) The second is the cylindrical vessed used in the first device, where few experiments are conducted. The detonation can be initiated at the center of the vessel by an exploding wire explosion source of possible different characteristics. This source can be considered as an ideal source of explosion and real energy E deposited in the gas can be measured (by shock wave characteristics measurements).

3. Experimental Results

A. Detonation Velocities

Generally speaking, measured self-sustained detonation velocities are very close to the CJ values. Their magnitudes are especially low with large dilutions by krypton (cf., Fig. 1).

Table 1 ZND and CJ parameters of detonations in $C_2H_2 + 2.5\ O_2 + Z\ I$ mixtures at $p_0 = 1$ atm and $T_0 = 293$ K with different inert : He(*), Ar(+) and Kr(•).

	M	p_{ZND}/p_0	T_{ZND}, K	p_{CJ}/p_0	T_{CJ}, K	D_{CJ}, m/s	a_0, m/s	a_{CJ}, m/s
Z=0	7.416	65.701	2236	34.47	4216	2426	327.1	1317
Z=3	6.472	51.518	2369	28.10	3938.2	2831.6(*) 2046.9(+) 1627.2(•)	437.5 316.2 251.5	1551 1122 891.9
Z=9	5.752	41.25	2402	23.30	3594.4	3277.9(*) 1810.5(+) 1332.2(•)	569.8 314.7 231.6	1824 1007 739.8
Z=12	5.553	38.501	2380	21.98	3465.5	3409.6(*) 1749.0(+) 1269.6(•)	613.9 314.9 228.6	1906 978 709.4
Z=15	5.397	36.379	2350	20.88	3350.3	3506(*) 1701.1(+) 1224(•)	649.5 315.3 226.7	1972 984 688

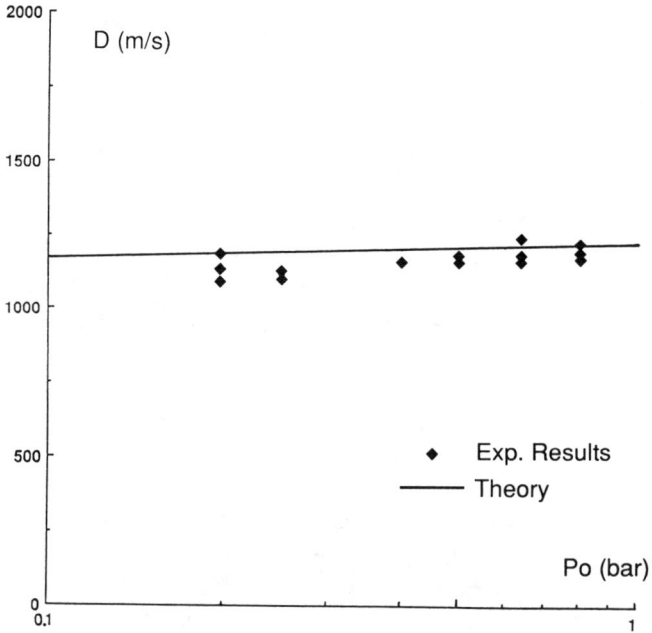

Fig. 1 Experimental and CJ detonation velocities in the $C_2H_2 + 2.5\, O_2 + 15\, Kr$ mixture at different initial pressures p_o.

B. λ_{CJ} Function of Dilution Z

For different initial pressure p_0, measurements of cell width, namely, λ_{CJ}, are displayed in Fig. 2 for $Z = 10.5$ and in Fig. 3 for $Z = 15$. The $Z = 0$ curve is given for comparison.

For any dilution the dependance of λ_{CJ} on p_0 follows the classical law:

$$\lambda_{CJ} = A\, p_0^{-1.2} \qquad (1)$$

Moreover, as it has been demonstrated for different chemical systems, λ_{CJ} varies as the induction length L_i calculated by assuming the ZND conditions of the mixture and represented in C_2H_2 / O_2 systems by

$$L_i \sim D\, \rho_0\, \rho_s^{-1}\, [O_2]^{-1}\, \exp \frac{Ea}{RT} \qquad (2)$$

where $Ea = 25$ kcal/mole. Predictions of λ_{CJ} (Z) for the three different inert gas at different pressure p_0, with a reference taken for the indiluted mixture $C_2H_2 + 2.5 O_2$ at $p_0 = 100$ Torr where $\lambda_{CJ} = 1.3$ mm, agree very well with all the experimental results (cf., Figs. 2 and 3). For the same dilution Z, the

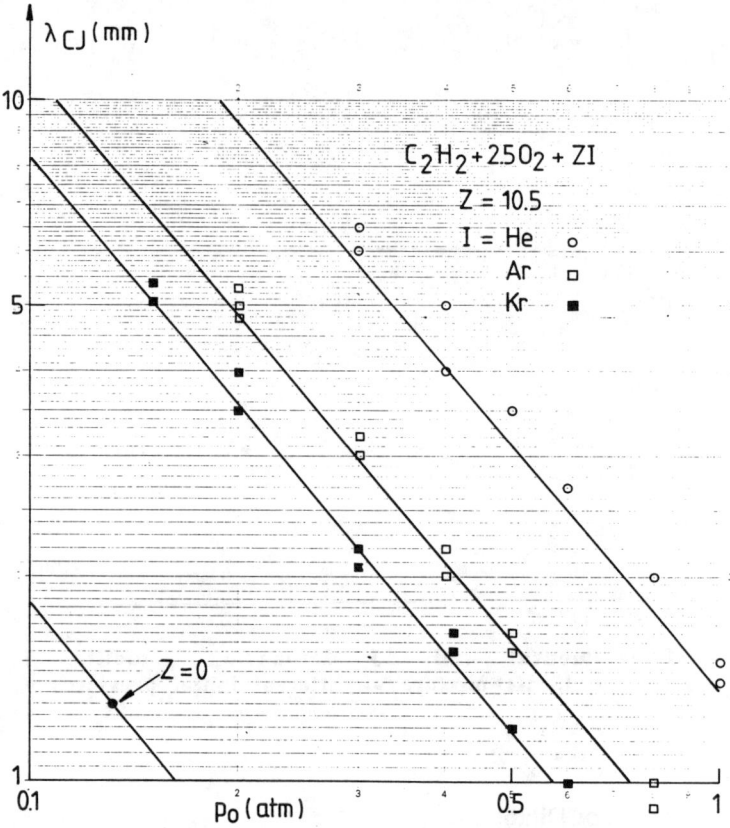

Fig. 2 Measured cell spacings λ_{CJ} in C_2H_2 + 2.5 O_2 + 10.5 I mixtures (for I = He, Ar, and Kr) at different initial pressures p_o; straight lines are drawn assuming the conservation of the ratio λ_{CJ} / L_i obtained at Z = 0, p_o = 100 Torr where λ_{CJ} = 1.3 mm.

magnitude of $\lambda_{CJ}(Z)$ depends indirectly on the inert atomic weight via the sound velocity a, as

$$\lambda_{CJ}(Z) \sim D_{CJ}(Z) \sim a_{CJ}(Z) \sim a_0(Z) \tag{3}$$

As shown in Table 1, for the same dilution Z and initial pressure p_0, D_{CJ} is inversely proportional to the square root of the initial density of the mixture. Then, the heavier the mixture (i.e., the monoatomic diluant) the lower the CJ detonation velocity is and the more stable the detonation wave.

C. Critical Diameter of Transmission d_c

For Z = 0 and as has been previously pointed out for Z up to 3.5 (Ref. 3), the classical rule $d_C = 13\lambda_{CJ}$ governs the transmission of the detonation from a

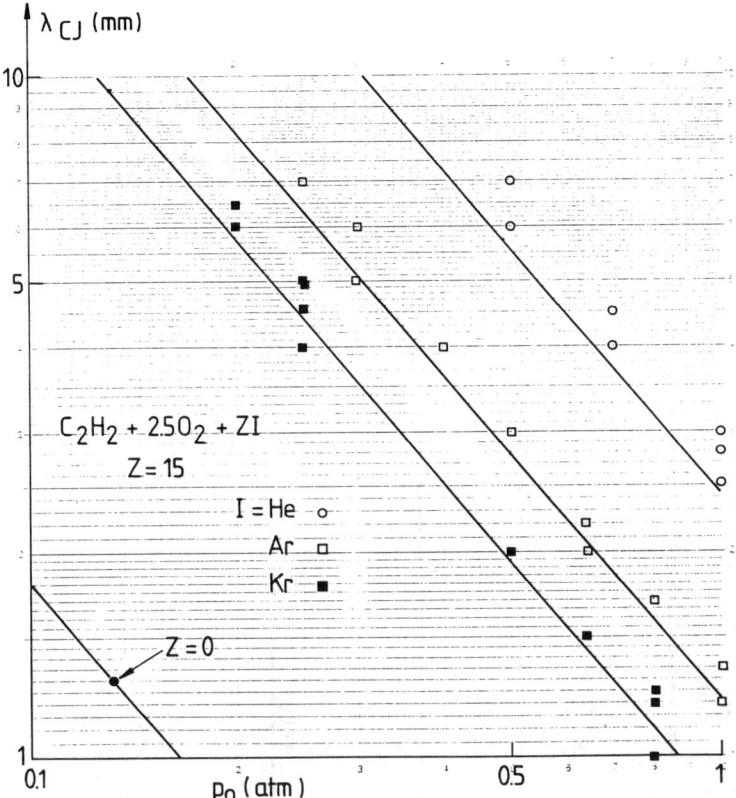

Fig. 3 Measured cell spacings λ_{CJ} in $C_2H_2 + 2.5\ O_2 + 15\ I$ mixtures (for I = He, Ar, and Kr) at different initial pressures p_o; straight lines are drawn assuming the conservation of the ratio λ_{CJ}/L_i obtained at Z = 0, p_o = 100 Torr where λ_{CJ} = 1.3 mm.

pipe into a large volume. The quasispherical predetonation region observed has a radius $R_C \cong 20\ \lambda_{CJ}$.

For Z = 10.5 and Z = 15, for the three inerts He, Ar, and Kr, the critical initial pressure p_{OC} of the mixture needed for the transmission of plane detonation is obsviously different and reported in Table 2. Also given are the ratios d_C/λ_{CJ} and R_C/λ_{CJ} observed in that mode of initiation of spherical detonations. The first ratio varies from 24 to 26 and the second from 40 to 45 approximatively, depending more on Z than on the inert.

D. Critical Energy of the Point Source E_C

Only mixtures for Z = 0 and Z = 10.5 are concerned with the direct initiation by the exploding wire source. For the same characteristics of the source, critical initial pressures p_{OC} for the successful initiation of the spherical detonation are

Table 2 Approximate critical conditions of transmission of a plane detonation in mixtures with different monatomic diluents : He, Ar and Kr

Mixture $C_2H_2 + 2.5 O_2 + ZI$	d_c, mm	p_{0c}, Torr	λ_{CJ}, mm	d_c/λ_{CJ}	R_c/λ_{CJ}
Z = 0	52	30-35	4-5	10-13	20
Z = 10.5	52	550	2.2	24	40
I = He					
Z = 15	-	1000	2	26	44
Z = 3.5	-	80	4	13	22
I = Ar					
Z = 10.5	-	325	2.2	24	40-42
Z = 15	-	530	2	26	42-45
Z = 10.5	52	220	2.2	24	40-42
I = Kr					
Z = 15	-	380	2	26	41-45

given in Table 3. This explosion source can be considered as ideal because, at low initial pressure in the $C_2H_2 + 2.5O_2$ mixtures, for instance, very fine cell structures appear with $\lambda \ll \lambda_{CJ}$ inside the sphere of radius R_* (Ref. 8). This radius defines the sphere where the chemical energy contained inside equals the energy E of the source, i.e.,

$$R_* = \left(\frac{E}{\frac{4}{3} \pi \rho_0 Q} \right)^{1/3} \quad (4)$$

The energy E of the source transferred in the medium is deduced from the decay of the amplitude of the shock wave in the function of the distance R from the center, generated by the explosion in an inert gas and is estimated here to be about 8-10 J. The observed critical predetonation radius R_c is given in Table 3 by the ratio R_c/λ_{CJ}. For $Z = 10.5$, this ratio takes the value of 40-44, and the standard value of 20 for $Z = 0$.

As it can seen, the detonability limit (initial pressure p_{oc}) depends on the weight of the inert diluent and following the general trend, the detonation is easier to obtain with the heavier diluent.

An estimation of the critical energy of initiation E_c can be given considering two models.

1) The first model is based on the experiments of Elsworth reported by Benedick et al.[13] on the initiation of spherical detonation in large-scale experiments in H_2-Air mixtures (which obey the $d_c = 13\lambda_{CJ}$ rule for successful transmission) initiated by different charges of tetryl and those of Desbordes[5] on the predetonation sphere observed systematically at critical initiation of detonation. On the hand, in the large range of equivalence ratio of the mixture, critical initiations are obtained when the explosion length R_o ($=(E/p_0)^{1/3}$) of the sphere of tetryl used equals approximatively $20\lambda_{CJ}$. In the other hand, the radius R_c of the quasi-spherical predetonation zone, for many systems and for different initiation sources (laser sparks, exploding wires, high explosives) and as observed in critical transmission of a plane detonation,[3] is included roughly 20 times the cell width. So, according to such experimental evidence

$$R_c \cong 1.6 \, d_c \cong R_o \quad (5)$$

i.e.,

$$E_{c1} = p_o R_c^3 \quad (6)$$

2) The second model is based on the observations, at criticality, that the size of the radius R_s where decoupling of the flame and the shock wave occurs during the expansion of the detonation (location where the cellular structure desappears) equals approximatively R_* (Refs. 14 and 15) and $0.2R_c$ (Ref. 8). Considering Eq. (4) and the approximate relationship $D_{CJ}^2 \cong 2(\gamma^2 - 1)Q$, the critical energy can be expressed by

$$E_{c2} = B \rho_o D_{CJ}^2 R_c^3 \quad (7)$$

Table 3 Approximate critical conditions of direct initiation by an exploding wire of $E = 8$-10 J in mixtures ($I \equiv$ He, Ar, and Kr) for $Z = 0$ and $Z = 10.5$

$C_2H_2 + 2.5 O_2 + ZI$		p_{oc}, Torr	$\rho_o{}^a$, kg/m³	D_{CJ}, m/s	λ_{CJ}, mm	R_c/λ_{CJ}	R_c/d_c	$E_{c1} = \rho_o R_c^3$, J	$E_{c2} = B\rho_o D_{CJ}^2 R_c^3$, J
$Z = 0$		35	1.26	2270	4.5	20	1.54	3.4	8.3
$Z = 10.5$	$I = $ He	1520	0.44	3360	0.7	40-44	1.6	4.4	8.3
	$I = $ Ar	760	1.565	1775	0.8	40-44	1.6	3.3	6.2
	$I = $ Kr	430	2.924	1280	1	40-44	1.6	3.6	6.6

[a] Standart values ($p_o = 1$ atm, $T_o = 293$ K).

where
$$B = \frac{4\pi}{750(\gamma^2 - 1)}, \quad (B \cong 3.8 \; 10^{-2})$$

and which gives values very close to those provided by the Vasiliev and Grigoriev[16] and Lee[17] models.

For our experiments, calculated values of E_{c1} and E_{c2} are reported in Table 3 concerning the direct initiation by exploding wire and give a reasonably good estimation of the energy of the source.

4. Discussion

The low-velocity (~ 1000 m/s) detonation wave can only be obtained with large dilution by an heavy inert of a chemical system (Q small : $D_{CJ} \sim \sqrt{Q}$) and generally the detonation regime cannot be obtained because of the too large chemical kinetic length. Especially, a monoatomic inert diluent, because of its weak heat capacity which remains constant even at high temperatures, considerably reduces the chemical length to make the detonation possible and even very stable for high dilutions.

The "dynamic parameters" of the detonation, as usually called now, d_c, R_{oc} ($\equiv E_c/p_0)^{1/3}$), and R_c, are generally linked to the cell size λ_{CJ} of the detonation structure by constant values widely recalled in this paper. With dilution by an inert as a monoatomic gas, classical correlations no longer remain valid. Nevertheless, the abovementioned correlations between R_c and d_c, on the one hand, and E_c and R_c(or d_c) on the other hand, hold. The experimental determination of the factor of proportionality k between one dynamic parameter (d_c, for instance, because the factor k is easier to obtain with a good accuracy) and λ_{CJ} is necessary to get the quantification of the others.

Increasing the dilution by a monoatomic inert from 0 and up to 81% in volume in the C_2H_2/O_2 stoichiometric mixture, the factor k (from $d_c = k\lambda_{CJ}$) changes from classical value of 13 to 26 which corresponds to an increase of the amplitude of the critical energy of initiation expected by an order of magnitude. On the one hand, the cell size λ_{CJ} varies in those reactive mixtures as for many others, like the one-dimensional chemical length behind the ZND shock wave ; and on the other hand, the direct initiation shows that the onset of the curved detonation needs a smaller curvature of the front in comparison with the cell size scale than generally observed (which represents, when, $R = R_c$, the competition at criticality between divergence and chemical production). All of these observations demonstrate that the cellular structure observed with large dilution by a monoatomic inert, which are surprisingly regular, represents a special case. The observations, furthermore, contribute to the emphasis of the role played by the transversal structure on the existence of the phenomena called detonation. Indeed, concerning the self-sustained propagation, these detonations are more dependant on the boundaries[9], and not as resistant to large curvature of the front. In general, from an initiation point of view, the near CJ multiheaded detonation wave cannot be simply reduced to the characteristic size of its cellular structure as demonstrated here. The intimate mechanisms that govern the cell and generally the onset of instabilities (main and higher frequencies) in the detonation phenomena and the role played by boundaries still are not yet really understood.

5. Conclusion

Critical direct initiation of spherical gaseous detonations in $C_2H_2 + 2.5O_2 + Z\,I$ systems has been investigated. Dilutions from $Z = 0$ up to $Z = 15$ by three different monoatomic inerts I of different atomic number, namely helium, argon and krypton are carried out. Spherical detonations are produced by two means: 1) transmission into a large volume of a plane CJ detonation wave propagating in a **d** i.d. tube or 2) explosion of a platinum wire considered as a point source of energy E. As dilution increases, and independent of the inert diluent, classical correlations generally observed between the dynamic parameters of the detonation and the cell size ($R_c = 20\,\lambda_{CJ}$, $d_c = 13\lambda_{CJ}$ and $E_c \sim \rho_0\,D_{CJ}^2\,\lambda_{CJ}^3$) vary substantially; particularly the following has been observed :

1) ratios d_c/λ_{CJ} and R_c/λ_{CJ} start from classical values for $Z = 0$ (up to 3.5) and grow monotonically with Z the same way and attain about twofold the classical value for $10.5 \leq Z \leq 15$.

2) Critical initiation energy E_c varies as R_c^3 (or d_c^3), taking into account the modification of the ratio R_c/λ_{CJ} (or d_c/λ_{CJ}) with Z.

3) Cell size λ_{CJ} of the mixture remains proportionnal to the chemical induction length, and therefore for the same dilution Z varies only like the CJ detonation velocity or the sound velocity of the initial mixture. As a consequence, the heavier is the monoatomic diluent, the more stable and the lower is the velocity of the self-sustained detonation wave.

A submillimetric cell size and near 1100 m/s quasi-CJ detonation is observed in stoichiometric C_2H_2/O_2 mixtures diluted by 81% of krypton (atomic weight of 83.8g) at ambiant conditions.

Very stable and sub-1000 m/s velocity detonation waves could be obtained with a very heavy monoatomic diluent such as Xe or Rd.

References

[1] Zeldovich, Y. B., Kogarko, S. M., and Simonov, N. M., "Etude expérimentale de la détonation sphérique dans les gaz," Z.E.T.P., Vol. 26, 1956, pp. 1744-1772.

[2] Matsui, H., and Lee, J. H., "On the Measure of the Relative Detonation Hazards of Gaseous Fuel Oxygen and Air Mixtures," 17th Symposium (International) on Combustion, The Combustion Institute, Pittsburgh, PA, 1979, pp. 1269-1280.

[3] Desbordes, D., and Vachon M. ,"Critical Diameter of Diffraction for Strong Plane Detonation," Progress in Astronautics and Aeronautics, Vol. 106, AIAA, New York, 1986, pp. 131-143.

[4] Knystautas, R., Guirao, C., Lee J. H., and Sulmistras, A., "Measurements of Cell Size in Hydrocarbon - Air Mixtures and Prediction of Critical Tube Diameter,

Critical Initiation Energy, and Detonability limits," Progress in Astronautics and Aeronautics, Vol. 94, AIAA, New York, 1984, pp. 23-37.

[5]Desbordes, D., "Correlation Between Shock-Flame Predetonation Zone Size and Cell Spacing in Critically Initiated Spherical Detonations," Progress in Astronautics and Aeronautics, Vol. 106, AIAA, New York, 1986, pp. 166-180.

[6]Desbordes, D., "Transmission of Overdriven Plane Detonations : Critical Diameter as a Functions of Cell Regularity and Size," Progress in Astronautics and Aeronautics, Vol. 114, AIAA, New York, 1988, pp.170-185.

[7]Baker, W. E., Cox, P. A., Westine, P. S., Kulesz, J. J. and Strehlow, R.A., Explosion Hazards and Evaluation, Elsevier Scientific, 1983.

[8]Desbordes, D., "Aspects stationnaires et transitoires de la détonation dans les gaz: relation avec la structure cellulaire du front," Thèse d'Etat de l'Université de Poitiers, France, July 1990.

[9]Moen, I. O., Sulmistras A., Thomas, G. O., Bjerketvedt, D. and Thibault P. A., "Influence of Cellular Regularity on the Behavior of Gaseous Detonations", Progress in Astronautics and Aeronautics, Vol. 106, AIAA, New York, 1986, pp. 220-243.

[10]Desbordes, D., Brisot, D., and Guerraud, C.,"Taille et régularité de la structure du front de la détonation dans les gaz. Corrélation avec le diamètre critique de transmission de la détonation," Annales de Physique, Paris, Vol. 14, 1989, pp. 629-635

[11]Manson, N. ,"Propagation des Détonations et des Déflagrations dans les Mélanges Gazeux," Edition ONERA IFP, Paris, 1947.

[12]Strehlow, R. A., "Multi-dimensional Detonation Wave Structure" Astro-Acta, Vol. 15, 1970, pp. 345-357.

[13]Benedick, W. B., Guirao, C. M., Knystauras, R., and Lee J. H., "Critical Charge for the Direct Initiation of Detonation in Gaseous Fuel-Air Mixtures," Progress in Astronautics and Aeronautics, Vol. 106, 1986, pp. 181-202.

[14]Bach, G.G., Knystautas, R., and Lee, J. H. , "Direct Initiation of Spherical Detonations in Gaseous Explosives" 12th Symposium (International) on Combustion. The Combustion Institute, Pittsburgh, PA, 1969, p. 853.

[15]Korobeinikov, V. P., Levin, V. A., Markov, V. V. and Chernyi, G.G., "Propagation of Blast Waves in a Combustible gas," Acta-Astronautica, Vol. 17, 1972, pp. 529-537.

[16]Vasiliev, A. A., and Grigoriev, V. V. , "Critical Conditions for Gas Detonation in Sharply Expanding Channels," F.G.I.V., Vol. 16, 1980, pp. 117-125.

[17]Lee, J. H., "Dynamic Parameters of Gaseous Detonations" Annual Review of Fluid Mechanics, Vol. 16, 1984, pp. 311-336.

Chapter IV.   Nonideal Detonations and Boundary Effects

Mechanisms of Detonation Propagation in a Porous Medium

A. Makris,* A. Papyrin,† M. Kamel,* G. Kilambi,‡ J. H. S. Lee,§ and R. Knystautas§
McGill University, Montreal, Quebec, Canada

Abstract

The mechanism of detonation propagation in two-dimensional porous media has been studied. Short cylindrical obstacles are used instead of three-dimensional particles to simulate the two-dimensional porous medium. Two thin rectangular channels (cross section of 3 × 127 mm and 25 × 101 mm) with optical windows are used for the experiment. Mixtures of $C_2H_2-O_2$, stoichiometric $C_3H_8-O_2$ and $C_2H_2-O_2$ diluted with argon are used. The range of initial pressures are $10 \leq p_o \leq 250$ Torr. High-speed framing Schlieren and open shutter photography are used to observe the detonation phenomena in the two-dimensional porous medium while streak photography or light detecting probes are used to monitor the propagation velocity. It is found that when the detonation cell size λ is very small compared to the void dimensions and obstacle size, measured propagation velocities approach the C-J values. However, when the cell size is of the order of the void dimension, the propagation mechanism is quite similar to that in very rough tubes.[1,2] The detonation

---

Copyright © 1992 by the American Institute of Aeronautics and Astronautics, Inc. All rights reserved.
* Graduate Student, Department of Mechanical Engineering.
† Professor, On sabbatical leave from the USSR Academy of Sciences, Institute of Pure and Applied Mechanics, Novosibirsk, USSR.
‡ Undergraduate Student, Department of Mechanical Engineering.
§ Professor, Department of Mechanical Engineering.

is attenuated by repeated diffractions around obstacles and reinitiated by shock reflections, resulting in a velocity much less than the C-J value. Similar to the rough tube case, the frequency of failure and reinitiation effects the measured propagation velocity. For the same mixture, higher velocities are measured in more porous arrangements of cylindrical obstacles compared to less porous ones. Furthermore, there are indications that cell regularity of the mixture can effect the velocity deficit experienced. Present results for the dependence of velocity deficits as measured by V/V_{CJ}, correlate well with $\lambda/$(characteristic length × porosity) when compared with other studies in porous media[3,4] and rough tubes.[1,2] Because of similarities in propagation mechanisms from photographic evidence and velocity deficit dependence with mixture sensitivity and geometrical characteristics, it is strongly suggested that detonations in porous media are only an extension of quasidetonations and detonations in rough tubes.

1. Introduction

The study of the propagation of gaseous detonation in porous media was initiated at McGill[5] in 1978 to explore the feasibility of bitumen extraction from oil sands by direct coking. Perhaps the most "surprising" result from the investigation was that for atmospheric pressure $C_2H_2-O_2$ mixtures, steady-state waves with velocities ranging from about 700 m/s to the full Chapman-Jouguet velocity of 2425 m/s were observed in a fine sand bed, with grain diameter of the order of 0.5 mm without being quenched. Subsequent studies were carried out by Kauffman et al.[6,7] where a wider range of fuels (propane, methane, and hydrogen) with pure oxygen and initial pressures 1-9 atm were used. Steel and ceramic spheres of diameters ranging from 19 to 38 mm were used in the porous bed. Similar results were obtained in that steady-state waves with velocity ranging from 0.4 times the C-J velocity to approximately the C-J velocity itself were observed. The two very different materials of the spheres which comprised the porous bed did not seem to have an influence on the detonation propagation, indicating that heat and momentum losses to the porous medium may not be the dominant mechanism to account for the velocity deficit.

Perhaps the most extensive study of detonations in porous media to date is by researchers at Novosibirsk.[3,4,8-10] They have investigated a range of fuel-air and fuel-oxygen mixtures (C_2H_2, H_2, C_3H_8)

at initial pressure ranging from 10^{-2} to 5 MPa. For the porous media, fine quartz sand with grain size 0.06 - 2.5 mm and larger grains including steel spheres with sizes ranging from 2 to 12 mm were used. They observed quasisteady velocities of different ranges with sharp transition from one regime to the other, similar to those reported by Lee et al.[11] for detonation propagation in rough tubes (i.e., quenching regime, turbulent flames, choking regime, quasidetonation, and C-J detonation). The experimental results indicate a strong similarity between detonation in porous media and detonation in rough tubes. In fact, the detailed photographic studies of Lyamin et al.[10] of a one-dimensional analog of a porous medium (i.e., a number of chambers interconnected by a narrow channel in a linear array) indicate that the propagation mechanism is almost identical to that of detonation in rough tubes. The study of Abdullin et al.[12] of the combustion in a linear system of interconnected vessels also yields results of a similar nature. Thus far, these investigations suggested the possibility of quasisteady wave propagation in a porous medium. From the studies in Refs. 3,4 and 8-10, it was demonstrated that the wave velocity increases with particle size, porosity, and mixture sensitivity. At high initial pressures, where the detonation cells are very small compared to the pore size, the propagation velocity becomes independent of the porosity and particle size.[9] In a recent study by Thomas et al.[13] detonations were able to propagate within tightly packed metal foil porous structures, even when the interfoil spacing was a fraction of the cell width. Despite the availability of experimental data, no meaningful correlations have been attempted to link the sensitivity of the explosive mixture to the characteristics of the porous medium itself.

Theoretically, a "random walk" model was first proposed by Donato et al.[5] to attempt to explain the observed velocity deficits, although no correlation of the detonation velocity with the characteristics of the mixture and porous medium was made. Later, a quasi-one-dimensional theoretical model based on heat and momentum losses in the reaction zone was proposed by Kauffman et al.[6,7] to describe the phenomenon. In a related study, Ikeh[14] transformed the porous medium into an equivalent array of capillary tubes and expanded the boundary-layer properties in terms of a small dimensionless loss parameter, to predict the detonation velocity. This model is limited in its application and does not consider flow in directions normal to the axis of capillary tubes, thus it does not account for the three-dimensional nature of the actual propagation phenomenon. It is well known from the study of quasidetonations in rough tubes[1,2] that energy and momentum losses do not provide the mechanisms to account for the observed velocity deficits. It is found that the frequency of detonation failure by diffraction past the obstacles and reinitiation by shock reflections is responsible for the velocity deficit in rough tubes. Experiments suggest that a porous medium should have a similar behavior. Thus, theoretical models based solely on heat and momentum losses may not describe the true mechanism of propagation in porous media.

Although the studies of detonation propagation in rough tubes have suggested that a similar phenomenon may occur in porous media, nevertheless, the influence of the randomness arrangement of the obstacles (as is the case for porous media) and its effect on the structure of the detonation front has not been investigated. The present study attempts to elucidate the fundamental mechanisms of propagation of a detonation in a porous medium. A two-dimensional porous medium is used to permit photographic observation of the wave obstacle interaction phenomena. The influence of the mixture sensitivity (i.e., cell size) as well as regularity and the geometrical properties of the porous medium is also investigated.

2. Experimental Details

The experiments for the present investigation were performed in a channel of rectangular cross section with inner dimensions of 1085 mm length, 101 mm width, and 25.4 mm thickness. Opposing side walls of the rectangular channel are made of glass to permit photographic observations. Short cylindrical obstacles with both diameter and length of 25.4 mm are used to produce the two-dimensional porous medium. Different arrangements of the cylinders determine the particular geometries (spacing and porosity) of the porous media to be tested. Fuel mixtures tested include stoichiometric and equimolar $C_2H_2-O_2$, stoichiometric $C_3H_8-O_2$, and stoichiometric $C_2H_2-O_2$ diluted with 75% argon. A schematic of the experimental setup and the relevant geometrical parameters of the two-dimensional porous media is provided in Fig. 1.

All tests are carried out in the range of initial pressures between 10 and 250 Torr. Direct initiation of detonation in the mixture at one end of the channel is effected by a high-energy spark (10 or 100-μF capacitor charged to 4 kV). The detonation propagates a distance of

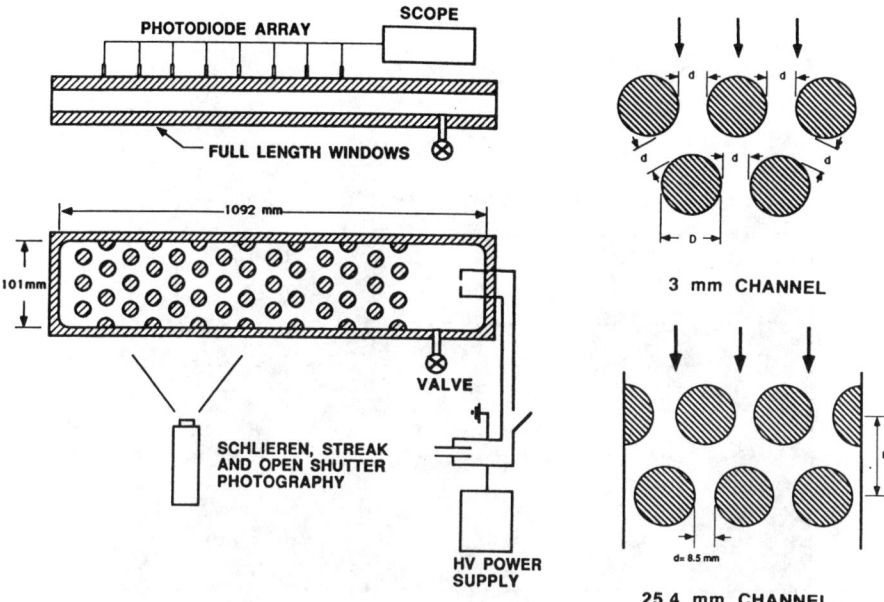

Fig. 1: Schematic of experimental setup and relevant geometric parameters of two-dimensional porous media.

at least 200 mm from the point of ignition prior to entering the first column of obstacles in the porous medium, to permit a relatively planar detonation wave to develop.

High-speed framing Schlieren photographs were taken using a Barr and Stroud camera. A stroboscopic laser Schlieren system was also used with a much shorter exposure time of 30 ns to facilitate the observation of the detailed structure of the detonation propagation in the porous media. Both Schlieren systems were of the double-pass type. The detonation velocity was monitored using an array of eight phototransistor detectors, connected in a four-channel computer scope with a total sampling capability of 1 MHz. A thinner 3-mm rectangular channel was also used for the open shutter photography of the transverse wave pattern as the detonation propagates in the two-dimensional porous medium. Detonation velocities in this 3-mm-thick channel were also measured using streak photography.

3. Results and Discussion

The propagation of a detonation wave in the simulated two-dimensional porous medium is illustrated in Fig. 2. Similar to a

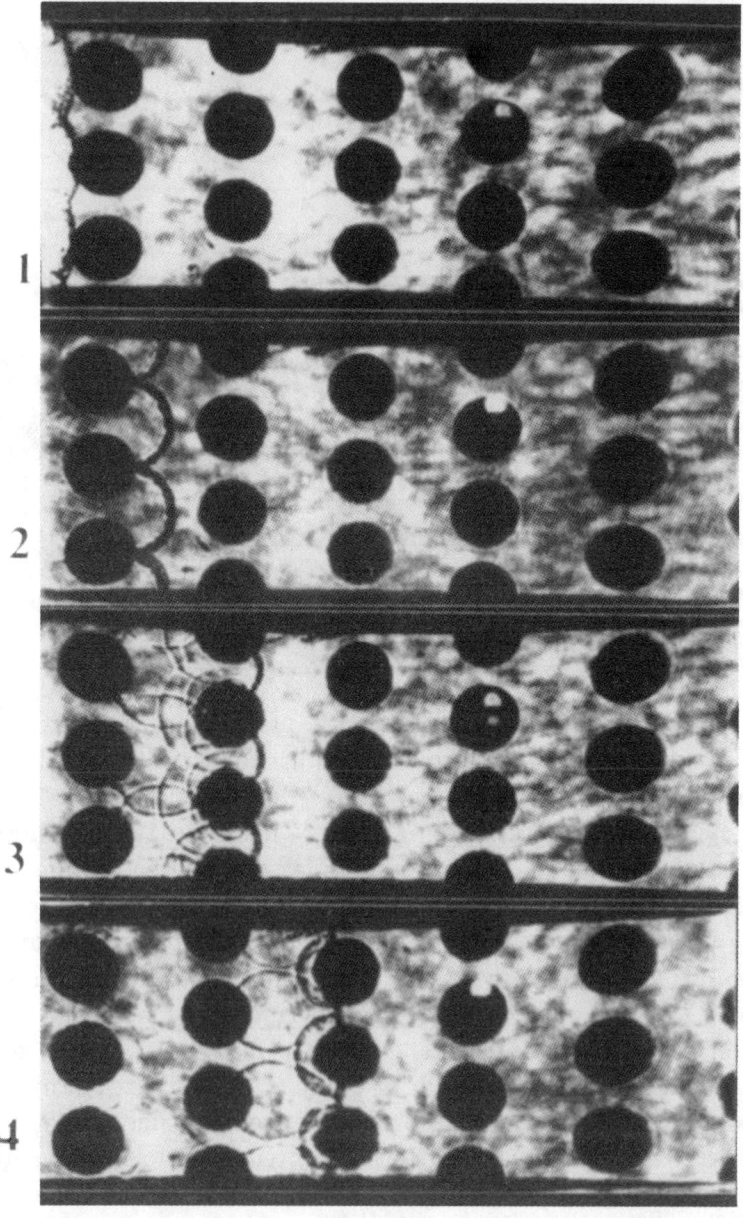

Fig. 2: Stroboscopic laser Schlieren photographs of detonation propagation in a two-dimensional porous medium, in $C_2H_2 + 2.5$ O_2 at $P_0 = 40$ Torr; $\Delta t = 18$ μs, $d = 8.5$ mm, $D = 25.4$ mm, $L = 50.8$ mm, porosity = 0.71.

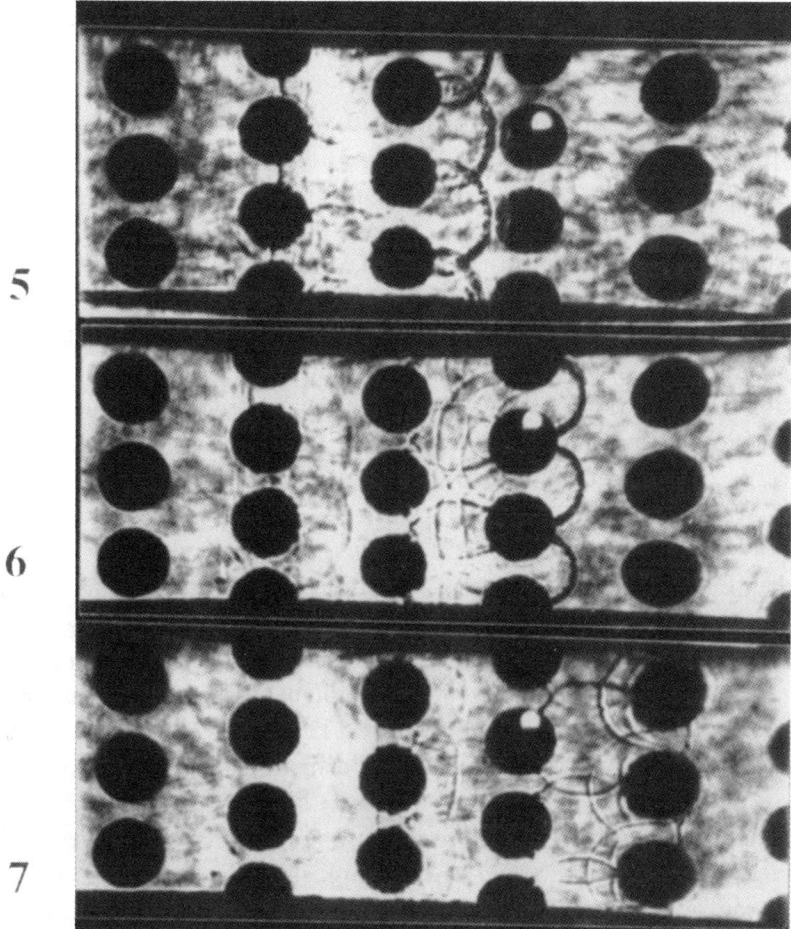

Fig. 2: (continued) Stroboscopic laser Schlieren photographs of detonation propagation in a two-dimensional porous medium, in $C_2H_2 + 2.5\ O_2$ at $P_0 = 40$ Torr; D = 18 m, d = 8.5 mm, D = 25.4 mm, L = 50.8 mm, porosity = .71.

critical tube, the waves diffract as they emerge from the openings between particles (frames 2,6). As the diffracted waves collide in the wake of each vertical column of particles, regular and then Mach reflections will occur which may reinitiate the wave and allow it to propagate farther. The repeated interaction of the detonation wave with the particles of the porous medium, even in the present two-dimensional array of obstacles of an artificial porous medium, is very complex. To elucidate the wave particle interaction processes (in particular, the diffraction and reinitiation mechanisms), a photographic study was undertaken using one obstacle and progressively increasing the complexity by adding one or more obstacles so as to generate different obstacle configurations. To avoid near-limit conditions, the initial pressures of the mixtures studied were chosen such that $\lambda \leq t \leq 1.5\lambda$, where t is the channel thickness and λ the detonation cell size of the mixture.

Figure 3a shows the interaction of a planar detonation wave with a single cylindrical particle. As the detonation wave propagates past the forward stagnation half of the cylinder, a reflected shock is generated and propagates back into the products. The part of the detonation wave near the cylinder first undergoes regular reflection and later Mach reflections when the incident angle increases. The detonation is overdriven at the Mach stem. When the detonation wave propagates past the downstream half of the cylinder, expansion waves are now generated which attenuate the overdriven detonation of the Mach stem, causing it to curve. A slight thickening of the reaction zone can now be observed in frame 13. As the diffracted waves collide at the rear stagnation point, a regular reflection is first obtained and later changed into a Mach reflection, when the colliding wave angle is greater than the critical value (frame 17). The reactive Mach stem which is formed is stronger than the diffracted incident wave, thus the detonative Mach stem is overdriven, as indicated by a finer cellular structure and a thinner reaction zone (frame 17 onward). The Mach stem eventually grows to encompass the incident wave and a planar detonation is restored across the entire channel. In Fig. 3b, the rear half of the cylinder is cut away to produce a more severe centered expansion fan. However, a similar strong reactive Mach stem is formed in the wake by frame 13, and a planar detonation wave is eventually restored.

The interaction of a detonation wave with an obstacle is governed by the ratio λ/D, where D is the diameter of the cylindrical obstacle. When the cell size $\lambda << D$, the wave can be considered as a discontinuity and will not be influenced by expansion. Ong[15] analyzed theoretically this problem of regular and Mach reflection of a thin detonation. When $\lambda >> D$, the obstacle simply perturbs the inner structure of a detonation cell and does not influence globally the propagation of the detonation wave. It is also evident that there is a range of λ/D where the expansion attenuates the diffracted wave to such a degree that when the waves do collide at the rear stagnation point of the cylinder, they fail to produce a Mach stem strong enough to reinitiate the detonation. The attenuation of the diffracted shock along the wall normal to the direction of propagation is given by Edwards et al.[16]

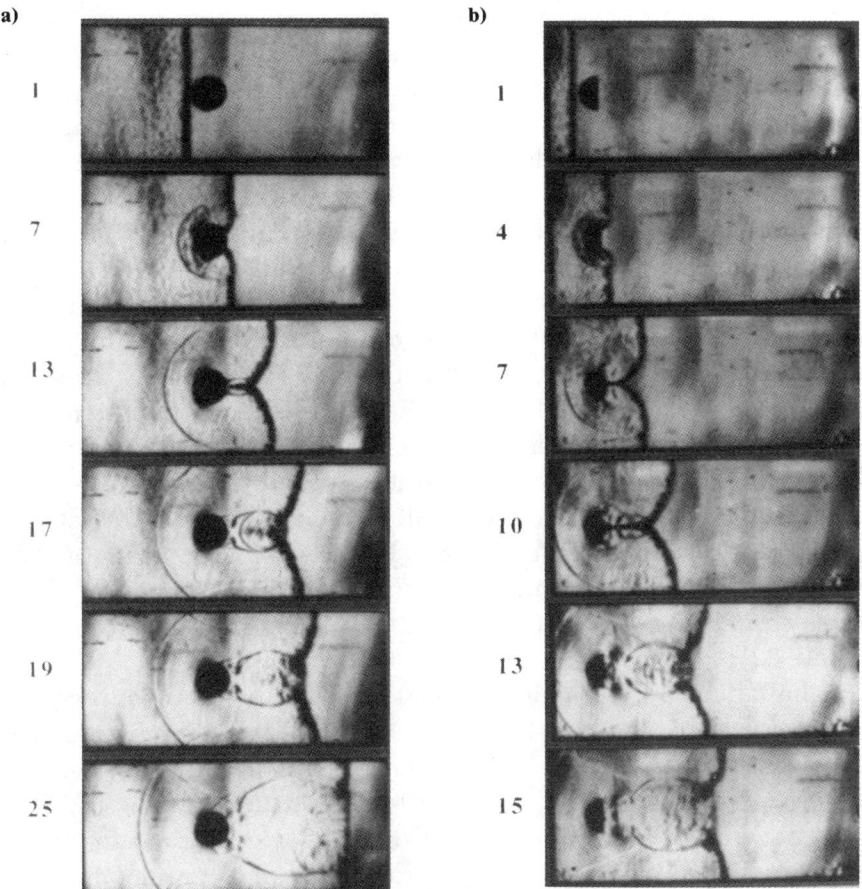

Fig. 3: Interaction of detonation with a) a single cylindrical obstacle in $C_3H_8 + 5\ O_2$ at $P_0 = 58$ Torr; $\Delta t = 1.64$ µs, $D = 25.4$ mm, $D/\lambda = 1.54$; and b) a semicylindrical obstacle in $C_3H_8 + 5\ O_2$ at $P_0 = 58$ Torr; $\Delta t = 3.64$ µs, $D = 25.4$ mm, $D/\lambda = 1.54$.

Consider next the configuration of a vertical column of semi-cylindrical obstacles of diameter D separated by a space d. From previous studies of the critical tube diameter problem,[16,17] we know that when $d/\lambda < 10$, the wave fails to self-initiate at the centerline of the channel. However, reinitiation of the detonation can occur via the collisions of the diffracted waves at the wake of each obstacle. If the diffracted waves at the time of collision are strong enough, then the Mach stems form an overdriven detonation which will eventually sweep out and reinitiate the entire wave front. However, if the mach stem formed is too weak to form a sufficiently overdriven detonation, then reinitiation of the entire wave will not occur. This is illustrated in Fig. 4a where $\lambda/d = 0.5$ and $D/\lambda = 1.5$, and the wave fails. The decoupling of the shock from the reaction front can clearly be observed when the detonation emerges from the openings (frame 7). Mach stems are formed by the diffracted waves at the rear stagnation point of the semi cylinders. These Mach stems are initially stronger than the incident waves and, hence, propagate faster to catch up with the diffracted detonation (frames 10 and 13). However, the Mach stems eventually decay and fail to reinitiate the entire detonation across the channel (frame 18). In Fig. 4b, the spacing d that separates the obstacles is increased so that $d/\lambda = 1.5$ and the Mach stems (frame 8) sweep out and eventually reinitiate the entire wave (frame 22).

The failure by diffraction through the opening between two obstacles (when $d/\lambda < 10$) and reinitiation by the overdriven Mach stems at the wake of the obstacles can be observed from the cellular pattern, illustrated in the open shutter photographs of Fig. 5. It is well known that for $C_2H_2-O_2$ mixture, the trajectories of the triple points can readily be recorded by open shutter photography in a thin channel. Since cellular structure is inherent in detonative combustion, the open shutter photographs reveal the relative fractions of detonative versus deflagrative combustion as the wave propagates in the porous medium (deflagrative combustion is indicated by the absence of cellular structure). For example, if $d/\lambda > 10$ and detonative combustion occurs throughout (Fig. 5a), cellular structure can be observed everywhere. The velocity in this case is essentially the C-J velocity of the mixture. For decreasing sensitivity (Fig. 5b), due to failure by diffraction, regions of deflagrative combustion appear and this results in a decreased overall burning rate. The average velocity for this case is $\frac{V}{V_{CJ}} \simeq 0.75$.

DETONATION IN POROUS MEDIA

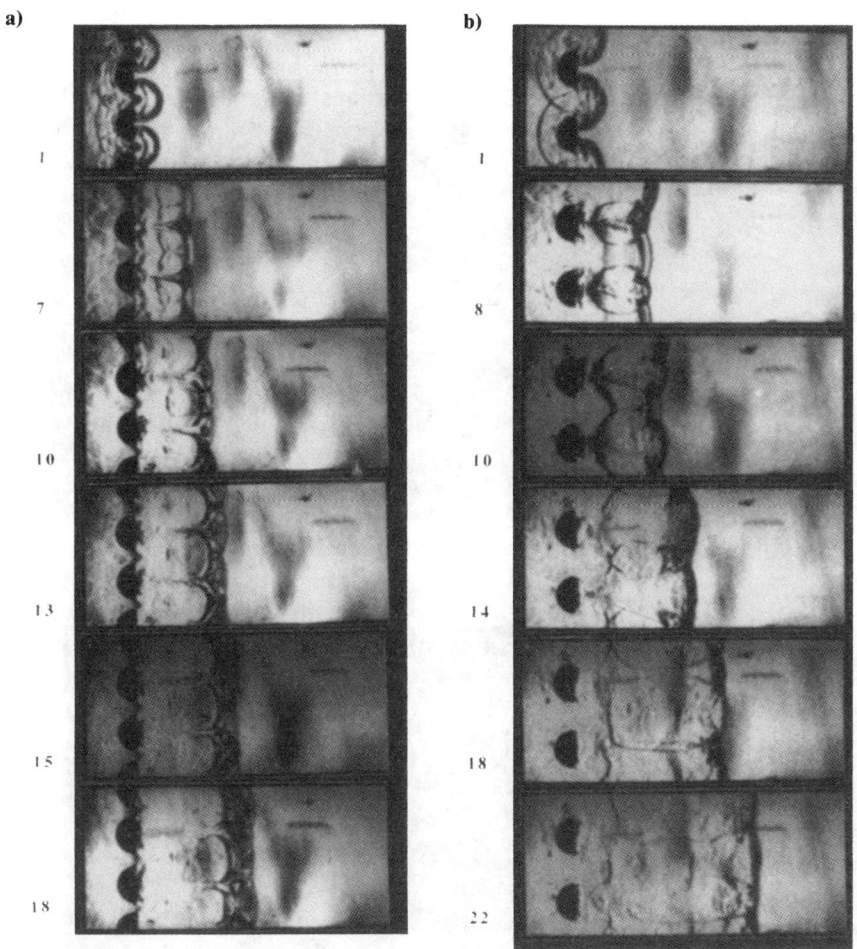

Fig. 4: a) Failure of detonation wave after passage across a single column of semicylindrical particles with a gap $d = 8.5$ mm, in $(C_2H_2 + 2.5 \, O_2)/75\%$ Ar at $P_0 = 56$ Torr, $\Delta t = 3.56$ μs, $d/\lambda = 0.5$, $D/\lambda = 1.5$; and b) reinitiation of detonation wave after passage across a single column of semicylindrical particles with a gap $d = 25.4$ mm, in $(C_2H_2 + 2.5 \, O_2)/75\%$ Ar at $P_0 = 56$ Torr, $\Delta t = 3.63$ μs, $d/\lambda = 1.5$, $D/\lambda = 1.5$.

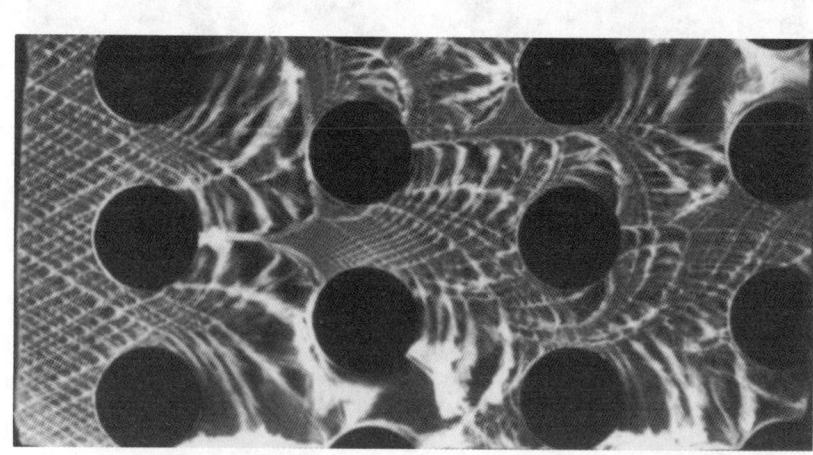

Fig. 5: Open shutter photographs from 3 mm thick channel, in $C_2H_2 + O_2$ mixture: a) $P_0 = 40$ Torr, $D = 31.75$ mm, and b) $P_0 = 26$ Torr, $D = 31.75$ mm.

Figure 6 shows the dependence of the average detonation velocity with the sensitivity of the mixture. Generally speaking, the detonation velocity is independent on cell size and is dependent only on the energetics of the mixture. However, from previous discussions we see that the dimension of the cell size relative to the various characteristic length scales of the porous medium are important in controlling the failure and reinitiation of the detonation (e.g., the relative frequency of failure and reinitiation controls the average velocity). For $C_2H_2 - O_2$ mixture, we note from Fig. 6a that V/V_{CJ} drops sharply at some critical value of the cell size, $\lambda \simeq 4$ mm. This is close to the thickness of the channel of 3 mm. The abrupt drop in the velocity corresponds to near-limit conditions in accord with the studies of Vasiliev et al.,[18] who found that limit conditions occur when the cell size λ is of the order of the channel thickness. For the case of propane mixture, it is noted that no abrupt drop in the detonation velocity occurs even for $\lambda \gtrsim t$. This could be due to the effect of cell regularity as demonstrated previously by Moen et al.[19] and Dupré et al.[20] Figure 6 also shows that for the same cell size (sensitivity), the average velocity V/V_{CJ} is higher for larger obstacle spacing d, or L. This is in agreement with the observation of Teodorczyk et al.[1,2] who found that the velocity of quasidetonation is higher when the obstacle spacing is larger. The duration in which the detonation propagates at its C-J velocity depends on the distance between obstacles, while failure by diffraction depends on the obstacles per unit length along the direction of propagation. Furthermore, when d/λ is increased by increasing the mixture sensitivity, the detonation can propagate with less failure and eventually V/V_{CJ} will approach unity when $d/\lambda \to \infty$. For a wider channel of 25 mm (Fig. 6b) we note that no abrupt decrease in the velocity occurs at a cell size of about 4 mm, thereby confirming Vasiliev's criterion of limit when $\lambda \simeq t$. For propane mixtures, the velocity deficits were much less than that for highly diluted mixtures of $C_2H_2 + 2.5\ O_2$ with argon (i.e., 75%). This again points to the effect of cell regularity on velocity deficit as demonstrated by Moen et al.[19] and Laberge et al.[21]

From the evidence provided, it is clear that the propagation mechanism of a detonation wave in two-dimensional porous media is essentially identical to that of a quasidetonation in rough tubes.[1,2] It is the frequency of failure and reinitiation determined by the obstacle configurations which accounts for the velocity deficits measured. Studies conducted in actual three-dimensional porous media[3,4,8–10]

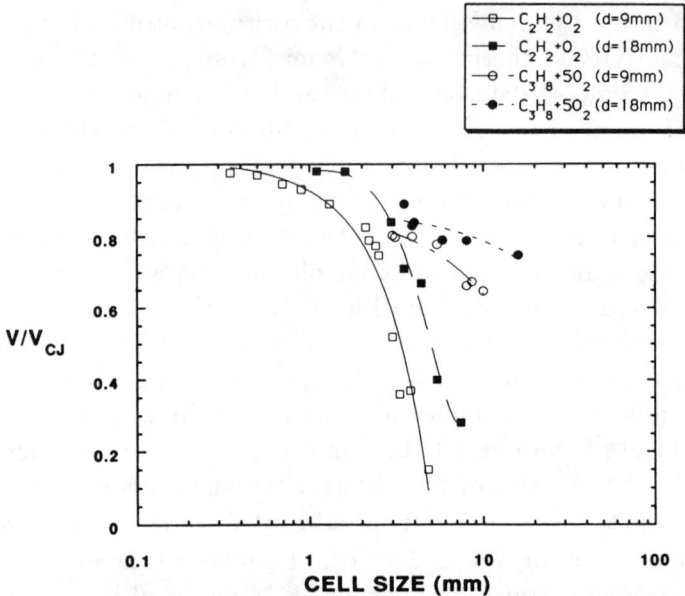

Fig. 6a: Variation of normalized velocity V/V_{CJ} with cell size λ in the 3-mm thick channel, for $C_2H_2 + O_2$ and $C_3H_8 + 5\,O_2$.

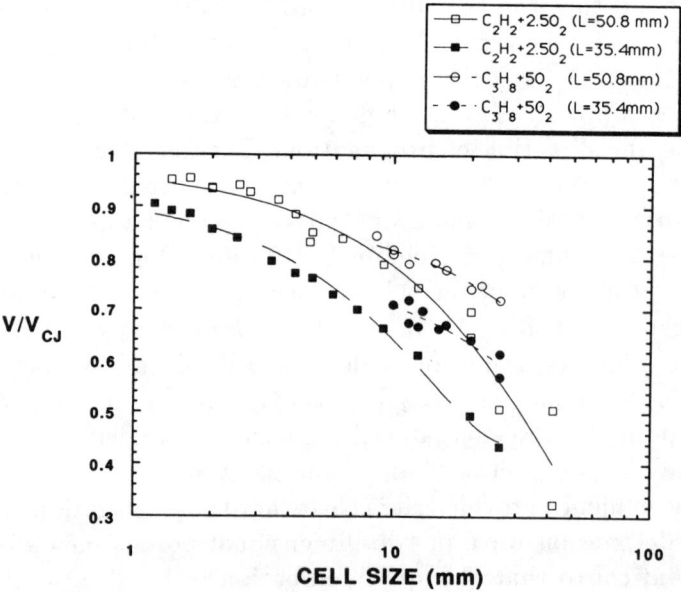

Fig. 6b: Variation of normalized velocity V/V_{CJ} with cell size λ in the 25.4-mm thick channel, for $C_2H_2 + O_2$ and $C_3H_8 + 5\,O_2$.

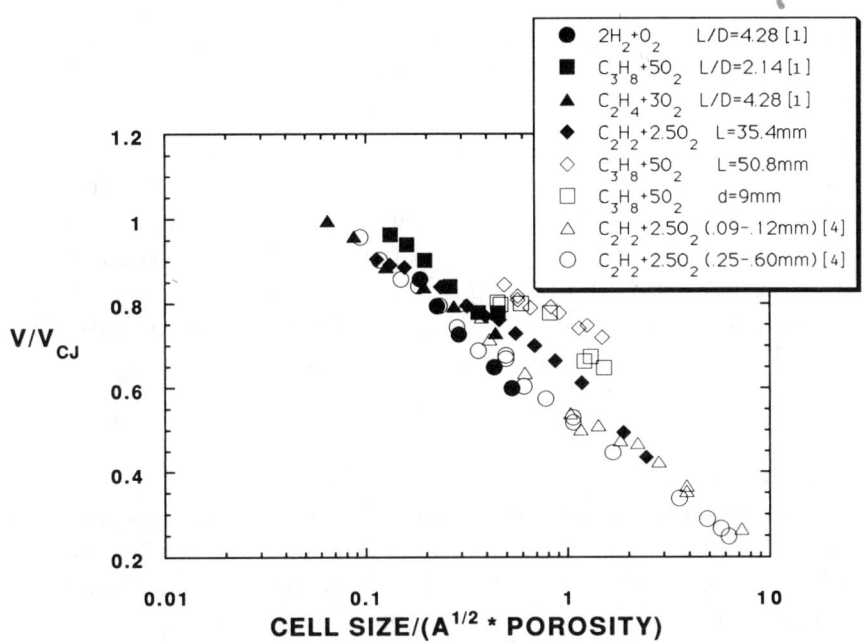

Fig. 7: Normalized velocities for porous media and rough tubes correlated with mixture sensitivity λ and geometry.

indicate that the dependence of the measured quasisteady velocities on mixture sensitivity (i.e., initial pressure) is similar to that observed by Lee et al.[11] for quasidetonations. Thus it appears possible to compare the two cases by appropriate normalized parameters (which include the cell size). It is found that a characteristics length scale, formed by taking the square root of the pore area (or cube root of the pore volume for a three-dimensional porous medium) in combination with the porosity and cell size, forms a dimensionless group that can correlate the data from actual porous media,[4] the present two-dimensional simulated porous media, and the quasidetonation data in rough tubes (Fig. 7). For cases where data for the cell size was not available, induction zone length data reported by Westbrook and Urtiew[22] and multiplied by a recommended correlating factor of 29 were used. For the correlation of the rough tube data, the porosity is taken as unity and the square root of the product between the unobstructed height d and the distance between obstacles in a rough tube L is used as the equivalent pore area. As indicated in Fig. 7, the correlation is quite good in spite of the diverse geometries, fuel mixtures, and range of initial pressures in which the velocity data is derived. This suggests that the propagation mechanism is similar for all the different cases. This fact compliments the photographic evidence in confirming that the mechanism of detonation propagation in porous media is one of quasidetonation. Figure 7 shows that V/V_{CJ} approaches unity for all geometries, when the detonation cell size becomes much smaller than the characteristic length scales of the porous medium. The data also suggests that cell regularity may play an important role in the propagation, since smaller velocity deficits are measured for mixtures of $C_3H_8 + 5\ O_2$, as compared to $C_2H_2 + 2.5\ O_2$ which has a more regular cellular pattern.

4. Conclusion

Photographic observation of detonations propagating in simulated two-dimensional porous media indicates that the propagation is one of repeated attenuation by diffraction and reinitiation by shock reflections, similar to what has been observed for quasidetonations in rough tubes.[1,2] The frequency of failure and reinitiation determines the average propagation velocity. Existing quasisteady one-dimensional models of the detonation process in a porous medium,[6,14] which consider heat and momentum losses in the reaction zone, do not appear to account for the true physical mechanism involved. The present study also demonstrated that cell regularity of the mixture may have an influence in the observed velocity deficits.

Acknowledgments

Gratitude is expressed to Richard Day for his valuable contributions in completing this study. Furthermore, the assistance of undergraduates Peter Kotsiopriftis, Walter Loch, and George Latouf is acknowledged and greatly appreciated.

References

[1]Teodorczyk, A., Lee, J. H. S., and Knystautas, R., "Propagation Mechanism of Quasidetonations," *Proceedings of the 22nd Symposium (International) on Combustion*, The Combustion Institute, Pittsburgh, PA, 1988, 1723-1731.

[2]Teodorczyk, A., Lee, J. H. S., and Knystautas, R., "Photographic Study of the Structure and Propagation Mechanisms of Quasidetonations in Rough tubes," *Dynamics of Detonations and Explosions: Detonations*, edited by A. L. Kuhl, J.-C. Leyer, A. A. Borisov, and W. A. Sirignano, Vol. 133, Progress in Astronautics and Aeronautics, AIAA, Washington, D.C., 1991, pp. 223-240.

[3] Lyamin, G. A., and Pinaev, A. V., "Study of Non-Ideal Gaseous Detonation and its Limits in Porous Media," *Dynamics of Multiphase Media*, Siberian Division of Academy of Sciences, Vol. 68, 1984, pp. 99-107.

[4] Lyamin, G. A., and Pinaev, A. V., "Supersonic (Detonation) Combustion in Gases in Inert Porous Media," *Doklady Akademie Nauk SSSR*, Fizika, Vol. 283, No. 6, 1985.

[5] Donato, L., Genadry, M., and Mavriplis, D., "Detonation Wave Propagation Through Porous Media," *Mechanical Laboratory II Report*, Dept. of Mechanical Engineering, McGill University, Montreal, May 1978.

[6] Kauffman, C. W., Chuanjun, Y., and Nicholls, J. A., "Gaseous Detonation in Porous Materials for Enhanced Fossil Fuel Utilization and Recovery," D.O.E. Rept. No DC/13407-1, Aug. 1982.

[7] Kauffman, C. W., Chuanjun, Y., and Nicholls, J. A., "Gaseous Detonations in Porous Media," *Proceedings of the 19th Symposium (International) on Combustion*, The Combustion Institute, Pittsburgh, PA, 1982, pp. 591-597.

[8] Lyamin, G. A., Pinaev, A. V., "Combustion Regimes for Gases in Inert Porous Material," translated from *Fizika Goreniya i Vzryva*, Vol. 22, No. 5, 1986, pp. 64-70.

[9] Pinaev, A. V., and Lyamin, G. A., "Fundamental Laws Governing Subsonic and Detonating Gas Combustion in Inert Porous Media," translated from *Fizika Goreniya i Vzryva*, Vol. 25, No. 4, Plenum, July-Aug. 1989, pp. 75-85.

[10] Lyamin, G. A., Mitrofanov, V. V., Pinaev, A. V., and Subbuton, V. A., "Propagation of Gas Explosion in Channels with Uneven Walls and in Porous Media," *Dynamic Structure of Detonation in Gaseous and Dispersed Media*, edited by A. Borisov, Kluwer Academ., Netherlands, 1991, pp. 51-75.

[11] Lee, J. H. S., Knystautas, R., and Chan, C. K., "Turbulent Flame Propagation in Obstacle-Filled Tubes," *Proceedings of the 20th Symposium (International) on Combustion*, The Combustion Institute, Pittsburgh, PA, 1984, pp. 1663-1672.

[12] Abdullin, R. H., Babkin, V. S., and Borisenko, A. V., "Regimes of Gas Combustion in the Linear Systems of Connected Vessels," *Institute of Chemical Kinetics*, Novosibirsk, USSR, 1990.

[13] Thomas, G. O., Edwards, D. H., and Jones, S. H. M., "Studies of the Mechanism of Detonation Propagation in Porous Structures," *Dynamics of Detonations and Explosions: Detonations*, edited by A. L. Kuhl, J.-C. Leyer, A. A. Borisov and W. A. Sirignano, Vol. 133, Progress in Astronautics and Aeronautics, AIAA, New York, 1991, pp. 257-267.

[14] Ikeh, M. O., "The Passage of Detonations Through Porous Media," Ph.D. Thesis, Dept. of Aerospace Engineering, The University of Michigan, 1981.

[15]Ong, R. S. B., "On the Interaction of a Chapman-Jouguet Detonation Wave with a Wedge," Ph.D. Thesis, Dept. of Physics, The University of Michigan, 1955.

[16]Edwards, D. H., Thomas, G. O., and Nettleton, M. A., "The Diffraction of a Planar Detonation Wave at an Abrupt Area Change," *Journal of Fluid Mechanics*, Vol. 95, No. 1, 1979, pp. 79-96.

[17]Liu, Y. K., Lee, J. H. S., and Knystautas, R., "Effect of Geometry on the Transmission of Detonation through an Orifice," *Combustion and Flame*, The Combustion Institute, Vol. 56, No. 2, 1984, pp. 215-225.

[18]Vasiliev, A. A., Mitrofanov, V. V., and Topchiyan, M. E., "Detonation Waves in Gases," translated from *Fizika Goreniya i Vzryva*, Vol. 23, No. 5, 1987, pp. 109-131.

[19]Moen, I. O., Sulmistras, A., Thomas, G. O., Bjerketvedt, D., and Thibault, P. A., "Influence of Cellular Regularity on the Behavior of Gaseous Detonations," *Dynamics of Explosions*, edited by R. R. Bowen, J.-C. Leyer, and R. I. Soloukhin, Vol. 105, Progress in Astronautics and Aeronautics, AIAA, New York, 1986, pp. 37-52.

[20]Dupré, G., Joannon, J., Knystautas, R., and Lee, J. H. S., "Unstable Detonations in the Near-Limit Regime in Tubes," *Proceedings of the 23rd Symposium (International) on Combustion*, The Combustion Institute, Pittsburgh, PA., 1990, pp. 1813-1820.

[21]Laberge, S., Atanasov, M., Knystautas, R., and Lee, J. H. S., "Propagation and Extinction of Detonation Waves in Tube Bundles," Progress in Astronautics and Aeronautics, edited by A. L. Kuhl, J.-C. Leyer, A. A. Borisov, and W. A. Sirignano, AIAA, (in press).

[22]Westbrook, C. K., and Urtiew, P. A., "Chemical Kinetic Prediction of Critical Parameters in Gaseous Detonations," *Proceedings of the 19th Symposium (International) on Combustion*, The Combustion Institute, Pittsburgh, PA, 1982, pp. 615-623.

Propagation and Extinction of Detonation Waves in Tube Bundles

S. Laberge,* R. Knystautas,† and J. H. S. Lee†
McGill University, Montreal, Quebec, Canada

Abstract

Experiments have been carried out to measure the detonation propagation and extinction in tube bundles composed of arrays of thin-walled and smooth-walled tubes stacked together wall-to-wall. A well-established detonation from a 44-mm-diam tube was made to enter a tube bundle, any one of which was 1.2 m long and made up of one size of glass tubes 1.6 mm, 5.46 mm, or 11.55 mm in internal diameter. Detonation velocity measurements in the tube bundles were via self-luminous streak photography and via photodetectors. Detonable mixtures with regular and irregular cellular structures were studied. The measured velocity deficit for detonable mixtures with a regular cellular structure follows the prediction of the Fay-Dabora theory up to a maximum value of $\Delta V/V_{CJ} \sim 15\%$. Beyond that, detonation extinction is observed corresponding to a critical sensitivity parameter $d_c/d \leq 25$. For irregularly structured detonable mixtures, the velocity deficit is much less than the theoretical prediction falling in the range $4\% < \Delta V/V_{CJ} < 13\%$. Detonation extinction occurs at $d_c/d \sim 100$. The trend of the results for velocity deficit and detonation extinction for the tube bundles is in accord with the findings of Moen et al.[10] for single smooth-walled tubes.

Copyright © 1992 by the American Institute of Aeronautics and Astronautics, Inc. All rights reserved.

*Graduate Student. Currently at l'Institut de Génie Energétique, Ecole Polytechnique, Montreal.

†Professor, Department of Mechanical Engineering.

Introduction

The propagation of detonation waves in slender smooth-walled tubes has been observed to exhibit significant velocity deficits[1-5] (measured velocities up to 10-15% below V_{CJ}) and even to undergo extinction under the right conditions. These effects have been attributed to the growth of the boundary layer in the wake of the detonation front. Theoretical models to account for this boundary-layer effect in terms of an effective streamline divergence behind the detonation wave have been developed.[4,6,7] Experimentally, velocity deficits up to maximum value of about 15% V_{CJ} have been observed. Beyond that, the detonation wave is invariably observed to fail.

The present paper re-examines the velocity attenuation aspect of smooth-walled detonation tubes. In particular, it aims to assess the detonation quenching capability of such tubes to potentially develop an effective detonation arrester. This would be done by assembling a number of thin-walled and smooth-walled slender tubes, stacked together wall-to-wall, into tight bundles. Such bundles could then be installed into the larger ducts of, say, petrochemical process equipment. In the event of development of detonation in such equipment, the tube bundle arresters would prevent the detonation from transmitting into vulnerable segments of the petrochemical plant that could otherwise result in severe and extensive damage or even loss of life.

A tube bundle detonation arrester has a number of advantages over other ideas for such devices. For example, a tube bundle insert would require minimal equipment modification and would involve small pressure losses and negligible blockage for material flow through under normal operating conditions. This is quite unlike other proposals for detonation arresters generally based on the diffraction process in which a detonation wave is forced to pass through a sudden area enlargement (the critical tube diameter situation), or a perforated plate, grid, screen, or porous medium. In such cases, major equipment modification may be involved (e.g., substantial increase of duct cross-sectional area) or may result in severe pressure losses, turbulence production, or flow impediment during normal operation.

The attractiveness of the tube bundle detonation arrester is, of course, predicated on its detonation quenching capability. To evaluate this aspect one needs to examine the mechanisms responsible for detonation extinction. It is well established that the steady-state propagation of a detonation wave at about the characteristic Chapman-Jouguet velocity is intimately linked to the existence of transverse waves of appropriate strength. Velocity deficits or detonation quenching arise when the transverse waves are attenuated or destroyed,[8,9] respectively. The survivability of a detonation wave would therefore appear to be linked to its self-regenerating capability of transverse waves. Recent articles[10-12] suggest that this capability appears to be significantly more pronounced in detonative media which are characterized by an "irregular" detonation structure. At present, "regularity" of structure remains a qualitative concept, although a tentative connection between detonation cell regularity and the dimensionless activation energy (E/RT) for the mixture has been proposed.[10-12] At present, there is no direct conclusive evidence to confirm such a link. However, evidence does exist which suggests that mixtures which possess an irregular detonation cellular structure exhibit better detonation survivability by being better able to cope with effects which attenuate the transverse wave structure. It has been postulated that this is because irregularly structured detonative media are better able to regenerate transverse waves. The reason is that a wider spectrum of transverse fluctuations exists in such a case which can then be sustained and amplified by the nonlinearly coupled chemical reactions.

The critical tube diameter phenomenon is a good example of where this effect was first detected. A universal correlation between the critical tube diameter and the detonation cell width λ, namely, $d_c \sim 13 \lambda$, has been established in the past decade.[13-17]. However, for mixtures with high ($\sim 75\%$) monatomic gas (Ar, He) dilution, which is known to result in a very regular cellular detonation structure, the critical tube diameter correlation is $d_c \sim 25\lambda - 28\lambda$. The implication is that in such regularly structured detonative media, the transverse waves are weak, the frequency spectrum of fluctuations narrow, and hence the self-regeneration capability of transverse waves greatly suppressed.

Recent though limited measurements of Moen et al.[10] in the three smooth-walled slender tubes also appear to show contrasting phenomenological consequences for regular and irregular detonation structures. They measured velocity deficits and detonation extinction under such conditions. For regularly structured detonative media, the velocity deficit results follow the trend predicted by the Fay-Dabora model, that is, a monotonically decreasing detonation velocity with decreasing mixture sensitivity to a maximum velocity deficit level $\Delta V/V_{CJ} \sim 15\%$. Less sensitive mixtures lead to detonation extinction which they have characterized by a critical sensitivity criterion $d_c/d < 25$. For irregularly structured detonable mixtures they observed negligible velocity deficits ($\Delta V/V_{CJ} < 2\%$) even up to very insensitive mixtures at which point the detonation was quenched corresponding to a quenching criterion of $d_c/d \sim 75\text{-}100$.

Clearly, these recent findings have direct and serious implications for detonation transmission in tube bundles. It is of interest, therefore, to extend these very limited observations to other very different chemical systems to determine if these recent findings are universal in nature. It is the purpose of the present paper to do this. The propagation and extinction of detonation waves in an array of thin-walled and smooth-walled tubes bound together in tight tube bundles has been systematically investigated for a range of detonable mixtures with regular and irregular cellular structures.

Experimental Details

All of the experiments were performed in a 4.8-m-long detonation tube which had a 44-mm inner diameter (see Fig. 1). The detonation tube was constructed of four standard lengths (1.2 m) of heavy-walled plexiglass, joined at the extremities by O-ring equipped sleeves in such a way as to maintain a constant inner diameter throughout. The first three lengths (3.6m) were clear and unobstructed and were used to achieve steady propagation of a well-established C-J detonation wave in the 44-mm-diam tube. The last segment (1.2 m) was filled with a tube bundle. Three tube sizes were used for the tube bundles composed of thin-walled smooth-walled glass tubing. The inner tube diameters for the tubes in the three tube bundles were d = 1.69, 5.46, and 11.55 mm. One size of tube was used for each tube bundle. The ends of each tube bundle were cast into a circular block of epoxy resin to leave open only the circular inner tube diameters while sealing the intertube spaces. The tube bundles were then slid and sealed inside the last segment of the 44-mm inner diameter detonation tube.

For any given experiment, the detonation tube was first evacuated to better than 0.1 mbar and then filled with premixed detonable mixture to the desired subatmospheric pressure. The safe range of initial pressures for the experiments was up to $p_o = 200$ Torr. Three

DETONATION WAVES IN TUBE BUNDLES 385

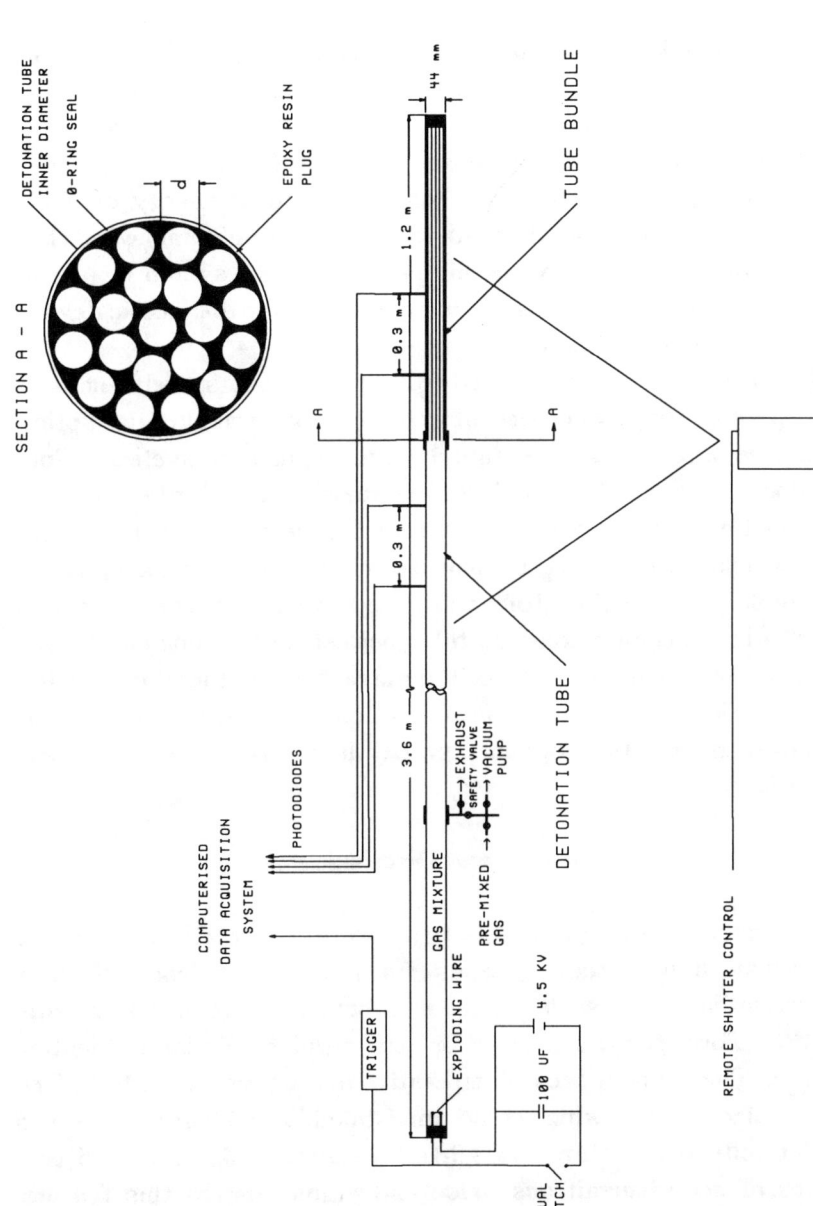

Fig. 1 Schematic diagram of experimental apparatus.

explosive mixtures were studied to simulate the range of experimental conditions desired. These were equimolar oxy-acetylene ($C_2H_2 + O_2$), stoichiometric methane-oxygen ($CH_4 + 2O_2$) and stoichiometric acetylene-nitrous oxide ($C_2H_2 + 5N_2O$) with 50% argon dilution. The gases were premixed by the method of partial pressures in reservoir bottles at 2 atm initial pressure.

Detonation initiation was achieved by an exploding wire driven by a low voltage (~4 KV) condenser bank with stored energy of about 1000 J. The energy was dumped through the exploding wire via a manual switch. In this way the high-frequency RF switch triggering noise was largely eliminated so as not to perturb the diagnostic devices and digital data acquisition system.

Diagnostics included smoked-foil measurements, self-luminous streak photography, and time-of-arrival measurements via optical photodiodes interfaced to a digital data acquisition system. Four photodiodes were used: one pair (0.3 m apart) in the third segment of the detonation tube to measure the near C-J detonation velocity, the other pair (also 0.3 m apart) to measure the detonation velocity in the tube bundle. The smoked foils were used to measure the detonation cell width in the main detonation tube just before the tube bundle and to observe the regularity of the cellular structure. Detonation velocity, velocity deficits, and detonation extinction in the tube bundle were determined from the optical probe and self-luminous streak observations.

Results and Discussion

The results of the present study indicate that mixtures involving the common hydrocarbon gaseous fuels (e.g., acetylene, ethylene, propane, methane, etc.) with oxygen at subatmospheric initial pressure (p_o < 200 Torr) possess a more or less regular cellular detonation structure. Therefore, it became imperative in the present study to find a fuel-oxidizer system which could unmistakably be identified with an irregular cellular structure. In 1981, Libouton et al.[18] reported that mixtures of acetylene-nitrous oxide and argon possess this feature. Figure 2 illustrates the two contrasting cellular detonation structures. The smoked foil on the left (Fig.2a) corresponds to a typical hydrocarbon fuel-oxygen regular structure, in this case equimolar $C_2H_2 - O_2$ at p_o = 6 Torr. The cellular structure on the right (Fig. 2b) is clearly

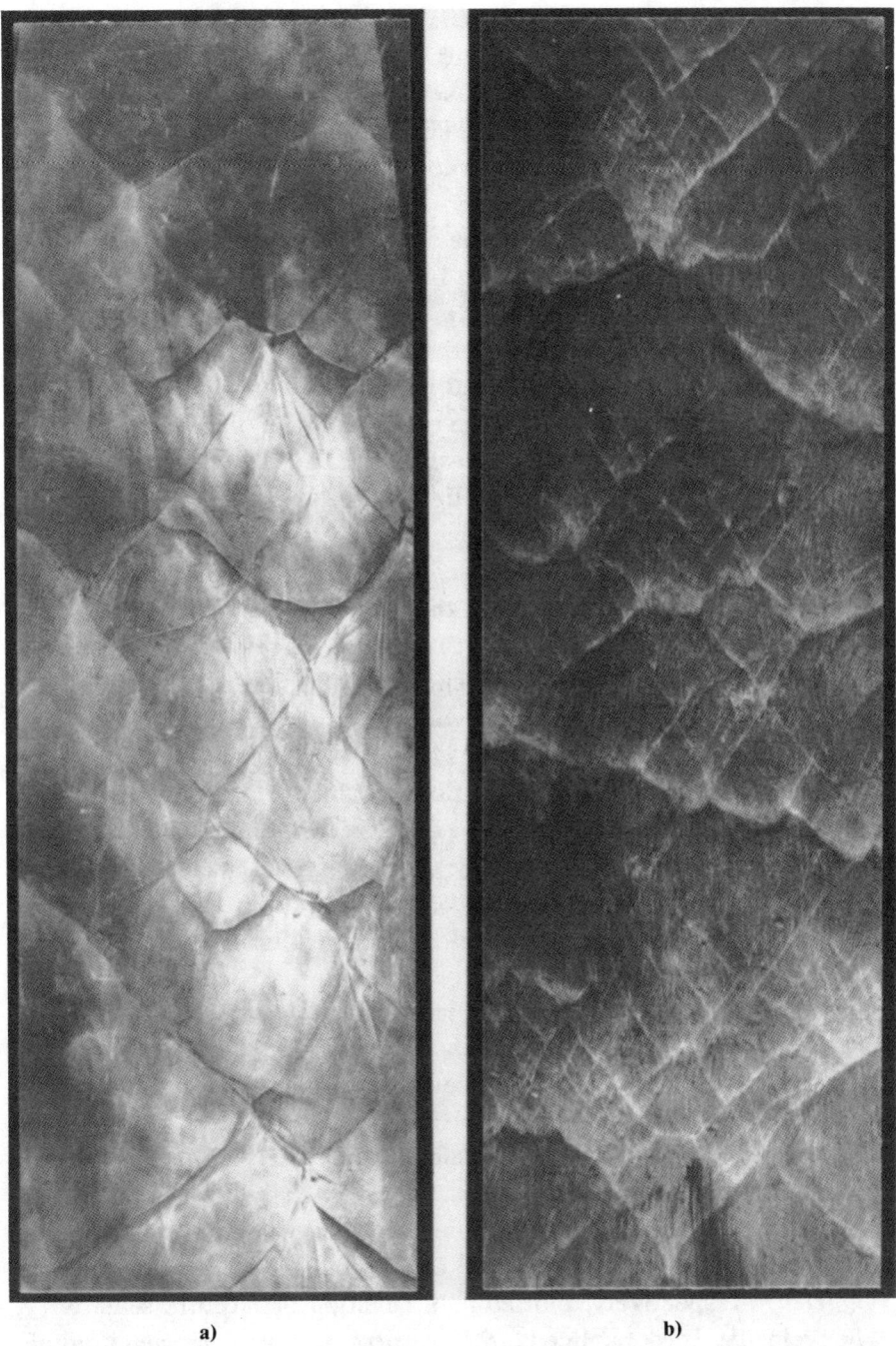

Fig. 2 Smoked-foil record illustrating cellular structure: a) regular and b) irregular.

irregular and corresponds to a mixture of stoichiometric acetylene-nitrous oxide with 50% argon dilution at p_o = 30 Torr.

Following Moen et al.[10] we have also chosen the critical tube diameter d_c as the characteristic detonation length scale to be used wherever a length parameter has to be normalized. This choice was motivated to avoid the subjectivity inherent in the measurement of the detonation cell size λ. Because the critical tube diameter d_c represents a length scale larger, typically by more more than an order of magnitude, than the cell size λ, its measurement requires a much larger scale of experiment. This drawback is mitigated by the fact that experimentally d_c can be measured precisely and unambiguously.

Figure 3 displays a compilation of critical tube diameter data based on earlier as well as current measurements. At atmospheric initial pressure, direct measurements of d_c are available from Knystautas et al.[14] Empirical critical tube diameter correlations from Matsui and Lee[20], namely $p_o = K d_c^{-\alpha}$ are represented in Fig. 3 by solid lines. Some very recent direct measurements that we have carried out for d_c in acetylene-nitrous oxide-argon mixtures for a range of initial pressures are also shown in Fig. 3. To extend the data base for d_c as much as possible, we have also taken the voluminous data available for λ and linked this data to d_c via the $d_c = 13\lambda$ correlation. It is therefore to be kept in mind that all the information presented in Fig. 3 comes from separate and distinct experiments. The agreement of all these separate data with each other is remarkably good. Even in the case of the very irregular structure (i.e. $C_2H_2 + 5N_2O$, 50% argon dilution), the $d_c = 13\lambda$ correlation is in good agreement with other results. This does suggest that measurement of λ is still a useful approach to determine the detonation sensitivity of a given mixture. The results displayed in Fig. 3 clearly indicate that the detonation sensitivity of the stoichiometric mixture of acetylene and nitrous oxide with 50% argon dilution is roughly comparable to that for the oxy-hydrogen system. The most sensitive mixture is equimolar oxy-acetylene and the least sensitive, stoichiometric methane-oxygen.

Figure 4 displays the detonation velocity measured in slender tubes and rectangular channels by Moen et al.[10] and Vandermeiren and Van Tiggelen,[21] respectively, plotted as a function of mixture sensitivity. The velocity is normalized with respect to the Chapman-Jouguet velocity for the mixture. The mixture sensitivity is characterized by the critical tube diameter rather than the cell size λ because the

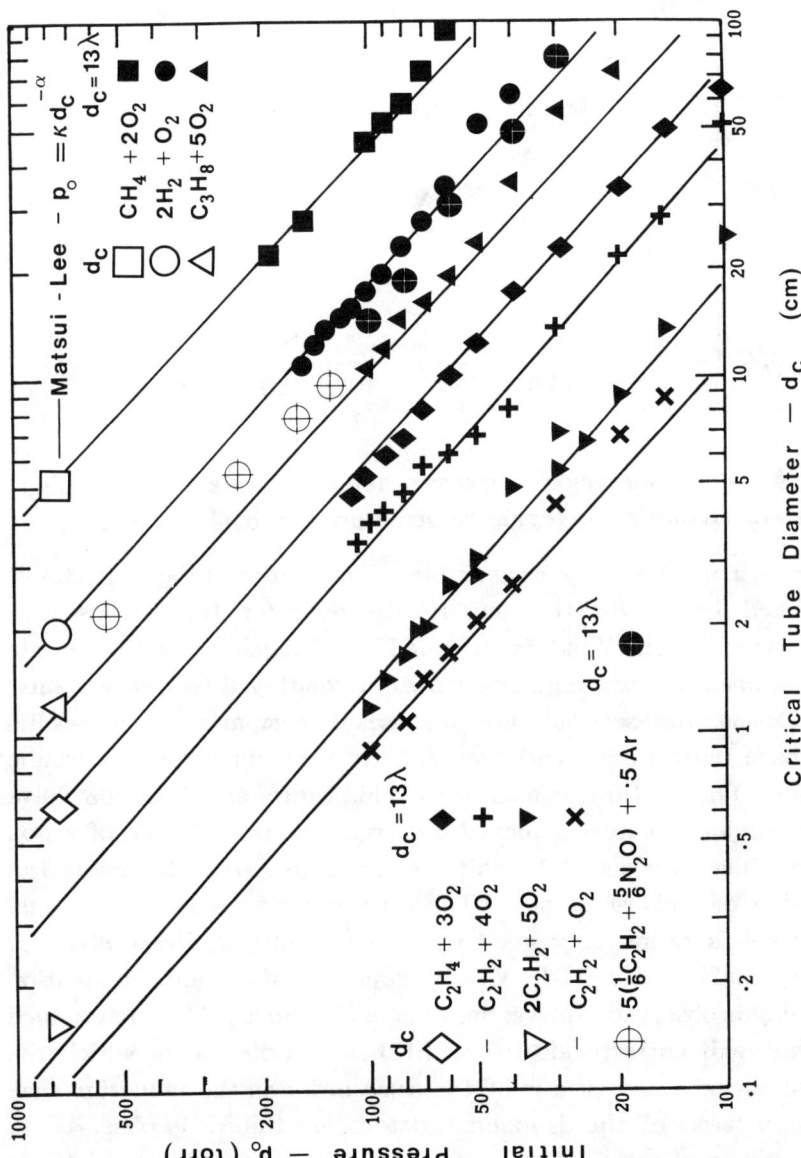

Fig. 3 Critical tube diameter results in fuel-oxidizer systems as a function of initial pressure.

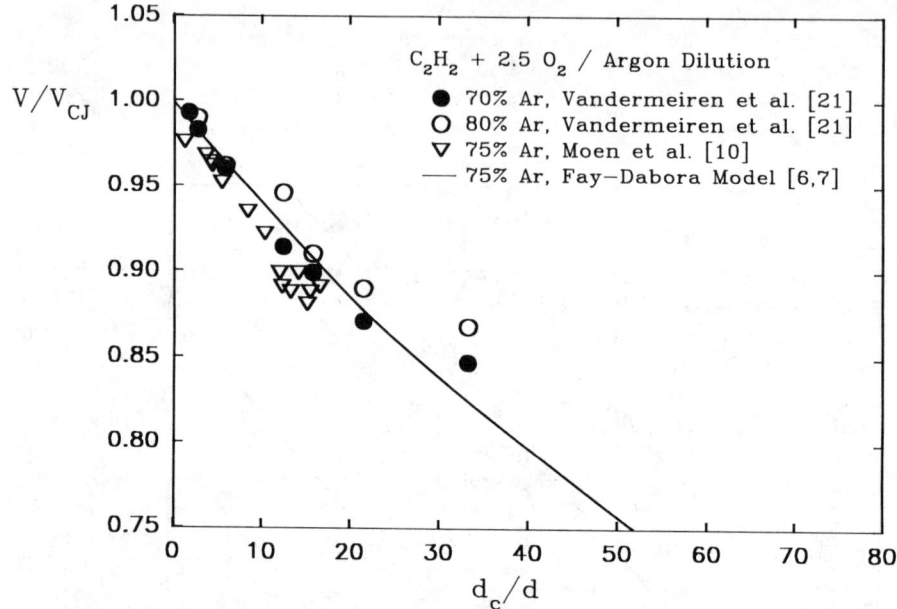

Fig. 4 Detonation velocity measurements in single smooth-walled tubes or channels for regularly structured detonable mixtures.

former is more precisely measurable. The critical tube diameter is normalized here with the actual tube diameter for purposes of comparison. For the Vandermeiren and Van Tiggelen results, the data were obtained in a rectangular channel of roughly 3 to 1 aspect ratio (32 x 92 mm cross-section). For purposes of comparison their results have been normalized with respect to the equivalent hydraulic diameter. The mixtures considered in this figure are those that have a very regular structure achieved by large amounts (70-80%) of argon dilution. The experimental results are correlated with the theoretical prediction of Dabora et al.[7]. The agreement between theory and experiment is remarkably good up to a velocity of about 85% V_{CJ} (velocity deficit $\Delta V = 15\%$ V_{CJ}). Beyond this point, detonation quenching is observed. Interestingly enough, Murray[22] has determined that this limit corresponds to the Shchelkin criterion in which the increase in induction time is of the same order as the induction time itself. In terms of the d_c/d limit, detonation failure is observed to correspond to about $d_c/d < 25$ for these very regularly structured media.

For detonation propagation in tube bundles, the results of the present study displayed in Fig. 5, show that for fairly regular

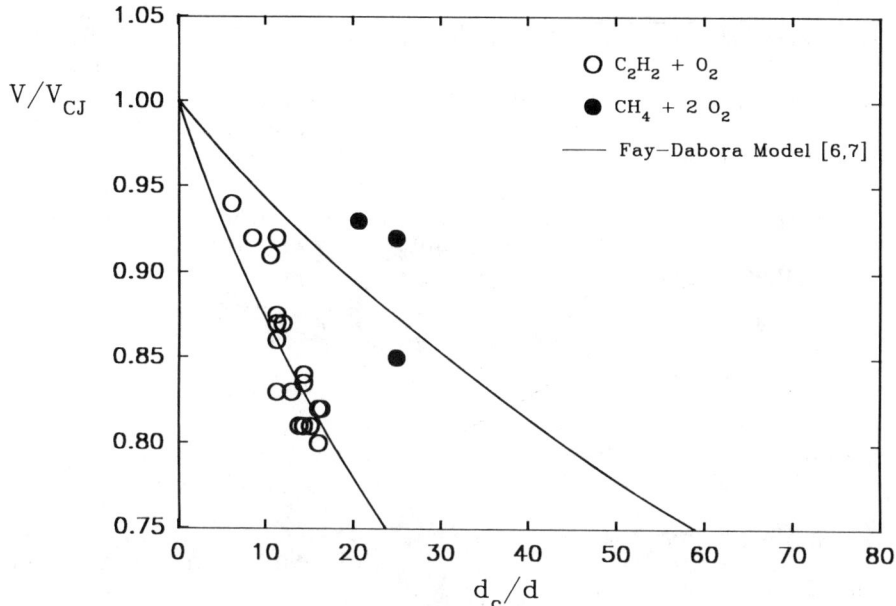

Fig. 5 Detonation velocity measurements in tube bundles for regularly structured detonable mixtures.

structured mixtures the detonation velocity deficit also follows the Dabora et al. theoretical curve. Again the maximum observed velocity deficit is in the range of about $\Delta V \approx 15\% \, V_{CJ}$. Beyond that, detonation quenching begins to occur. By this we mean that in the case of the tube bundle, the quenching limit defined in terms of initial pressure is not abrupt in the sense that detonation does not fail in all the tubes of the bundle at the first incipient condition. Rather, there is a range of about 5-6 Torr in initial pressure over which detonation fails progressively in more and more of the tubes in the bundle as the initial pressure decreases. At the lower initial pressure limit, detonation fails in all the tubes in the bundle. This progressive nature of detonation failure is clearly evident from both the streak records and the photodiode observations. In the streak record, a bifurcation of the self-luminous trace is observed which arises and vanishes at the beginning and the end of this pressure range. The bifurcation corresponds to a detonation propagating in some of the tubes in the bundle and a deflagration in the others. At the lower end of the pressure range, detonation is quenched in all of the tubes and a deflagration is the only self-luminous trace. In the case of the

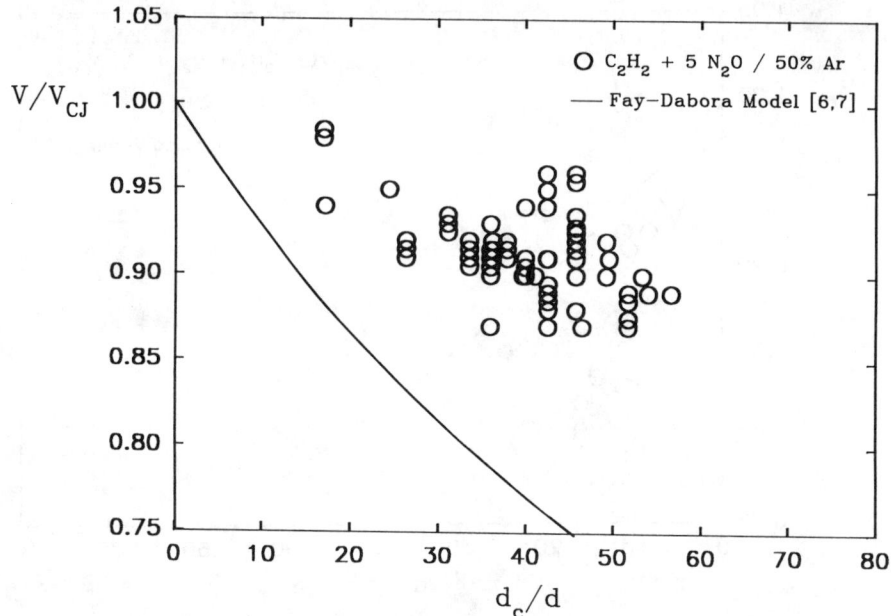

Fig. 6 Detonation velocity measurement in tube bundles for an irregularly structured detonable mixture.

photodiode time-of-arrival measurements, dual signals are detected at a given position in the tube bundle. The first (earlier) signal rise is due to the detonation detected in some of the tubes of the bundle. The later signal rise is due to the deflagration in the rest of the tubes. As the initial pressure is reduced, the detonation signal amplitude decreases while the deflagration signal amplitude rises. This is because as more tubes in the bundle experience detonation quenching the intensity of self-luminous emission shifts with increasing amplitude to the later time signal corresponding to deflagration mode of combustion in these tubes. Detonation failure corresponding to the results of Fig. 5 occurs for $18 < d_c/d < 24$ covering this 5-6 Torr range of initial pressure over which detonation fails totally in the tube bundle. This is in very good agreement with the failure criterion of $d_c/d < 25$ measured by Moen et al.[10] for mixtures with regular cellular detonation structure.

Figure 6 shows detonation velocity measurements in tube bundles for a detonable mixture with a very irregular cellular structure. The mixture in question is stoichiometric acetylene-nitrous oxide with 50% argon dilution. As before, the velocity results are correlated with the

Fay-Dabora velocity deficit theory.[7] In this case again the measured detonation velocity in the tube bundles is normalized with respect to the Chapman-Jouguet velocity and the mixture sensitivity characterized by d_c, the critical tube diameter, is normalized with the bundle tube diameter d. Figure 6 clearly shows that there is no agreement between the Fay-Dabora theory and experimental measurements. First, there is large fluctuation of experimental measurements for a particular d_c/d. Second, the velocity deficit measured is much less than that predicted by the Fay-Dabora model. The lower range of velocity deficit $\Delta V/V_{CJ}$ is up to 4% and does not increase significantly with decreasing mixture sensitivity (i.e., larger d_c/d values). This is very reminiscent of the Moen et al.[10] findings for irregularly structured mixtures of hydrocarbon-air in which the velocity deficit was typically no more than $\Delta V/V_{CJ} < 2\%$ regardless of mixture sensitivity. However, we do observe fluctuation in our measured velocity deficit results, with the maximum $\Delta V/V_{CJ} \sim 13\%$. This maximum measured value is typically about half that predicted by the Fay-Dabora theory. Detonation failure in this case is observed to occur for $d_c/d \sim 100$ which is in accord with the Moen et al.[10] findings in single tubes for mixtures with irregular cellular structure. For tube bundles, however, detonation failure occurs over a range of initial pressures in a manner already described in Fig. 5 for regularly structured media.

Conclusions

The present study has investigated the propagation and extinction of detonation waves in tube bundles composed of arrays of thin-walled and smooth-walled tubes stacked together wall-to-wall. For detonable mixtures with regular cellular structure, velocity deficits are observed which are well correlated by the Fay-Dabora theory. Maximum velocity deficits up to $\Delta V/V_{CJ} \sim 15\%$ are observed. Beyond this level, detonations are observed to fail. Failure occurs at a sensitivity criterion, characterized by the critical tube diameter d_c, corresponding to $d_c/d < 25$. This is in accord with similar measurements by Moen et al.[10] in single tubes. However, for tube bundles detonation failure occurs progressively over a band of initial pressures rather than abruptly at one initial pressure as is the case for a single tube.

For irregularly structured detonative media, wide fluctuations are observed in the velocity deficit measurements with the overall range extending from $4\% \leq \Delta V/V_{CJ} \leq 13\%$. These values are much less than that predicted by the Fay-Dabora theory, by more than a factor of two. Detonation extinction occurs for mixture sensitivities corresponding to $d_c/d \sim 100$ which is again in accord with the Moen et al. findings for single tubes.

Acknowledgments

This work was supported by the Natural Sciences and Engineering Research Council of Canada under Grants A-3347 and A-7091 and by the Department of Defense of Canada (DRES) under Contract 01SG.W7702-6-2521-A. The authors would like to acknowledge the assistance of Ms. Maria Atanasov in carrying out some of the experiments as the measurement of the critical tube diameter.

References

[1] Edwards, D. H., Jones, T. G., and Price, B., "Observations on Oblique Shock Waves in Gaseous Detonations," *Journal of Fluid Mechanics*, Vol. 17, 1963, pp. 21-30.

[2] Brochet, C., "Contributions à l'Étude des Détonations Instables dans les Mélanges Gaseux," Ph.D Thesis, Université de Poitiers, Poitiers, France, 1966.

[3] Renault, G., "Propagation des Détonations dans les Mélanges Gazeux Contenus dans des Tubes de Section Circulaire et de Section Rectangulaire: Influence de l'État de la Surface Interne des Tubes," Ph.D Thesis, Université de Poitiers, Poitiers, France, 1972.

[4] Dove, J. E., Scroggie, B. J., and Semerjian, H., "Velocity Deficits and Detonability Limits of Hydrogen-Oxygen Detonations," *Acta Astronautica*, Vol. 1, 1974, pp. 345-359.

[5] Dupré, G., Knystautas, R., and Lee, J. H. S., "Near-Limit Propagation of Detonation in Tubes," Vol. 106, Progress in Astronautics and Aeronautics, 1986, pp. 244-259, AIAA, New York.

[6] Fay, J. A., "Two-Dimensional Gaseous Detonations: Velocity Deficit," *Physics of Fluids*, Vol. 2, 1959, pp. 283-289.

[7] Dabora, E. K., Nicholls, J. A., and Morrison, R. B., "The Influence of a Compressible Boundary on the Propagation of Gaseous Detonations," *Proceedings of the 10th Symposium (International) on Combustion*, The Combustion Institute, Pittsburgh, PA, 1965, pp. 817-830.

[8]Evans, M. W., Given, F. I., and Richeson, W. F., "Effects of Attenuating Materials on Detonation Induction Distances Gases," *Journal of Applied Physics*, Vol. 26, 1955, pp. 1111-1113.

[9]Dupré, G., Peraldi, O., Lee, J. H. S., and Knystautas, R., "Propagation of Detonation Waves in an Acoustic Absorbing Walled Tube," Vol. 114, Progress in Astronautics and Aeronautics, 1988, pp. 248-263, AIAA, Washington, D.C.

[10]Moen, I. O., Sulmistras, A., Thomas, G. O., Bjerketvedt, D., and Thibault, P. A., "Influence of Cellular Regularity on the Behavior of Gaseous Detonations," Vol. 106, Progress in Astronautics and Aeronautics, 1986, pp. 220-243, AIAA, New York.

[11]Dupré, G., Joannon, J., Knystautas, R., and Lee, J. H. S., "Unstable Detonations in the Near-Limit Regime in Tubes," *Proceedings of the 23rd Symposium (International) on Combustion*, The Combustion Institute, Pittsburgh, PA, 1991.

[12]Lee, J. H., S., "Dynamic Structure of Gaseous Detonation," Third International conference 'Lavrentyev Readings on Mathematics, Mechanics and Physics,' Novosibirsk, USSR, 1990.

[13]Moen, I. O., Funk, J. W., Ward, S. A., Rude, G. M., and Thibault, P. A., "Detonation Length Scales for Fuel-Air Explosives," Vol. 94, Progress in Astronautics and Aeronautics, 1984, pp. 55-79, AIAA, New York.

[14]Knystautas, R., and Lee, J. H. S., and Guirao, C., "The Critical Tube Diameter for Detonation Failure in Hydrocarbon-Air Mixtures," *Combustion and Flame*, Vol. 48, 1982, pp. 63-83.

[15]Knystautas, R., and Lee, J. H. S., "Detonation Parameters for the Hydrogen-Chlorine System," Vol. 114, Progress in Astronautics and Aeronautics, 1988, pp. 32-44, AIAA, Washington, D.C.

[16]Pedley, M. D., Bishop, C. V., Benz, F. J., Bennett, C. A., McClenagan, R. D., Fenton, D. L., Knystautas, R., Lee, J. H. S., Peraldi, O., Dupré, G., and Shepherd, J. E., "Hydrazine Vapor Detonations," Vol. 114, Progress in Astronautics and Aeronautics, 1988, pp. 45-63, AIAA, Washington, D.C.

[17]Desbordes, D., "Transmission of Overdriven Plane Detonations: Critical Diameter as a Function of Cell Regularity and Size," Vol. 114, Progress in Astronautics and Aeronautics, 1988, pp. 170-185, AIAA, Washington, D.C.

[18]Libouton, J. C., Jacques, A., and Van Tiggelen, P.J., "Cinétique, Structure et Entretien des Ondes des Detonations," *Colloque International Berthelot-Veille-Mallard-Le Chatelier Proceedings*, Université de Bordeaux, France, Vol. II, 1981, pp. 437-442.

[19]Moen, I. O., Murray, S. B., Bjerketvedt, D., Rinnan, A., Knystautas, R., and Lee, J. H. S., "Diffraction of Detonation from Tubes into Large Fuel-Air Explosive Cloud," *Proceedings of the 19th Symposium (International) on Combustion*, The Combustion Institute, Pittsburgh, PA, 1982, pp. 635-644.

[20]Matsui, H., and Lee, J. H. S., "On the Measure of the Relative Detonation Hazards of Gaseous Fuel-Oxygen and Air Mixtures," *Proceedings of the 17th Symposium (International) on Combustion*, The Combustion Institute, Pittsburgh, PA, 1979, pp. 1269-1280.

[21]Vandermeiren, M., and Van Tiggelen, P. J., "Cellular Structure of Detonation of Acetylene-Oxygen Mixtures," Vol. 94, Progress in Astronautics and Aeronautics, 1984, pp. 104-117, AIAA, New York.

[22]Murray, S. B., "The Influence of Initial and Boundary Conditions on Gaseous Detonation Waves," Ph.D. Thesis, McGill University, Department of Mechanical Engineering, Montreal, Canada, 1984, pp. 102-119.

Simultaneous Strong and Quasi-Chapman-Jouguet Detonation Wave Propagation

Roger Chéret*
Commissariat à l'Energie Atomique, Paris, France

Abstract

Usual explosive devices (sticks or tubes) detonate according to the so-called Chapman-Jouguet regime, whereas appropriate priming conditions (for example, endowed with spherical converging symmetry) may induce a strong detonation regime. On the fringe of these traditional circumstances, an axisymmetric experiment has been designed, where the two regimes propagate simultaneously in the one and same explosive. Whereas already published analysis has focused on temperature measurements and EOS calibrations, the present work deals with the whole wave pattern; its geometric and optical structures are completely interpreted in terms of the detonation layer theory.

Introduction

In terms of the boundary conditions, it can be said that the traditional "plane" detonation wave experiment is characterized by a priming boundary occupying all or part of the entry face (of the tube or stick whose axis is **i**) whereas the lateral surface and the output face have free boundaries.

What happens when this distribution between priming boundary and free boundaries is different? In particular, what

---
Copyright © 1989 by the American Institute of Aeronautics and Astronautics, Inc. All rights reserved.
*Research Director, CEA/DAM, 33 rue de la Fédération.

happens if a stress progresses on the lateral surface with a constant velocity higher than the velocity D_* which the detonation will have in the traditional configuration ? This question was raised by Chéret in Ref 1, and partly answered by three series of experiments reported by Krishnan et al.,[2] Sellam et al.,[3] and Hamada et al.,[4].

First, the experimental set-up and main features of the propagation are recalled. A detailed account of the optical measurements is then given. Finally an interpretation is produced which calls on the detonation layer theory[5]. (N.B. Specific terms of this theory are marked with a dagger †, and briefly explained at the end of the paper, in the Appendix).

Main Experimental Features

The principle of the experiment is to create the required lateral stress by careful use of a "fast" explosive whose Chapman-Jouguet velocity D'_* is greater than that D_* of the explosive forming the principal "core" charge. Fig.1, taken from Ref.3, shows the device obtained when the annular chamber is filled with astrolite in aqueous solution, and the central tube with commercial grade nitromethane.

The dimensions and materials used were obtained from preliminary research which showed that, when the tube is rather long compared to the thickness of the annular chamber, terminal propagation consists of 1) in the astrolite, a quasi-C-J (†) detonation wave moving with velocity D'i in the laboratory frame; and 2) in the nitromethane, a complex wave pattern Ω

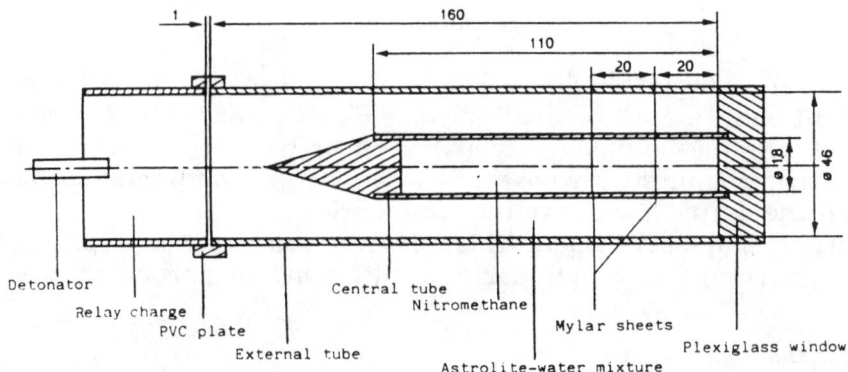

Fig.1 Experimental device for the generation of a strong axisymmetric detonation in nitromethane (from Ref.3). Dimensions are given in millimeters.

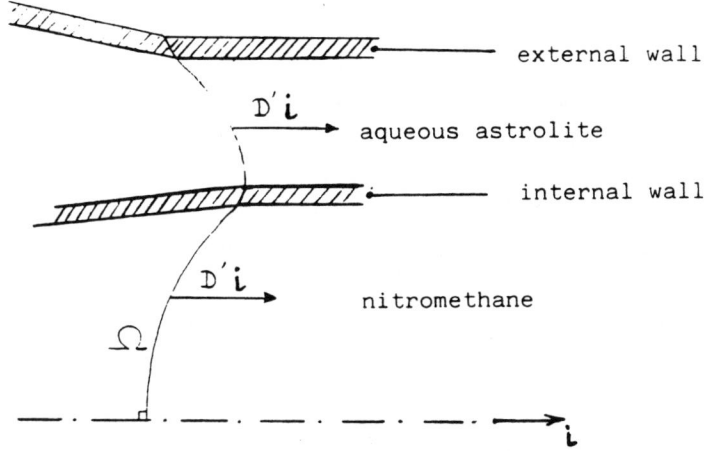

Fig.2 Detonation of the structure shown in Fig.1. Scheme of the wave pattern.

moving by translation with the same velocity D'i; as schemed in Fig.2 ($D_* < D' < D'_*$). Simple adjustment of the concentration of the aqueous astrolite solution allows D' to be varied between 7 and 8 mm/μs, an interval which compares with the value 6.29 mm/μs of the standard C-J velocity of the nitromethane.

Whereas the already published work focuses on temperature measurements and EOS assessment in the detonation products, present analysis deals with the geometric and optical structures of the whole pattern Ω.

Optical Measurements

On its end section, the nitromethane core is in contact with a plexiglass "window" whose shock impedance is less than that of the nitromethane detonation products. Interaction of Ω with the window reduces the emissivity of the detonation products together with the plexiglass transparency. As a result, on the film of a slit scanning camera, whose optical axis coincides with the assembly axis and whose entry lens sees a diameter of the output face, there appear significant traces of arrival times on the window of the frontal parts of Ω: "vertical" discontinuity of optical density on the right of Fig.3a. Starting with these traces, and using the precautions given above to ensure that Ω is stationary, it is possible to "reconstruct" Ω in the neighborhood of the window.

This "end" observation can be complemented (using the same camera) by that of the interaction of Ω with two mylar sheets

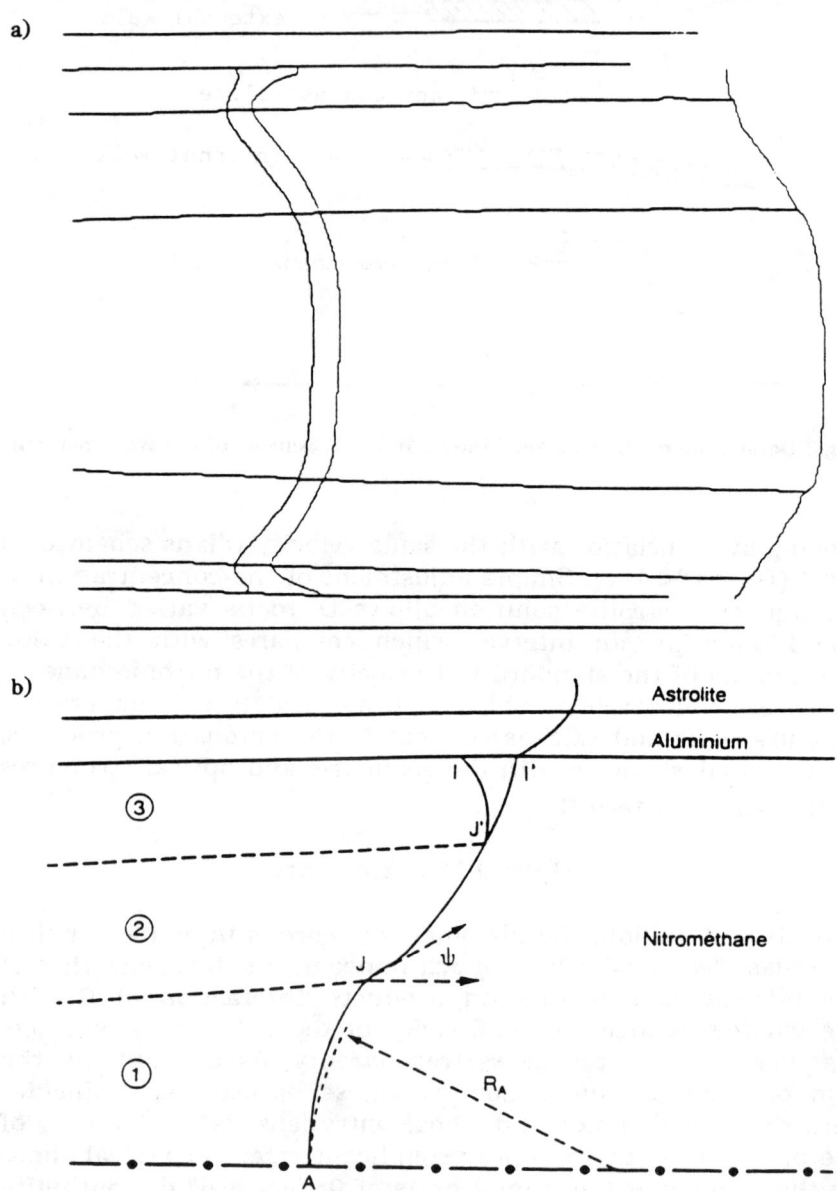

Fig.3 Detonation of experimental set-up shown in Fig.1, typical record: a) result of the densitometric analysis of the recording on the film of a streak camera (time increases from left to right); b) wave pattern Ω reconstructed from the recording.

(0.1 mm thick, 10 mm wide) placed normally to i, at 20 ± 0.2 mm and 40 ± 0.2 mm, respectively from the window. This interaction causes a momentary extinction of detonation and thus an interruption in the transmission of light towards the camera. The dark traces recorded on the film thus show the time of arrival on the mylar of emissive parts of Ω: "vertical" lines on the left of Fig.3a. These "intermediate" traces can be reconstructed in the same way as the "end" traces.

Many recordings have been achieved while varying the natures and thicknesses of central and external tubes (P.V.C., aluminium, brass, or stainless steel). Each of them shows three zones (1, 2, 3 in Fig.3b) clearly differentiated by their optical density: only their relative radial extension depends on the nature and thickness of the walls. These three zones are limited in the scanning direction by two arcs AJ and JI (cf., in Fig.2b):

1) Arc AJ is completely situated in the most luminous zone; it is concave towards the window; and it is normal in A to i. Its curvature in A increases as D' increases (see Table 1).

2) Arc JI extends from the interface between zones 1 and 2 up to the wall of the tube: it is concave towards the detonation products; in J it is inclined on i by an angle ψ whose sine is little different from D_*/D' (refer Table 1), which means that the local normal component of the upstream velocity vector does not differ significantly from the nitromethane C-J velocity D_* (Table 1). Some arc J'I' starts from the interface between zones 2 and 3, and makes an angle in I' with i close to that made by the shock wave induced in the nitromethane across the central tube by the detonating astrolite.

Table 1 exhibits some values taken from a series of experiments where the central tube is made of brass 1mm thick and the external tube is made of aluminium.

Table 1 Values of angle ψ and radius R_A defined on Fig.3 as a function of velocity D' (see Ref.3).

Volume % of water is astrolite	20	15	10	5	0
D', mm/μs	7.3	7.51	7.72	7.87	8
R_A, mm	44.3	31.20	25.50	18.50	19
$(\frac{D'}{D_*} \sin \psi) - 1$		$9\ 10^{-3}$	$17\ 10^{-3}$	$9\ 10^{-3}$	$12\ 10^{-3}$

Interpretation

As to the nature of the propagation wave pattern Ω in nitromethane, the important experimental features are the sign of the algebraic curvature of the meridian and the nullity of $[(D'/D_*) \sin\ ]-1$ at point J. Indeed, the rules of propagation (†) demonstrated by CHERET[5] show that AJ is the meridian of a lenticular strong detonation and JI is the meridian of an annular quasi C-J detonation.

These conclusions allow interpretation of the discontinuity of illumination between zones 1 and 2 as the manifestation of the sharp change in the temperature profile when moving from one regime to the other. As a matter of fact, the "detonation layer" (†) theory[5] shows that the temperature profile T (Z) in the vicinity of the downstream state (subscript 1 in the strong detonation case and subscript * in the quasi-C-J detonation case) is :

$$\frac{T(Z)}{T_1} = 1 + G_1 \frac{2|u_1| w_1}{a_1^2 - w_1^2} \frac{Z}{r_m}, \quad r_m < 0 \qquad (1)$$

along a streamline starting from the arc AJ and

$$\frac{T(Z)}{T_*} = 1 + G_* \frac{2|u_*|}{a_*} \sqrt{\beta_* Z/r_m}, \quad r_m > 0 \qquad (2)$$

along a streamline starting from the arc JI'; where the following notation is used:

G   = Gruneisen coefficient
u   = normal absolute velocity
w   = normal relative velocity
r_m = mean algebraic curvature

$$\beta_* = (\rho w)_*^3 \left[|u| \frac{\partial^2 p}{\partial v^2} (v,s) \right]_*^{-1}$$

The above conclusions also allow interpretation of the difference in illumination between zones 2 and 3 and the significance of the arc J'I':

1) Across the central tube, the detonation of the astrolite induces a shock which curves itself and intensifies from I' to J', but remains too weak to supply directly the ignition temperature to the nitromethane.

2) Ignition is obtained only at the cost of subsequent compressions due to "the accumulation of shocks" in the tube itself, in such a way that the different points of IJ' correspond to quasi-C-J detonation of thin streams of precompressed, preheated nitromethane and not to nitromethane at rest; the corresponding C-J temperatures are found to be lower, as is the emissivity of the detonation products.

Conclusion

The configuration which we have described and interpreted is much less "natural" than the traditional configuration. Nevertheless it is of great interest in so far as it provides the following: 1) an example of strong stationary detonation, the velocity of which may be chosen over a continuum, 2) an exemple of the co-existence of strong and quasi-C-J detonations and of the absence of continuous transition from one to the other; 3) an example of the co-existence of "instantaneous" and "delayed" quasi-C-J detonations.

Appendix : apropos of detonation layer

The nature and the contents of the "detonation layer" theory is best understood after recalling Prandtl "boundary-layer" theory on one hand, and Weyl "shock-layer" theory on the other hand.

At the beginning of the century, Prandtl boundary-layer theory exhibits a brilliant and intuitive way of matching the viscous fluid flow near the edge of a foil with the perfect fluid flow far from it. However the basic principles of the method remain hidden and unformulated until Hermann Weyl put them forward in a rigorous and convincing way, in connection with the problem of the shock wave, where dissipative effects are to be neglected everywhere in the flow except in a narrow zone: the shock layer. In particular, Weyl shows that the key to matching the "internal" dissipative fluid flow with the "external" perfect fluid flow is a dimensionless perturbation parameter ϵ: the range of dissipative effects divided by the main dimension of the system. This theory is rather straightforward inasmuch as any experimental value of the normal component of the upstream velocity vector (i.e., normal shock velocity) allows one and only one solution flow.

When applying Weyl's method (known to-day as the "matched asymptotic expansions" method) to the detonation wave[5], a new and specific difficulty arises, due to the chemical change: any experimental value of the normal component of the upstream velocity vector does not allow one and only one flow solution (of course, this difficulty bears some relation with the intersection between Rayleigh line and Crussard curve being two points, one point, or none). This difficulty is circumvented by considering that the detonation velocity to be inserted into the theory is not the experimental value but the first term $D^{(0)}$ of an expansion in the form: $D^{(0)} + \epsilon\, D^{(1)} + O(\epsilon)$. When $D^{(0)}$ equals C-J value D_*, then the wave is quasi-C-J; when $D^{(0)}$ is larger than D_*, then the wave is strong. The "detonation layer" theory enables to settle some propagation rules. Among these, one is of prime importance here: when a detonation is axisymmetric and concave, it is strong. Other rules call on the notion of "autonomous wave" which lies beyond the scope of this Appendix.

References

[1] Brun, L., Chéret, R. and Vacellier, J. "Considérations sur les détonations fortes", *Proceedings of the Symposium High Dynamic Pressures*, Paris, France, 1978, p.269.

[2] Krishnan, S., Brochet, C., and Chéret, R., "Mach Reflexion in Condensed Explosives", *Propellants and explosives*, vol.6, 1981, p.170.

[3] Sellam, M., Presles, H.-N., Brochet, C., and Chéret, R., "Characterization of Strong Detonation Waves in Nitromethane", *Proceedings of the 8th Symposium on detonation*, Albuquerque, NM, 1985, p.425.

[4] Hamada, L., Presles, H.-N., Brochet, C., Bouriannes, R., and Chéret, R., "Characterization of an Overdriven Detonation State in Nitromethane", Vol.94, *Progress in Aeronautics and Astronautics*, AIAA, New York, 1985, p.343.

[5] Chéret, R., "La détonation des Explosifs Condensés", *Masson éditeur*, Paris, France, 1989.

Structure and Velocity Deficit of Gaseous Detonation in Rough Tubes

A. Teodorczyk*
Warsaw University of Technology, Warsaw, Poland

Abstract

Detailed parametric analysis of the structure and velocity deficit of nonideal gaseous detonation waves is reported. The analysis applies to detonations, with wall friction and heat transfer effects taken into account, in rough tubes. The bulk heat release model due to self-ignition of the gas behind the leading shock wave with wall losses has been used. The flow in the reaction zone is assumed to be one-dimensional and steady. The reaction zone profiles of pressure, temperature, density and velocity for the wide range of activation energies, heats of combustion, reaction orders, tube diameters, as well as wall skin-friction coefficients and heat transfer to the wall coefficients, have been calculated. The parametric analysis of the reaction zone length and unburned fraction at C-J plane is also reported. The calculations were performed for the following magnitudes of the parameters: wall friction coefficient $c_f = 0$–0.5, heat transfer coefficient $c_h = 0$–0.05, activation energy $E/R = 12000$–24000 K, heat of reaction $Q = 3 \cdot 10^6$–$15 \cdot 10^6$ J/kg, tube diameter $d = 0.01$–0.1 m, reaction order $m = 0, 1, 2$. Extensive

---

Copyright © 1993 by the American Institute of Aeronautics and Astronautics, Inc. All rights reserved.
*Associate Professor, Institute of Heat Engineering

calculations of velocity deficit were performed and its dependence on problem parameters are reported. The velocity deficit in all calculations has not exceeded $0.16 D_{C-J}$.

Nomenclature

a = speed of sound
A = cross-sectional area
b = channel perimeter
C_f = friction drag coefficient
C_h = heat transfer coefficient
c_p = specific heat at constant pressure
d = hydraulic diameter
D = detonation velocity
E = activation energy
H = enthalpy
k = pre-exponential factor
m = order of chemical reaction
p = pressure
q = heat flux to the wall
Q = heat of reaction
R = gas constant
T = temperature
T_c = stagnation temperature
T_w = wall temperature
w = gas velocity with respect to wall
v = specific volume
x = coordinate
β = unburned mixture fraction
γ = ratio of specific heats
ρ = density
σ = viscous shear stress at the wall

Introduction

In recent years nonideal detonation regimes have been extensively studied[1-5]. This is because in most real explosion hazard problems ideal detonations do not occur. However, the destructions caused by nonideal detonations and their ability to propagate for long distances show that they are comparable to classical waves.

DEFICITS OF NONIDEAL GASEOUS DETONATIONS

In nonideal detonation, energy and momentum losses to the walls can change the wave structure and its velocity of propagation. Rough channel surface changes the ignition conditions behind the leading shock wave. When the temperature behind the shock is not high enough to cause fast ignition, local mixture ignitions, caused by multiple shock reflections from surface roughness, are still possible in the wall vicinity. Another important factor is that wall roughness intensively turbulizes the flow. Depending on the initial and boundary conditions, two different steady combustion waves can possibly occur: 1) a very fast turbulent flame propagating in a group with one or more shock waves in front of it and stabilized by ignition centers close to the wall (Fig. 1a); or 2) a nonideal multiheaded detonation with propagation velocity below C-J detonation velocity (Fig. 1b).

In the second regime the mechanism of wave propagation is similar to classical C-J detonation with ignition behind the shock wave and a multiheaded cell structure. However, a velocity deficit with respect to C-J value is observed, which is caused by energy and momentum losses to the walls.

Zeldovich[1,6] was the first to name this process as quasidetonation. He also believed that the first

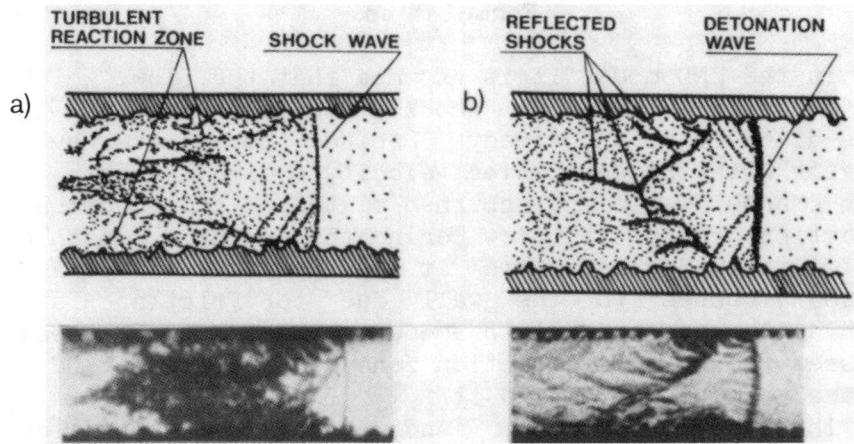

Fig. 1 Schematics and schlieren photographs of two regimes of very fast combustion in rough tubes: a) fast deflagration; b) nonideal detonation.

regime with shock separated from the flame can be classified as low-velocity quasidetonation. Detailed studies performed by Lee[2] and Teodorczyk[4,5,7] have not confirmed this hypothesis; they have shown that this is a deflagration wave with the propagation velocity not exceeding the sound speed in combustion products.

Quasidetonation, like galloping detonation, is highly unstable; it exists in cyclic reinitiations and failures of the multiheaded detonation wave, with phases of overdriven detonation and very fast turbulent deflagration[4,5]. As a result, the average propagation velocity over a long distance is much lower than C-J value (up to 50%).

Thus, restricting the name of nonideal detonation to the case of stable propagation velocity in the length scale of several cell dimensions is proposed. In this regime, the detonation wave preserves all characteristic features of classical, ideal C-J wave with a small velocity deficit.

This paper is a detailed parametric study of the structure and velocity of the gaseous detonation wave in this nonideal regime. The well-known Zeldovich-Von Neuman-Doring (ZND) detonation wave model, modified for heat and momentum losses to the walls, has been used for the calculations[6].

Formulation

In the ZND model it is assumed that the flow, which is shown in Fig. 2 in shock-fixed coordinates, is one-dimensional and steady. Behind the shock wave, moving with the detonation velocity D with respect to the channel walls and unburned mixture, the chemical reaction occurs. The flow behind the shock is subsonic until the C-J plane where it reaches the speed of sound. Losses to the wall due to friction and convection are taken into account. Because of these losses the chemical reaction zone does not end at C-J plane but extends behind it.

The mass, momentum, and energy conservation equations are as follows:

$$\rho(D - w) = \rho_o D \qquad (1)$$

DEFICITS OF NONIDEAL GASEOUS DETONATIONS

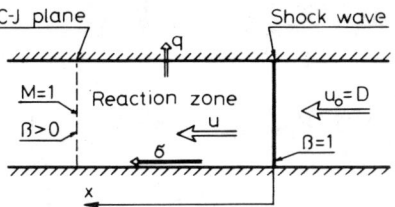

Fig. 2 Schematic of ZND model of nonideal detonation.

$$\frac{d}{dx}[p + \rho(D - w)^2] = \frac{b}{A}\sigma \qquad (2)$$

$$\frac{d}{dx}\left\{\rho(D - w)[H + \frac{(D - w)^2}{2}]\right\} = \frac{b}{A}(-q + \sigma D) \qquad (3)$$

The shear stress at the wall is

$$\sigma = C_f \rho \, w \, \frac{|w|}{2} \qquad (4)$$

and the heat flux is

$$q = C_h c_p \rho \, w \, (T_c - T_w) \qquad (5)$$

The gas enthaply is given by the formula

$$H = Q\beta + \frac{\gamma p}{\rho(\gamma - 1)} \qquad (6)$$

An unburned mixture fraction is described by the chemical kinetics equation

$$\frac{d\beta}{dx} = -\frac{k\beta^m}{D - w}\exp\left(-\frac{E}{RT}\right) \qquad (7)$$

After differentiation and rearrangement, Eqs. (2) and (3) take the form

$$\frac{dp}{dx} = \frac{b}{A}C_f \rho_o D \frac{w\,|w|}{2(D-w)} + \rho_o D \frac{dw}{dx} \qquad (8)$$

$$\frac{dw}{dx} = \left[S_1 - \rho_o D Q \frac{d\beta}{dx}\right] / S_2 \qquad (9)$$

where S_1 and S_2 are

$$S_1 = \frac{b}{A}\left[C_f \frac{w}{2} |w| \rho_o \frac{D^2}{(D-w)} - C_f \rho_o \frac{w}{2} |w| D \frac{\gamma}{(\gamma-1)} \right.$$
$$\left. - C_h c_p w \rho_o \frac{D}{(D-w)}\left(T + \frac{w^2}{2c_p} - T_w\right)\right] \qquad (10)$$

$$S_2 = \frac{\rho_o D (D-w) \gamma}{(\gamma-1)} - \frac{\gamma p}{\gamma-1} - \rho_o D (D-w) \qquad (11)$$

It is assumed that the gas is ideal, and so the equation of state is

$$\frac{p}{\rho} = RT \qquad (12)$$

The solutions of Eqs. (7), (8) and (9) must satisfy the following boundary conditions:

for $x = 0$

$$\beta = 1$$
$$\rho = \rho_o \frac{\gamma+1}{\gamma-1}$$
$$p = \frac{2D^2 \rho_o}{\gamma+1} \qquad (13)$$
$$w = \frac{2D}{\gamma+1}$$

for $x \to \infty$

$$\rho \to \rho_o$$
$$p \to p_o \qquad (14)$$

In the formulation of boundary conditions it was assumed that the chemical reaction does not change the number of moles of substance. The conditions for $x=0$

were calculated from the equations for the parameters behind the shock front.

To close the model the C_f and C_h coefficient must be defined. In general, C_f and C_h are the functions of the distance behind the shock wave. Detailed methods for their calculations are given in Refs. 8 and 9. In this parametric study it was assumed that C_f and C_h have constant values averaged along the distance from the shock wave to the C-J plane.

The integration of Eqs. (7)-(9) (together with boundary conditions and auxiliary relationships) allows for calculations of detonation wave structure provided that the detonation velocity is known.

The model can also be used for calculations of detonation velocity if the heat and momentum losses to the walls are specified. In these calculations the singularity of model differential equations must be taken into the account. To get a continuous solution from shock conditions to the state when combustion products are totally cooled down and stopped it is necessary to satisfy two conditions at singular point[1,3,6]:

$$D - w = a \qquad (15)$$

$$-Q \frac{d\beta}{dx} = \frac{b}{A} \left[-\frac{q}{\rho(D-w)} - \sigma\left(\frac{1}{\rho(\gamma-1)} - \frac{1}{\rho_o} + \frac{1}{\rho}\right) \right] \qquad (16)$$

The first condition states that the gas velocity withrespect to the detonation wave is equal to the sound speed. The second condition, which expresses the balance between energy generation and its losses, is the equivalent of the tangency condition between Reyleigh and Hugoniot lines in ideal detonation. It follows that at the position where D-w=a there is still some mixture unburned (because $d\beta/dx \neq 0$).

Method of Solution

The detonation velocity has been calculated with the use of an iterative procedure. The velocity was assumed initially, and then calculations of the reaction zone structure were performed to the point where one of the conditions ((15) or (16)) was satisfied. Then a new value of D was assumed and calculations repeated until two conditions were satisfied simultaneously.

For larger values of C_f close to the limit of the detonation existence the model gives two solutions for the velocity. The bigger value is identified as the real detonation velocity, and the lower one as the unstable detonation velocity. This fact was shown previously by Zeldovich et al.[1]. However, there is the limiting value of C_f for which the velocity curve has the sadle with one solution. This point defines the maximum velocity deficit for nonideal detonation.

The ordinary differential equations ((7)-(9)) were integrated numerically with the use of DVERK subroutine (from IMSL).

Computational Results

The calculations were carried out for the following ranges of parameters:

Friction drag coefficient

$$C_f = 0 - 0.5$$

Heat transfer coefficient

$$C_h = 0 - 0.05$$

Hydraulic diameter

$$d = 0.01 - 0.1 \text{ m}$$

Order of chemical reaction

$$m = 0, 1, 2$$

Heat of reaction per 1-kg of mixture

$$Q = 2 \cdot 10^6 - 16 \cdot 10^6 \text{ J/kg}$$

Activation energy of chemical reaction

$$E/R = 12000 - 24000 \text{ K}$$

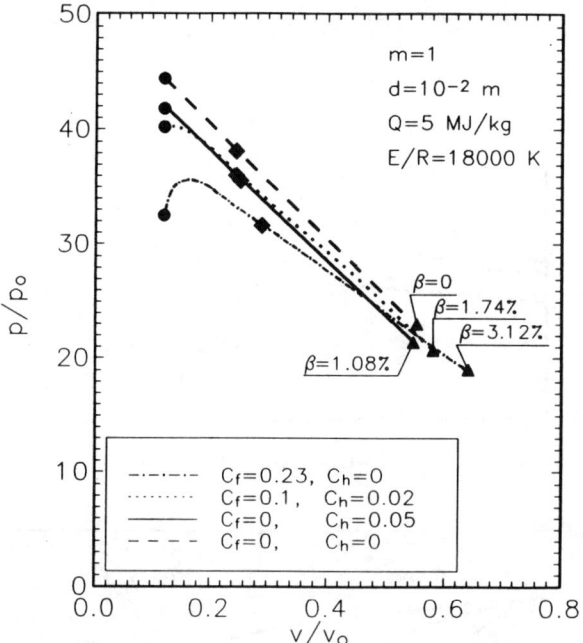

Fig. 3 Calculated Rayleigh lines for different nonideal detonations: ● - initial state behind the leading shock wave, ▲ - final states at CJ plane, ■ - 50% burnout of the mixture.

Pre-exponential factor

$$k = 10^{10} \ 1/s$$

Specific heat ratio

$$\gamma = 1.3$$

Generally, in all calculated cases detonation velocity deficit never exceeded $\Delta D = D_{CJ} - D = 0.16 D_{CJ}$, where $D_{CJ} = \sqrt{2(\gamma^2 - 1)Q}$ is the C-J detonation velocity.

Figure 3 shows the calculated Rayleigh lines for typical values of parameters: $Q = 5 \cdot 10^6$ J/kg, $E/R = 18000$ K, $m = 1$, and $d = 0.01$ m. This figure illustrates the changes in the path of Rayleigh lines with the change of C_f and C_h. Solid circles describe the initial conditions of the combustible mixture directly behind the shock wave where $\beta = 1$. Triangles stand for the respective final states, and diamonds describe the

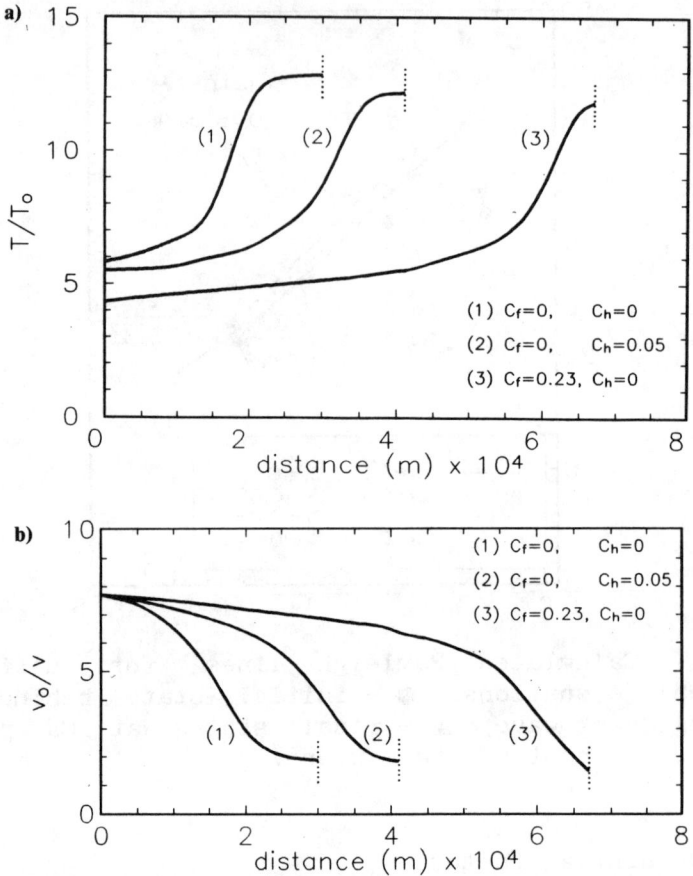

Fig. 4 Variation of nondimensional temperature (a) and specific volume (b) in the reaction zone of nonideal detonation; $m=1$, $d=0.01$ m, $E/R=18000$ K, $Q=5$ MJ/kg.

conditions when 50% of the mixture is burned out. For each final point the unburnt mixture fraction is given in the figure. It is evident from the plot that heat and momentum losses decrease the pressure in the reaction zone. The momentum losses separately cause curving of the Rayleigh line, and they move the point of maximum pressure (von Neuman spike) from the shock wave inside the reaction zone. Also, with increasing C_f, the unburnt mixture fraction β increases at C-J plane.

Figures 4 and 5 show the profiles of pressure, temperature, and specific volume related to the

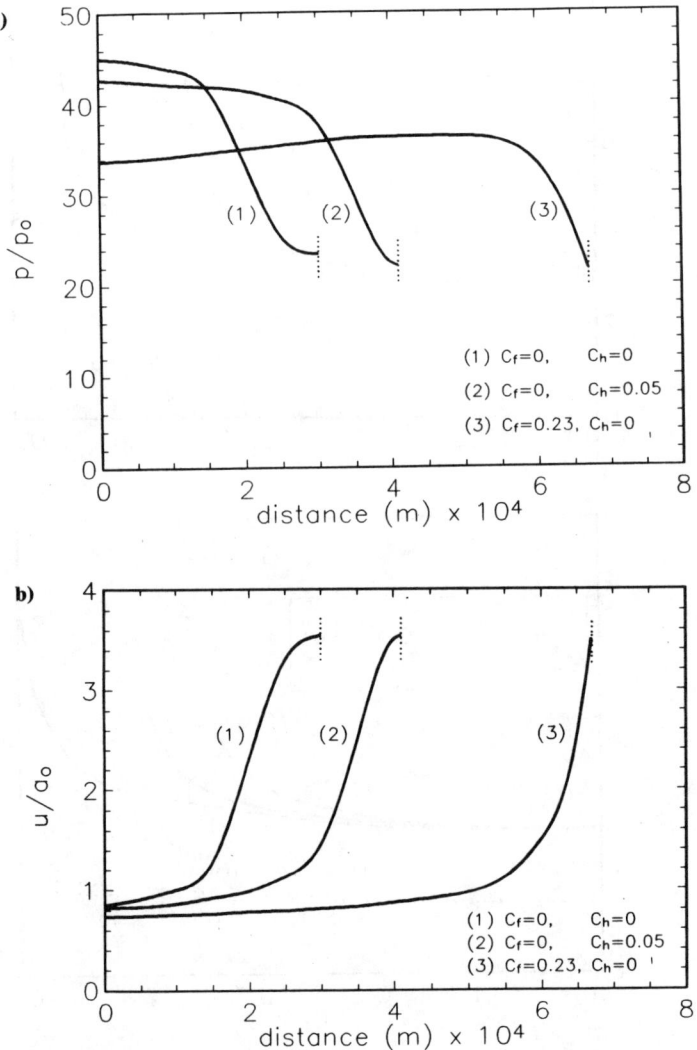

Fig. 5 Variation of nondimensional pressure (a) and gas velocity with respect to the wave (b) in the reaction zone of nonideal detonation; $m=1$, $d=0.01$ m, $E/R=18000$ K, $Q=5$ MJ/kg.

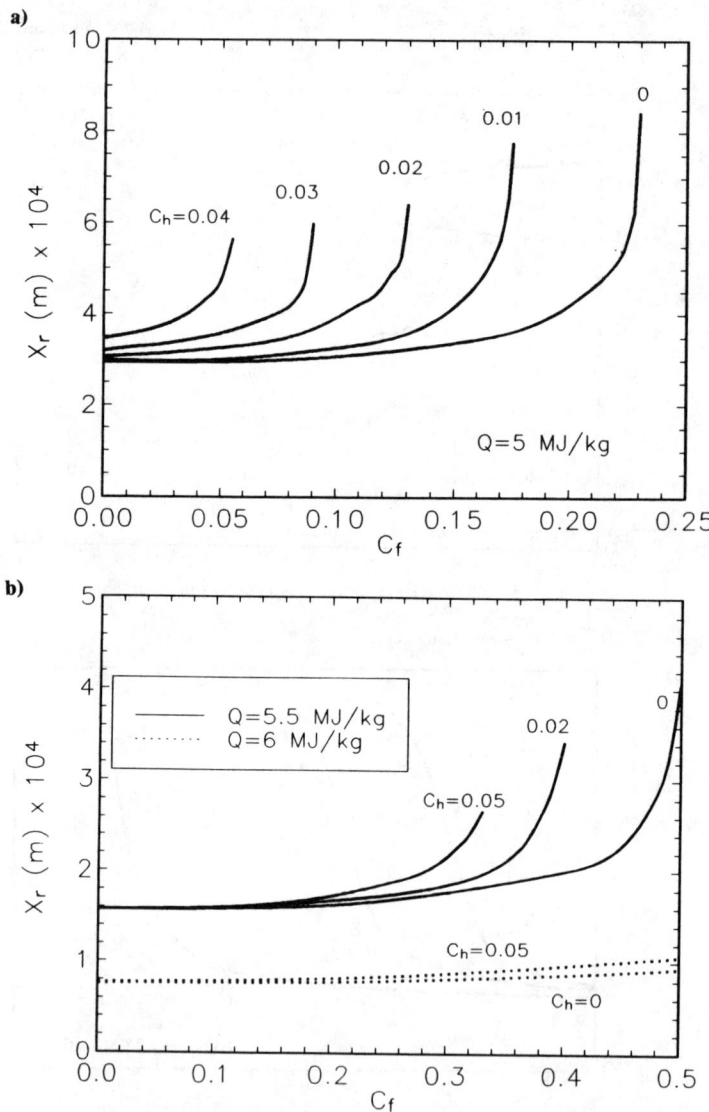

Fig. 6 Dependence of reaction zone length on drag coefficient for several values of heat transfer coefficient and three values of heat of reaction; $m=1$, $d=0.01$ m, $E/R=18000$ K.

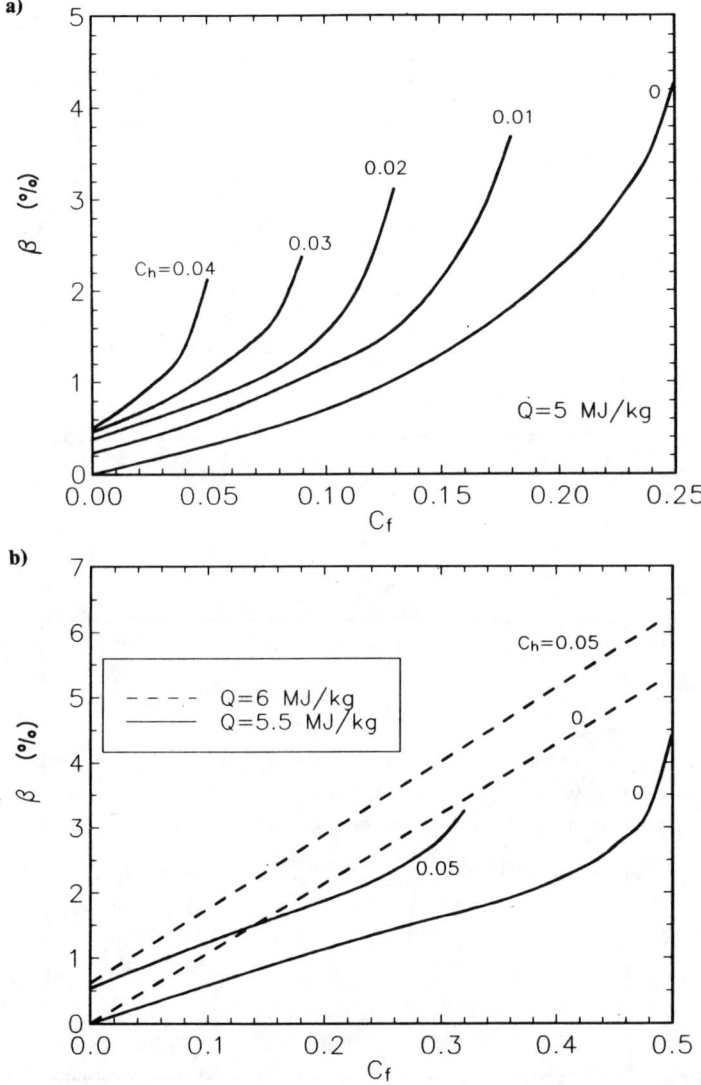

Fig. 7 Unburnt mixture fraction at C-J point versus drag coefficient for several values of heat transfer coefficient and three values of heat of reaction; $m=1$, $d=0.01$ m, $E/R=18000$ K.

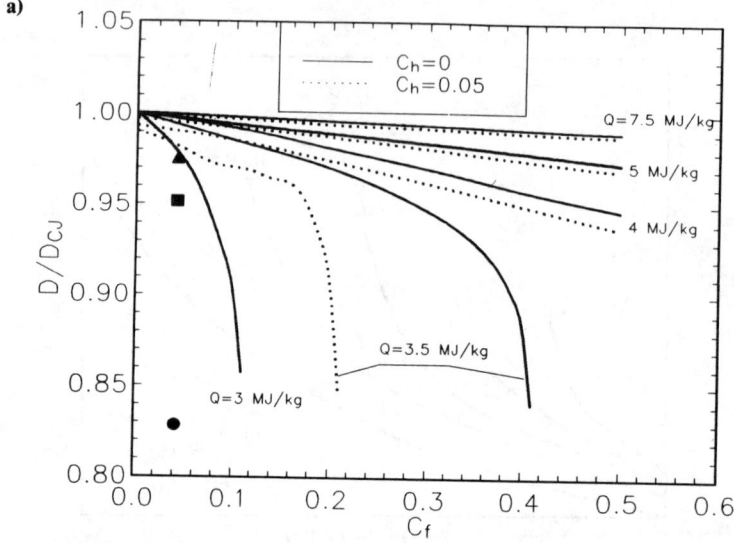

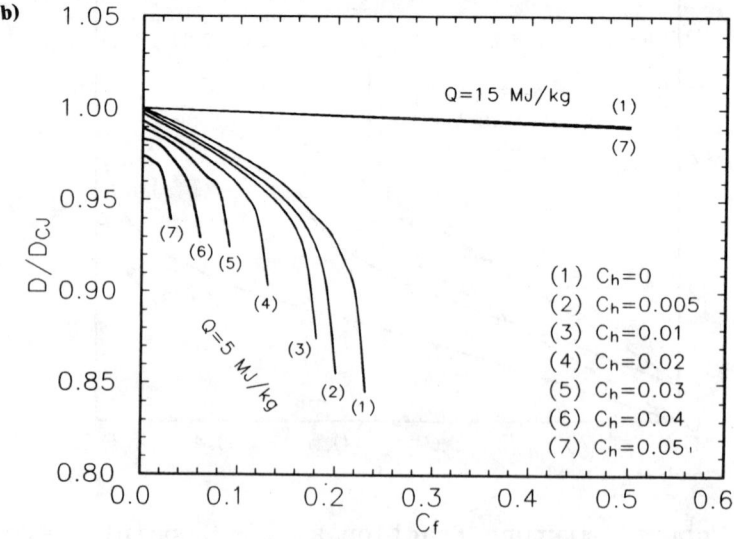

Fig. 8 Nondimensional detonation velocity versus drag coefficient for different values of heat transfer coefficient, heat of reaction and two values of activation energy; m=1, d=0.01 m; a) E/R=12000 K, ●, ■, ▲ - experimental points of Lee[12] for stoichiometric acetylene-air mixture; b)and c)E/R=18000 K.

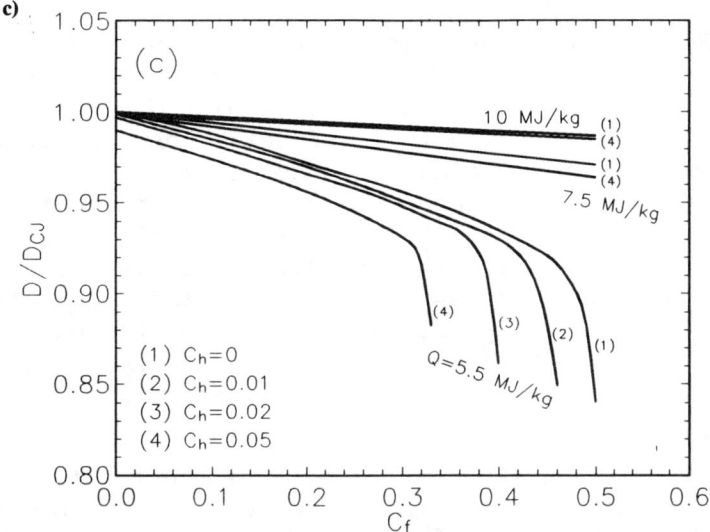

Fig.8 (continued) Nondimensional detonation velocity versus drag coefficient for different values of heat transfer coefficient, heat of reaction and two values of activation energy; m=1, d=0.01 m; a) E/R=12000 K, ●, ■, ▲, - experimental points of Lee[12] for stoichiometric acetylene-air mixture; b) and c) E/R = 18000 K.

initial values and gas velocity with respect to detonation wave related to initial sound speed in the reaction zone. The values of the parameters Q, E/R, m, and d are the same as in Fig. 3. Dot vertical lines show the positions of C-J plane. These figures demonstrate the change in detonation wave structure with C_f and C_h. The most important changes in comparison with the ideal case are the increase of the length of reaction zone and decrease of absolute values of pressure and temperature.

Figure 6 shows the dependence of the length of reaction zone on C_f for different values of C_h and three values of Q. For constant C_f the increase in heat transfer results in the increase of reaction zone length while the increase of heat of reaction gives the opposite result.

Figure 7 shows the unburned mixture fraction β in terms of C_f for different values of C_h and three values of Q. The detonation parameters are the same as in Fig. 6. The lines in Fig. 7 demonstrate that, neglecting heat losses, the unburned mixture fraction at the C-J plane changes linearly with C_f. Heat losses additionally increase this effect. The increase of heat of reaction Q also increases β.

Figure 8 demonstrates the nondimensional detonation velocity in terms of C_f for different values of Q and two values of activation energy. It is clearly visible that in some cases at particular conditions, detonation velocity decreases very fast for particular C_f. This phenomenon is commonly observed experimentally in the studies of detonability limits [2,10,11,12].

The limiting value of C_f for which detonation still exists increases with the increase of Q and decreases with the increase of E. For bigger values of E the influence of Q is stronger. Heat losses influence detonation velocity only close to the limits.

Figure 9 shows nondimensional detonation velocity as a function of C_f for two values of the order of reaction and several hydraulic diameters. For a higher order of reaction, the lower detonation velocity is observed with narrower limits. The same effect is observed for the hydraulic diameter.

Figure 10 shows the nondimensional detonation velocity in terms of heat of reaction for three values of the ratio C_f/d. The lines that are shown in this plot are qualitatively consistent with experimental dependencies of detonation velocity on mixture composition or initial pressure obtained by Matsui et al.[10] and the author. Figure 10 also demonstrates an interesting feature that which is very important from a practical point of view; it shows the universal character of the C_f/d parameter which makes it possible to apply experimental data obtained in a laboratory scale to large-scale real processes.

Calculations have shown that, far from the detonation limits, the influence of tube diameter is small and can be neglected. This conclusion is in agreement with results of Rybanin[11] and Zeldovich[1]. Thus, the results from small-scale experiments can be transformed to large-scale situations provided that the wall friction drag coefficient C_f will be conserved.

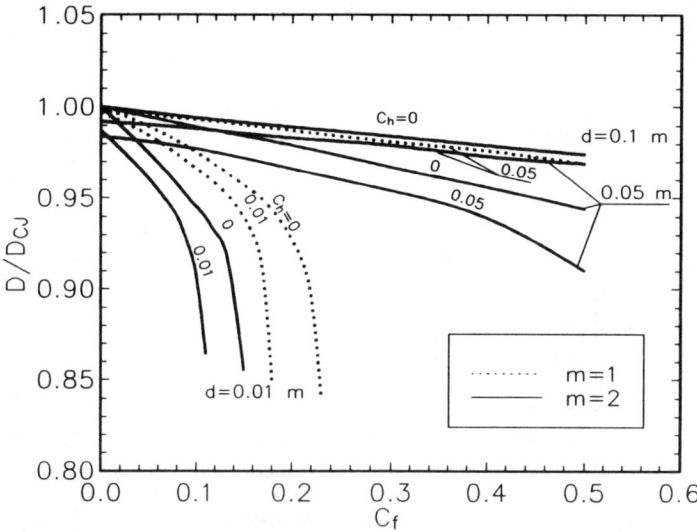

Fig. 9 Nondimensional detonation velocity versus drag coefficient for several values of heat transfer coefficient, of hydraulic diameter, and two values of chemical reaction order; E/R =18 000 K, Q=5 MJ/kg.

Experimental Validation

There is a general lack of experimental data concerning the velocity deficit of steady multiheaded gaseous detonations in rough tubes and channels. Only very limited data are available for some particular experimental conditions. This is mainly because the very small velocity deficit, which exists in most of the experiments with nonideal detonations, is difficult to record without special techniques. Using ordinary experimental methods this deficit is usually within experimental error.

The first data were reported by Shchelkin[13] for detonation of stoichiometric hydrogen-oxygen mixtures in the tubes of 7 and 18 mm in diameter, each containing a metal spiral. The geometrical parameters of Shchelkin's experiments were recently recalculated by Frolov et al.[14] to give the average drag coefficient for both cases equal to C_f = 0.036. The velocity deficit in the smaller tube was D/D_{CJ} = 0.845. These two experimental points are shown in Fig. 10; they obviously do not coincide with theoretical results.

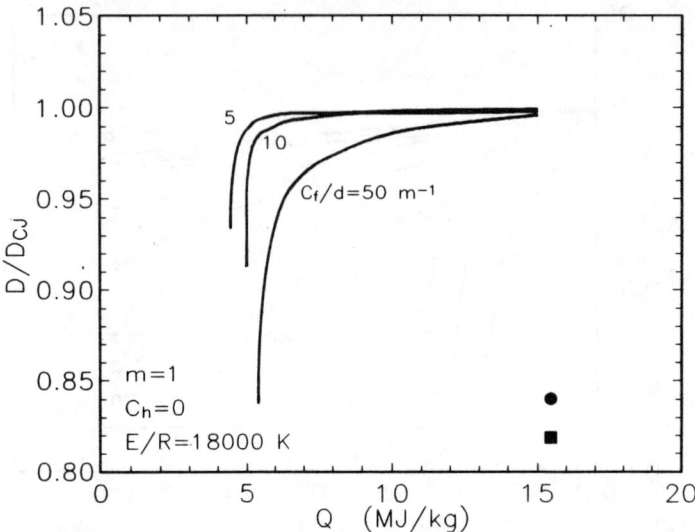

Fig. 10 Nondimensional detonation velocity versus heat of reaction for three values of C_f/d parameter; ● , ■ - experimental points of Shchelkin[13] for stoichiometric hydrogen-oxygen mixture.

Other data were reported by Lee[12] for the stoichiometric mixture of acetylene and air. The nonideal detonations were studied in tubes with three diameters: 50, 150 and 300 mm containing baffle plates with different blockage ratios. The total momentum losses in the reaction zone were similar in all cases according to the analysis of Lee's data by Gelfand and Frolov[15]. The kinetic parameters for this mixture are: $k=10^{10}$ 1/s, $E/R=12000$ K, $m=1$, and $Q=2.9$ MJ/kg. Three of Lee's experimental points are shown in Fig. 8a. Although the theoretical results are for the smaller tube diameters, it can be concluded that experimental points lay in the region of possible real solutions; they do not generally disagree with theoretical predictions.

As was stated, more experimental data are needed for the analysis of this practical and important regime of detonation; this is planned by the author.

References

[1] Zeldovich, Y. B., Borisov, A. A., Gelfand B. E., Frolov, S. M., and Mailkov, A. E., "Nonideal Detonation Waves in Rough Tubes", *Dynamics of Reactive Systems*, edited by A. L. Kuhl, J. R. Bowen, J.-C. Leyer, and A. Borisov, Vol. 113, Progress in Astronautics and Aeronautics, AIAA, Washington, DC, 1988, pp. 211-231.

[2] Peraldi, O., Knystautas, R., and Lee, J. H., "Criteria for Transition to Detonation in Tubes", Twenty-First Symposium (International) on Combustion, The Combustion Institute, Pittsburgh, PA, 1986.

[3] Gelfand, B. E., Frolov, S. M., and Polenov, A. N., "Specific Features of Detonation in Systems with Losses of an Arbitrary Type", *Archivum Combustionis*, Vol. 7, No. 1-2, 1987, pp. 197-214.

[4] Teodorczyk, A., Lee, J. H. S., and Knystautas, R., "Propagation Mechanism of Quasi-Detonations", Twenty-Second Symposium (International) on Combustion, The Combustion Institute, Pittsburgh, PA, 1988, pp. 1723-1731.

[5] Teodorczyk, A., Lee, J. H. S., and Knystautas, R., "Photographic Studies of the Structure and Propagation Mechanisms of Quasi-Detonations in a Rough Tube", *Proceedings of the 12th International Colloquium on the Dynamics of Explosions and Reactive Systems*, Ann Arbor, MI, 1989.

[6] Zeldovich, Y. B., and Kompaneets, A. S., *Theory of Detonation*, Academic Press, New York, 1960.

[7] Teodorczyk, A., Lee, J. H. S., and Knystautas, R., "The Structure of Fast Turbulent Flames in Very Rough Obstacle Filled Channels", Twenty-Third Symposium (International) on Combustion, The Combustion Institute, Pittsburgh, 1990.

[8] Dalle Donne, M., and Meyer, L., "Turbulent Convective Heat Transfer from Rough Surfaces with Two-Dimensional Reactangular Ribs", *International Journal of Heat and Mass Transfer*, Vol. 20, 1977, pp. 583-620.

[9] Kuhl, A. L., and Fink, S. F., IV, "Numerical Calculations of Shock Decay in Rib Walled Ducts", TRW Rept. 78.4735.9-14, 1978.

[10] Matsui, H., and Lee, J. H., "On the Measure of the Relative Detonation Hazards of Gaseous Fuel-Oxygen and Air Mixtures", Seventeenth Symposium (Int.)on Combustion, The Combustion Institute, Pittsburgh, PA, 1978, pp.1269-1280.

[11] Rybanin, S. S., "Detonation Theory in Rough Tubes", *Physics of Combustion and Explosions*, (in Russian), Vol. 5, 1969, pp.395-403.

[12] Lee, J. H. S., "The Propagation of Turbulent Flames and Detonations in Tubes", *Advances in Chemical Reaction Dynamics*, edited by P. M. Rentzepis and C.Capellos, D.Reidel Publishing Company, 1986, pp.345-378.

[13] Shchelkin, K. I., "Decrease of Detonation Velocity in Rough Tubes", *Acta Phys.-Chem. USSR*, Vol.20, 1945.

[14] Gelfand, B. E., and Frolov, S. M., "Similarity Criteria for Detonative Systems", *Chemical Physics (USSR)*, (in Russian), Vol.7, No.3, 1988, pp.397-405.

[15] Frolov, S. M., Polenov, A. N., Gelfand, B. E., and Borisov, A. A., "The Characteristics of Detonations in Systems with Losses of an Arbitrary Type", *Chemical Physics (USSR)*, (in Russian), Vol.5, No.7, 1986, pp.987-988.

Possible Method for Quenching of Gaseous Detonations

J. Bakken* and O. K. Sønju†
Norwegian Institute of Technology, Trondheim, Norway
D. Bjerketvedt‡
Christian Michelsen Institute, Bergen, Norway
and
T. Engebretsen§
Norwegian Defense Construction Service, Oslo, Norway

Abstract

Experimental results for quenching of gaseous detonations propagating in acetylene/air and ethylene/air mixtures at atmospheric pressure using water are presented, and mechanisms for quenching are discussed. The apparatus consisted of an 8-m-long square tube with an internal dimension of 125 mm. A tube bundle consisting of 36 square tubes with an internal dimension of 17 mm was mounted inside the large tube. Tests were performed with and without a 2-mm layer of water in the bottom of each tube. Three different tube bundle lengths were used: 0.3, 0.5, and 1.0 m. Without water in the tubes the detonation did not reinitiate when the acetylene concentration was less than 4.9% for the 0.3-m tube bundle length and less than 5.5% for the 1.0-m tube bundle length. However, radar Doppler measurements clearly showed that a flame was still propagating downstream of the tube bundle. With a layer of water in the tubes the detonation did not reinitiate after the tube bundle when the acetylene concentration was less than 6.2, 6.4, and 6.6% for the 0.3-, 0.5-, and 1.0-m tube bundle lengths, respectively. The $S = d_c/13$

Copyright © 1992 by the American Institute of Aeronautics and Astronautics, Inc. All rights reserved.
*Research Engineer, Thermal Energy Division.
† Professor, Thermal Energy Division.
‡ Senior Scientist, Department of Science and Technology.
§ Research Scientist.

correlation (S is the transverse cell width, d_c is the critical tube diameter) using the data of Moen et al. gives a cell width for 6.4% acetylene/air that is approximately equal to the internal dimension of the small tubes. With ethylene/air mixtures and a layer of water in the tubes the detonation reinitiated in only one test. The minimum cell width for ethylene/air mixtures is at least 30% greater than the internal dimensions of the tubes in the tube bundle. The results indicate that when $d_c/13$ divided by the internal dimension of the tubes is approximately equal to or greater than 1, the detonation will be quenched when a water layer is used in the tubes. The velocity of the shock front measured between two pressure transducers downstream of the tube bundle was reduced to less than 750 m/s, and the pressure was reduced to less than 4 bars from the CJ conditions in front of the tube bundle when there was a layer of water in the tubes.

Introduction

This paper presents experimental results for quenching of gaseous detonations in acetylene/air and ethylene/air mixtures at atmospheric pressure using water. Quenching of detonation waves using water have been studied by several investigators. Thomas et al.[1] investigated the use of water sprays for quenching of detonation waves. They used a 10% velocity deficit from the predicted CJ velocity as the criterion taken for quenching in C_2H_2-O_2-Ar, C_2H_2-O_2-N_2, H_2-O_2-Ar, C_2H_6-O_2, and C_2H_6-O_2-N_2 mixtures. They observed that detonations with regular wave structures were more easily quenched than detonations with irregular structures. Borisov et al.[2] showed that a thin layer of water on the bottom wall of the tube was enough to quench a detonation in C_3H_8-O_2-N_2 under certain conditions. Jenssen[3] was able to stop detonations in acetylene/air and even quench the flame when water was dispersed by detonating a high explosive charge inside a water container. These large-scale tests were performed at Raufoss in 1985 and 1986. Jenssen[3] also successfully used tube bundles to decouple the shock wave and the flame front of a detonation. He used a 3-m-long tube bundle mounted inside a tube with internal diameter of 0.64 m. The tubes were 15 mm in diameter. A detonation in a 7.0% acetylene/air mixture was initially decoupled, but reinitiated after the tube bundle.

In the present study the possibility of applying the two concepts of a tube bundle and a water layer has been investigated. The concept investigated was based on the idea that for a certain cell width the detonation would decouple in a square tube. Based on the tests by Engebretsen et al.,[4] this cell width was expected to be equal to the width/height of the tubes in the tube bundle for acetylene/air mixtures. However, the cell width is not a unique dimension.[5,6] In this report the cell

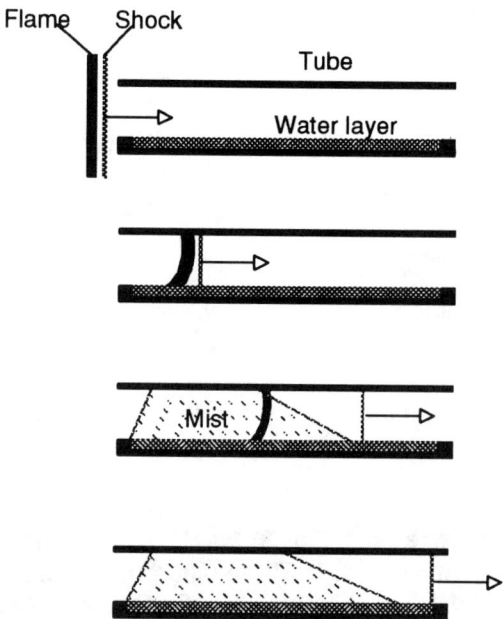

Fig. 1 Illustration of the quenching process in the tubes.

width based on $d_c = 13 \cdot S$ by Moen et al.[6] will be used for acetylene/air and ethylene/air mixtures. When the shock front and the reaction zone have been decoupled the high velocity flow behind the shock wave would strip the water surface and form a micromist between the shock and the flame front. This micromist would then be able to quench the flame (see Fig. 1).

In a practical situation, using water sprays as a quenching device has clear disadvantages since it is an active system that has to be triggered prior to arrival of the detonation wave. However, passive devices for quenching of detonations often represent an unacceptable impedance to the flow under normal operating conditions (i.e., large pressure drop). There is, therefore, a need for a passive system with low impedance such as investigated in the present work.

Experimental Setup

Figure 2 shows the experimental apparatus. The test section consists of a square tube with an internal dimension of 125 mm and a length of 8 m. The test section can be divided into three sections by two slide valves. Section 3 can be lengthened with a 1.15-m tube with windows on both sides, as shown in Fig. 2. The three sections can be filled separately with different gas mixtures. In this study the gas composition was the same in all sections.

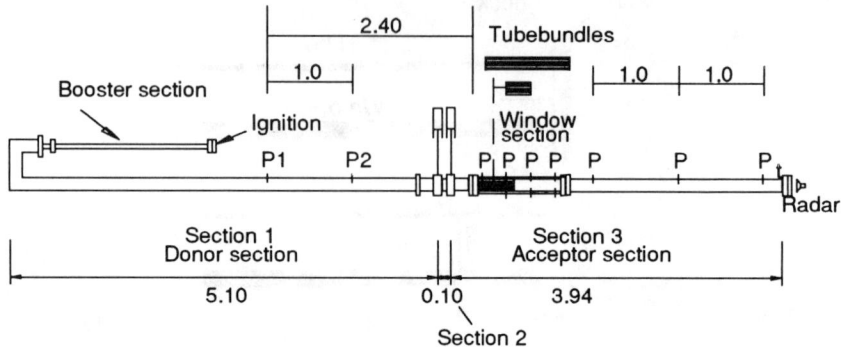

Fig. 2 Experimental setup showing position of tube bundle and pressure transducers.

A tube bundle (see Fig. 3) was placed in the acceptor section (section 3) as shown in Fig. 2. Three different tube bundle lengths were used: 0.3, 0.5, and 1.0 m. The tube bundle consisted of 36 square tubes with an external dimension of 20 mm and an internal dimension of 16 mm for the 0.5-m tubes, and 17 mm for the 0.3- and 1.0-m tubes. In the bottom of each tube it was possible to have a 2-mm-thick layer of water, which gives a total amount of 0.37, 0.58, and 1.22 liters of water for the three tube bundles, respectively. The thickness of the water layer was chosen to be 2 mm to make sure that the whole surface was covered with water and still minimize the amount of water, and to make the open area ratio (OAR) as large as possible. The 1.15-m section was used in all the tests with the 1.0-m tube bundle and in some tests with the 0.3- and 0.5-m tube bundles.

The gas mixture was produced by using bottles of compressed air and commercial grade acetylene and ethylene. The gas was premixed prior to entering the test section. The tube was purged with the premixed gas for at least 10 tube volumes prior to the experiment. The inlet gas composition was continuously monitored by a Wilkins Miran 1A IR-analyzer (+/- 0.05%).

A booster section filled with acetylene/oxygen was used to initiate the detonation wave in the donor section (section 1). Prior to the initiation, the pressure in the apparatus was vented to atmospheric pressure.

The diagnostic system consisted of pressure transducers (Kistler 603B) and an X-band radar Doppler unit (M/A-COM type MA86656-D with a wavelength of 30 mm). Two pressure transducers (P1 and P2) were mounted in the donor section, upstream of the tube bundle. When the 0.3- and 0.5-m tube bundles were used, two pressure transducers were mounted in the acceptor section after the tube bundle. When the 1.0-m tube bundle was used and the acceptor section was lengthened with 1.15

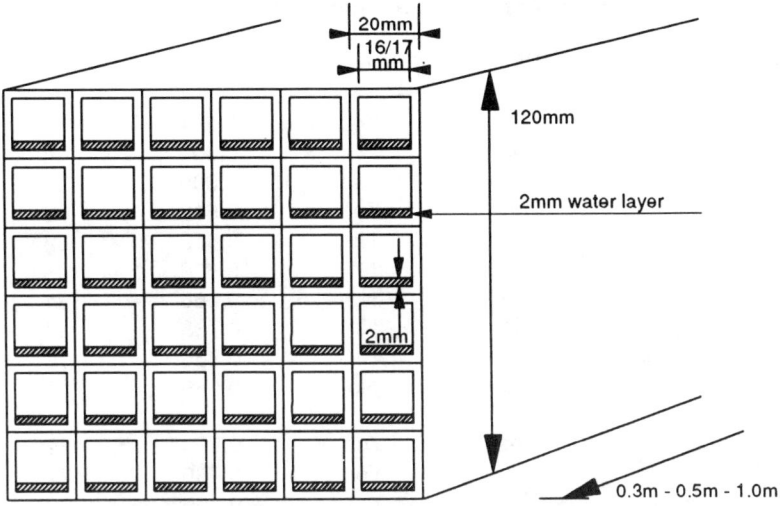

Fig. 3 Tube bundle with 2-mm water layer.

m, three pressure transducers were mounted in the acceptor section after the tube bundle. The Doppler unit was located at the end of the acceptor section. A high-density polyethylene plate was used as a "window" for transmission of microwaves into the tube. However the Doppler unit was unable to "see" through the tube bundle because the wavelength of the microwaves was too large. With the pressure transducers the establishment of a CJ detonation in the donor section was verified, and the propagation of the shock wave in the acceptor section after the tube bundle was monitored. The Doppler unit and the pressure transducers were used to observe whether the detonation reinitiated or not after the tube bundle.

The diagnostic signals were recorded on a transient data logger with sampling rate up to 10^7 samples/s. The data logger was triggered by time of arrival of the detonation at the pressure transducer P1. The data logger was connected to a personal computer (PC) for arming, transferring, and presentation of data.

Experimental Results

The experimental conditions and the results are given in Tables 1-3. Mainly acetylene/air mixtures were used in this study, but several tests where also carried out with ethylene/air mixtures. Ethylene/air mixtures with the smallest cell width was tested. The calculated values of $d_c/13$ and the equivalence ratio Φ are given .

Table 1 Experimental results, tube bundle 0.3 m

Gas	Water	Quenching	Φ^a	$d_c/13$ mm
4.50% C_2H_2/air	No	Yes	0.561	70
4.80% C_2H_2/air	No	Yes	0.600	54
4.80% C_2H_2/air	No	Yes	0.600	54
4.90% C_2H_2/air	No	Yes	0.613	52
4.90% C_2H_2/air	No	No	0.613	52
5.00% C_2H_2/air	No	No	0.626	45
5.30% C_2H_2/air	No	No	0.666	34
5.80% C_2H_2/air	No	No	0.733	24
6.80% C_2H_2/air	No	No	0.868	14
7.50% C_2H_2/air	No	No	0.965	10
5.50% C_2H_2/air	Yes	Yes	0.693	29
6.00% C_2H_2/air	Yes	Yes	0.760	21
6.05% C_2H_2/air	Yes	Yes	0.766	20
6.10% C_2H_2/air	Yes	Yes	0.773	20
6.20% C_2H_2/air	Yes	Yes	0.787	19
6.20% C_2H_2/air	Yes	Yes	0.787	19
6.20% C_2H_2/air	Yes	No	0.787	19
6.30% C_2H_2/air	Yes	Yes	0.800	18
6.30% C_2H_2/air	Yes	Yes	0.800	18
6.30% C_2H_2/air	Yes	No	0.800	18
6.30% C_2H_2/air	Yes	No	0.800	18
6.40% C_2H_2/air	Yes	Yes	0.814	16
6.40% C_2H_2/air	Yes	No	0.814	16
6.40% C_2H_2/air	Yes	No	0.814	16
6.40% C_2H_2/air	Yes	Yes	0.814	16
6.50% C_2H_2/air	Yes	No	0.827	16
6.50% C_2H_2/air	Yes	No	0.827	16
6.60% C_2H_2/air	Yes	Yes	0.841	15
6.70% C_2H_2/air	Yes	No	0.855	14
6.80% C_2H_2/air	Yes	No	0.868	14
6.90% C_2H_2/air	Yes	Yes	0.882	13
6.90% C_2H_2/air	Yes	No	0.882	13
7.00% C_2H_2/air	Yes	No	0.896	13
7.50% C_2H_2/air	Yes	No	0.965	11
5.50% C_2H_4/air	No	Yes	0.831	53
5.90% C_2H_4/air	No	Yes	0.895	43
6.20% C_2H_4/air	No	Yes	0.944	38
6.40% C_2H_4/air	No	Yes	0.976	35
6.55% C_2H_4/air	No	No	1.009	33
6.80% C_2H_4/air	No	Yes	1.042	31
7.00% C_2H_4/air	No	Yes	1.075	29.
7.50% C_2H_4/air	No	Yes	1.158	26

(Table continued on next page)

Table 1 (cont.) Experimental results, tube bundle 0.3 m

Gas	Water	Quenching	Φ^a	d_c/13 mm
7.80% C_2H_4/air	No	Yes	1.208	25
7.90% C_2H_4/air	No	Yes	1.225	25
8.00% C_2H_4/air	No	No	1.242	25
8.20% C_2H_4/air	No	Yes	1.276	24
8.40% C_2H_4/air	No	Yes	1.310	24
8.50% C_2H_4/air	No	No	1.327	24
8.60% C_2H_4/air	No	No	1.344	24
8.70% C_2H_4/air	No	No	1.361	24
8.80% C_2H_4/air	No	No	1.378	24
8.80% C_2H_4/air	No	No	1.378	24
8.90% C_2H_4/air	No	Yes	1.395	24
8.95% C_2H_4/air	No	No	1.404	24
9.00% C_2H_4/air	No	Yes	1.412	24
6.55% C_2H_4/air	Yes	Yes	1.001	34
6.80% C_2H_4/air	Yes	Yes	1.042	31
7.00% C_2H_4/air	Yes	Yes	1.075	29
7.50% C_2H_4/air	Yes	Yes	1.158	26
8.00% C_2H_4/air	Yes	Yes	1.242	25
8.50% C_2H_4/air	Yes	Yes	1.327	24
9.00% C_2H_4/air	Yes	Yes	1.412	24

^{a} Φ = Equivalence ratio

For tests without water in the tube bundle, the acetylene concentration ranged from 4.5-7.5%. When the 0.3-m tube bundle was used, the detonation did not reinitiate when the acetylene concentration was less than 4.9%. With the 1.0-m tube bundle, this limit was increased to 5.5%. With ethylene/air the concentration ranged from 5.5-9.0%. When the 0.5-m and 1.0-m tube bundle was used, the detonation did not reinitiate for any concentration. With the 0.3-m tube bundle the quenching limit was 8.0-8.5%. However, the detonation reinitiated also in one test with 6.55% ethylene. In all the tests without water in the tube bundle and when the detonation did not reinitiate, the radar Doppler signals clearly showed a flame behind the shock wave. Figure 4a shows the pressure record from one of the tests where the detonation reinitiated after the tube bundle. Figure 4b shows the voltage signal from the radar Doppler unit from the same experiment, and Fig. 4c shows the velocity calculated from the radar signal. Figures 5a-5c shows the experimental records from one of the tests where the detonation did not reinitiate. When the detonation reinitiated, it reinitiated a short distance from the outlet of the tube bundle. The test results indicated that the tube bundle without water was able to

Table 2 Experimental results, tube bundle 0.5 m

Gas	Water	Quenching	Φ^a	$d_c/13$ mm
4.50% C_2H_2/air	No	Yes	0.561	70
5.00% C_2H_2/air	No	Yes	0.626	45
5.20% C_2H_2/air	No	Yes	0.653	37
5.40% C_2H_2/air	No	Yes	0.679	31
5.50% C_2H_2/air	No	No	0.693	29
5.50% C_2H_2/air	No	Yes	0.693	29
5.60% C_2H_2/air	No	Yes	0.706	27
5.70% C_2H_2/air	No	Yes	0.719	25
5.80% C_2H_2/air	No	Yes	0.733	24
5.90% C_2H_2/air	No	Yes	0.746	23
5.90% C_2H_2/air	No	Yes	0.746	23
6.00% C_2H_2/air	No	No	0.760	21
6.00% C_2H_2/air	No	Yes	0.760	21
6.00% C_2H_2/air	No	No	0.760	21
6.10% C_2H_2/air	No	No	0.773	20
6.20% C_2H_2/air	No	Yes	0.787	19
6.20% C_2H_2/air	No	No	0.787	19
6.30% C_2H_2/air	No	No	0.800	18
6.40% C_2H_2/air	No	No	0.814	16
6.50% C_2H_2/air	No	No	0.827	16
7.00% C_2H_2/air	No	No	0.896	13
4.50% C_2H_2/air	Yes	Yes	0.561	70
4.50% C_2H_2/air	Yes	Yes	0.561	70
5.00% C_2H_2/air	Yes	Yes	0.626	45
5.00% C_2H_2/air	Yes	Yes	0.626	45
5.20% C_2H_2/air	Yes	Yes	0.653	37
6.00% C_2H_2/air	Yes	Yes	0.760	21
6.10% C_2H_2/air	Yes	Yes	0.773	20
6.20% C_2H_2/air	Yes	Yes	0.787	19
6.30% C_2H_2/air	Yes	Yes	0.800	18
6.30% C_2H_2/air	Yes	Yes	0.800	18
6.30% C_2H_2/air	Yes	Yes	0.800	18
6.40% C_2H_2/air	Yes	Yes	0.814	16
6.40% C_2H_2/air	Yes	No	0.814	16
6.40% C_2H_2/air	Yes	Yes	0.814	16
6.50% C_2H_2/air	Yes	No	0.827	16
6.50% C_2H_2/air	Yes	No	0.827	16
6.90% C_2H_2/air	Yes	No	0.882	13
7.70% C_2H_2/air	Yes	No	0.993	10
7.75% C_2H_2/air	Yes	Yes	1.000	10

(Table continued on next page)

Table 2 (cont.) Experimental results, tube bundle 0.5 m

Gas	Water	Quenching	Φ^a	d_c/13 mm
6.00% C_2H_4/air	No	Yes	0.912	41
6.50% C_2H_4/air	No	Yes	0.993	34
7.00% C_2H_4/air	No	Yes	1.075	29
7.50% C_2H_4/air	No	Yes	1.158	26
8.00% C_2H_4/air	No	Yes	1.242	25
8.50% C_2H_4/air	No	Yes	1.327	24
8.70% C_2H_4/air	No	Yes	1.361	24
9.00% C_2H_4/air	No	Yes	1.412	24
7.50% C_2H_4/air	Yes	Yes	1.158	26
7.85% C_2H_4/air	Yes	Yes	1.216	25
8.00% C_2H_4/air	Yes	Yes	1.242	25
8.50% C_2H_4/air	Yes	Yes	1.327	24
9.00% C_2H_4/air	Yes	Yes	1.412	24

^{a} Φ = Equivalence ratio

decouple the shock and the reaction zone. But for more sensitive mixtures (i.e., higher concentration of fuel), reinitiation of detonation occurred after the tube bundle as shown above.

In the next series of tests a 2-mm layer of water was placed in the bottom of each tube of the tube bundle. The acetylene and ethylene concentration in these tests ranged from 4.5-7.75%, and 6.55-9.0%, respectively. In all these tests the radar Doppler signals showed that the shock wave and the reaction zone were decoupled after the tube bundle. The acetylene concentration for which the detonation reinitiated was found to be 6.2% with the 0.3-m tube bundle, 6.4% with the 0.5-m tube bundle, and 6.6% with the 1.0-m tube bundle. For all weaker acetylene/air mixtures, the detonation did not reinitiate. However, when the concentration of acetylene was higher, the detonation also failed to reinitiate in some tests. With ethylene/air mixtures, the detonation was queched for all concentrations, except 7.5% when the 0.3-m tube bundle was used. The reason for this might be to little water in the tube bundle. The quenching limits found are summarized in Table 4.

When water was used in the tube bundles and the detonation did not reinitiate, the radar Doppler signals were weak. However, a velocity of 100-500 m/s could be measured. This velocity was probably the velocity of the water droplets and the water vapour. When the detonation did not reinitiate, the pressure measured by the last pressure transducer was reduced to less than 4 bar when the 0.3-m tube bundle was used, and less than 3.0 bar when the 1.0-m tube bundle was used. This is shown in

Table 3 Experimental results, tube bundle 1.0 m

Gas	Water	Quenching	Φ^a	$d_c/13$ mm
4.50% C_2H_2/air	No	Yes	0.561	70
5.00% C_2H_2/air	No	Yes	0.626	45
5.30% C_2H_2/air	No	Yes	0.666	34
5.50% C_2H_2/air	No	Yes	0.693	29
5.50% C_2H_2/air	No	No	0.693	29
5.60% C_2H_2/air	No	No	0.706	27
5.70% C_2H_2/air	No	No	0.719	25
6.00% C_2H_2/air	No	No	0.760	21
4.50% C_2H_2/air	Yes	Yes	0.561	70
6.00% C_2H_2/air	Yes	Yes	0.760	21
6.00% C_2H_2/air	Yes	Yes	0.760	21
6.00% C_2H_2/air	Yes	Yes	0.760	21
6.20% C_2H_2/air	Yes	Yes	0.787	19
6.30% C_2H_2/air	Yes	Yes	0.800	18
6.40% C_2H_2/air	Yes	Yes	0.814	16
6.50% C_2H_2/air	Yes	Yes	0.827	16
6.50% C_2H_2/air	Yes	Yes	0.827	16
6.60% C_2H_2/air	Yes	Yes	0.841	15
6.60% C_2H_2/air	Yes	No	0.841	15
6.70% C_2H_2/air	Yes	No	0.855	14
6.70% C_2H_2/air	Yes	No	0.855	14
6.80% C_2H_2/air	Yes	No	0.868	14
7.00% C_2H_2/air	Yes	No	0.896	13
6.50% C_2H_4/air	No	Yes	0.993	34
7.00% C_2H_4/air	No	Yes	1.075	29
7.50% C_2H_4/air	No	No	1.158	26
8.00% C_2H_4/air	No	Yes	1.242	25
8.50% C_2H_4/air	No	Yes	1.327	24
9.00% C_2H_4/air	No	Yes	1.412	24
7.50% C_2H_4/air	Yes	Yes	1.158	26
8.00% C_2H_4/air	Yes	Yes	1.242	25
8.30% C_2H_4/air	Yes	Yes	1.293	24
8.50% C_2H_4/air	Yes	Yes	1.327	24
9.00% C_2H_4/air	Yes	Yes	1.412	24

[a] Φ = Equivalence ratio

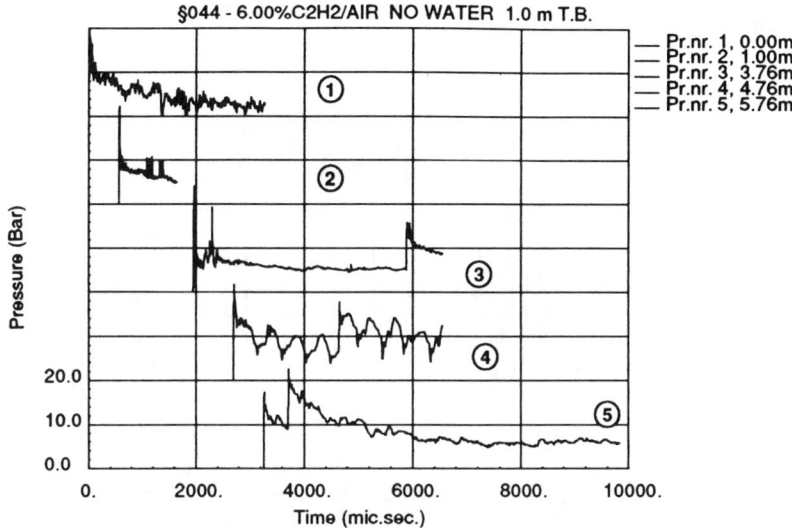

a) Pressure record (each transducer shifted 20 bar on the axes).

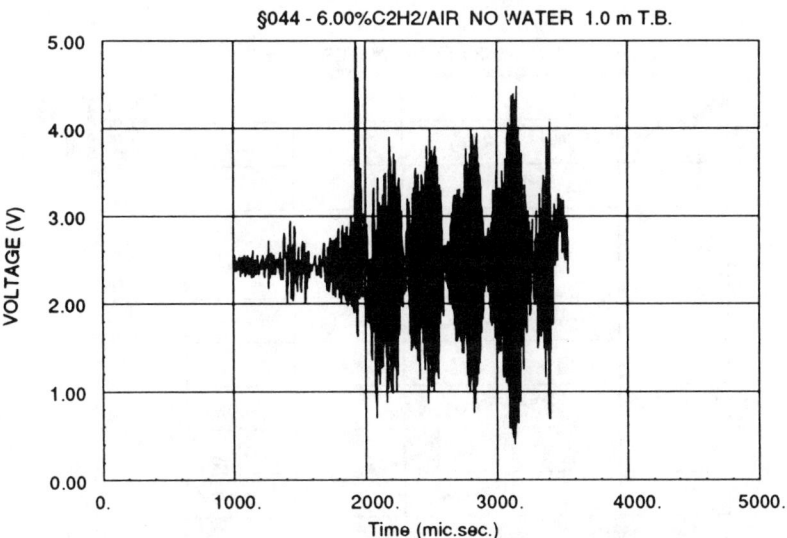

b) Voltage signal from radar Doppler unit.

Fig. 4 Typical example of a detonation that reinitiated after the tube bundle, no water in the tubes: 6.0% C_2H_2/air, 1.0-m tube bundle.

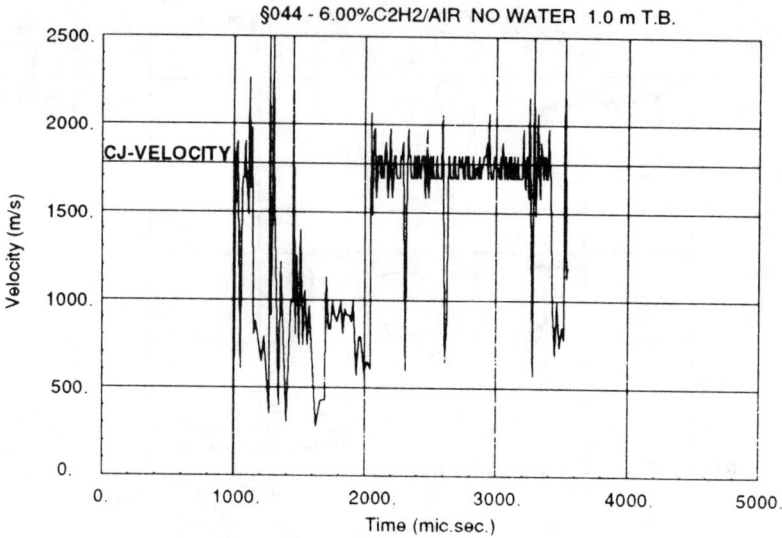

c) Velocity calculated from radar signal.

Fig. 4 (continued) Typical example of a detonation that reinitiated after the tube bundle, no water in the tubes: 6.0% C_2H_2/air, 1.0-m tube bundle.

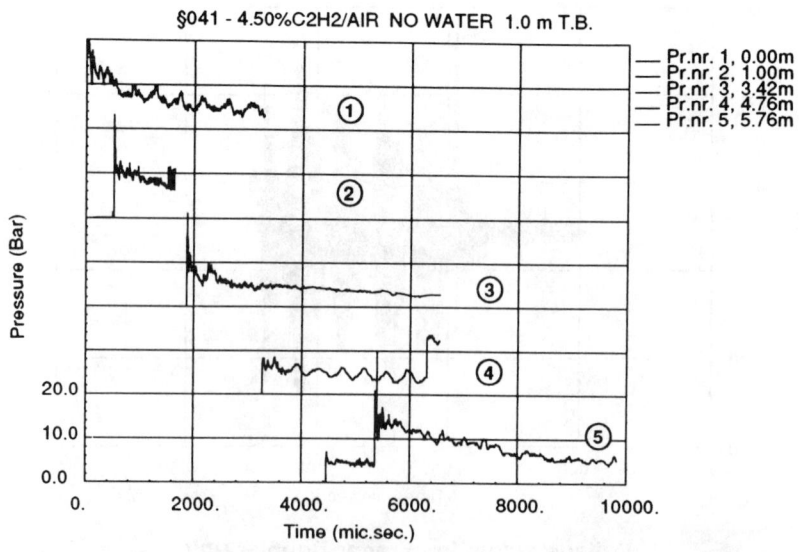

a) Pressure record (each transducer shifted 20 bar on the axes).

Fig. 5 Typical example of a detonation that is not reinitiated after the tube bundle, no water in the tubes: 4.5% C_2H_2/air, 1.0-m tube bundle.

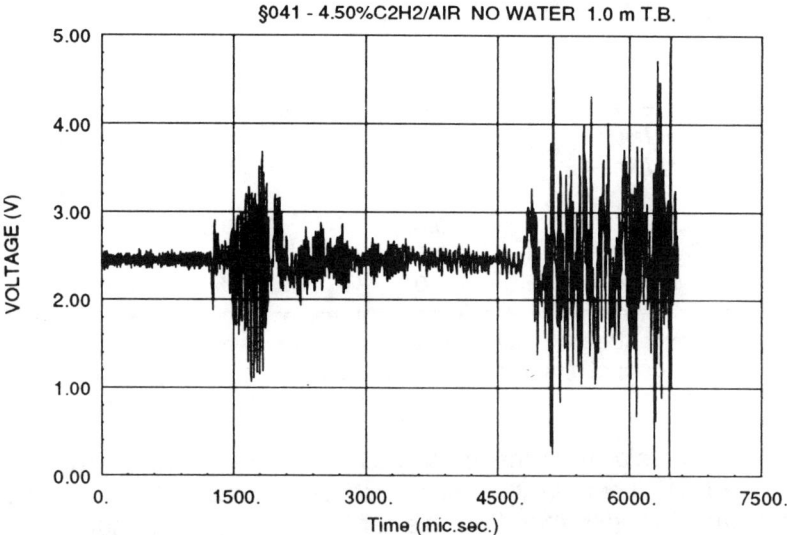

b) Voltage signal from radar Doppler unit.

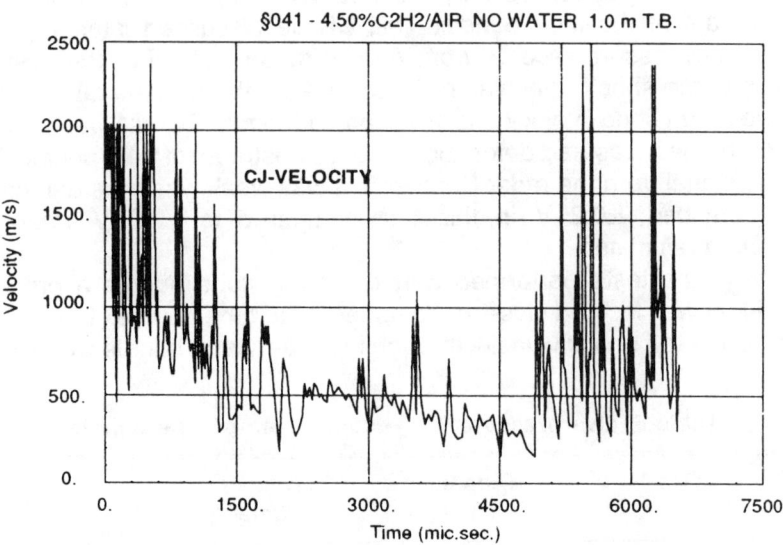

c) Velocity calculated from radar signal.

Fig. 5 (continued) Typical example of a detonation that is not reinitiated after the tube bundle, no water in the tubes: 4.5% C_2H_2/air, 1.0-m tube bundle.

Table 4 Quenching limits, vol%

Gas	Tube bundle length, m	Without water	With water
C_2H_2/air	0.3	4.9	6.2
C_2H_2/air	0.5	4.5-5.5	6.4
C_2H_2/air	1.0	5.5	6.6
C_2H_4/air	0.3	8.0-8.5	All[a]
C_2H_4/air	0.5	All[b]	All[b]
C_2H_4/air	1.0	All[b]	All[b]

[a] All concentrations tested were quenched except 7.5%
[b] All concentrations tested were quenched

Table 5. The velocity of the shock wave measured between the last two pressure transducers in the test section downstream of the tube bundle was reduced to less than 750 m/s for the 0.3-m tube bundle, and less than 670 m/s for the 1.0-m tube bundle. Table 6 shows the reduction of the shock velocity downstream of the tube bundle in percentage of the measured detonation velocity upstream of the tube bundle. The main reduction of the shock velocity did not occur in the tube bundle, but downstream of the tube bundle. This will be discussed later. Figure 6 shows the pressure records from one of these tests. For this test the velocity of the shock front was reduced by 22% through the tube bundle, and the mean shock velocity downstream of the tube bundle was reduced by 63% of the measured detonation velocity upstream of tube bundle. The voltage signal from the radar Doppler unit was weak when the detonation was quenched, ±0.2 V in this test compared to ±2.0 V when the detonation reinitiated.

In the tests performed with the 1.0-m tube bundle a pressure transducer was located close to the outlet of the tube bundle (11 cm). The reduction of the shock velocity through the 1.0-m tube bundle when water

Table 5 Overpressure [bar] downstream of tube bundle[a]

Gas	Tube bundle length, m	Without water	With water
C_2H_2/air	0.3	< 8.0	< 4.0
C_2H_2/air	0.5	< 8.0	< 3.5
C_2H_2/air	1.0	< 7.0	< 3.0
C_2H_4/air	0.3	< 7.0	< 4.0
C_2H_4/air	0.5	< 7.0	< 3.0
C_2H_4/air	1.0	< 7.0	< 1.0

[a] Pressure measured at the last pressure transducer in the test section

Table 6 Reduction of shock velocity downstream of tube bundle[a]

Gas	Tube bundle length, m	Without water, %	With water, %
C_2H_2/air	0.3	45-47	59-63
C_2H_2/air	0.5	44-56	60-64
C_2H_2/air	1.0	51-54	63-72
C_2H_4/air	0.3	46-56	64-67
C_2H_4/air	0.5	51-52	63-67
C_2H_4/air	1.0	45-51	68-75

[a] Velocity measured between the last two pressure transducer in the test section compared to the measured detonation velocity upstream of the tube bundle

was used in the tubes and the detonation did not reinitiate was approximately 18-23%, compared to less than 14% when the detonation reinitiated. The shock velocity, however, very rapidly decreased after the tube bundle. In the ethylene/air tests at similar conditions, the reduction of the shock velocity was approximately 50% through the tube bundle. Figure 7 shows the pressure record from one of these tests. In the acetylene/air tests without water in the tube bundle, the reduction of the shock velocity was less than 10%, regardless of whether the detonation reinitiated or not. These observations could not be seen from the tests performed with the 0.3-m tube bundle and the 0.5-m tube bundle since the

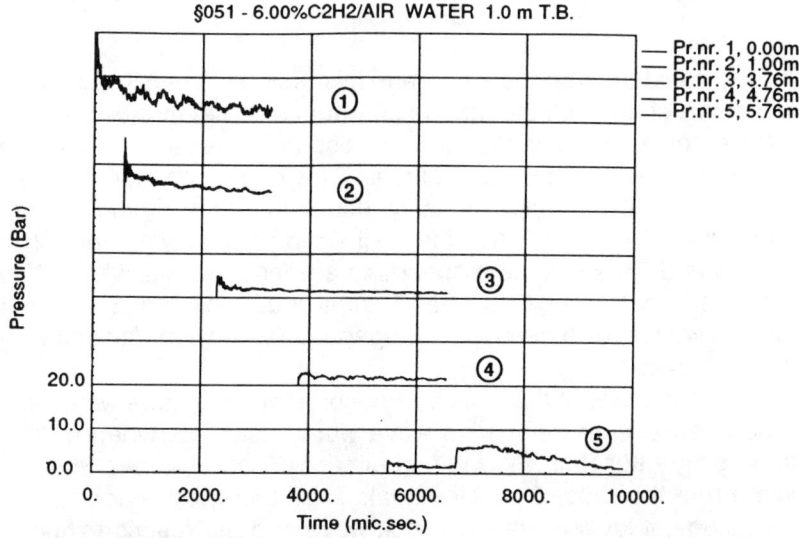

Fig. 6 Typical pressure record of a detonation in C_2H_2/air that is not reinitiated after the tube bundle, water in the tubes: 6.0% C_2H_2/air, 1.0-m tube bundle, each transducer shifted 20 bar on the axes.

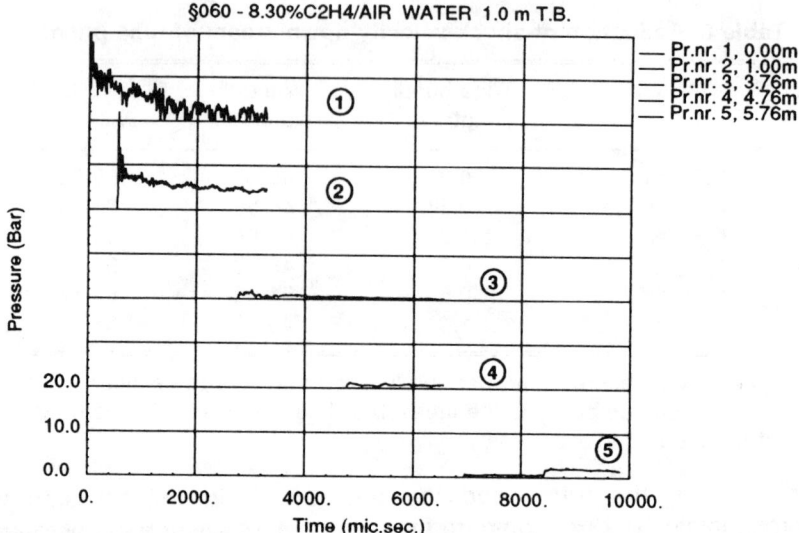

Fig. 7 Typical pressure record of a detonation in C_2H_4/air that is not reinitiated after the tube bundle, water in the tubes: 8.3% C_2H_4/air, 1.0-m tube bundle, each transducer shifted 20 bar on the axes.

first pressure transducer after the tube bundle was located further downstream of the outlet of the tube bundle in these tests than in the tests with the 1.0-m tube bundle.

Discussion

Detonation propagation in lean mixtures has been investigated by Engebretsen et al.[4] with the same apparatus as used in these tests. The limit for detonation propagation at atmospheric pressure in a 125-mm square tube was 4.0% for acetylene/air. This corresponds to an average transverse cell width approximately equal to the width/height of the channel when $S = d_c/13$ is used for cell width.[6] For ethylene/air mixtures the limit was 5.0% which corresponds to a transverse cell width of about 0.6 times the width/height of the channel indicating that three or four transverse waves were needed for propagation of ethylene/air detonations in square tubes.

On the basis of the tests by Engebretsen et al.[4] one would expect that the acetylene/air detonation wave would decouple when $d_c/13$ was equal to or greater than the internal dimension of the tubes in the tube bundle. From the radar Doppler signals in all the tests performed in the present study, it looked like the shock wave and the reaction zone were always decoupled when leaving the tube bundle, as well as when the cell width was less than the internal dimension of the tubes. This was not as expected. However, re-establishment of the detonation after the tube bundle will not occur instantaneously, and can then explain why the radar Doppler signals showed that the detonation wave was decoupled at the outlet of the tube bundle even when the cell width was less than the internal dimension of the tubes. The shock velocity through the tube bundle indicates, in contradiction to the radar Doppler signals, that the acetylene/air detonations were not decoupled in the tube bundle, at least not in all the tubes. But since the radar Doppler unit was unable to "see" what happened inside the tube bundle, one can not tell for certain that the detonation did not decouple somewhere inside the tube bundle. Thomas et al.[1] use the Shchelkin instability criterion, and they conclude that when the velocity of a plane CJ detonation falls approximately 10% below the CJ value, the temperature will decrease by about 20%. If the induction zone is assumed to be controlled by Arrhenius kinetics, the reaction will be decoupled from the shock front and the detonation wave decays. The Shchelkin instability criterion states that, for a one-dimensional wave front, if an increase δt_i occurs in the steady-state induction-reaction time t_i such that

$$\frac{\delta t_i}{t_i} \geq 1 \qquad (1)$$

failure of the detonation wave front occurs. Applying the result of Thomas et al.[1] of 10% reduction in the CJ velocity to the present tests, decoupling of the wave front in the tube bundle is very likely in all the cases where water is used in the tube bundle.

However, the detonation might be able to propagate in some of the tubes because of the very irregular structure of the detonation front in acetylene/air mixtures. When the detonation hits the tube bundle, the cellular structure of the front is destroyed and marginal detonations continue to propagate only in some of the tubes. Depending on the length of the tubes, some of these marginal detonations will decouple in the tube bundle. This can explain the difference in the quenching limits found for the different tube bundle lengths for acetylene/air mixtures without water in the tubes.

Since there are significant differences between the quenching limits with and without water in the tube bundle, it is obvious that the water layer itself has an effect on the quenching process. Borisov et al.[2] showed that a thin layer of water on the bottom wall of the tube was able to reduce the detonation velocity in C_3H_8-O_2-N_2 mixtures from about 1800 m/s to about 1300 m/s over a distance of 300 mm. They observed that the detonation wave decay started near the water surface and that the heat release zone grew progressively until the chemical reaction and the shock front decoupled completely. The detonation waves were quenched by the water layer when the cell size was greater than one-third of the channel width. This does not agree with the present results for acetylene/air mixtures. However, the regularity of the cell structure is different. This difference in regularity might explain the difference in the results between the present results and the results of Borisov et al.[2] When water was used in the tube bundle, the measured reduction of the shock velocity through the tube bundle indicates that the decay of the detonation wave starts later in acetylene/air mixtures, or is a much slower process, than observed by Borisov et al.[2] The cell width was, however, greater than half the width/height of the tubes in all the acetylene/air tests. The results of Borisov et al.[2] can not be compared to the ethylene/air tests in this study since the cell width always was greater than the width/height of the tubes and the detonation very likely decoupled in all the tests.

A few experiments were also carried out with 4.5% C_2H_2 in the donor section and air in section 3 (i.e., no combustion in section 3). The detonation then stopped 0.5 m upstream of the tube bundle and only a shock wave propagated through the tube bundle. The 0.5-m tube bundle was used in these tests. When there was no water in the tube bundle, the radar Doppler signals were weak, but a velocity fluctuating between 100-500 m/s was measured. With water in the tube bundle, the radar Doppler signals were stronger, and a velocity of about 100-150 m/s was measured. The pressure record from the test with water in the tube bundle and air in section 3 is shown in Fig. 8. Figure 9 shows the pressure record from a test with the same gas mixture in the whole test section. These two

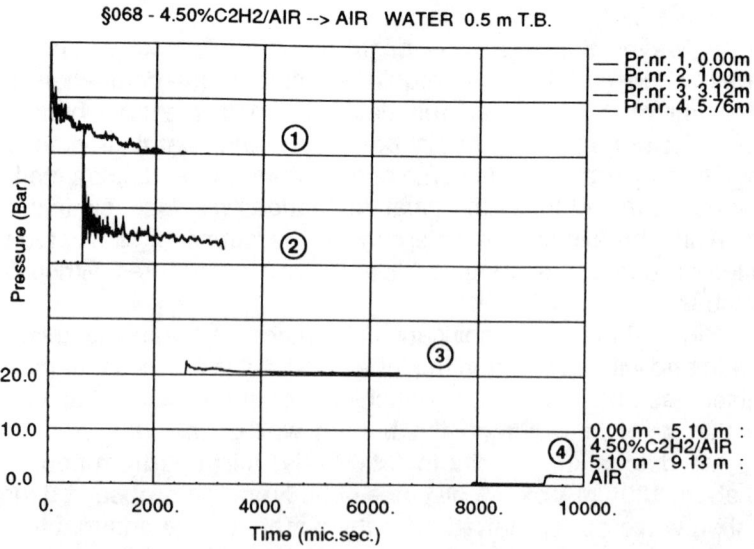

Fig. 8 Pressure record of a detonation in 4.50% C_2H_2/air that propagates into air in section 3: water in the tubes, 0.5-m tube bundle, each transducer shifted 20 bar on the axes.

pressure records show no significant difference in pressure or shock velocity between these two tests. The fact that the detonation propagated 0.5 m longer before it stopped when the whole test section was filled with the acetylene/air mixture can explain the small differences in pressure and shock velocity. The radar Doppler signals and the radar Doppler velocity from these two tests were also nearly identical. This tends to support the theory that the velocity measured by the radar Doppler unit when the detonation was quenched was the velocity of the water droplets and the water vapour.

Some tests with high-speed film of shock waves over water layers showed a very rapid entrainment of water from the surface due to the high velocity gas flow (in less than 100 µs). This agrees with the observations by Borisov et al.[2] who observed that the first lift up of water occurred about 10 µs after the arrival of the shock front. The correlation presented by van Rossum[7] shows that the high velocity gas atomizes the water. For the water to be able to quench the detonation, it must be lifted from the surface and form a micromist that covers the whole cross section of the tubes. The distance between the shock wave and the flame front, therefore, must be large enough for such entrainment of water.

If one assumes that the shock wave and the flame are decoupled immediately after entering the tube bundle, which is very likely for the ethylene/air tests, and that the average shock velocity in the tubes is about 1500 m/s and the average flame speed is about 1000 m/s, one

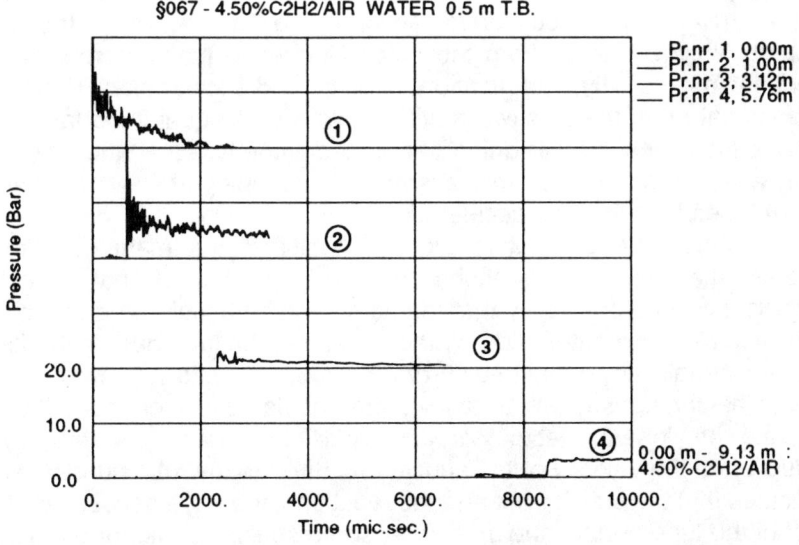

Fig. 9 Pressure record of a detonation in 4.50% C_2H_2/air, water in the tubes, 0.5-m tube bundle, each transducer shifted 20 bar on the axes.

obtains a distance of 17 cm between the shock wave and the flame front in the 0.5-m tube bundle. When the shock wave leaves the tube bundle, the time between the shock and the flame front will be 170 µs. This indicates that the time needed to form a micro-mist covering the whole cross section of the tubes between the shock wave and the flame should be sufficient. The largest droplets suspended in the gas behind the shock will also be further reduced in size by stripping liquid from the droplet surface by the high velocity gas flow.[8,9] A "plug" of small water droplets and water vapour will then follow the shock wave. The length of this plug will depend on the amount of water in the tube bundle. When the flame reaches the micromist, the mist will evaporate very quickly and, probably, quench the flame as observed by Jenssen.[3] Thomas et al.[1] have estimated the evaporation time to be 7 µs for a 1-µm droplet and 670 µs for a 10-µm droplet. Applying the correlation of Pilch and Erdman,[10] the time needed for total breakup of a 10-µm droplet is approximately 2.5 µs when the gas velocity is 1000 m/s and approximately 18 µs for a 100-µm droplet. This indicates that only very small water droplets (i.e., micromist) will be present in the reaction zone if the detonation wave has decoupled.

Thomas et al.[1] discuss mechanisms for water droplets to quench a detonation. The leading shock front does not directly influence the droplet breakup. The viscous action of the high-speed gas flowing behind the shock wave strips liquid from the droplet surface and forms a micromist in the wake. This is in agreement with the observations from the high-speed films in the present study of shock waves propagating over a water layer. This means that the droplets do not have to be present prior to the arrival of the shock wave. According to Thomas et al.[1] the water droplets absorb energy from the detonation wave mainly through three mechanisms, namely, vaporization, heating, and acceleration. Thomas et al.[1] estimate that the first two mechanisms are of equal importance and that the last mechanism is one order of magnitude less than the other two. They also estimate the energy absorbed by stripping of water droplets to be 10 times less than by acceleration.

When the detonation did not reinitiate, the main question is whether the flame was quenched or not. If the flame is not quenched there is always a possibility that the flame can be accelerated and cause transition to a new detonation. Without water in the tube bundle the flame was for certain not quenched. From the present tests with water in the tubes one can not say for certain whether the flame was quenched or not. For the ethylene/air tests this is very likely because of the severe reduction of the velocity of the shock front through the tube bundle, which indicates that the shock front and the reaction zone were decoupled at the inlet of the tube bundle and thereby ensured sufficient time for breakup of the water layer and establishment of a micromist between the shock front and the reaction zone. For the acetylene/air tests this is, however, more uncertain. This depends on whether the detonation was decoupled in the tube bundle or whether it managed to pass through some of the tubes in the tube bundle. The pressure behind the shock wave is so low that reinitiation due to shock compression is very unlikely.

The criterion for quenching of detonations in acetylene mixtures using a tube bundle with a water layer in the tubes is found to be

$$\frac{d_c/13}{H} > 1 \qquad (2)$$

where H is the width/height of the tubes. This equation is expected to be valid for other gas mixtures also, and even be a conservative criterion, since acetylene/air is a very sensitive gas mixture with a very irregular structure. However, the length of the tube bundle or the amount of water is not taken into account in this equation. In the present test the length of the tube bundle seems to play a minor role on whether the detonation reinitiates or not. Using the equation above, one find that a detonation in stoichiometric acetylene/air could be quenched if the internal dimension of the tubes was approximately 10 mm. Reducing the internal dimension of the tubes also means that the time needed for covering the whole cross section of the tubes with mist is reduced. On the other hand, increasing the internal dimension of the tubes increases this time.

Conclusions

The experiments in this study shows that a tube bundle without water is able to deccuple the shock and the flame front. Whether this happens in the tube bundle or at the exit of the tube bundle is not clear, but probably depends of the sensitivity of the gas mixture. Under certain conditions the detonation will not reinitiate after the tube bundle, but a shock wave followed by a deflagration will continue to propagate. These conditions are, for example, the sensitivity of the gas mixture and the geometry of the tube after the tube bundle (i.e., wall roughness and obstacles).

The high velocity gas flowing behind the shock front will rapidly break up the water layer in the tubes and form a micromist in the tubes in less than 100 μs.

When water is used in the tube bundle, the detonation will not reinitiate after the tube bundle when $d_c/13$ is approximately equal to or greater than the internal dimension of the tubes in the tube bundle for acetylene/air mixtures. This is also valid for other gas mixtures, since less sensitive gas mixtures are more easily quenched than sensitive gas mixtures.

When water is used in the tube bundle and the detonation does not reinitiate, the pressure is reduced to less than 4 bar and the velocity of the shock front is reduced to less than 750 m/s downstream of the tube bundle.

From the present experiments it is not clear whether the flame was quenched or not; for the ethylene/air tests this was very likely, but for the acetylene/air tests it was more uncertain.

References

[1]Thomas, G. O., Edwards, M. J., and Edwards, D. H., "Studies of Detonation Quenching by Water Sprays,". *Combustion Science and Technology*, Vol. 71, No. 4-6, 1990, pp. 233-245.

[2]Borisov, A. A., Mailkov, A. E., Kosenkov, V. V., and Aksenov, V. S., "Propagation of Detonations over Liquid Layers," Paper presented at the 12th International Colloquium on Dynamics of Explosions and Reactive Systems, Ann Arbor, MI, July, 1989.

[3]Jenssen, A., private communication, Norwegian Defence Construction Service Oslo, Norway, May, 1991.

[4]Engebretsen, T., Bjerketvedt, D., and Sønju, O. K., "Propagation of Gaseous Detonations Through Regions of Low Reactivity," AIAA Paper PS4300, 13th International Colloquium on Dynamics of Explosions and Reactive Systems, Nagoya, Japan, July, 1991.

[5]Knystautas, R., Guirao, C., Lee, J. H., and Sulmistras, A., "Measurements of Cell Size in Hydrocarbon-Air Mixtures and Predictions of Critical Tube Diameter, Critical Initiation Energy, and Detonability Limits," *Dynamics of Shock Waves, Explosions, and Detonations: AIAA Progress in Astronautics and Aeronautics* (edited by Bowen, Manson, Oppenheim, and Soloukhin), Vol. 94, AIAA, New York, 1984, pp. 23-37.

[6]Moen, I. O., Funk, J. W., and Ward, S. A., "Detonation Length Scales for Fuel-Air Explosives," *Dynamics of Shock Waves, Explosions, and Detonations: AIAA Progress in Astronautics and Aeronautics* (edited by Bowen, Manson, Oppenheim, and Soloukhin), Vol. 94, AIAA, New York, 1984, pp. 55-79.

[7]van Rossum, J. J., "Experimental Investigation of Horizontal Liquid Films," *Chemical Engineering Science*, Vol. 11, No. 1, 1959, pp. 35-52.

[8]Ranger, A. A., and Nicholls, J. A., "Aerodynamic Shattering of Liquid Drops," *AIAA Journal*, Vol. 7, No. 2, 1969, pp. 285-290.

[9]Lane, W. R., "Shatter of Drops in Steams of Air," *Industrial and Engineering Chemistry*, Vol. 43, No. 6, 1951, pp. 1312-1317.

[10]Pilch, M., and Erdman, C. A., "Use of Breakup Time Data and Velocity History Data to Predict the Maximum Size of Stable Fragments for Acceleration-Induced Breakup of a Liquid Drop," *International Journal of Multiphase Flow*, Vol. 13, No. 6, 1987, pp. 741-757.

Effect of Losses on the Existence of Nonideal Detonations in Hybrid Two-Phase Mixtures

B. A. Khasainov*
Russian Academy of Sciences, Moscow, Russia
and
B. Veyssière†
Ecole Nationale Supérieure de Mécanique et d'Aérotechnique, Poitiers, France

Abstract

The ranges of existence of steady propagation regimes of nonideal detonations in hybrid two-phase mixtures are studied numerically for different values of mean solid particle size and shock tube diameter. Three steady regimes are examined in mixtures of aluminum particles with detonable gases: pseudogas (or "practically" gas) detonation, double-front detonation and single-front detonation. In the pseudogas detonation (PGD), the aluminum particles are not ignited upstream of the CJ plane. Single-front detonations (SFD) are supported by heat release from both gaseous reactions and solid particles-gas reactions. In the double-front detonation (DFD), a secondary detonation wave (supported by reactions between particles and gases) propagates at a constant time delay τ downstream of the leading detonation wave (supported by heat release from gaseous reactions). In some particle concentration ranges, both PGD and SFD may propagate but the velocity of SFD is higher than that of PGD. When particle size diminishes or tube diameter increases, the domain of SFD increases, whereas for coarser particles or small shock tube diameter, PGD is the dominant detonation regime. DFD may propagate in a relatively narrow range of particle sizes and shock tube diameters where PGD exists and SFD does not. For a given composition of the gas mixture, there are optimal values of particle size and shock tube diameter for which the domain of DFD is the largest. The growth of particle size (or

---

Copyright © 1992 by the American Institute of Aeronautics and Astronautics, Inc. All rights reserved.
*Senior Researcher, Institute of Chemical Physics.
†Chargé de Recherches, Centre National de la Recherche Scientifique, Laboratoire d'Enérgetique et de Détonique.

reduction of tube diameter) increases the delay τ between the two fronts of DFD and finally DFD disappears being replaced by PGD. With decrease of particle size (or increase of tube diameter) SFD propagation becomes more and more likely. With bimodal particle size distribution, it is shown that the presence in the aluminum powder of relatively large particles may stabilize the nonideal detonation regime even in the absence of wall losses. Thus, due to the presence of coarser particles in the dispersed powders, propagation of unconfined PGD, SFD and DFD in real two-phase hybrid mixtures should follow the same trends as those displayed in the study of wall losses effects.

Introduction

Previously we derived a simple model of two-phase steady detonations in hybrid mixtures of aluminum particles with reactive gaseous mixtures. This Khasainov-Veyssière (KV) model[1-3] provides reasonable agreement with experimental observations[4-5] of double-front detonations (DFD) in a number of different gaseous mixtures. The main feature of DFD is the secondary detonation wave propagation downstream of the leading detonation front with a constant time delay between the two fronts. This secondary wave is due to the reaction of aluminum particles with reaction products of the leading detonation.

The KV model also predicts the existence of steady pseudogas (or "practically" gas) detonation (PGD) and steady single-front detonation (SFD) in some particle concentration ranges.[6] From a practical point of view, SFD generates the highest detonation velocity in comparison with DFD or PGD. The PGD regime in hybrid mixtures corresponds to the case when detonation wave propagation is supported mainly by gaseous reactions, that is aluminum particles burn mostly downstream of the CJ plane of PGD wave. This regime of PGD is similar to gas-inert particle detonations, i.e., aluminum particles are not ignited and remain inert in the normal PGD wave: this regime exists up to particle concentration σ_{qi} at which gas detonation is quenched by inert particles. Conversely, SFD wave propagation is supported by gaseous reactions together with burning of aluminum particles and the SFD velocity is higher than that of PGD.

The dependence of the structure and concentration limits of different detonation regimes upon the composition of the gaseous mixture is quite complicated[6] due to the nonmonotonic character of heat release behind the shock front. For example, for lean gaseous compositions, the time delay τ between the two detonation fronts of DFD decreases rapidly with particle concentration σ in a narrow range of a relatively small particle concentrations. For near-stoichiometric gas mixtures, at larger σ a second branch of steady DFD appears for which τ grows with σ, and for rich mixtures the dependence $\tau(\sigma)$ may be continuous and nonmonotonic due to coincidence of the two different branches of DFD (see Ref. 6). The possibility of nonuniqueness of detonation regimes in hybrid mixtures (that is, at a given particle concentration different detonation regimes may propagate) was also demonstrated earlier[6] and is displayed elsewhere.[7]

PGD, SFD, and DFD are nonideal detonations because following the theory of nonideal detonations,[8] their existence and structure depend on the balance between the chemical heat release rate (dq_+/dt) and the effective heat losses rate (dq_-/dt) in the sonic point of the flow behind the detonation front (equivalent CJ condition[8] means that $dq/dt = dq_+/dt - dq_-/dt = 0$ and $M = (D - u)/c = 1$). From a mathematical point of view, the PGD solution corresponds to a first appearance of $dq/dt = 0$ ($d^2q/dt^2 < 0$) and $M = 1$ behind the front; SFD corresponds to a second appearance of those conditions; and for DFD, the CJ condition must be satisfied in two points.

The effective heat losses rate includes the impulse and energy losses due to the interaction of the gas flow behind the detonation front with shock tube walls and those due to the interaction between solid particles and gases. All of those losses are taken into account by the KV model. In the case of polymodal particle size distribution, the presence of coarse particles (the ignition of which may be strongly delayed in comparison with finer particles) may provide losses even in the absence of shock tube walls. This was shown by analysis of the detonation wave structure in a suspension of monopropellant particles with bimodal particle size distribution.[9] In the case of aluminum suspensions, the endothermicity of aluminum burning at temperatures exceeding the "boiling" temperature of aluminum oxide[10] may also affect the structure of the detonation wave; this proposal was stated in Ref. 11 and was confirmed numerically for aluminum - air and oxygen bimodal suspensions.[12]

Because the presence of coarse particles in the real aluminum powders is quite likely, it is interesting to check numerically the possibility of steady propagation of unconfined detonations in hybrid mixtures for the case when losses are only due to the presence of coarse particles. For this purpose, we used the KV model to study the detonation structures for the case of bimodal aluminum particle size distribution in the absence of shock tube wall losses. To make the results clearer it is worthwhile to begin with the study of the influence of losses on the existence and structure of the different nonideal detonation regimes in the hybrid mixtures of aluminum particles with reactive gases and to investigate the sensitivity of these detonation regimes to the governing physical parameters, namely, to the mean particle size and shock tube diameter. Surely the change of particle size not only alters heat losses rate but also provides relative change of heat release rate from particles.

KV Model and Input Parameters

The KV model and main input parameters were described earlier.[1-3] It gives reasonable agreement with the structure of DFD observed experimentally[4,5] for aluminum particle concentrations up to 200 g/m^3 in a number of different gaseous explosive mixtures. The KV model was also used successfully to analyze the structure and detonability of aluminum-air and oxygen mixtures.[12] Good agreement with experimental data was obtained due to modification of the original KV model by assuming that $\gamma = C_p/C_v$ is not constant and depends on temperature and composition of the gaseous mixture behind the shock wave front. This assumption was important mainly at high σ (about 3000 g/m^3). Here, we use the previous assumption γ = const. Hence, at high σ the results shown below should be considered as a first approximation. However, a series of computations with γ variable has shown that the results presented later remain valid with a lowering of the upper values of mass particle concentration σ.

The computations were performed for a near-stoichiometric gaseous mixture of hydrogen with oxygen and nitrogen for a richness coefficient 1.06. The effect of the gaseous composition was considered previously.[6] Let us study first the effect of mean particle size and then the effect of tube diameter on the domains of existence and detonation velocities of the different detonation regimes.

Results for Monomodal Particles in the Presence of Shock Tube Walls

Effect of Particle Size

Figure 1 shows the calculated time delay τ of DFD vs particle concentration σ for the case of the "nominal" shock tube used in experiments[4,5] (internal diameter 6.9 cm and prescribed "technical" roughness 15 μm, value chosen as being suitable for realistic modeling[13]) at different mean diameters of aluminum particles d_p. DFD does not exist for particles with diameters $d_p \leq 9$ μm and $d_p \geq 19$ μm. The domain of DFD is the largest when particle size is ≈15 μm.

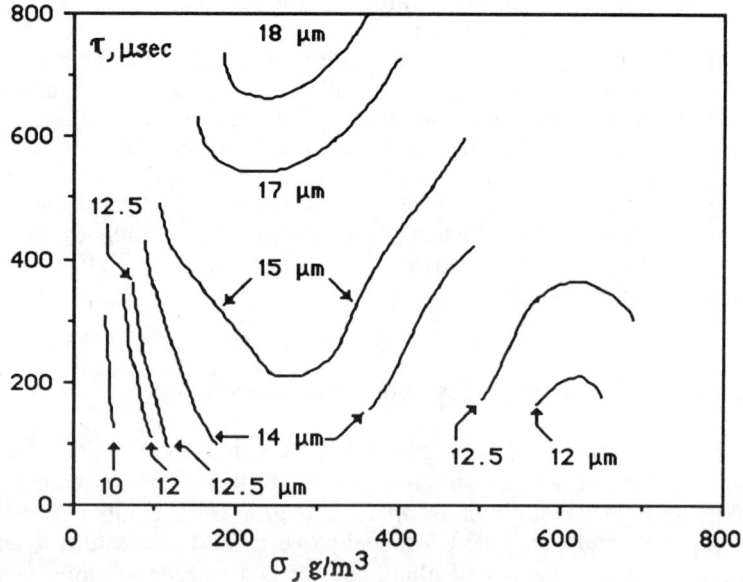

Fig. 1 The dependence of time delay τ between the two fronts of double-front detonation upon particle concentration σ for different values of mean particle size d_p.

Hence, the optimal size of aluminum particles for DFD to exist for the present gaseous mixture and shock tube diameter is about 15 μm.

For relatively large particle diameters with d_p from 15 to 18 μm, the dependence $\tau(\sigma)$ is nonmonotonic with a passage through a minimum. The growth of particle diameter increases τ and narrows the domains of DFD. When $d_p > 19$ μm, the DFD domain collapses, because burning of particles becomes so slow (burning time of particles is proportional to d_p^2) that the heat release rate from particles downstream of the CJ plane of the leading detonation cannot provide the secondary detonation wave formation at the present level of heat losses.

For smaller particles (10 μm $\leq d_p <$ 15 μm) the nonmonotonic dependence $\tau(\sigma)$ separates into two different branches: the left one where τ decreases when σ increases and the right one at larger σ, with opposite behavior of τ (the right DFD branch may be nonmonotonic as in the case of 12.5- and 12-μm particles). At $d_p = 10$ μm the right DFD branch had collapsed and the dependence $\tau(\sigma)$ is represented only by the left branch. Thus, decrease of d_p below optimal particle size shifts the left domain of existence of DFD to smaller σ, the right branch is shifted to larger σ, at the same time the ranges of σ where DFD exists diminish, then the right DFD branch disappears and, finally, the left DFD branch vanishes as well at $d_p < 10$ μm.

Figure 2 shows the dependencies of the velocity D of PGD and SFD on σ at different particle sizes (note that DFD has the same velocity as PGD in domains of DFD existence but the structures of DFD and PGD are different). The lower concentration limit of PGD is $\sigma_{PGD}* = 0$. For the sake of comparison, calculations have been performed under the assumption that aluminum particles are "inert." Comparison of different curves $D(\sigma)$ corresponding to the same particle size but to reactive and inert particles shows that up to a particle concentration σ_{qi} where fast decrease of detonation velocity occurs, those curves are identical. Hence, the PGD propagation regime at $0 < \sigma < \sigma_{qi}$ corresponds to inert particles; and it means that aluminum particles are not ignited along the normal branch of PGD curve until the "quenching" concentration σ_{qi}. The value of σ_{qi} diminishes rapidly when particle size decreases ($\sigma_{qi} \approx$ 1225, 750, 575, and 275 g/m^3 for d_p=20, 13, 10, and 5 μm, respectively, compare Figs. 2a-2d) due to increase of losses, caused by acceleration and heating of particles. Though in the case of relatively large particles (d_p = 20 μm, Fig. 2a and d_p = 13 μm, Fig. 2b) at $\sigma > \sigma_{qi}$ quenching of detonation really takes place in the case of inert particles, at smaller diameters of inert particles, that is, at greater losses, a kind of low velocity detonation may propagate at $\sigma > \sigma_{qi}$ (Fig. 2c and Fig. 2d). This result is in agreement with predictions made by Menga[14] for inert particles.

Regarding the problem of DFD it is important to emphasize that the velocity of DFD exactly equals the velocity of PGD. Comparison of Fig. 1 and Fig. 2 shows that DFD cannot propagate at $\sigma > \sigma_{qi}$ where D changes from a normal to a low velocity value. The DFD domains never intersect with the SFD domains

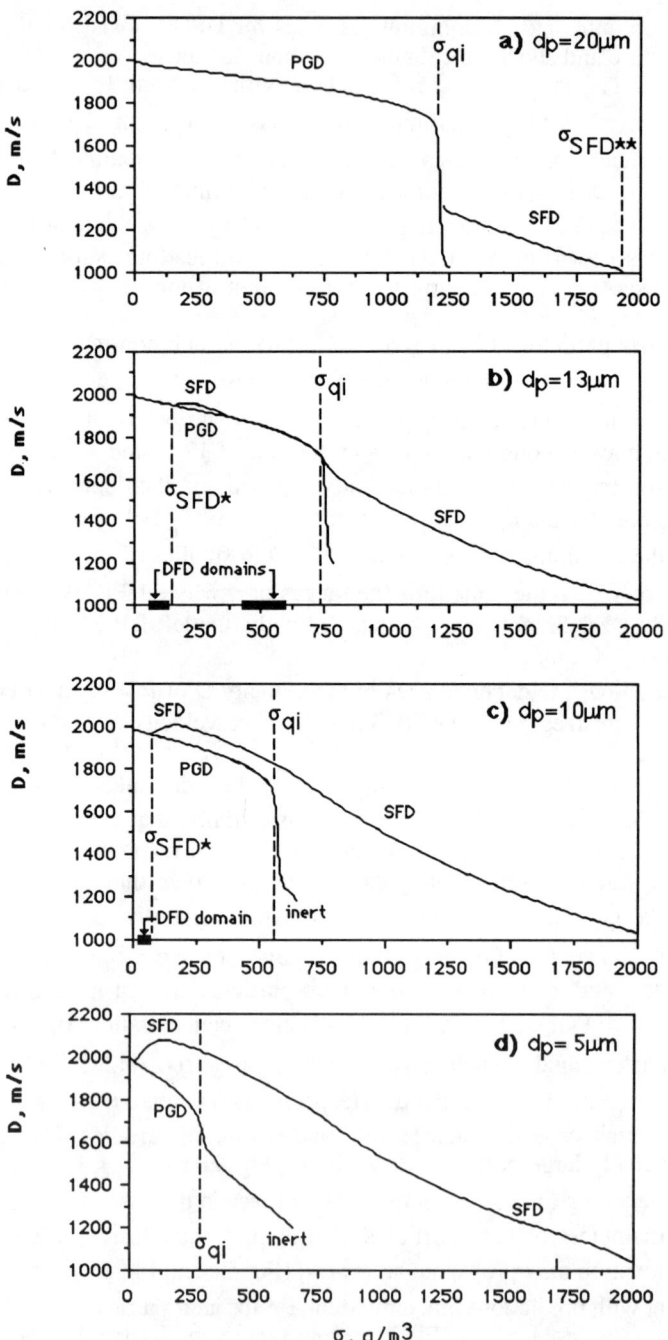

Fig. 2 The dependence of the velocity of SFD and PGD on particle concentration σ at different particle diameters: a) $d_p = 20$ μm, b) $d_p = 13$ μm, c) $d_p = 10$ μm, and d) $d_p = 5$ μm.

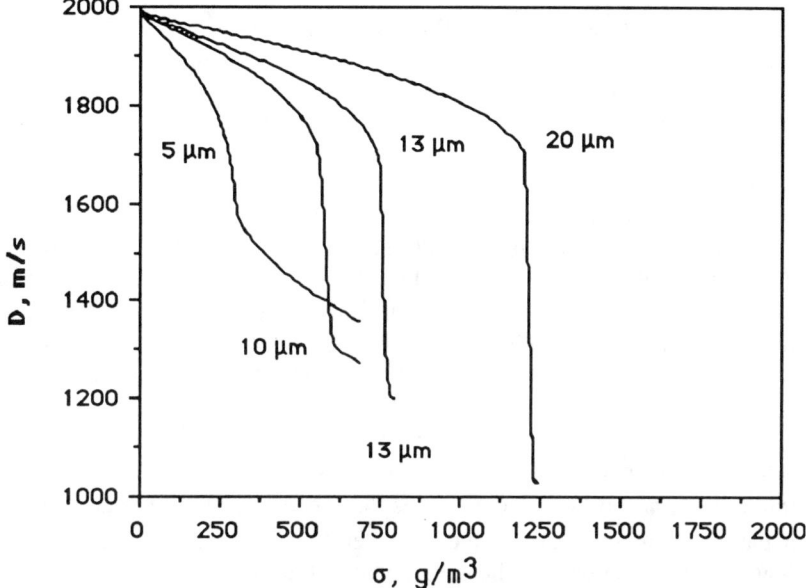

Fig. 3 Comparison of PGD velocities at different particle sizes.

(see Fig. 2b for 13-μm particles and Fig. 2c for 10-μm particles where domains of DFD are shown along the σ axis).

The dependence of concentration limits of DFD on particle size is easier to understand if one refers to SFD concentration limits, because DFD and SFD cannot propagate at the same σ. Figure 2 shows the dependence of SFD velocities on σ. At relatively large particle diameters, SFD can propagate only at σ > σ_{qi} (remember that DFD does not exist in the present conditions when d_p < 10 μm and d_p > 19 μm). But when d_p diminishes and the dependence τ(σ) transforms from a single nonmonotonic branch to two separate ones, the SFD propagation becomes possible in a gap between the left and right branches of DFD (Fig. 2b). Further decrease of particle size results in widening of SFD domains and coincidence of left and right domains of SFD. This is why the DFD domain collapses with a decrease of particle size. When d_p diminishes, the lower concentration limit of SFD (σ_{SFD}*) decreases. The sensitivity of upper limit of SFD (σ_{SFD}**) to d_p is quite weak.

Figure 3 shows the dependence of PGD velocities on σ for different d_p. As was mentioned above, the increase of losses due to the reduction of size of "inert" particles decreases the domain of PGD regime.

Figure 4 reports SFD velocities for different particle diameters. The maximum detonation velocity of SFD becomes closer and closer to the ideal detonation velocity when d_p decreases and the heat release rate from particles grows. The greater is d_p, the smaller are the domains of SFD. Comparison of

Fig. 4 with Fig. 1 shows that DFD is completely replaced by SFD for sufficiently small particles. Nevertheless, even for fine particles at $\sigma < \sigma_{qi}$ propagation of two different regimes, PGD and SFD is possible at the same particle concentration (compare Fig. 2c and Fig. 2d).

Effect of Tube Diameter

Figure 5 shows the effect of tube diameter on the dependence $\tau(\sigma)$ for the case of nominal 13-μm particles. Roughness of the tube h was changed simultaneously with the internal tube diameter D_{tube} to maintain the roughness of the tube at the same "technical" level[2,3] as formerly, i.e., $h/D_{tube}=1/2300$. Numbers near the curves indicate the factor by which the tube diameter was changed. Figure 5 looks similar to Fig. 1 but the effect of tube diameter is opposite to that of particle diameter, because the effective heat release rate in the burning zone of particles grows when d_p *decreases* and D_{tube} *increases*.

Comparison of Fig. 5 with Fig. 1 shows that the sensitivity of $\tau(\sigma)$ to D_{tube} is important but not to the same extent as the sensitivity of $\tau(\sigma)$ to d_p. It is seen again that there is an optimal tube diameter (6.44 cm, which is close to the nominal tube diameter used in the experimental studies of DFD[4,5]) for which the domain of DFD is the largest for the considered hybrid mixture and nominal mean particle diameter. The shift to higher tube diameters (corresponding to smaller losses) decreases DFD domain because in this case propagation of SFD becomes more likely. In tubes with diameters smaller than 3.4 cm neither DFD nor SFD can propagate at $\sigma < \sigma_{qi}$, though SFD can propagate at higher σ.

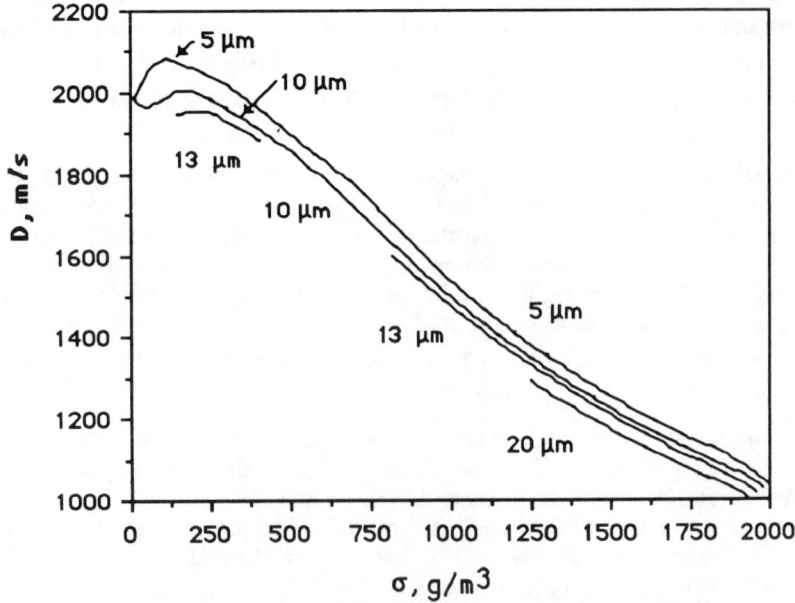

Fig. 4 Comparison of SFD velocities at different particle sizes.

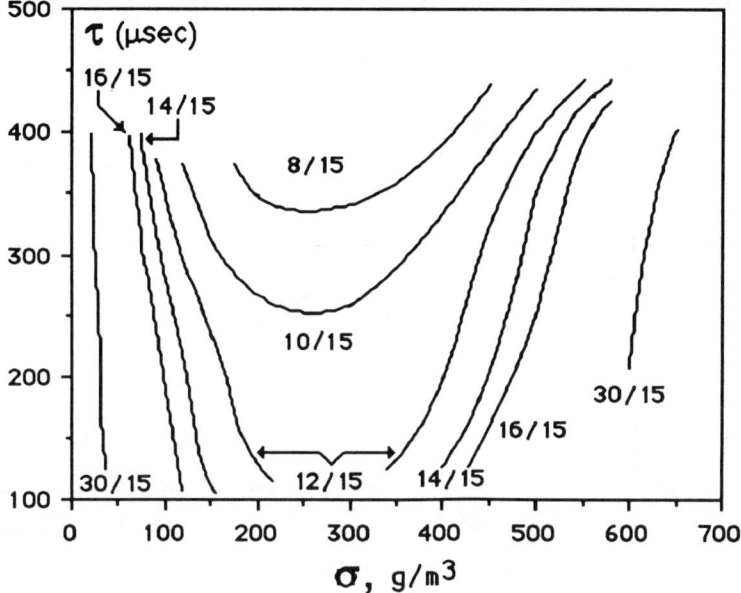

Fig. 5 Influence of shock tube diameter D_{tube} on the dependence of time delay τ between the two fronts of double-front detonation on particle concentration σ.

Figure 6 demonstrates that the decrease of D_{tube} diminishes the value of the maximum SFD velocity and decreases the domain of SFD at $\sigma < \sigma_{qi}$. The sensitivity of σ_{qi} to D_{tube} is less important than to d_p (compare the lower concentration limits of SFD at $\sigma > \sigma_{qi}$). Thus, with decrease of losses due to increase of D_{tube}, SFD becomes the prevailing regime of propagation; for larger losses, PGD is the dominant regime.

It is worthwhile to note that the distance between the shock front and the position x_{CJ} of CJ point for all three detonation regimes (in the case of DFD we mean the position of the second CJ point) grows when the level of heat losses decreases (in the case of ideal detonations ZND model implies that x_{CJ} is infinite[8]).

Bimodal Distribution of Particles

Now, since the process through which the losses affect the domains of existence of PGD, SFD, and DFD in hybrid two-phase mixtures is clearer, it is possible to expect that the addition of a given quantity of coarser particles to fine aluminum monodispersed powder may supply the losses provided by the shock tube walls, because acceleration and heating of these coarse particles should absorb a certain part of the energy from the flow. This mechanism was shown previously to be valid for bimodal suspension of monopropellant particles even

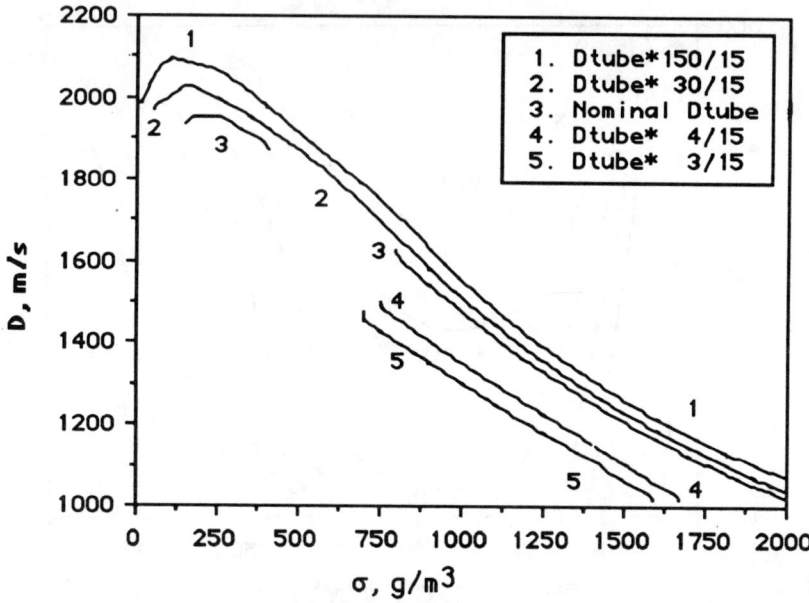

Fig. 6 Comparison of SFD velocities at different tube diameters.

in the case of single-velocity two-phase flow.[9] In the case of aluminum particles, the distribution of particles into two different fractions may also stabilize the nonideal detonation due to the fact that burning of aluminum becomes endothermic at temperatures above the boiling temperature of aluminum oxide T_{ox}. It may appear that at the moment of ignition of coarse particles, the temperature of the gas phase may already have reached T_{ox} due to burning of the fine particles only, hence the "burning" of coarser particles should absorb the energy from the flow.[12] Let us consider now the results of preliminary calculations using the KV code for the case of bimodal aluminum particles in the absence of losses to shock tube walls.

Results for Bimodal Particles

Bimodal particle size distribution is characterized by the diameters of fine and coarse particles d_{p1} and d_{p2} respectively, and by mass fractions of fine and coarse particles f_1 and f_2 ($f_1 + f_2 = 1$). Figure 7 shows the dependence of time delay τ between the two detonation fronts of DFD on σ in the case when the size of coarse particles d_{p2} is 130 or 260 µm and the size of main fraction of aluminum particles d_{p1} is assumed to be 13 µm ($d_{p2} = 10$ or $20 d_{p1}$). Curve 1 corresponds to the case when the mass fraction f_2 of coarse 130-µm particles is 0.2; curve 2 is computed for the same f_2 but for $d_{p2} = 260$ µm; and curve 3 differs from curve 2 by a higher value of f_2 ($f_2 = 0.5$). All of the curves correspond to the right branch of DFD; the left DFD branches are located at very small σ (about a few g/m^3) and have a very narrow domain of DFD existence.

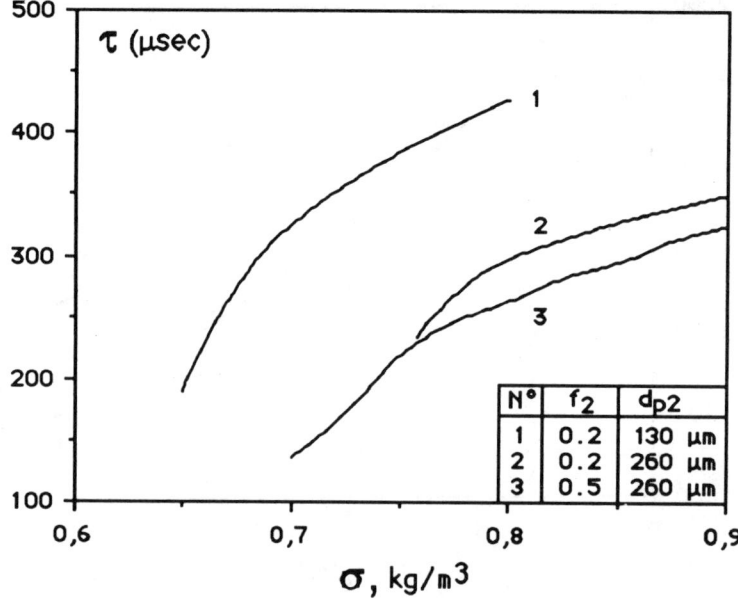

Fig. 7 Time delay between the two fronts of unconfined double-front detonation in the case of bimodal particle size distribution: 1) $f_2 = 0.20$ and $d_{p2} = 130$ μm, 2) $f_2 = 0.20$ and $d_{p2} = 260$ μm, and 3) $f_2 = 0.50$ and $d_{p2} = 260$ μm.

Comparison with the monomodal case (Fig. 5) shows that the right branches of $\tau(\sigma)$ dependence in the case of bimodal particle size distribution are shifted to higher σ. This is due partially to the fact that the concentration of smaller particles, which mainly control the heat release rate in the detonation wave is only $\sigma_1 = \sigma(1 - f_2)$ and is less than σ, and partially to the fact that larger particles in bimodal mixtures provide smaller heat losses rate than shock tube walls. Curve 1 corresponds to larger effective losses than curve 2 because particles with smaller d_{p2} are accelerated and heated by the gas flow faster than coarser ones. Comparison of relative positions of curves 1 and 2 shows that increase of losses shifts the right branch of $\tau(\sigma)$ dependence to smaller concentrations in agreement with the results obtained above. The dashed curves in Fig. 8 correspond to PGD velocities. It is seen that the difference between PGD curves 1a and 2a is very small; that is, in agreement with previous results, PGD concentration limit σ_{qi} depends only slightly on losses to coarse particles.

Curve 3 with $f_2 = 0.50$ should correspond to more important losses than curve 2 with $f_2 = 0.20$. But it appeared that the increase of mass fraction of coarse particles shifts the dependence $\tau(\sigma)$ to higher σ (Fig. 7). This effect is due mainly to the fact that, for the curve 3, $\sigma_1 = \sigma(1 - f_2) = \sigma/2$ is less than the concentration of smaller particles σ_1 in the case of the curve 2 when $\sigma_1 = \sigma (1 -$

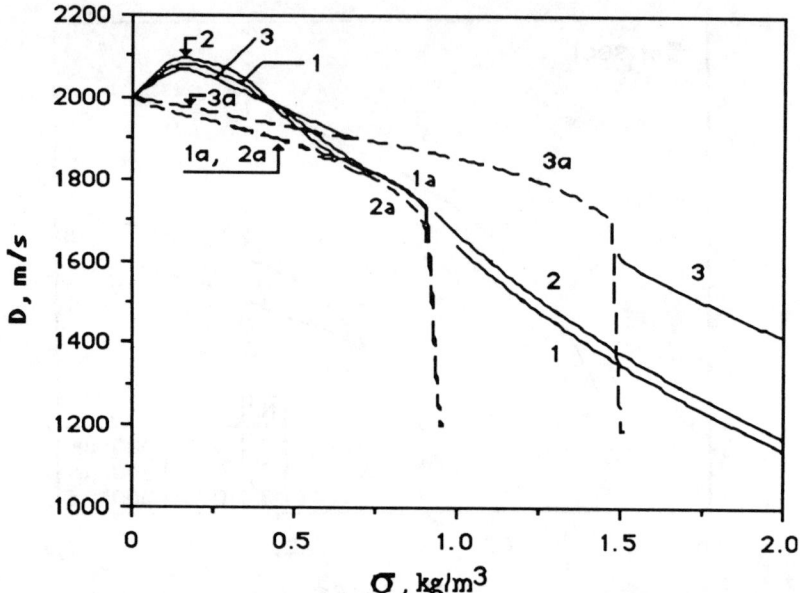

Fig. 8 Velocities of different detonation regimes in the case of bimodal particle size distribution: 1) $f_2 = 0.20$ and $d_{p2} = 130$ μm, 2) $f_2 = 0.20$ and $d_{p2} = 260$ μm, and 3) $f_2 = 0.50$ and $d_{p2} = 260$ μm; dashed lines correspond to PGD.

f_2) = σ*0.8. Indeed, from Fig. 2b for monomodal 13-μm particles, it follows that $\sigma_{qi} \approx 750$ g/m^3 and for the curve 3 with $f_2 = 0.50$ (Fig.8), $\sigma_{qi} \approx 1475$ g/m^3; hence, the ratio of those σ_{qi} is about 1/2 and corresponds approximately to a change of only σ_1. The same comparison for curves 1 or 2 with $f_2 = 0.20$ and $\sigma_{qi} \approx 925$ g/m^3 (Fig. 8) shows that at this concentration of particles $\sigma_1 = \sigma(1 - f_2) = 925*0.8 = 740$ g/m^3 which again approximately equals a quenching concentration 750 g/m^3 for monomodal particles. Thus quenching concentration σ_{qi} in bimodal suspensions in the absence of walls depends mainly on mass fraction of smaller particles and only slightly on losses to coarse ones.

Figure 8 shows also the dependence of SFD velocities on σ for the same bimodal particle size distributions as in Fig. 7. Because increase of f_2 increases the value of σ_{qi}, the dependence D(σ) shifts to higher σ in comparison with monomodal particles.

The position x_{CJ} of CJ point in the case of bimodal particle size distribution is essentially higher than in the presence of shock tube walls. This increase of x_{CJ} becomes much more important when coarse particle size d_{p2} is decreased to smaller values slightly exceeding d_{p1}.

Discussion

The results shown above indicate that in comparison with SFD and PGD, DFD exists in a quite narrow range of governing physical parameters such as mean particle size and shock tube diameter. Really, if one decreases the size of aluminum particles to such a small value that their reactivity becomes of the same order of value as that of gas, mainly SFD propagation must take place in this case (Fig. 2). PGD propagation is likely in the other extreme case of coarse aluminum particles which do not ignite at all or burn so slowly that the heat release from particles is not sufficient to result in the secondary detonation wave formation (though the formation of the secondary compression waves behind the CJ point of PGD may be possible). The conditions of DFD existence depend strongly not only on the particle size but on the shock tube diameter and roughness because each of these parameters influences the way the balance between the heat release and losses rate is reached at the CJ point.

Because of the complexity of the problem, the dependence of PGD, DFD, and SFD structure on governing parameters may be obtained only numerically, but results shown above clearly indicate the following:

1) DFD may be observed only for particle concentrations below the quenching limit by "inert" particles, σ_{qi}. DFD is replaced by SFD when the level of losses decreases. In the other extreme case of relatively strong losses (large particles or narrow shock tubes) DFD is replaced by PGD.

2) SFD has concentrations limits different from those of DFD; SFD and DFD cannot propagate at the same particle concentration.

3) At a given particle concentration the following possibilities can be met:

a) Only PGD can propagate (at $\sigma < \sigma_{qi}$) if SFD or DFD do not exist in this concentration range.

b) PGD and DFD can propagate at the same particle concentration ($\sigma < \sigma_{qi}$).

c) PGD and SFD can propagate (at $\sigma < \sigma_{qi}$).

d) Only SFD can propagate (this case may be realized only for large particle concentrations: $\sigma_{qi} < \sigma < \sigma_{SFD**}$).

e) No steady nonideal detonation can propagate at $\sigma > \sigma_{SFD**}$.

As was shown above, bimodal particle size distribution really can result in the behavior of PGD, SFD, and DFD similar to that in the case of monomodal particles and the presence of shock tube walls. Namely, even the DFD propagation is possible for the considered cases. The dependences $D(\sigma)$ and $\tau(\sigma)$ are shifted to higher σ in comparison with the monomodal case mainly due to decrease of concentration of smaller particles. The effectiveness of losses due to the presence of coarse particles is lower than that due to the presence of shock tube walls. This fact is confirmed by the behavior of the position x_{CJ} of the CJ point: in the case of bimodal particle size distribution, x_{CJ} is essentially higher than in the presence of shock tube walls. The shift of CJ point to higher values becomes much more important when mass fraction of coarse particles decreases or the size of coarse particles is closer to the size of the main fraction. In the last case the endothermicity of aluminum burning at high temperatures becomes the main mechanism of stabilizing the nonideal detonation. As a result, in bimodal powders in the absence of wall losses, nonideal detonation zone thickness is generally much greater than in the case when wall losses are present. But this increase of x_{CJ} does not practically change the detonability of the hybrid mixtures.[12] It is due to the two-stage nature of heat release from aluminum particles: the initial zone of fast heat release (in which gas temperature rapidly reaches the aluminum oxide boiling temperature T_{ox}) is followed by slow heat release zone (in which gas temperature is close to T_{ox}) and the chemical energy of aluminum is released mainly in the initial heat release zone from particles.

Because of the effect of shock tube walls on the concentration limits of different detonation regimes in hybrid mixtures, it would be interesting to know what may happen if one tries to detonate a hybrid mixture in unconfined conditions. If the experimental conditions are really ideal (no boundary effects, strictly monomodal particle size distribution, etc.) then, following the results shown above, DFD cannot exist and the only stable propagation regime can be PGD or SFD, depending on particle size. (In the last case at $\sigma < \sigma_{qi}$, PGD may propagate at the initial stage of the initiation process if the initiation source is not very strong and the duration of this stage must grow with particle size.) One can wonder whether this situation may be achieved in actual conditions. Surely, the user prefers to limit the scattering of particle size distribution of the powder, but this distribution is always at least a Gaussian one. However, the experimental conditions may be not ideal due, for example, to coagulation of particles during the dispersion of the powder or, more likely, due to the presence of coarse particles in the aluminum powder (note that the numeric fraction of larger particles is proportional to $f_2(d_{p1}/d_{p2})^3$ and may be essentially lower than mass fraction f_2). Hence, steady plane DFD may exist even in real cases.

If one considers the general problem of detonations supported by nonmonotonic heat release, it is possible to imagine that there may exist numerous situations where coarser particles (other than aluminum) are able to provide heat losses and play a role comparable with that of walls.

Conclusions

The effect of mean particle size and shock tube diameter on aluminum particle concentration limits of pseudogas (PGD), double-front (DFD), and single-front (SFD) detonations in hybrid mixtures of aluminum particles with detonable gases has been studied numerically. The pseudogas detonation regime has been shown to exist up to the quenching limit σ_{qi} of gas detonation by "inert" particles.

At $\sigma > \sigma_{qi}$, a single-front detonation (SFD) may propagate which is supported by gas reactions and heat release from particles. The upper concentration limit of SFD is $\sigma_{SFD**} \geq \sigma_{qi}$. If particle size is small enough, SFD also may propagate at $\sigma < \sigma_{qi}$. When particle size diminishes, the second region of existence of SFD increases and may coincide with the first one. The lower concentration limit of SFD, σ_{SFD*}, increases with particle size. If one wants to increase the detonation velocity of gaseous mixtures by adding aluminum particles it is necessary to use as fine aluminum particles as possible.

It is shown that double-front detonations (DFD) can propagate steadily only in a relatively narrow range of particle sizes and shock tube diameters, where normal PGD exists, but SFD cannot propagate. For a given composition of the gas mixture and shock tube diameter, there is an optimal particle size for which the domain of DFD is the largest. The increase of particle size (or reduction of shock tube diameter) results in an increase of time delay τ between the two detonation fronts of DFD and, finally, DFD disappears. For smaller particles or greater diameters of the shock tube, DFD is replaced by SFD.

It is shown also that the presence in the aluminum powder of relatively large particles may stabilize a propagation regime of nonideal detonation even in the absence of shock tube walls. Thus, the domains of existence of unconfined PGD, SFD, and DFD should follow the same trends as in the case of detonation propagation in shock tubes.

References

[1] Khasainov, B. A., and Veyssière, B., "Analysis of the Steady Double-Front Detonation Structure for a Detonable Gas Laden with Aluminium Particles," *Archivum Combustionis,* Vol.7, No. 3-4, 1987, pp. 333-352.

[2] Khasainov, B. A., and Veyssière, B., "Steady, Plane, Double-Front Detonations in Gaseous Detonable Mixtures Containing a Suspension of Aluminum Particles." *Dynamics of Explosions,* Vol. 114, Progress in Astronautics and Aeronautics, AIAA, New York, 1988, pp. 284-299.

[3] Veyssière, B., and Khasainov B. A., "A Model for Steady, Plane, Double-Front Detonations (DFD) in Gaseous Explosives Mixtures with Aluminium Particles in Suspension," *Combustion and Flame,* Vol.85, No.1-2, 1991, pp. 241-253.

[4] Veyssière, B., and Manson, N., "Sur l'Existence d'un Second Front de Détonation des Mélanges Biphasiques Hydrogène-Oxygène-Azote-Particules d'Aluminium," *Comptes Rendus à l'Académie des Sciences,* Vol. 295, NoII, 1982, pp. 335-338.

[5] Veyssière, B., "Structure of the Detonations in Gaseous Mixtures Containing Aluminium Particles in Suspension," *Dynamics of Explosions,* Vol.106, Progress in Astronautics and Aeronautics, AIAA, New York, Vol. 106, 1986, pp. 522-544.

[6] Veyssière, B., and Khasainov, B. A., "The Propagation Regimes of Non-ideal Detonations in Combustible Gaseous Mixtures with Reactive Solid Particles," *Khimicheskaya Fizika* , Vol.10, No.11, 1991, pp. 1533-1544.

[7] Veyssière, B., and Khasainov, B. A., "Nonideal Detonation Regimes in Aluminum Suspensions in Reacting Gases," XIIIth *International Colloquium on Dynamics and Explosions of Reactive Systems,* Poster No. PAB12/B3, Nagoya, Japan, August 1991.

[8] Zel'dovich, Ya. B., and Kompaneets, A. S., *Theory of Detonations,* Academic Press, New York, 1960.

[9] Borisov, A. A., Ermolaev, B. S., and Khasainov, B. A., "Non-Ideal Detonation in a Mixture of Gas with Bimodal Monopropellant Particles," *Khimicheskaya Fizika,* Vol. 2, No 8, 1983, pp. 1129-1133.

[10] Brewer, L., and Searcy, A. W., "The Gaseous Species of the $Al-Al_2O_3$ System" *J.Am.Chem.Soc.,* Vol. 73, No. 11, 1951, pp. 5408-5414.

[11] Kopotev, V. A., and Kuznetsov, N. M., "On the Problem of Existence of Steady Double-Front Detonation Waves," *Detonatsia i udarnye volny,* Proceedings of the VIIIth Soviet Symposium on Combustion and Explosions, Tashkent, 1986, pp. 139-142.

[12] Borisov, A. A., Khasainov, B. A., Veyssière, B., Saneev, E. L., Khomik, S. V., Fomin, I. B., "On Detonation of Aluminum Suspensions in Air and Oxygen," *Khimicheskaya Fizika* , Vol. 10, No.2, 1991, pp. 250-272.

[13] Schlichting, H., *Boundary Layer Theory,* Mc Graw Hill, New York, 1968.

[14] Menga, H., "Contribution à l'Etude de l'Influence des Pertes Thermomécaniques sur la Détonation dans les Mélanges Hétérogènes," *Thèse de Docteur Ingénieur,* University of Poitiers, France, 1981.

Effect of Hollow Heterogeneities on Nitromethane Detonation

C. Gois,* H. N. Presles,† and P. Vidal*
Ecole Nationale Supérieure de Mécanique et d'Aérotechnique, Poitiers, France

Abstract

Mixtures of nitromethane (NM) and polymethylmetacrylate (PMMA) containing physical heterogeneities exhibit detonation properties more like solids than liquids because under shock conditions the heterogeneities produce hot-spots which increase the energy release rate in the NM. Part of an overall study to understand this behavior and its dependence on the size, concentration, nature, and shape of the heterogeneities is presented here. In this part of the study, an experimental program was undertaken to determine 1) the relationship between steady-state detonation velocity and reciprocal charge size (the diameter effect curve) and 2) the critical diameter, for mixtures of NM-PMMA containing different concentrations of glass microballoons (GMBs) with well-defined diameters in the range 37-50 μm. Sets of experiments with mixtures containing a given concentration of GMBs up to 4.36% were performed in steel and plastic tubes to determine how the diameter effect curve and critical diameter of these mixtures depend on the concentration of GMBs and the confinement. The critical diameter measurements in plastic tubes show that the hot-spots produced by the GMBs decrease the critical diameter of our NM-PMMA mixture by up to 80% and are effective in increasing the energy release rate in NM. The measurements of the diameter effect curve show that their shapes are sensitive to the GMB concentration and the confinement. In steel confinement, the diameter effect curves are linear for tube diameter ≥ 2 mm (but may not be linear for tubes with diameters < 2 mm). In plastic confinement, the shape of the diameter effect curve for the mixture containing 1% of GMBs is typical of a heterogeneous explosive, but as the concentration of GMBs is increased the diameter effect curves become more linear.

I. Introduction

Explosives can be classified as homogeneous or heterogeneous according to the different curvatures and velocity decrements exhibited by their detonation

---

Copyright © 1992 by the American Institute of Aeronautics and Astronautics, Inc. All rights reserved.
*Research Scientist, Detonation, Centre National de la Recherche Scientifique, Laboratoire d'Energétique et de Détonique.
†Senior Research Scientist, Detonation, Centre National de la Recherche Scientifique, Laboratoire d'Energétique et de Détonique.

velocity vs reciprocal charge diameter curves. For these diameter effect curves, the velocity decrement is the difference between the detonation velocity in a charge with an "infinite" diameter and the detonation velocity in a charge with the critical diameter. For homogeneous explosives, the diameter effect curve is linear and has a small velocity decrement of about 1%, but for heterogeneous explosives it is strongly concave downward and has a larger velocity decrement of about 10%.

These significant differences between the diameter effect curves for homogeneous and heterogeneous explosives can be attributed to their different chemical energy release rates. In a homogeneous explosive, the chemical energy is released globally in shocked material by thermal explosion after an induction time determined by the bulk temperature. In a heterogeneous explosive, however, the chemical energy is released at local sites called "hot-spots," where the temperatures are significantly higher than the bulk shock temperature in the surrounding explosive. Consequently, a significant amount of chemical energy can be released before the global induction time and used to support wave propagation.

Many mechanisms were pointed out to account for hot-spot formation (gas compression in pores,[1] friction between grains,[1] impact of the upstream surface of a cavity on the downstream side;[2] plastic work in the vicinity of collapsing cavities,[3] and shear[4]) but up to now it is not possible to state which is the dominant physical process.

Our approach to obtain a better understanding of the role hot-spots play in the detonation process is to determine how the diameter effect curve of a liquid explosive is altered by incorporating a well-defined field of physical heterogeneities into this liquid explosive.

Nitromethane (NM) is convenient for performing such an experimental study because its shock initiation and detonation properties are known to be significantly changed by the presence of imperfections and heterogeneities. More specifically, Campbell et al.,[5] thirty years ago, showed that the shocked initiating process in NM containing small imperfections is more typical of a heterogeneous than a homogeneous explosive. Engelke,[6] ten years ago, showed that the steady detonation process in NM containing silica particles is more like that in a heterogeneous explosive than in a homogeneous explosive. Fauquignon and Moulard[7] also found that the distance between hot-spots in nitromethane plays a major role in reducing the initiation distance to detonation.

A few years ago, we incorporated glass microballoons (GMBs) into NM to investigate how its steady-state detonation velocity and pressure are influenced by hot-spots produced by physical inhomogeneities.[8] Stable mixtures of NM and GMBs for this investigation were made by the following procedure. We first added 3% of polymethylmetacrylate (PMMA) to the NM to increase its viscosity and form a homogeneous NM-PMMA, and then added 5-40% (mass fraction) of the GMBs.

Although the experiments performed with these mixtures are not completely definitive because the size distribution of the GMBs was too large, the influence of GMBs on the detonation process in NM is clearly shown by the following experimental result. NM-PMMA mixtures containing GMBs confined in plastic tubes with diameter as small as 10 mm were found to support steady detonation even though the critical diameter of the NM-PMMA mixture is greater than 30 mm.

This pronounced sensitization effect is well known and is the reason for the use of GMBs in industrial explosives such as emulsions.[9,10] No definitive conclusion about the influence of the GMBs concentration on the shape of the diameter effect curves could be reached, however, because the diameter of the tubes used in the experiments were always greater than the critical diameters.

To extend our previous study and obtain more definitive results, we started an experimental program to determine the diameter effect curves for NM-PMMA mixtures containing different concentrations of GMBs with a well-defined size distribution. Because such a program allows us to determine how the steady-state velocities and critical diameters of such mixtures depend on the size, concentration, and nature of physical heterogeneities, we expect our experimental results to provide a better understanding of some aspects of the role hot-spots play in the detonation process.

II. Experimental Conditions

1. Mixtures

GMBs supplied by Minnesota Mining and Manufacturing (No. C15-250) were sieved into 12 sizes, and those with diameters in the range 37-50 μm (90% of GMBs have a diameter ≤ 50.3 μm and 10% have a diameter ≤ 37.1 μm) were used to prepare our mixtures. These GMB had a wall thickness of about 1 μm and an apparent density 0.208 g/cm^3. To obtain stable mixtures containing lower mass fraction of GMBs than those used previously, it was necessary to increase the viscosity of the base NM-PMMA mixture by increasing its mass fraction of PMMA from 3 to 4%. A particular NM/PMMA-GMB mixture was defined by the mass fraction X of GMBs it contained. The variation of the measured initial density ρ_0 of our mixtures with respect to X is given in Table 1.

2. Detonation Tubes

We used two kinds of confinement, steel tubes and PMMA tubes. The internal diameter ϕ of the steel tubes ranged from 2 to 21 mm and their wall thickness was always 2 mm. The PMMA tubes had internal diameters between 3 and 20 mm and, depending on the diameter, had wall thickness between 2 and 3 mm.

All of the tubes were equipped with three ionization coaxial gauges located every 10 cm (accuracy ± 0.05 mm). The distance between the plane of initiation and the first probe was equal to 10 times the charge size. Electrical impulses delivered by the gauges were used to trigger an electronic counter (Thomson 632M32, accuracy ± 1 ns).

Table 1 Initial density of the mixtures

X, %	1	2	3	4.36
ρ_0	1.094	1.037	1.004	0.984

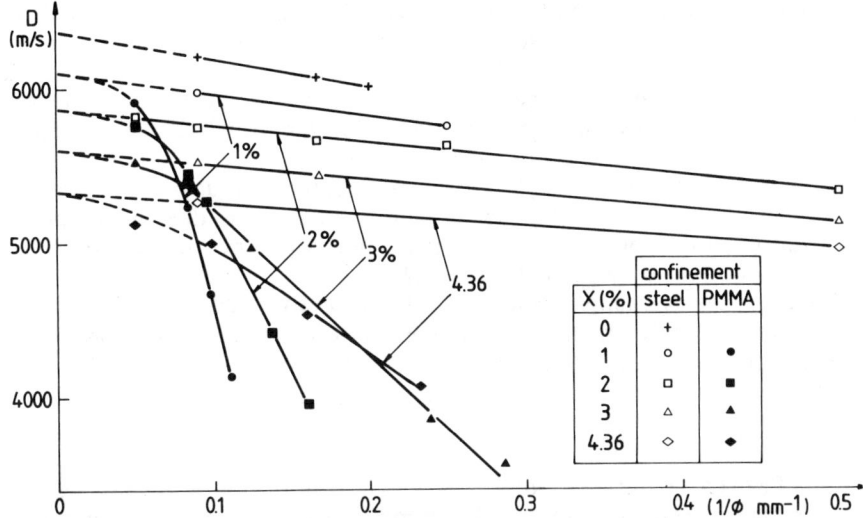

Fig. 1 Detonation velocity vs the reciprocal charge diameter.

The mixtures can be initiated with a No. 8 detonator, but to get a better defined surface of initiation we used a booster made of a plastic explosive whose external diameter was equal to that of the detonation tube.

III. Experimental Results

1. Detonation Velocity

We used the same procedure for each experiment. The amount of the mixture necessary to fill a tube was prepared just before the experiment and each mixture was stirred slowly until it was poured or injected in the tube according to its size. The charge was then fired in less than 5 min; this time was short enough to keep the mixture uniform. All the experiments were performed at an initial temperature of $6 \pm 2°C$.

Streak camera records were taken in some experiments with mixtures contained in transparent tubes to verify that detonation velocities were constant. The experimental results plotted in Fig. 1 show the relationship between the detonation velocity and the reciprocal charge diameter for the mixtures. At first glance the nature of the confinement is a very important parameter.

1.A. Steel Confinement

Five diameter effect curves are shown in Fig. 1 for our mixtures, confined in steel tubes, with values of X varying from 0 to 4.36%. These diameter effect curves are linear and characteristic of homogeneous explosives. They also show that the inverse diameter range for self-sustaining detonation in our homogeneous NM-PMMA mixture confined in steel tubes is significantly increased by the presence of GMBs. In fact, the critical diameter for the NM-PMMA mixture (X = 0) was found to be 5 mm, while the critical diameters for the mixtures with X = 2, 3, and 4.36% were found to be less than 2 mm. We thus conclude that any heterogeneous behavior to be found in steel confinement must be exhibited by our mixtures confined in tubes with diameters greater than their critical diameters but less than 2 mm.

Detonation velocities of explosive charges confined in tubes with diameters less than 2 mm are difficult to measure because it is difficult to install ionization gauges on such narrow tubes, but procedures to overcome this difficulty will be developed in the near future.

Values of the detonation velocity D_∞ for charges with an "infinite" diameter were obtained by extrapolating the diameter effect curves backward to the detonation velocity axis where $\phi = \infty$. These extrapolated values for D_∞ given in Table 2, as our previous results,[11] show that the variation of D_∞ with X is nearly linear.

1.B. PMMA Confinement

Four diameter effect curves are shown in Fig. 1 for our mixtures, confined in PMMA tubes, with values of X varying from 1 to 4.36%. No diameter effect curve is shown for our base NM-PMMA mixture (X = 0) because we were unable to initiate self-sustaining detonation in this mixture confined in PMMA tubes with diameters up to 30 mm. These four diameter effect curves for PMMA confinement, in contrast to those for steel confinement, are concave downward and thus characteristic of heterogeneous explosives. However, as the mass fraction X increases, the velocity decrement increases from 35 to 45%, but the diameter effect curves become more linear. With respect to curvature, the heterogeneous nature of our mixtures decreases as X increases from 1%, and from the decreasing curvature of the diameter effect curves shown in Fig. 1, we expect that the diameter effect curve will become linear when $X \approx 7\%$. This is an interesting prediction, when we recall for mixtures with unsieved GMBs, that the diameter effect curves are linear for values of $X \geq 5\%$.

Further examination of the four diameter effect curves for plastic confinement shows that our NM/PMMA-GMB mixtures have another interesting property. Because the velocity decrement increases but the curvature decreases as X increases, the curves intersect. Consequently, two different mixtures confined in the same diameter PMMA tube can have the same detonation velocity. Comparing the relative shapes of two of these diameter effect curves leads us to the following conclusions. The mixture with the lower value of X contains more explosive per unit volume than the mixture with the higher value of X and thus has a higher value of D_∞.

Table 2 D_∞ vs GMB concentration

X, %	0	1	2	3	4.36
D_∞ m/s	6360	6100	5860	5600	5330

After the diameter effect curves intersect, the mixture with the higher value of X has a higher energy release rate than the mixture with the lower value of X because it contains more heterogeneities that produce more hot-spots. An increase of the GMB concentration thus reduces the chemical energy available to support wave propagation but gives rise to a larger number of hot-spots which increase the energy release rate at smaller diameters.

2. Critical Diameter

As shown in Fig. 2, the value of the critical diameter of our NM/PMMA-GMB system decreases rapidly as the value of X increases. For the base NM-PMMA mixture, the critical diameter has a value > 30 mm, for the mixture with X = 1% a value of ~ 8.5 mm corresponding to a reduction of at least 70%, and for the mixture with X = 4%, a value of ~ 3 mm corresponding to a reduction of at least 85%. Thus the concentration of GMBs has a very strong influence on the critical diameter of our NM/PMMA-GMB system.

Engelke[12] used glass microspheres (GMSs) in diameters in the range 1-4 μm as physical inhomogeneities in a NM-guar-gum mixture, and showed that there is a linear relationship between the interparticles distance and the critical diameter of such a system. We observe a similar dependence in the present work (Fig. 3), when the interparticles distance calculated from the GMB concentration is in the 60-110 μm range. Such a linear relationship is valid for only a limited range of interparticles distances because the critical diameter becomes asymptotic to that of the base NM-PMMA mixture as the

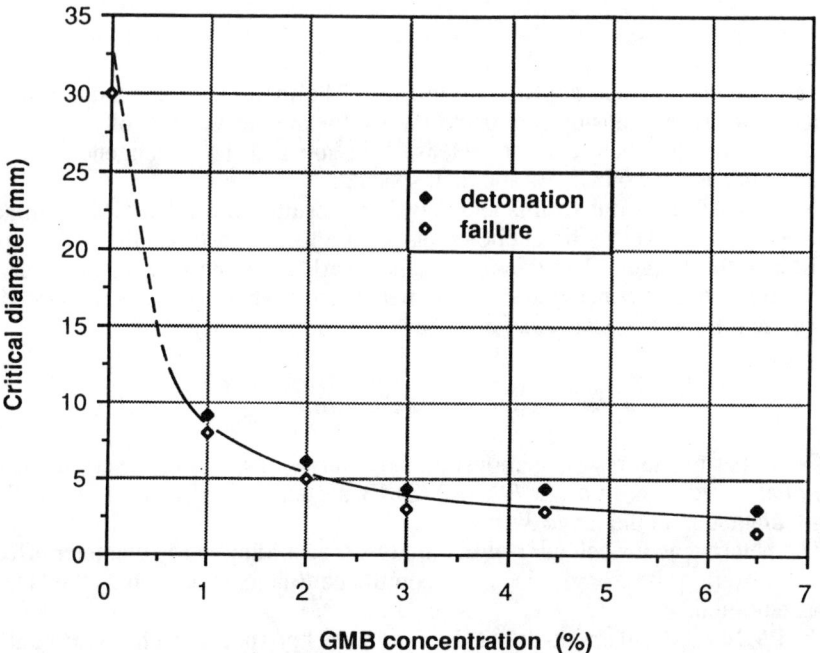

Fig. 2 Critical diameter of the mixtures confined in PMMA tubes.

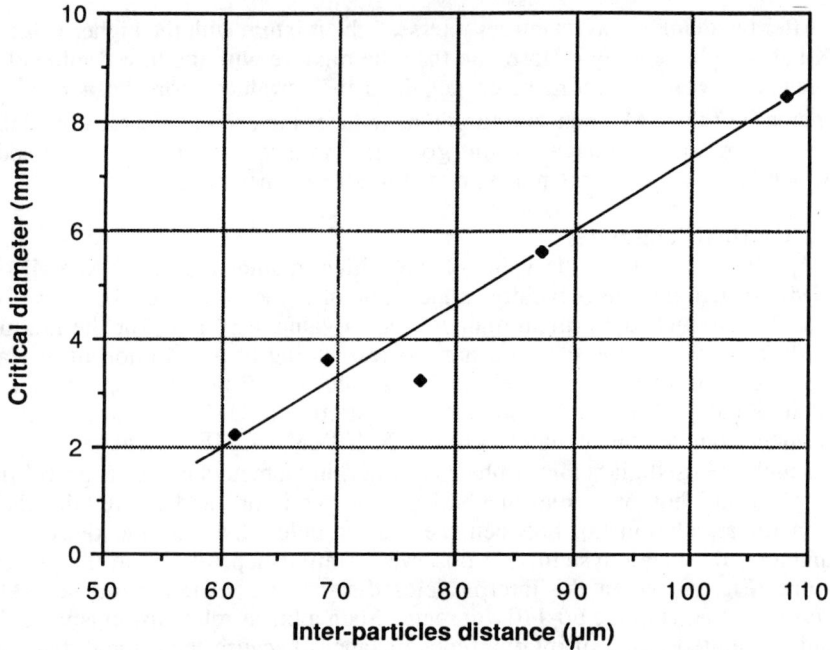

Fig. 3 Critical diameter of the mixtures confined in PMMA tubes vs the interparticles distance.

interparticles distance approaches infinity. The theoretical understanding of such a linear relationship requires a theory for the dependence of the critical diameter on the chemical heat release function and the dependence of this energy release rate on GMBs and their spacing.

A comparison of our results and Engelke's results shows that GMBs unlike GMSs produced very efficient hot-spots. Consequently, because the GMBs and GMSs in these studies had the same shape, nearly the same size, and were used in similar volume concentrations, the mechanisms whereby GMBs and GMSs increase the energy release rate must be significantly different.

IV. Conclusions

Our experimental results demonstrate how various aspects of the steady state detonation process in a NM/PMMA-GMB system are influenced by GMBs with diameters in the range 37-50 µm.

The heterogeneous behavior of the mixtures, according to the diameter effect curve criterion, is strongly dependent on the confinement as well as the GMB concentration.

In PMMA confinement the GMBs produce hot-spots which significantly increase the chemical energy release rate, because 1) mixtures containing GMBs support steady detonation in charges with much smaller diameter than the critical diameter of the base NM-PMMA mixture, and 2) the critical diameter decreases as the concentration of GMBs is increased from 1 to 6.5%.

We chose not to comment on the mechanisms involved in hot-spot formation because our experimental study of the dependence of the detonation process on GMBs with a well-defined size distribution is limited to distribution (37-50 μm), and because the detonation process is influenced by the integrated effect of chemical reaction initiated locally at many sites. Accordingly, we plan to extend our data base by performing similar experiments with our base NM-PMMA mixture containing GMBs with other well-defined size distributions and also containing GMSs with corresponding well-defined size distributions.

We hope that the additional data obtained from these planned experiments will provide the information required to develop a better physical understanding of the dependence of the detonation process on physical heterogeneities.

Acknowledgment

The authors are grateful to M. Cowperthwaite for helpful discussions.

References

[1] Bowden, F. P., and Yoffe, A. D., *Initiation and Growth of Explosion in Liquids and Solids*, Cambridge University Press, Cambridge, London, New York, 1952, pp. 12-55.

[2] Mader, C., "Initiation of Detonation by the Interaction of Shock with Density Discontinuities," *Physics of Fluids*, Vol. 8, No. 10, 1965, pp. 1811-1816.

[3] Khasainov, B. A., Borisov, A. A., Ermolaev, B. S., and Korotkov, A. I., "Two Phase Visco-Plastic Model of Shock Initiation of Detonation on High Density Pressed Explosives," *Seventh Symposium on Detonation*, NSWC 82-334, White Oak, 1981, pp. 435-447.

[4] Frey, R. B., "The Initiation of Explosive Charges by Rapid Shear," *Eight Symposium on Detonation*, NSWC 86-194, White Oak, 1985, pp. 32-64.

[5] Campbell, A. W., Davis, W. C., and Travis, J. R., "Shock Initiation of Detonation in Liquid Explosives," *The Physics of Fluids*, Vol. 4, No. 4, 1961, pp. 498-510.

[6] Engelke, R., "Effect of a Physical Inhomogeneity on Steady-State Detonation Velocity," *Physics of Fluids*, Vol. 22, No. 9, 1979, pp. 1623-1630.

[7] Fauquignon, C., Moulard, H., "Shock Sensitivity of Nitromethane with Well Defined Hot-Spots Distribution," *Acta Astronautica*, Vol. 5, No. 11-12, 1978, pp. 1035-1040.

[8] Presles, H. N., Campos, J., Heuzé, O., and Bauer, P., "Effects of Microballoons Concentration on the Detonation Characteristics of Nitromethane - PMMA Mixtures," *Ninth Symposium on Detonation*, NAVSWC, Silver Spring, Vol. 2, 1989, pp. 925-929.

[9] Yoshida, M., Iida, M., Tanaka, K., Fujiwara, S., Kusakabe, M., and Shiino, K., "Detonation Behavior of Emulsion Explosives Containing Glass Micro-Balloons," *Eight Symposium on Detonation*, NSWC 86-194, White Oak, 1985, pp. 993-1000.

[10] Lee, J., and Persson, P. A., "Detonation Behavior of Emulsion Explosives," *Propellants Explosives Pyrotechnics*, Vol. 15, No. 5, 1990, pp. 208-216.

[11]Presles, H. N., Campos, J., Heuzé, O., and Bauer, P., "Détonation de Mélanges Nitrométhane - PMMA Billes de Verre Creuses," *Third Symposium on Behaviour of Dense Media under High Dynamic Pressures*, Association Française de Pyrotechnie, Paris, 1989, pp. 123-126.

[12]Engelke, R., "Effect of the Number and Density of Heterogeneities on the Critical Diameter of Condensed Explosives," *Physics of Fluids*, Vol. 26, No. 9, 1983, pp. 2420-2424.

Author Index

Abid, S. 162
Akbar, R. 78
Aksamentov, S. M. 112
Aminallah, M. 203
Bakken, J. 425
Bjerketvedt, D. 324, 425
Boris, J. P. 241
Borisov, A. A. 312
Bourlioux, A. 43
Brossard, J. 203
Chéret, R. 397
Chue, R. S. 270
Dabora, E. K. 3
Desbordes, D. 347
Dremin, A. N. 105
Dupré, G. 162
Engebretsen, T. 324, 425
Frolov, S. M. 298
Frost, D. L. 182
Gelfand, B. E. 298
Gois, C. 462
Guerraud, C. 347
Hamada, L. 347
Huang, Z. W. 132
Inada, M. 253
Jones, D. A. 241
Kailasanath, K. 64, 231
Kamel, M. 363
Khasainov, B. A. 447
Khomik, S. V. 312
Kilambi, G. 363
Knystautas, R. ....
.... 182, 253, 270, 363, 381

Kosenkov, V. V. 312
Laberge, S. 381
Lee, J. H. 95, 253, 270
Lee, J. H. S. 182, 363, 381
Lee, J. J. 182
Lefebvre, M. H. 64, 144
Li, C. 231
Mailkov, A. E. 312
Majda, A. J. 43
Makris, A. 363
Manson, N. 3
Manzhaley, V. I. 112
Meltzer, J. 78
Mikhalkin, V. N. 312
Mitrofanov, V. V. 112
Nzeyimana, E. 144
Oran, E. S. 64, 231, 241
Paillard, C. 162
Papyrin, A. 270, 363
Presles, H. N. 347, 462
Sabet, A. 78
Scarinci, T. 270
Shepherd, J. E. 78
Sichel, M. 241
Sønju, O. K. 324, 425
Takano, Y. 283
Teodorczyk, A. 405
Van Tiggelen, P. J. 64, 132, 144
Vasiliev, A. 203
Veyssière, B. 447
Vidal, P. 462
Yoshikawa, N. 95

PROGRESS IN ASTRONAUTICS AND AERONAUTICS SERIES VOLUMES

*1. **Solid Propellant Rocket Research** (1960)
Martin Summerfield
Princeton University

*2. **Liquid Rockets and Propellants** (1960)
Loren E. Bollinger
Ohio State University
Martin Goldsmith
The Rand Corp.
Alexis W. Lemmon Jr.
Battelle Memorial Institute

*3. **Energy Conversion for Space Power** (1961)
Nathan W. Snyder
Institute for Defense Analyses

*4. **Space Power Systems** (1961)
Nathan W. Snyder
Institute for Defense Analyses

*5. **Electrostatic Propulsion** (1961)
David B. Langmuir
Space Technology Laboratories, Inc.
Ernst Stuhlinger
NASA George C. Marshall Space Flight Center
J.M. Sellen Jr.
Space Technology Laboratories, Inc.

*6. **Detonation and Two-Phase Flow** (1962)
S.S. Penner
California Institute of Technology
F.A. Williams
Harvard University

*Out of print.

*7. **Hypersonic Flow Research** (1962)
Frederick R. Riddell
AVCO Corp.

*8. **Guidance and Control** (1962)
Robert E. Roberson,
Consultant
James S. Farrior
Lockheed Missiles and Space Co.

*9. **Electric Propulsion Development** (1963)
Ernst Stuhlinger
NASA George C. Marshall Space Flight Center

*10. **Technology of Lunar Exploration** (1963)
Clifford I. Cummings
Harold R. Lawrence
Jet Propulsion Laboratory

*11. **Power Systems for Space Flight** (1963)
Morris A. Zipkin
Russell N. Edwards
General Electric Co.

*12. **Ionization in High-Temperature Gases** (1963)
Kurt E. Shuler, Editor
National Bureau of Standards
John B. Fenn,
Associate Editor
Princeton University

*13. **Guidance and Control—II** (1964)
Robert C. Langford
General Precision Inc.
Charles J. Mundo
Institute of Naval Studies

*14. **Celestial Mechanics and Astrodynamics** (1964)
Victor G. Szebehely
Yale University Observatory

*15. **Heterogeneous Combustion** (1964)
Hans G. Wolfhard
Institute for Defense Analyses
Irvin Glassman
Princeton University
Leon Green Jr.
Air Force Systems Command

*16. **Space Power Systems Engineering** (1966)
George C. Szego
Institute for Defense Analyses
J. Edward Taylor
TRW Inc.

*17. **Methods in Astrodynamics and Celestial Mechanics** (1966)
Raynor L. Duncombe
U.S. Naval Observatory
Victor G. Szebehely
Yale University Observatory

*18. **Thermophysics and Temperature Control of Spacecraft and Entry Vehicles** (1966)
Gerhard B. Heller
NASA George C. Marshall Space Flight Center

*19. **Communication Satellite Systems Technology** (1966)
Richard B. Marsten
Radio Corporation of America

*20. **Thermophysics of Spacecraft and Planetary Bodies: Radiation Properties of Solids and the Electromagnetic Radiation Environment in Space** (1967)
Gerhard B. Heller
NASA George C. Marshall Space Flight Center

*21. **Thermal Design Principles of Spacecraft and Entry Bodies** (1969)
Jerry T. Bevans
TRW Systems

*22. **Stratospheric Circulation** (1969)
Willis L. Webb
Atmospheric Sciences Laboratory, White Sands, and University of Texas at El Paso

*23. **Thermophysics: Applications to Thermal Design of Spacecraft** (1970)
Jerry T. Bevans
TRW Systems

24. **Heat Transfer and Spacecraft Thermal Control** (1971)
John W. Lucas
Jet Propulsion Laboratory

25. **Communication Satellites for the 70's: Technology** (1971)
Nathaniel E. Feldman
The Rand Corp.
Charles M. Kelly
The Aerospace Corp.

26. **Communication Satellites for the 70's: Systems** (1971)
Nathaniel E. Feldman
The Rand Corp.
Charles M. Kelly
The Aerospace Corp.

27. **Thermospheric Circulation** (1972)
Willis L. Webb
Atmospheric Sciences Laboratory, White Sands, and University of Texas at El Paso

28. **Thermal Characteristics of the Moon** (1972)
John W. Lucas
Jet Propulsion Laboratory

*29. **Fundamentals of Spacecraft Thermal Design** (1972)
John W. Lucas
Jet Propulsion Laboratory

30. **Solar Activity Observations and Predictions** (1972)
Patrick S. McIntosh
Murray Dryer
Environmental Research Laboratories, National Oceanic and Atmospheric Administration

31. **Thermal Control and Radiation** (1973)
Chang-Lin Tien
University of California at Berkeley

32. **Communications Satellite Systems** (1974)
P.L. Bargellini
COMSAT Laboratories

33. **Communications Satellite Technology** (1974)
P.L. Bargellini
COMSAT Laboratories

*34. **Instrumentation for Airbreathing Propulsion** (1974)
Allen E. Fuhs
Naval Postgraduate School
Marshall Kingery
Arnold Engineering Development Center

35. **Thermophysics and Spacecraft Thermal Control** (1974)
Robert G. Hering
University of Iowa

36. **Thermal Pollution Analysis** (1975)
Joseph A. Schetz
Virginia Polytechnic Institute
ISBN 0-915928-00-0

37. **Aeroacoustics: Jet and Combustion Noise; Duct Acoustics** (1975)
Henry T. Nagamatsu, Editor
General Electric Research and Development Center
Jack V. O'Keefe, Associate Editor
The Boeing Co.
Ira R. Schwartz, Associate Editor
NASA Ames Research Center
ISBN 0-915928-01-9

38. **Aeroacoustics: Fan, STOL, and Boundary Layer Noise; Sonic Boom; Aeroacoustics Instrumentation** (1975)
Henry T. Nagamatsu, Editor
General Electric Research and Development Center
Jack V. O'Keefe, Associate Editor
The Boeing Co.
Ira R. Schwartz, Associate Editor
NASA Ames Research Center
ISBN 0-915928-02-7

39. **Heat Transfer with Thermal Control Applications** (1975)
M. Michael Yovanovich
University of Waterloo
ISBN 0-915928-03-5

*40. **Aerodynamics of Base Combustion** (1976)
S.N.B. Murthy, Editor
J.R. Osborn,
Associate Editor
Purdue University
A.W. Barrows
J.R. Ward,
Associate Editors
Ballistics Research Laboratories
ISBN 0-915928-04-3

41. **Communications Satellite Developments: Systems** (1976)
Gilbert E. LaVean
Defense Communications Agency
William G. Schmidt
CML Satellite Corp.
ISBN 0-915928-05-1

42. **Communications Satellite Developments: Technology** (1976)
William G. Schmidt
CML Satellite Corp.
Gilbert E. LaVean
Defense Communications Agency
ISBN 0-915928-06-X

*43. **Aeroacoustics: Jet Noise, Combustion and Core Engine Noise** (1976)
Ira R. Schwartz, Editor
NASA Ames Research Center
Henry T. Nagamatsu,
Associate Editor
General Electric Research and Development Center
Warren C. Strahle,
Associate Editor
Georgia Institute of Technology
ISBN 0-915928-07-8

*44. **Aeroacoustics: Fan Noise and Control; Duct Acoustics; Rotor Noise** (1976)
Ira R. Schwartz, Editor
NASA Ames Research Center
Henry T. Nagamatsu,
Associate Editor
General Electric Research and Development Center
Warren C. Strahle,
Associate Editor
Georgia Institute of Technology
ISBN 0-915928-08-6

*45. **Aeroacoustics: STOL Noise; Airframe and Airfoil Noise** (1976)
Ira R. Schwartz, Editor
NASA Ames Research Center
Henry T. Nagamatsu,
Associate Editor
General Electric Research and Development Center
Warren C. Strahle,
Associate Editor
Georgia Institute of Technology
ISBN 0-915928-09-4

*46. **Aeroacoustics: Acoustic Wave Propagation; Aircraft Noise Prediction; Aeroacoustic Instrumentation** (1976)
Ira R. Schwartz, Editor
NASA Ames Research Center
Henry T. Nagamatsu,
Associate Editor
General Electric Research and Development Center
Warren C. Strahle,
Associate Editor
Georgia Institute of Technology
ISBN 0-915928-10-8

47. **Spacecraft Charging by Magnetospheric Plasmas** (1976)
Alan Rosen
TRW Inc.
ISBN 0-915928-11-6

48. **Scientific Investigations on the Skylab Satellite** (1976)
Marion I. Kent
Ernst Stuhlinger
NASA George C. Marshall Space Flight Center
Shi-Tsan Wu
University of Alabama
ISBN 0-915928-12-4

49. **Radiative Transfer and Thermal Control** (1976)
Allie M. Smith
ARO Inc.
ISBN 0-915928-13-2

50. **Exploration of the Outer Solar System** (1976)
Eugene W. Greenstadt
TRW Inc.
Murray Dryer
National Oceanic and Atmospheric Administration
Devrie S. Intriligator
University of Southern California
ISBN 0-915928-14-0

51. **Rarefied Gas Dynamics, Parts I and II** (two volumes) (1977)
J. Leith Potter
ARO Inc.
ISBN 0-915928-15-9

52. **Materials Sciences in Space with Application to Space Processing** (1977)
Leo Steg
General Electric Co.
ISBN 0-915928-16-7

53. **Experimental Diagnostics in Gas Phase Combustion Systems** (1977)
Ben T. Zinn, Editor
Georgia Institute of Technology
Craig T. Bowman, Associate Editor
Stanford University
Daniel L. Hartley, Associate Editor
Sandia Laboratories
Edward W. Price, Associate Editor
Georgia Institute of Technology
James G. Skifstad, Associate Editor
Purdue University
ISBN 0-015928-18-3

54. **Satellite Communications: Future Systems** (1977)
David Jarett
TRW Inc.
ISBN 0-915928-18-3

55. **Satellite Communications: Advanced Technologies** (1977)
David Jarett
TRW Inc.
ISBN 0-915928-19-1

56. **Thermophysics of Spacecraft and Outer Planet Entry Probes** (1977)
Allie M. Smith
ARO Inc.
ISBN 0-915928-20-5

57. **Space-Based Manufacturing from Nonterrestrial Materials** (1977)
Gerard K. O'Neill, Editor
Brian O'Leary, Assistant Editor
Princeton University
ISBN 0-915928-21-3

58. **Turbulent Combustion** (1978)
Lawrence A. Kennedy
State University of New York at Buffalo
ISBN 0-915928-22-1

59. **Aerodynamic Heating and Thermal Protection Systems** (1978)
Leroy S. Fletcher
University of Virginia
ISBN 0-915928-23-X

60. **Heat Transfer and Thermal Control Systems** (1978)
Leroy S. Fletcher
University of Virginia
ISBN 0-915928-24-8

61. **Radiation Energy Conversion in Space** (1978)
Kenneth W. Billman
NASA Ames Research Center
ISBN 0-915928-26-4

62. **Alternative Hydrocarbon Fuels: Combustion and Chemical Kinetics** (1978)
Craig T. Bowman
Stanford University
Jorgen Birkeland
Department of Energy
ISBN 0-915928-25-6

63. **Experimental Diagnostics in Combustion of Solids** (1978)
Thomas L. Boggs
Naval Weapons Center
Ben T. Zinn
Georgia Institute of Technology
ISBN 0-915928-28-0

64. **Outer Planet Entry Heating and Thermal Protection** (1979)
Raymond Viskanta
Purdue University
ISBN 0-915928-29-9

65. **Thermophysics and Thermal Control** (1979)
Raymond Viskanta
Purdue University
ISBN 0-915928-30-2

66. **Interior Ballistics of Guns** (1979)
Herman Krier
University of Illinois at Urbana-Champaign
Martin Summerfield
New York University
ISBN 0-915928-32-9

*67. **Remote Sensing of Earth from Space: Role of "Smart Sensors"** (1979)
Roger A. Breckenridge
NASA Langley Research Center
ISBN 0-915928-33-7

68. **Injection and Mixing in Turbulent Flow** (1980)
Joseph A. Schetz
Virginia Polytechnic Institute and State University
ISBN 0-915928-35-3

69. **Entry Heating and Thermal Protection** (1980)
Walter B. Olstad
NASA Headquarters
ISBN 0-915928-38-8

70. **Heat Transfer, Thermal Control, and Heat Pipes** (1980)
Walter B. Olstad
NASA Headquarters
ISBN 0-915928-39-6

*71. **Space Systems and Their Interactions with Earth's Space Environment** (1980)
Henry B. Garrett
Charles P. Pike
Hanscom Air Force Base
ISBN 0-915928-41-8

72. **Viscous Flow Drag Reduction** (1980)
Gary R. Hough
Vought Advanced Technology Center
ISBN 0-915928-44-2

SERIES LISTING

73. **Combustion Experiments in a Zero-Gravity Laboratory** (1981)
Thomas H. Cochran
NASA Lewis Research Center
ISBN 0-915928-48-5

74. **Rarefied Gas Dynamics, Parts I and II** (two volumes) (1981)
Sam S. Fisher
University of Virginia
ISBN 0-915928-51-5

75. **Gasdynamics of Detonations and Explosions** (1981)
J.R. Bowen
University of Wisconsin at Madison
N. Manson
Université de Poitiers
A.K. Oppenheim
University of California at Berkeley
R.I. Soloukhin
Institute of Heat and Mass Transfer, BSSR Academy of Sciences
ISBN 0-915928-46-9

76. **Combustion in Reactive Systems** (1981)
J.R. Bowen
University of Wisconsin at Madison
N. Manson
Université de Poitiers
A.K. Oppenheim
University of California at Berkeley
R.I. Soloukhin
Institute of Heat and Mass Transfer, BSSR Academy of Sciences
ISBN 0-915928-47-7

77. **Aerothermodynamics and Planetary Entry** (1981)
A.L. Crosbie
University of Missouri-Rolla
ISBN 0-915928-52-3

78. **Heat Transfer and Thermal Control** (1981)
A.L. Crosbie
University of Missouri-Rolla
ISBN 0-915928-53-1

79. **Electric Propulsion and Its Applications to Space Missions** (1981)
Robert C. Finke
NASA Lewis Research Center
ISBN 0-915928-55-8

80. **Aero-Optical Phenomena** (1982)
Keith G. Gilbert
Leonard J. Otten
Air Force Weapons Laboratory
ISBN 0-915928-60-4

81. **Transonic Aerodynamics** (1982)
David Nixon
Nielsen Engineering & Research, Inc.
ISBN 0-915928-65-5

82. **Thermophysics of Atmospheric Entry** (1982)
T.E. Horton
University of Mississippi
ISBN 0-915928-66-3

83. **Spacecraft Radiative Transfer and Temperature Control** (1982)
T.E. Horton
University of Mississippi
ISBN 0-915928-67-1

84. **Liquid-Metal Flows and Magnetohydrodynamics** (1983)
H. Branover
Ben-Gurion University of the Negev
P.S. Lykoudis
Purdue University
A. Yakhot
Ben-Gurion University of the Negev
ISBN 0-915928-70-1

85. **Entry Vehicle Heating and Thermal Protection Systems: Space Shuttle, Solar Starprobe, Jupiter Galileo Probe** (1983)
Paul E. Bauer
McDonnell Douglas Astronautics Co.
Howard E. Collicott
The Boeing Co.
ISBN 0-915928-74-4

86. **Spacecraft Thermal Control, Design, and Operation** (1983)
Howard E. Collicott
The Boeing Co.
Paul E. Bauer
McDonnell Douglas Astronautics Co.
ISBN 0-915928-75-2

87. **Shock Waves, Explosions, and Detonations** (1983)
J.R. Bowen
University of Washington
N. Manson
Université de Poitiers
A.K. Oppenheim
University of California at Berkeley
R.I. Soloukhin
Institute of Heat and Mass Transfer, BSSR Academy of Sciences
ISBN 0-915928-76-0

88. **Flames, Lasers, and Reactive Systems** (1983)
J.R. Bowen
University of Washington
N. Manson
Université de Poitiers
A.K. Oppenheim
University of California at Berkeley
R.I. Soloukhin
Institute of Heat and Mass Transfer, BSSR Academy of Sciences
ISBN 0-915928-77-9

89. **Orbit-Raising and Maneuvering Propulsion: Research Status and Needs** (1984)
Leonard H. Caveny
Air Force Office of Scientific Research
ISBN 0-915928-82-5

90. **Fundamentals of Solid-Propellant Combustion** (1984)
Kenneth K. Kuo
Pennsylvania State University
Martin Summerfield
Princeton Combustion Research Laboratories, Inc.
ISBN 0-915928-84-1

91. **Spacecraft Contamination: Sources and Prevention** (1984)
J.A. Roux
University of Mississippi
T.D. McCay
NASA Marshall Space Flight Center
ISBN 0-915928-85-X

92. **Combustion Diagnostics by Nonintrusive Methods** (1984)
T.D. McCay
NASA Marshall Space Flight Center
J.A. Roux
University of Mississippi
ISBN 0-915928-86-8

93. **The INTELSAT Global Satellite System** (1984)
Joel Alper
COMSAT Corp.
Joseph Pelton
INTELSAT
ISBN 0-915928-90-6

94. **Dynamics of Shock Waves, Explosions, and Detonations** (1984)
J.R. Bowen
University of Washington
N. Manson
Université de Poitiers
A.K. Oppenheim
University of California at Berkely
R.I. Soloukhin
Institute of Heat and Mass Transfer, BSSR Academy of Sciences
ISBN 0-915928-91-4

95. **Dynamics of Flames and Reactive Systems** (1984)
J.R. Bowen
University of Washington
N. Manson
Université de Poitiers
A.K. Oppenheim
University of California at Bereley
R.I. Soloukhin
Institute of Heat and Mass Transfer, BSSR Academy of Sciences
ISBN 0-915928-92-2

96. **Thermal Design of Aeroassisted Orbital Transfer Vehicles** (1985)
H.F. Nelson
University of Missouri-Rolla
ISBN 0-915928-94-9

97. **Monitoring Earth's Ocean, Land, and Atmosphere from Space — Sensors, Systems, and Applications** (1985)
Abraham Schnapf
Aerospace Systems Engineering
ISBN 0-915928-98-1

98. **Thrust and Drag: Its Prediction and Verification** (1985)
Eugene E. Covert
Massachusetts Institute of Technology
C.R. James
Vought Corp.
William F. Kimzey
Sverdrup Technology AEDC Group
George K. Richey
U.S. Air Force
Eugene C. Rooney
U.S. Navy Department of Defense
ISBN 0-930403-00-2

99. **Space Stations and Space Platforms — Concepts, Design, Infrastructure, and Uses** (1985)
Ivan Bekey
Daniel Herman
NASA Headquarters
ISBN 0-930403-01-0

100. **Single- and Multi-Phase Flows in an Electromagnetic Field: Energy, Metallurgical, and Solar Applications** (1985)
Herman Branover
Ben-Gurion University of the Negev
Paul S. Lykoudis
Purdue University
Michael Mond
Ben-Gurion University of the Negev
ISBN 0-930403-04-5

101. **MHD Energy Conversion: Physiotechnical Problems** (1986)
V.A. Kirillin
A.E. Sheyndlin
Soviet Academy of Sciences
ISBN 0-930403-05-3

102. **Numerical Methods for Engine-Airframe Integration** (1986)
S.N.B. Murthy
Purdue University
Gerald C. Paynter
Boeing Airplane Co.
ISBN 0-930403-09-6

103. **Thermophysical Aspects of Re-Entry Flows** (1986)
James N. Moss
NASA Langley Research Center
Carl D. Scott
NASA Johnson Space Center
ISBN 0-930403-10-X

104. **Tactical Missile Aerodynamics** (1986)
M.J. Hemsch
PRC Kentron, Inc.
J.N. Nielsen
NASA Ames Research Center
ISBN 0-930403-13-4

105. **Dynamics of Reactive Systems Part I: Flames and Configurations; Part II: Modeling and Heterogeneous Combustion** (1986)
J.R. Bowen
University of Washington
J.-C. Leyer
Université de Poitiers
R.I. Soloukhin
Institute of Heat and Mass Transfer, BSSR Academy of Sciences
ISBN 0-930403-14-2

106. **Dynamics of Explosions** (1986)
J.R. Bowen
University of Washington
J.-C. Leyer
Université de Poitiers
R.I. Soloukhin
Institute of Heat and Mass Transfer, BSSR Academy of Sciences
ISBN 0-930403-15-0

107. **Spacecraft Dielectric Material Properties and Spacecraft Charging** (1986)
A.R. Frederickson
U.S. Air Force Rome Air Development Center
D.B. Cotts
SRI International
J.A. Wall
U.S. Air Force Rome Air Development Center
F.L. Bouquet
Jet Propulsion Laboratory, California Institute of Technology
ISBN 0-930403-17-7

108. **Opportunities for Academic Research in a Low-Gravity Environment** (1986)
George A. Hazelrigg
National Science Foundation
Joseph M. Reynolds
Louisiana State University
ISBN 0-930403-18-5

109. **Gun Propulsion Technology** (1988)
Ludwig Stiefel
U.S. Army Armament Research, Development and Engineering Center
ISBN 0-930403-20-7

110. **Commercial Opportunities in Space** (1988)
F. Shahrokhi
K.E. Harwell
University of Tennessee Space Institute
C.C. Chao
National Cheng Kung University
ISBN 0-930403-39-8

111. **Liquid-Metal Flows: Magnetohydrodynamics and Applications** (1988)
Herman Branover,
Michael Mond, and
Yeshajahu Unger
Ben-Gurion University of the Negev
ISBN 0-930403-43-6

112. **Current Trends in Turbulence Research** (1988)
Herman Branover,
Michael Mond, and
Yeshajahu Unger
Ben-Gurion University of the Negev
ISBN 0-930403-44-4

113. **Dynamics of Reactive Systems Part I: Flames; Part II: Heterogeneous Combustion and Applications** (1988)
A.L. Kuhl
R & D Associates
J.R. Bowen
University of Washington
J.-C. Leyer
Université de Poitiers
A. Borisov
USSR Academy of Sciences
ISBN 0-930403-46-0

114. **Dynamics of Explosions** (1988)
A.L. Kuhl
R & D Associates
J.R. Bowen
University of Washington
J.-C. Leyer
Université de Poitiers
A. Borisov
USSR Academy of Sciences
ISBN 0-930403-47-9

115. **Machine Intelligence and Autonomy for Aerospace** (1988)
E. Heer
Heer Associates, Inc.
H. Lum
NASA Ames Research Center
ISBN 0-930403-48-7

116. **Rarefied Gas Dynamics: Space-Related Studies** (1989)
E.P. Muntz
University of Southern California
D.P. Weaver
U.S. Air Force Astronautics Laboratory (AFSC)
D.H. Campbell
University of Dayton Research Institute
ISBN 0-930403-53-3

117. **Rarefied Gas Dynamics: Physical Phenomena** (1989)
E.P. Muntz
University of Southern California
D.P. Weaver
U.S. Air Force Astronautics Laboratory (AFSC)
D. Campbell
University of Dayton Research Institute
ISBN 0-930403-54-1

118. **Rarefied Gas Dynamics: Theoretical and Computational Techniques** (1989)
E.P. Muntz
University of Southern California
D.P. Weaver
U.S. Air Force Astronautics Laboratory (AFSC)
D.H. Campbell
University of Dayton Research Institute
ISBN 0-930403-55-X

119. **Test and Evaluation of the Tactical Missile** (1989)
Emil J. Eichblatt Jr.
Pacific Missile Test Center
ISBN 0-930403-56-8

120. **Unsteady Transonic Aerodynamics** (1989)
David Nixon
Nielsen Engineering & Research, Inc.
ISBN 0-930403-52-5

121. **Orbital Debris from Upper-Stage Breakup** (1989)
Joseph P. Loftus Jr.
NASA Johnson Space Center
ISBN 0-930403-58-4

122. **Thermal-Hydraulics for Space Power, Propulsion and Thermal Management System Design** (1989)
William J. Krotiuk
General Electric Co.
ISBN 0-930403-64-9

123. **Viscous Drag Reduction in Boundary Layers** (1990)
Dennis M. Bushnell
Jerry N. Hefner
NASA Langley Research Center
ISBN 0-930403-66-5

124. **Tactical and Strategic Missile Guidance** (1990)
Paul Zarchan
Charles Stark Draper Laboratory, Inc.
ISBN 0-930403-68-1

125. **Applied Computational Aerodynamics** (1990)
P.A. Henne
Douglas Aircraft Company
ISBN 0-930403-69-X

126. **Space Commercialization: Launch Vehicles and Programs** (1990)
F. Shahrokhi
University of Tennessee Space Institute
J.S. Greenberg
Princeton Synergetics Inc.
T. Al-Saud
Ministry of Defense and Aviation Kingdom of Saudi Arabia
ISBN 0-930403-75-4

127. **Space Commercialization: Platforms and Processing** (1990)
F. Shahrokhi
University of Tennessee Space Institute
G. Hazelrigg
National Science Foundation
R. Bayuzick
Vanderbilt University
ISBN 0-930403-76-2

128. **Space Commercialization: Satellite Technology** (1990)
F. Shahrokhi
University of Tennessee Space Institute
N. Jasentuliyana
United Nations
N. Tarabzouni
King Abulaziz City for Science and Technology
ISBN 0-930403-77-0

129. **Mechanics and Control of Large Flexible Structures** (1990)
John L. Junkins
Texas A&M University
ISBN 0-930403-73-8

130. **Low-Gravity Fluid Dynamics and Transport Phenomena** (1990)
Jean N. Koster
Robert L. Sani
University of Colorado at Boulder
ISBN 0-930403-74-6

131. **Dynamics of Deflagrations and Reactive Systems: Flames** (1991)
A. L. Kuhl
Lawrence Livermore National Laboratory
J.-C. Leyer
Université de Poitiers
A. A. Borisov
USSR Academy of Sciences
W. A. Sirignano
University of California
ISBN 0-930403-95-9

132. **Dynamics of Deflagrations and Reactive Systems: Heterogeneous Combustion** (1991)
A. L. Kuhl
Lawrence Livermore National Laboratory
J.-C. Leyer
Université de Poitiers
A. A. Borisov
USSR Academy of Sciences
W. A. Sirignano
University of California
ISBN 0-930403-96-7

133. **Dynamics of Detonations and Explosions: Detonations** (1991)
A. L. Kuhl
Lawrence Livermore National Laboratory
J.-C. Leyer
Université de Poitiers
A. A. Borisov
USSR Academy of Sciences
W. A. Sirignano
University of California
ISBN 0-930403-97-5

134. **Dynamics of Detonations and Explosions: Explosion Phenomena** (1991)
A. L. Kuhl
Lawrence Livermore National Laboratory
J.-C. Leyer
Université de Poitiers
A. A. Borisov
USSR Academy of Sciences
W. A. Sirignano
University of California
ISBN 0-930403-98-3

135. **Numerical Approaches to Combustion Modeling** (1991)
Elaine S. Oran
Jay P. Boris
Naval Research Laboratory
ISBN 1-56347-004-7

136. **Aerospace Software Engineering** (1991)
Christine Anderson
U.S. Air Force Wright Laboratory
Merlin Dorfman
Lockheed Missiles & Space Company, Inc.
ISBN 1-56346-005-5

137. **High-Speed Flight Propulsion Systems** (1991)
S. N. B. Murthy
Purdue University
E. T. Curran
Wright Laboratory
ISBN 1-56347-011-X

138. **Propagation of Intensive Laser Radiation in Clouds** (1992)
O. A. Volkovitsky
Yu. S. Sedunov
L. P. Semenov
Institute of Experimental Meteorology
ISBN 1-56347-020-9

139. **Gun Muzzle Blast and Flash** (1992)
Günter Klingenberg
Fraunhofer-Institut für Kurzzeitdynamik, Ernst-Mach-Institut (EMI)
Joseph M. Heimerl
U.S. Army Ballistic Research Laboratory (BRL)
ISBN 1-56347-012-8

140. **Thermal Structures and Materials for High-Speed Flight** (1992)
Earl A. Thornton
University of Virginia
ISBN 1-56347-017-9

141. **Tactical Missile Aerodynamics: General Topics** (1992)
Michael J. Hemsch
Lockheed Engineering & Sciences Company
ISBN 1-56347-015-2

142. **Tactical Missile Aerodynamics: Prediction Methodology** (1992)
Michael R. Mendenhall
Nielsen Engineering & Research, Inc.
ISBN 1-56347-016-0

143. **Nonsteady Burning and Combustion Stability of Solid Propellants** (1992)
Luigi De Luca
Politecnico di Milano
Edward W. Price
Georgia Institute of Technology
Martin Summerfield
Princeton Combustion Research Laboratories, Inc.
ISBN 1-56347-014-4

144. **Space Economics** (1992)
Joel S. Greenberg
Princeton Synergetics, Inc.
Henry R. Hertzfeld
HRH Associates
ISBN 1-56347-042-X

145. **Mars: Past, Present, and Future** (1992)
E. Brian Pritchard
NASA Langley Research Center
ISBN 1-56347-043-8

146. **Computational Nonlinear Mechanics in Aerospace Engineering** (1992)
Satya N. Atluri
Georgia Institute of Technology
ISBN 1-56347-044-6

147. **Modern Engineering for Design of Liquid-Propellant Rocket Engines** (1992)
Dieter K. Huzel
David H. Huang
ISBN 1-56347-013-6

148. Metallurgical Technologies, Energy Conversion, and Magnetohydrodynamic Flows (1993)
Herman Branover
Yeshajahu Unger
Ben-Gurion University of the Negev
ISBN 1-56347-019-5

149. Advances in Turbulence Studies (1993)
Herman Branover
Yeshajahu Unger
Ben-Gurion University of the Negev
ISBN 1-56347-018-7

150. Structural Optimization: Status and Promise (1993)
Manohar P. Kamat
Georgia Institute of Technology
ISBN 1-56347-056-X

151. Dynamics of Gaseous Combustion (1993)
A. L. Kuhl
Lawrence Livermore National Laboratory
J.-C. Leyer
Université de Poitiers
A. A. Borisov
Russian Academy of Sciences
W. A. Sirignano
University of California
ISBN 1-56347-060-8

152. Dynamics of Heterogeneous Combustion and Reacting Systems (1993)
A. L. Kuhl
Lawrence Livermore National Laboratory
J.-C. Leyer
Université de Poitiers
A. A. Borisov
Russian Academy of Sciences
W. A. Sirignano
University of California
ISBN 1-56347-058-6

153. Dynamic Aspects of Detonations (1993)
A. L. Kuhl
Lawrence Livermore National Laboratory
J.-C. Leyer
Université de Poitiers
A. A. Borisov
Russian Academy of Sciences
W. A. Sirignano
University of California
ISBN 1-56347-057-8

154. Dynamic Aspects of Explosion Phenomena (1993)
A. L. Kuhl
Lawrence Livermore National Laboratory
J.-C. Leyer
Université de Poitiers
A. A. Borisov
Russian Academy of Sciences
W. A. Sirignano
University of California
ISBN 1-56347-059-4

(Other Volumes are planned.)